Environmental and Technological Aspects of Redox Processes

Gheorghe Duca
Institute of Chemistry, Moldova State University, Moldova

Ashok Vaseashta
International Clean Water Institute, USA

A volume in the Advances in Environmental Engineering and Green Technologies (AEEGT) Book Series

IGI Global
PUBLISHER of TIMELY KNOWLEDGE

Published in the United States of America by
IGI Global
Engineering Science Reference (an imprint of IGI Global)
701 E. Chocolate Avenue
Hershey PA, USA 17033
Tel: 717-533-8845
Fax: 717-533-8661
E-mail: cust@igi-global.com
Web site: http://www.igi-global.com

Library of Congress Cataloging-in-Publication Data

Names: Duca, Gheorghe, 1952- editor. | Vaseashta, A. (Ashok), editor.
Title: Environmental and technological aspects of redox processes / edited by Gheorghe Duca and Ashok Vaseashta.
Description: Hershey PA : Engineering Science Reference, [2023] | Includes bibliographical references and index. | Summary: "This book is a valuable resource for professionals and researchers in materials science, materials engineering, chemical engineering, ecological chemistry, and water professionals, as well as individuals working in related disciplines such as environment, chemistry, toxicology, administrative sciences and management, education, sociology, and subjects related to public policies and decision support. Moreover, the book provides insights and support to executives concerned with the management of expertise, knowledge, information, and policy development in different types of science and educational communities"-- Provided by publisher.
Identifiers: LCCN 2023022605 (print) | LCCN 2023022606 (ebook) | ISBN 9798369305126 (hardcover) | ISBN 9798369305133 (softcover) | ISBN 9798369305140 (ebook)
Subjects: LCSH: Oxidation-reduction reaction.
Classification: LCC QD716.O95 .H36 2023 (print) | LCC QD716.O95 (ebook) | DDC 541/.393--dc23/eng20230919
LC record available at https://lccn.loc.gov/2023022605
LC ebook record available at https://lccn.loc.gov/2023022606

This book is published in the IGI Global book series Advances in Environmental Engineering and Green Technologies (AEEGT) (ISSN: 2326-9162; eISSN: 2326-9170)

British Cataloguing in Publication Data
A Cataloguing in Publication record for this book is available from the British Library.

All work contributed to this book is new, previously-unpublished material. The views expressed in this book are those of the authors, but not necessarily of the publisher.

For electronic access to this publication, please contact: eresources@igi-global.com.

Advances in Environmental Engineering and Green Technologies (AEEGT) Book Series

Sang-Bing Tsai
Zhongshan Institute, University of Electronic Science and Technology of China, China & Wuyi University, China
Ming-Lang Tseng
Lunghwa University of Science and Technology, Taiwan
Yuchi Wang
University of Electronic Science and Technology of China Zhongshan Institute, China

ISSN:2326-9162
EISSN:2326-9170

Mission

Growing awareness and an increased focus on environmental issues such as climate change, energy use, and loss of non-renewable resources have brought about a greater need for research that provides potential solutions to these problems. Research in environmental science and engineering continues to play a vital role in uncovering new opportunities for a "green" future.

The **Advances in Environmental Engineering and Green Technologies (AEEGT)** book series is a mouthpiece for research in all aspects of environmental science, earth science, and green initiatives. This series supports the ongoing research in this field through publishing books that discuss topics within environmental engineering or that deal with the interdisciplinary field of green technologies.

Coverage

- Water Supply and Treatment
- Green Transportation
- Radioactive Waste Treatment
- Electric Vehicles
- Contaminated Site Remediation
- Sustainable Communities
- Green Technology
- Biofilters and Biofiltration
- Alternative Power Sources
- Cleantech

IGI Global is currently accepting manuscripts for publication within this series. To submit a proposal for a volume in this series, please contact our Acquisition Editors at Acquisitions@igi-global.com or visit: http://www.igi-global.com/publish/.

The Advances in Environmental Engineering and Green Technologies (AEEGT) Book Series (ISSN 2326-9162) is published by IGI Global, 701 E. Chocolate Avenue, Hershey, PA 17033-1240, USA, www.igi-global.com. This series is composed of titles available for purchase individually; each title is edited to be contextually exclusive from any other title within the series. For pricing and ordering information please visit http://www.igi-global.com/book-series/advances-environmental-engineering-green-technologies/73679. Postmaster: Send all address changes to above address.

Titles in this Series

For a list of additional titles in this series, please visit: www.igi-global.com/book-series

Handbook of Research on Improving the Natural and Ecological Conditions of the Polesie Zone
Anatoliy Rokochinskiy (National University of Water and Environmental Engineering, Ukraine) Lyudmyla Kuzmych (Institute of Water Problems and Land Reclamation of the National Academy of Agrarian Sciences, Ukraine) and Pavlo Volk (National University of Water and Environmental Engineering, Ukraine)
Engineering Science Reference • © 2023 • 479pp • H/C (ISBN: 9781668482483) • US $270.00

Multidisciplinary Approaches in AI, Creativity, Innovation, and Green Collaboration
Ziska Fields (University of Johannesburg, South Africa)
Engineering Science Reference • © 2023 • 405pp • H/C (ISBN: 9781668463666) • US $270.00

Opportunities and Challenges in Climate-Friendly Clean Water and Energy Technologies
J. Sadhik Basha (Department of Process Engineering, National University of Science & Technology, Oman) Jimoh Kayode Adewole (Department of Process Engineering, National University of Science & Technology, Oman) Anteneh Mesfin Yeneneh (Department of Process Engineering, National University of Science & Technology, Oman) and Luqman Kolawole Abidoye (Department of Process Engineering, National University of Science & Technology, Oman)
Engineering Science Reference • © 2023 • 344pp • H/C (ISBN: 9781668473030) • US $250.00

Transcending Humanitarian Engineering Strategies for Sustainable Futures
Yiannis Koumpouros (University of West Attica, Greece) Angelos Georgoulas (University of West Attica, Greece) and Georgia Kremmyda (University of Warwick, UK)
Engineering Science Reference • © 2023 • 338pp • H/C (ISBN: 9781668456194) • US $270.00

Food Sustainability, Environmental Awareness, and Adaptation and Mitigation Strategies for Developing Countries
Ahmad Ni'matullah Al-Baarri (Universitas Diponegoro, Indonesia) and Diana Nur Afifah (Universitas Diponegoro, Indonesia)
Engineering Science Reference • © 2023 • 293pp • H/C (ISBN: 9781668456293) • US $110.00

Climate Change and the Economic Importance and Damages of Insects
Ahmed Karmaoui (Moulay Ismail University of Meknes, Morocco & Moroccan Center for Culture and Science, Morocco)
Engineering Science Reference • © 2023 • 279pp • H/C (ISBN: 9781668448243) • US $215.00

IGI Global
PUBLISHER of TIMELY KNOWLEDGE

701 East Chocolate Avenue, Hershey, PA 17033, USA
Tel: 717-533-8845 x100 • Fax: 717-533-8661
E-Mail: cust@igi-global.com • www.igi-global.com

Editorial Advisory Board

Table of Contents

Detailed Table of Contents

Determination of the important wastewater parameter chemical oxygen demand (COD) in international standard procedures is still based on the use of mercury sulfate and potassium dichromate, which are hazardous to health and the environment (REACH). In order to overcome this environmental paradox, alternative oxidizing agents were compared, and a new wet-chemical method using Mn(III) as an oxidant was developed that works without Cr(VI) and Hg(II). Oxidation of potassium hydrogen phthalate as a reference substance was reproducible in a COD range of 20–500 mg/L with a mean recovery of 88.7% compared with the standard Cr(VI)-method. A high correlation (R2 = 0.9935) to the standard Cr(VI)-method with a mean recovery of 78.1% (± 5.2%) was found for a series of industrial and municipal wastewater samples. Although the results of the new method were not 100% of the standard method, its high correlation with the latter and its reproducibility offer an environmentally benign alternative wet chemical method.

This chapter presents a synthesis of the typical mechanisms and patterns of catalytic redox processes that take place in the natural aquatic environment with the participation of oxygen, its reactive species, hydrogen peroxide, OH radicals, and various forms of copper and iron transition metals. There are presented the redox subsystems (slow or fast) which form the redox character of the environment; the mechanisms and conditions for activating oxygen for its involvement in redox processes; biotic and abiotic sources and pathways of the hydrogen peroxide degradation, as well as its role in the formation, maintenance, and regulation of the environment redox state; and formation and peculiarities of hydroxyl radicals' participation in the processes inside the water basins.

This chapter studied the influence of TGA and TU on the self-purification processes of aquatic systems on model systems, whose composition was modeled to be as close as possible to the natural waters. As a source of irradiation was used the DRT-400 lamp and the Solar Simulator, Oriel model 9119X, the emission spectra of which overlap with the emission spectrum of the Sun. The results of the research show that on irradiation, TGA can regenerate an additional amount of OH radicals, thus contributing to the intensification of water self-purification processes. But in the presence of transition metals, such as Cu (II) and Fe (III), it has been established that the intensity of self-purification processes decreases due to the formation of the mercaptides, whose catalytic activity is low. Unlike TGA, TU leads to decreasing the intensity of the self-purification processes of aquatic systems, through the intense consumption of OH radicals, and at the same time on the interaction with transition metals, such as Cu (II) and Fe (III), form stable complexes with low catalytic activity.

The chemical composition of water is determined by the environment with which it comes into contact, with the establishment of a dynamic balance, conditioned by energy and substance flows, expressed by redox equivalents. The valuable biological state of natural waters as a living environment is oxidizing, determined by the presence of hydrogen peroxide. For the characterization of the redox state of the

Despite the major efforts that have been made over recent years to clean up the environment, pollution remains a major problem and poses continuing risks to health. Thus, the determination of contaminants in environmental samples is a challenging and critical issue. Highly accurate analytical techniques provide the possibility to assess and map the organic and inorganic pollutants in complex atmospheric, water, and soil matrices qualitatively and quantitatively. The present chapter brings together the wealth of information in the current literature, covering both the fundamentals and applications of these methodologies. The main attention is given to neutron activation analysis, inductively coupled plasma mass spectrometry, and inductively coupled plasma atomic emission spectroscopy. The principles, main strengths, and weaknesses of each technique are discussed. Several examples of techniques applied in air, water, and soil sample analysis are presented.

This chapter contains information related to the oxidation-reduction potential (ORP) in aqueous solutions, aiming to extend some aspects from theoretical electrochemistry to practical environmental monitoring and engineering applications. A more comprehensive and precise continuous monitoring of water quality for groundwater, surface water, wastewater, and drinking water could be achieved by deducing relations between ORP, the concentration of the oxidants and reductants in waters, pH, and temperature, delivering valuable information for the better comprehension of the requirements necessary to design an automated water quality monitoring system with real-time transmitted data, assuring better water governance. Such a system is presented as a case study related to an early warning of preliminary ecotoxicity. The case study was conducted by installing water pollution monitoring equipment on a passenger ship sailing on the Sulina Channel in the Danube Delta, Romania.

The utilization of nanopesticides is a young and very promising direction in agriculture. Their use can significantly improve many parameters of the used drugs. Nevertheless, there are many disputes and hypotheses about the impact of such pesticides on the environment. There are many examples of nanopesticides in the literature with an inorganic active substance. They are widely used in practice and are sufficiently described in the literature. In this chapter, emphasis will be placed on the use of nanopesticides with photosensitive organic active substances. Methods for their preparation, effect on the target pathogen, and degradation pathways will be described.

This chapter is related to the general characteristics of vegetable oils; the oxidative stability and the mechanism of lipid oxidation; the factors affecting the mechanism of lipid oxidation; the impact of lipid oxidation on the sensory properties of complex food products; and characteristics of the antioxidant complex from local horticultural products. Oil oxidation also destroys essential fatty acids and produces toxic compounds and oxidized polymers. It also affects the taste, nutritional quality, and toxicity of edible oils. Different chemical mechanisms, autoxidation, and photooxidation, are responsible for the oxidation of edible oils during processing and storage, depending on oxygen types. Two oxygen types can react with edible oils: atmospheric triplet oxygen, 3O2, and singlet oxygen, 1O2.

Ferrite treatment of wastewater is a promising approach, making it possible to resolve environmental and economic problems, due to the high degree of water purification and final formation of chemically stable sludge with crystal oxide structure and ferromagnetic properties. This chapter discusses the specifics of ferrite treatment of wastewater containing heavy metal ions and its advantages with regard to other physical-chemical technologies. The role of ζ-potential of microparticles Fe3O4 in the formation of spinel structures is discussed. The mechanism of ferritization reactions is connected with the catalytic nature of the running processes and the probability of the intermediate active radical's formation. To make the ferrite treatment more flexible and effective, it can be combined with photocatalytical treatment, which allows acceleration of the oxidation processes during the ferrite sludge formation, and at the same time promotes the destruction of organic contaminants due to the formation of free radicals in treated water.

Chapter 14

Various physico-chemical and biological methods have been developed for the removal of organic and inorganic pollutants from wastewater. The application of these techniques, their combination, and the development of hybrid technologies to remove the persistent pollutants from water are presented in this chapter. The advanced oxidation processes (AOPs) such as photocatalysis, modified Fenton reactions, electro- and photoelectrochemical oxidation are evidenced as the main applied technologies. The combination and effects of direct and indirect oxidation methods and of the alternative greener technologies with traditional methods are presented as a function of pollutants' nature, volume, and concentration. For each method, results were obtained for various pollutants such as pesticides, hydrocarbons, dyes, surfactants, aromatic compounds, and pharmaceutical residues. The main objective of this chapter is the application of redox technologies in wastewater treatment by correlating redox processes with working conditions, pollutants, and oxygen reactive species.

Chapter 15

The redox processes are an unavoidable and very complex part of winemaking. The chapter examines the main redox processes that take place during grape subsequent technological processes, maturation, and aging of wine. Some of them lead to the ennobling of wines and their stabilization, while others cause the alteration, and degradation of wines, and the appearance of numerous defects of wines. The consecutive, parallel, and spontaneous development of multiple redox processes, and their dependence on pedoclimatic, agro-technical, and technological conditions influence the structure, appearance, aroma/bouquet, taste, and stability of wines. The knowledge of the production mechanisms, the influencing factors, of the particularities allows the realization of precision vinification, characterized by the reduction of the additive and subtractive technological interventions, of the quantities of adjuvants, and the production of ecological wines. Monitoring redox processes during alcoholic fermentation through oxygen management is essential for the quality of wines.

Chapter 16

Plate-like cadmium sulfide (CdS) quantum dots were prepared by using β-cyclodextrin as reductants. Photoluminescence and UV-Visible diffused reflectance spectroscopic techniques are used to investigate the optical property of CdS. The blue-shifted emission of CdS at 469nm than the bulk particle is observed. The calculated band gap of the CdS quantum dot is 2.62 eV showing a difference of 0.2 eV energy greater than that of bulk, indicating the quantum size effect. The synthesized quantum dot has a size in the range of 3.5nm with a zinc blend crystallite structure. The plate-like nanoclusters of CdS are inferred from

Foreword

The problems related to the environment and the search for solutions to solve them are fundamental in our society. For a better quality of life and protection of our surrounding environment, we must be aware of all kinds of risks against them, including anthropogenic, to prevent, mitigate, and restore damages produced by various human activities. This is possible only by a deep understanding of the chemical processes that take place for and against the protection of the environment. In this context, the book edited by Professors Duca and Vaseashta is welcome as it offers an up-to-date overview of the redox processes involved in ecological chemistry.

There are 18 chapters authored by various scientists and researchers who have led significant contributions to the field. The book is divided into two parts. The first part (consisting of nine chapters) is devoted to a broad overview of the redox processes occurring in the environment, with a special emphasis on natural waters. The efficiency of the solutions found for environmental protection depends on the analytical methods employed to identify and characterize the various sources of pollution. The logical construction of this book is completed by the presentation of technological aspects of the redox processes in the second part. Indeed, the second part presented in the following nine chapters offers some current topics, such as biomass digestion, catalytic oxidation of carbon monoxide, wastewater treatment, and degradation of dyes using quantum dots technology. Environmental-friendly solutions in grape and wool processing are described in this section, as well.

The book is interesting and well-written and is addressed to a broad readership. Students will discover research topics that are exciting and timely, and scientists will find a valuable tool for developing research projects. The book is also useful for teaching, providing teachers and students with essential and recent achievements in ecological chemistry. I congratulate the editors and authors for this excellent work!

Marius Andruh
University of Bucharest, Romania

Foreword

The study of redox processes has become increasingly important as the world continues to face complex environmental, biomedical, and technological challenges. Redox reactions involve the transfer of electrons from one species to another, resulting in changes in the oxidation states of the reactants. The species that loses electrons is said to be oxidized, while the species that gains electrons is said to be reduced. These processes involve the transfer of electrons from one molecule to another and play a critical role in a wide range of chemical reactions in nature and the laboratory. The study of redox reactions has led to the development of many technologies for treatment and environmental remediation. These technologies rely on the use of chemical, physical, and biological processes to reduce or eliminate contaminants from the environment.

The *book of Environmental and Technological Aspects of Redox Processes* by editors Gheorghe Duca and Ashok Vaseashta provides a comprehensive overview of the latest research in this field. The book covers a wide range of topics related to redox processes, including the environmental impacts of redox reactions, the role of redox processes in biological systems, and the use of redox chemistry in technological applications. The book consists of two parts written by many very well-known scientists and professionals. The first and second parts are dedicated to the applications of redox reactions in the environment and technology. Almost all contributors are eminent scientists in their respective fields who offer insights into the fundamental principles of redox processes, as well as their practical applications.

Overall, the *book of Environmental and Technological Aspects of Redox Processes* is an invaluable resource of knowledge for researchers, students, and professionals who are interested in understanding the fundamental principles and practical applications of redox processes. It provides a comprehensive overview of this rapidly evolving field and highlights several ways redox chemistry shapes our surrounding world. Furthermore, redox reactions play a crucial role in the functioning of many biological processes such as respiration, photosynthesis, and digestion. These reactions also play an important role in industrial processes such as metallurgy, electroplating, and battery technology. Some common examples of redox technologies include, for example, water treatment which also includes electrocoagulation, electrochemical oxidation, photocatalysis, and biological treatment. These technologies are effective in treating a wide range of contaminants, including organic pollutants, heavy metals, and pathogens. We foresee many applications to be added to what is already provided in the book. Since the field is rapidly developing, this handbook will serve as a reference for many young researchers.

In summary, redox reactions play a critical role in both biological and industrial processes and have led to the development of many technologies for treatment and environmental remediation. These technologies provide a sustainable and effective way to treat and protect the contaminated environment. We congratulate the editors and chapter contributors for their excellent contributions and we hope that this book is successful in its aim to further the field of redox process in the environment, biomedical, and contamination remediation technologies.

Janusz Lipkowski
Cardinal Stefan Wyszynski University, Poland

Foreword

Redox reactions, or oxidation-reduction reactions, are one of the most fundamental types of chemical reactions. They play crucial roles both in biology, where they underlie fundamental processes like respiration, photosynthesis, and metabolism, and in technological applications, where they are used in such various processes as wastewater treatment, corrosion prevention, and as the fundamental basis for energy storage in batteries. The rich and diverse landscape of these reactions has prompted generations of chemists, biochemists, and engineers to explore their secrets and harness their power for the elucidation of fundamental processes and exploration of their technological potential.

This *book of Environmental and Technological Aspects of Redox Processes* aims to provide a comprehensive overview of current progress and applications of redox chemistry in diverse fields. Its 18 chapters, written by experts in their respective fields, are organized into two sections: Fundamental Aspects of Redox Processes in Environmental Aspects of Redox Processes, and Technology and Redox Processes. The first section provides a comprehensive analysis of the role of redox reactions in water bodies and their assessment for the monitoring of their environmental health. The chapters in this section cover a wide range of topics, including the biotic and abiotic sources of radical oxygen species and their roles in processes in water basins, analytic methods to determine chemical oxygen demand and other environmental health parameters, the mechanisms underlying redox self-purification mechanism of natural waters, the influence of environmental thiol compounds on the redox state of natural waters, and the application of bio-diagnostic methods in environmental chemistry. The final section, Redox Process, and Technology focus on the technological applications of redox chemistry. The chapters in this section provide a detailed analysis of the mechanisms of redox use of redox technologies in wastewater treatment, waste biomass digestion, wool processing, and wine production, as well as novel experimental results on the photochemical degradation of dyes, removal of contaminating antibiotics from wastewater and oxidation of carbon monoxide by atmospheric oxygen.

The quality and diversity of the contributions presented in this research *book on Environmental and Technological Aspects of Redox Processes* provide a comprehensive overview of the fundamental principles of redox chemistry and their applications in wide-ranging fields. We, therefore, expect this book to become an essential resource for researchers, academics, and students interested in the fascinating diversity of redox processes in the fields of chemistry, biochemistry, environmental science, and materials science.

Pedro Silva
Universidade Fernando Pessoa, Portugal

Preface

Processes involving the transfer of electrons between oxidants and reducing agents are called Redox processes and are essential in both natural and technological systems. Redox reactions are ubiquitous in nature, occurring in environmental processes and in many technological applications such as energy conversion, chemical synthesis, and pollution remediation. The environmental aspects of redox processes are diverse and complex and include a sequence of natural phenomena. Human activities have also had a profound impact on the redox balance of natural systems through the emission of greenhouse gases and the release of pollutants. Although, to date, several studies have been carried out in this area, the problem is far from solved. The study of redox environmental processes is essential for understanding and mitigating the environmental impact of anthropogenic activities.

The technological applications of redox processes are numerous and varied. Redox chemistry is also essential in chemical synthesis and manufacture since it is used to catalyze reactions, eliminate impurities, and produce high-value products. Therefore, the study of redox processes is essential both for understanding natural systems and for developing technological solutions to societal challenges. The field of redox chemistry continues to advance, with new discoveries and innovations that promise sustainable development and environmental management. Below are some examples of redox processes in environmental and technological contexts:

Qualitative state of the environment - Redox reactions can provide important information about the qualitative state of the environment. The presence of certain redox-sensitive chemicals, such as dissolved oxygen, hydrogen peroxide, nitrate, iron, copper, and manganese, may indicate whether an aquatic environment is healthy or polluted. In aerobic environments, dissolved oxygen levels are high, indicating a healthy ecosystem that can support a diversity of aquatic life. However, in anaerobic environments, such as polluted sediments or stagnant ponds, dissolved oxygen levels are low, and other redox-sensitive chemicals may be present in high concentrations, indicating poor water quality. Similarly, in soils, the presence of certain redox-sensitive chemicals, such as iron and sulfur, can provide information about soil fertility and the availability of nutrients for plant growth. Additionally, Hydrogen peroxide plays an important role in the environment. It is a natural compound that is formed in the atmosphere by the action of sunlight on water vapor and is also produced by many living organisms as a metabolic byproduct. Hydrogen peroxide is used by microorganisms to break down organic matter in water and is used to remedy contaminated soils by oxidizing organic pollutants, such as petroleum hydrocarbons and pesticides, into less toxic compounds., and serves as a stimulant for plant growth. Thus, Hydrogen peroxide contributes to the formation of the redox state of natural waters.

In terms of technological contexts - Redox processes are widely used in wastewater treatment, air pollution control, and solid waste treatment. In wastewater treatment, microorganisms use redox reactions to break down organic pollutants such as sewage and industrial effluents. Redox reactions are also used to produce chemicals in a more environmentally friendly way. Also, the use of hydrogen peroxide as an oxidizing agent in chemical synthesis can replace traditional oxidizing agents such as chlorine and bromine, which can generate harmful by-products. In addition, the use of electrochemical methods, such as electrolysis, can facilitate redox reactions that produce chemicals without the need for harsh chemicals or high temperatures.

This book aims to close the existing knowledge gaps and discusses the latest topics and other related topics from several leading experts, covering interdisciplinary studies on a wide range of chemical and physical methods, including studies from different fields of chemistry, in particular food chemistry, ecological chemistry, and related scientific disciplines.

The book presents the following points that are not covered in detail elsewhere:

Hydrogen peroxide in the environment
Redox processes in natural waters
Redox technologies in wastewater treatment
Redox Processes in wastewater treatment with ferritization method
Photochemical processes in the environment with the involvement of thiols
Redox processes in the decomposition of pesticides in natural waters
Redox processes in wine processing
Redox processes on the protection of oils in the corners
Redox wool processing
Redox processes in essential oil processing

The *book of Environmental and Technological Aspects of Redox Processes* provides a comprehensive overview of contemporary progress and potential applications of redox chemistry in different fields. The book consists of 18 chapters, written by experts in their respective fields, organized into two sections: Environmental Aspects of Redox Processes and Redox Technology and Processes. The chapters containing the results of research related to natural and wastewater studies are an important part of this book. Thus, Gladchi, Bunduchi, Duca, and Romanciuc summarized the typical mechanisms and patterns of redox catalytic processes in natural waters. Kolb and Bahadir were dedicated to determining one of the most important quality parameters of wastewater – chemical oxygen demand (COD). The involvement of reducing compounds – thioglycollic acid and thiourea in the mechanism of redox self-purification in natural water is reported by researchers, including Lis, Gladchi, Duca, Bunduchi, and Blonschi. A similar topic is discussed by Blonschi, Gladchi, Duca, Bunduchi, and Lis, which demonstrates the effect of reducing thiol compounds on the redox state of natural waters. A team of researchers, Schvydkiy, Travin, Schtamm, Cheremnykh and Semenyak demonstrate, in accordance with chemical parameters, that biodiagnostic methods can more clearly reveal the state of the natural environment of water. A case study of the investigation of redox processes in a lake in Georgia is presented by Kupatadze in a chapter. Zincovscaia highlights the advantages of neutron activation analysis in the investigation of environmental compartments. Another approach for the study of water quality, namely the evaluation of ecotoxicity with the help of redox measurements, is presented in a chapter by Cretescu, Borodaev,

Lutic, Soreanu, Duca, Ratnaweera. A new method of treating liquid waste from the agro-industrial sector in the presence of bioactive additives is described in a chapter by Duca, Sturza, Covaliova, Covaci, Romanciuc, and Tasca. A method of treating air emissions using the new catalyst was developed by Kiose, Rakitskaya, Ennan, and Truba. A fairly new direction in the application of pesticides using nano pesticides is discussed in a chapter by Sucman and Macaev. The studies of vegetable oils, their stability, and their oxidation mechanisms are presented in a chapter by Popovici, Sturza and Gnedov-Moşanu. A more complex approach to the combined treatment of multicomponent wastewater containing organic pollutants and heavy metal ions is described by Duca, Covaliova, Covaliov, Romanciuc, and Vaseashta. The final part of this book introduces several chapters describing some redox technological processes, modified in order to improve their efficiency and expand their practical applications. The multiple redox processes are known to determine the quality of wine in winemaking and were subjects of a meticulous study conducted by Sturza, Scutaru, and Duca. Another type of industrial technology, the processing of wool and the preparation of wool-based materials has been modified essentially due to the use of hydrogen peroxide, as described in a chapter by Condurache, Samoila, and Harabadjiu.

The expected result is the accumulation of new knowledge on the essence of oxidation-reduction processes, the theoretical disclosure and justification of key mechanisms, and the discovery of new aspects of redox reaction control methods, all of which have a major interest in technology, food quality, as well as the state of the environment, especially for water, since clean water resources continue to deplete at a rapid rate. The target audience of this book will consist of professionals and researchers working in the fields of Materials Science, Chemical Engineering, Ecological Chemistry, Water Professionals, and various other related disciplines, for example, Environment, Chemistry, Toxicology, Administrative Sciences and Management, Education, Sociology and topics related to public policy and decision support. In addition, the book will provide information and support to executives involved in managing expertise, knowledge, information, and policy development in different types of scientific and educational communities. In addition, the book will be very useful for many researchers, undergraduate /postgraduate students, scientists, policymakers, industry professionals, practitioners, and some chapters can be adopted as text supplements for independent study. We hope that the information provided by the authors of this book guides them in their research, new discoveries, and many directions, but undiscovered.

Finally, we would like to recognize many contributors for their excellent contributions, as they have been working with passion in this field for a long time. We also thank our reviewers who have tirelessly helped the book with their expert and objective comments. Such observations are essential to ensure that any technical gaps are closed and chapters appear on an ongoing basis. Publishers also recognize the publisher of IGI for his guidance and help with the book. We certainly hope that many researchers, students, and scientists benefit from these chapters for their ongoing investigations to seek answers from these chapters. We also look forward to the opportunity for many readers to become contributors to our later editions. Of course, efforts to prepare books invariably consume a lot of time away from our families and pets. One of the editors (AV) owes a lot of missed playtimes to his very cute companion dog, Isabella, who is an avid frisbee and soccer ballplayer, and another editor (GD) owes a lot of time to his nephew George and the beautiful flower garden, vegetables, garden fruits, and berries - features of his country. We would like to thank some authors from the Republic of Moldova who carried out the research within the national scientific program no.20.80009.5007.27.

Disclaimer: The opinions, recommendations, and proposals presented in different chapters represent the findings, views, and opinions of their authors and may not coincide with the position(s) of the editors. The editors do not promote, endorse, or express their dissent against any product and/or company mentioned in the chapters. Although due precautions are exercised before publication, the editors, however, do not take any responsibility for any infringement of any copyrighted content by the authors of either their own or from elsewhere.

Gheorghe Duca
Institute of Chemistry, State University of Moldova, Chisinau, Republic of Moldova

Ashok Vaseashta Acad
International Clean Water Institute, Manassas, USA, & Transilvania University of Brasov, Brasov, Romania

Section 1
Environmental Aspects of Redox Processes

Chapter 1
Redox Reactions for Chemical Oxygen Demand in the Light of Health, Environmental Protection, and Technical Practicability

Marit Kolb
https://orcid.org/0000-0002-3350-6409
Technische Universitaet Braunschweig, Germany

Müfit Bahadir
Technische Universitaet Braunschweig, Germany

ABSTRACT

Determination of the important wastewater parameter chemical oxygen demand (COD) in international standard procedures is still based on the use of mercury sulfate and potassium dichromate, which are hazardous to health and the environment (REACH). In order to overcome this environmental paradox, alternative oxidizing agents were compared, and a new wet-chemical method using Mn(III) as an oxidant was developed that works without Cr(VI) and Hg(II). Oxidation of potassium hydrogen phthalate as a reference substance was reproducible in a COD range of 20–500 mg/L with a mean recovery of 88.7% compared with the standard Cr(VI)-method. A high correlation (R2 = 0.9935) to the standard Cr(VI)-method with a mean recovery of 78.1% (± 5.2%) was found for a series of industrial and municipal wastewater samples. Although the results of the new method were not 100% of the standard method, its high correlation with the latter and its reproducibility offer an environmentally benign alternative wet chemical method.

DOI: 10.4018/979-8-3693-0512-6.ch001

INTRODUCTION

Pollutants in wastewater (WW) are various and of inorganic nature like heavy metals, toxic ions, and fertilizing compounds (nitrate and phosphate), and organic ones like persistent organic compounds, pesticides, detergents, personal care products, etc. Whilst some of these pollutants are regulated as individual chemicals, many others are difficult to analyze one by one due to practical reasons. Therefore, some group parameters are defined in water legislations and regulations, among which the Chemical Oxygen Demand (COD) is the most important one (Duca et al, 2021). COD is used worldwide to determine the organic load in wastewater, especially for the characterization of the WW and the operation of wastewater treatment plants (WWTP). Furthermore, in the case of mainly industrial WW producers, the taxation of the discharged pollution into the public sewerage is based often on COD.

The determination of the COD as a group parameter is based worldwide since the 1970s on the total oxidation of organic compounds to carbon dioxide and water with the help of strong oxidants after the following redox reactions, e.g., for oxalic acid and for ethanol:

$$\text{Oxalic acid: } C_2H_2O_4 + 0.5\ O_2 \rightarrow 2\ CO_2 + H_2O \tag{1}$$

$$\text{Ethanol: } C_2H_6O + 3\ O_2 \rightarrow 2\ CO_2 + 3\ H_2O \tag{2}$$

If one wants to quantify the amounts of these two organic compounds in WW, there are two opportunities to measure: (i) the amount of oxygen needed to completely oxidize the compounds (COD), and (ii) the total organic carbon that is oxidized to and collected as the amount of CO_2 (TOC). But as can be seen in Eq. 1 and Eq. 2, the two different chemical compounds produce the same amount of CO_2 due to their different oxidation states. Contrary to TOC, the amount of oxygen needed is different: (i) 0.5 mol O_2 for oxalic acid (Eq. 1) and (ii) 3 mol O_2 for ethanol (Eq. 2). This is the reason that TOC is unsuitable and, therefore, usually not used to measure the organic load in WW.

The ratio of COD/TOC depends on the average oxidation state of the organic compounds in WW. Additionally, in the case of COD, phosphorus and sulfur in the molecules are oxidized and contribute to the amount of oxygen needed for total oxidation. In fact, the theoretical oxygen demand for the total oxidation of an organic molecule is based on Eq. 3:

$$C_nH_mO_eX_kN_jS_iP_h + bO_2 \rightarrow nCO_2 + [(m - k - 3j - 3h)/2]H_2O + kHX + jNH_3 + iH_2SO_4 + hH_3PO_4 \tag{3}$$

with X = halogen atoms; $b = n + (m - k - 3j - 2i - 3h)/2 - e/2 + 2i + 2h$

In daily practice, the ratio COD/TOC can vary in a range of 2 to 8, why TOC is not suitable to replace the COD, although it was investigated often to do so due to toxic chemicals needed for the determination of COD after international standards (DIN 38409-41 (1980), ISO 6060 (1986)). Further alternatives to the established wet chemical COD method could be electrochemical, thermal, and photocatalytic oxidation. Here also some devices are already on the market (e.g., from LAR, Mantech, UVT). These devices are offered for online monitoring but have not yet found their way into monitoring laboratories. An important reason for this is the high acquisition costs, especially for smaller laboratories. Another disadvantage is that suspended solids (SS), which are an important component of the total COD, cannot be detected as they can clog tubes and other equipment parts.

THE STANDARD REDOX METHOD FOR THE CHEMICAL OXYGEN DEMAND

Since the end of the 1970ies, the redox reaction with potassium dichromate is the internationally used standard method for COD for instance in ISO 6060 (International Standard Organization, 1986), ASTM D1252-06 (American Society for Testing of Materials, 2012), APHA SM 5220 B (American Public Health Organization, 2011), DIN 38409-41 (Deutsches Institut für Normung, 1980), NEN 6633 (Nederlands Normalisatie Instituut, 2006), SANS 6048 (South African National Standard, 2010). The method requires potassium dichromate ($K_2Cr_2O_7$) in a strong sulfuric acid medium as an oxidizing agent and silver sulfate (Ag_2SO_4) as a catalyst in order to enhance the oxidation of refractory organic substances. A known excess amount of $K_2Cr_2O_7$ in 50% of the sulfuric acid solution containing Ag_2SO_4 is used to oxidize organic substances in water samples under reflux, according to the following reaction (Eq.4) (Bilanovic et al., 1997):

$$C_xH_yO_z + kCr_2O_7^{-2} + kH^+ \rightarrow x\ CO_2 + (y+8x)/2\ H_2O + 2kCr^3, \text{ where } k= 2x/3 + y/6 + z/3 \quad (4)$$

After oxidation, non-consumed $K_2Cr_2O_7$ is titrated with $(NH_4)_2Fe(SO_4)_2$ using Ferroin as an indicator (Jeffery et al., 1989) according to Eq.5.

$$Cr_2O_7^{2-} + 6\ Fe^{2+} + 14\ H^+ \rightarrow 2\ Cr^{3+} + 6\ Fe^{3+} + 7\ H_2O \quad (5)$$

From the $Cr_2O_7^{2-}$ consumption, the equivalent of O_2, which is needed for oxidation, is calculated in mg/L. However, many wastewater samples contain high levels of chloride (Cl^-) in the g/L range in addition to their organic loads. Since Cl^- is also oxidized by $Cr_2O_7^{2-}$ and thus leads to increased COD values, complexation of the chloride ions by mercury ions adding $HgSO_4$ was developed as a very efficient method to remedy this interference (Figure 1).

The principle of the German standard method DIN 38409-H41 (1980) for the determination of COD in the range from 15 to 300 mg/L is shown in Figure 2. For the other standard methods mentioned above, the procedures are very similar.

ENVIRONMENTAL AND HEALTH ISSUES OF THE STANDARD COD

For the measurement of an environmentally relevant parameter, relatively large amounts of waste harmful to the environment and health are generated in this method. According to the DIN 38409-H41 (1980), for a single COD determination in a wastewater sample, 0.80 g $HgSO_4$ is used to complex Cl^- and 0.058 g $K_2Cr_2O_7$ in concentrated sulfuric acid is used for oxidation. For a laboratory such as that of a large German wastewater association, which analyzes wastewater from sewage treatment plants for 3.6 million inhabitants and from various industrial and commercial enterprises, the annual consumption of hazardous compounds for COD determination is, e.g., 8.0 kg $HgSO_4$ and 588 g $K_2Cr_2O_7$ (Korber, 2014). The resulting laboratory waste is classified as hazardous waste and must be collected and disposed of separately due to its high toxicity. This causes high disposal costs and corresponding risks for humans and the environment during use and disposal.

Figure 1. COD determination of a potassium hydrogen phthalate solution (PHP) with a theoretical COD of 200 mg/L according to the German standard dichromate method (DIN 38409-41) using Ag_2SO_4 as catalyst, with and without $HgSO_4$ as complexing agent; 1 mg Cl^- results in 0.226 mg COD/L; (*) measuring range was exceeded

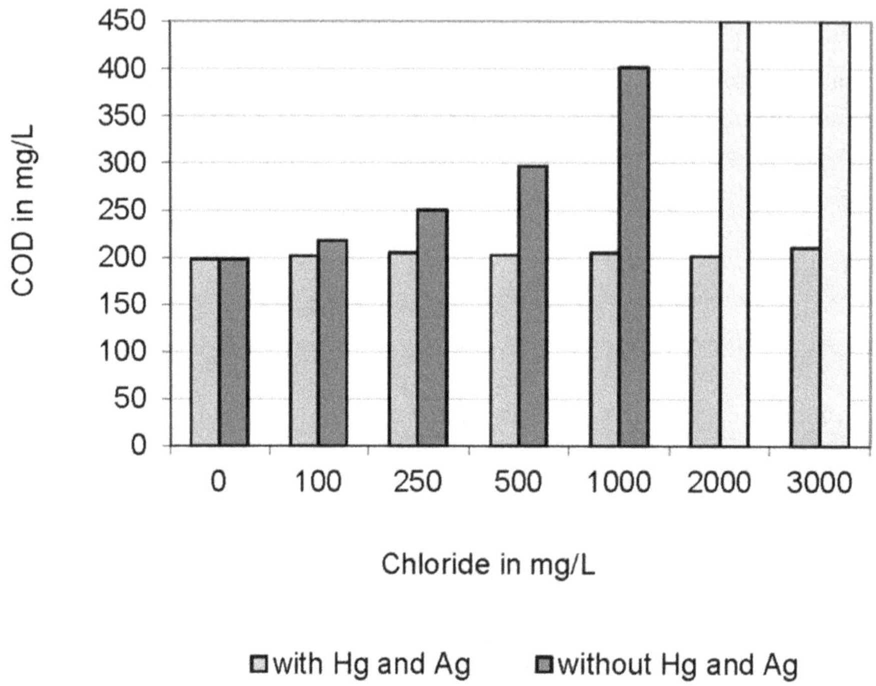

Hg is one of the most toxic non-essential heavy metals (Drasch et al. 2004). $HgSO_4$ is classified as very toxic and environmentally hazardous (LD_{50} oral rat 57 mg/kg (IFA 2022)). In 2008, Hg was classified as a priority hazardous substance (SVHC) by the EU WFD (EU (2008)), resulting in a "phasing out". Therefore, mercury and its compounds were included in REACH Annex XVII restricted substances list (ECHA 2016) to reduce Hg emissions and restrict its use. According to the restriction conditions, Hg compounds may no longer be placed on the market or used if they are intended for use as antifouling, wood preservatives, impregnation of heavy-duty textiles and yarns, and treatment of industrial waters. The EU RoHS 2 (Directive 2011/65/EU) restricts Hg and its compounds in all electrical and electronic products. In the frame of the global Minamata Convention on Mercury (2017) human health and the environment shall be protected from adverse effects by Hg, e.g., by the ban of new mercury mines, phasing out of existing mercury mines, as well as import and export restrictions. An exemption has been made so far for quantities used for laboratory-scale research or as reference standards and for some products, where no feasible mercury-free alternatives for replacement are available. Storage and disposal of Hg-containing wastes are subject to strict conditions. More than 140 countries have ratified the Convention and more restrictions might be expected.

Figure 2. Flow diagram of COD determination according to the German standard method DIN 38409-H41 (1980)

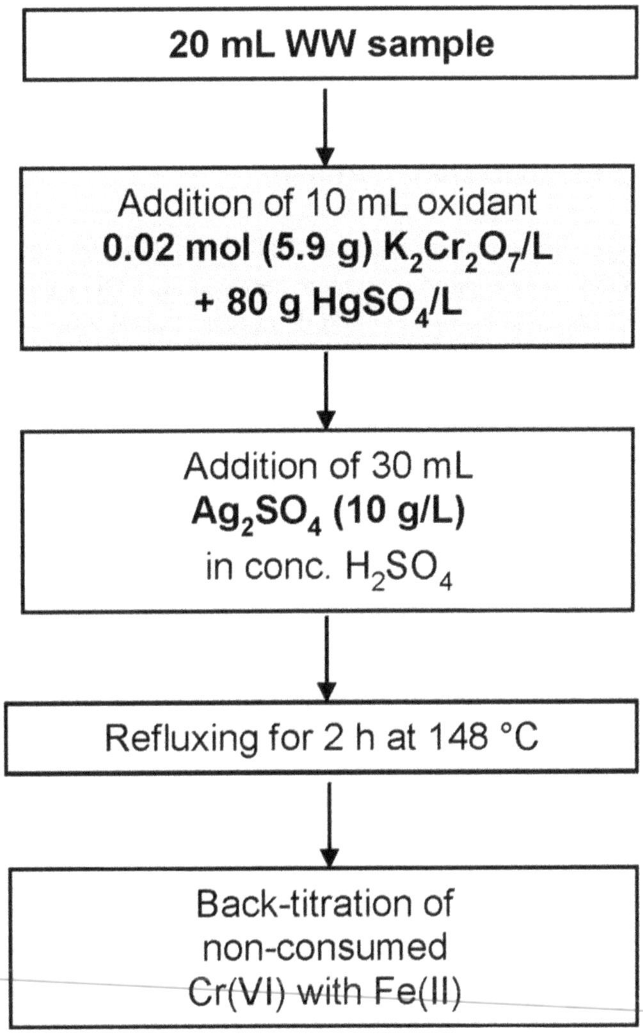

Cr(VI) inhibits plant growth, has high fish (96 h LC_{50} 51.1 mg/L (median), IFA 2022), crustacean (48 h LC_{50} 0.12 mg/L (median), IFA 2022), and algal toxicity (48 h LC_{50} 0.61 mg/L (median), IFA 2022), produces skin and respiratory irritation, and is classified as a lung carcinogen by the International Agency for Research on Cancer (Stoecker 2004); there are reasonable grounds to suspect reproductive toxicity and mutagenicity (IFA 2022). Hence, $K_2Cr_2O_7$ was added by ECHA (2010) to the list of substances of very high concern due to its CMR (carcinogenic, mutagenic, and toxic for reproduction) properties in 2010 and to the authorization list under REACH in 2013 (EU 2013). Therefore, companies that want to continue using $K_2Cr_2O_7$ after the sunset date of 21.09.2017 (01.03.2023 for some exceptions) needed to prepare an application for authorization and to have a positive authorization decision by the European Commission.

Since both substances Hg and $K_2Cr_2O_7$ are classified as toxicologically of very high concern and their future use is, therefore, very restrictively handled by the EU, alternatives to the classic dichromate method for COD determination are highly desirable.

ALTERNATIVES TO THE STANDARD Cr(VI)/Hg-COD DETERMINATION

Small Scale Sealed Tube COD Determination

The ISO 15705 (2002) was the first international standard for a small-scale sealed tube COD determination. In tests, it was shown that there is good agreement between the COD values obtained according to ISO 6060 (1989) and those obtained by this method. However, it cannot be assumed that this method will give results comparable to those obtained according to ISO 6060 in all cases without exhaustive testing. Furthermore, it is difficult to obtain a representative 2 mL sample out of a high volume of WW especially in the case of WW with a high content of suspended solids. However, as a result of such a low sample volume, there are big differences in needed quantities of potassium dichromate (0.006 g/L), silver sulfate (0.021 g/L), and mercury sulfate (0.088 g/L) used per analysis compared to the classical open reflux and titration method. Commercially available COD cell test uses around 90% less of toxic chemicals compared to the classical methods (Decker et al. 2015). The amount of dichromate consumed for oxidation of the sample is determined by measuring the Cr(III) ions formed at a wavelength of 600 nm ± 20 nm for a mass concentration up to 1000 mg/L. The absorbance is measured in the digestion tube that serves as the cuvette; the value is converted using a calibration line to a COD value. For a smaller calibration range up to 150 mg/l, an alternative measurement can be made at a wavelength of 440 nm ± 20 nm. At this wavelength, the absorbance of the non-consumed Cr(VI)-ions is measured. For turbid and atypically colored digested samples, also a titration is carried out with a standardized ammonium Fe(II) sulfate solution as in ISO 6060 (1989) or DIN 38409-H41 (1980). However, if toxic chemicals such as Hg and Cr(VI) are to be avoided in the future, the use of less problematic reagents for oxidation the organics and masking the chloride ions will be required.

Alternative Oxidizing Agents

Alternative oxidizing agents for $K_2Cr_2O_7$ should have at least the same or a greater redox potential than Cr(VI) with E^0= 1,36 V, and they should be less hazardous. Such oxidation agents are listed in Table 1.

However, kinetic limitations can prevent oxidation in reality despite theoretically suitable redox potential. For this reason, MnO_4^- proved to be an unsuitable oxidizing agent for the determination of COD (Cameron and Moore, 1957). Phthalic acid, which is oxidized to 100% with Cr(VI) in the standard COD procedure, could only be oxidized to 1% with MnO_4^- (Janicke, 1983). Oxidizing agents such as Fenton's reagent H_2O_2/(Fe^{2+}) and $S_2O_8^{2-}$, on the other hand, are known as very strong oxidizing agents and can completely oxidize organic compounds in WW (e.g., Rivas et al. 2001, Kilic et al. 2014).

$S_2O_8^{2-}$ is successfully used for the quantitative determination of TOC in cuvette test due to its high oxidation potential (e.g., LCK 380, Hach Lange, Düsseldorf, Germany). Here, the organic ingredients are oxidized to CO_2 with an excess of $S_2O_8^{2-}$, and the CO_2 formed is quantified by a color reaction. However, these reagents are unstable in solution and decompose relatively quickly even without the addition of reactants (blank), so a quantitative determination of the redox reaction and thus of the COD is not possible. This fact is demonstrated by the example of the oxidation of potassium hydrogen phthalate (PHP) with $S_2O_8^{2-}$ in Figure 3.

Thus, the oxidation reactions with $S_2O_8^{2-}$ resulted in very high blank values (reagent without PHP) in each case, which was in the order of magnitude of the respective theoretical oxygen demand of the samples.

Table 1. Overview of oxidizing agents

Redox pair	E^{0} a)	Remarks
$Cr_2O_7^{2-}$, H^+/Cr^{3+}	1,36 V	Catalysis by Ag is required for the oxidation of many recalcitrant organic compounds.
MnO_4^-, H^+/Mn^{2+}	1,51 V	The reaction rate is relatively low, many organic compounds are only incompletely oxidized.
Mn^{3+}/Mn^{2+}	1,54 V	Comproportionation of MnO_4^- and Mn^{2+} produces Mn^{3+}, stable in sulfuric acid solution with excess Mn^{2+}, no catalyst required.
Ce^{4+}/Ce^{3+}	1,72 V	Similar oxidation rates as MnO_4^-, H^+/Mn^{2+}
$S_2O_8^{2-}/SO_4^{2-}$ $S_2O_8^{2-}$, H^+/SO_4^{2-}	2,01 V 2,1 V	Sulfate radical provides the actual contribution to the standard potential, and the formation of the radical is rate-determining, autocatalytic decomposition at low pH
H_2O_2 (Fe^{2+})/ OH^-, OH•	2,8 V	Low stability of H_2O_2

a) CRC Handbook of Chemistry and Physics (2013-2014), W.M. Haynes (ed.), 94th edition.

Figure 3. COD-Value of potassium hydrogen phthalate (PHP, theoretical values of 20, 100, 200 and 300 mg COD/L) using $S_2O_8^{2-}$ as oxidizing agent under different reaction conditions. Blank: reagent without PHP.

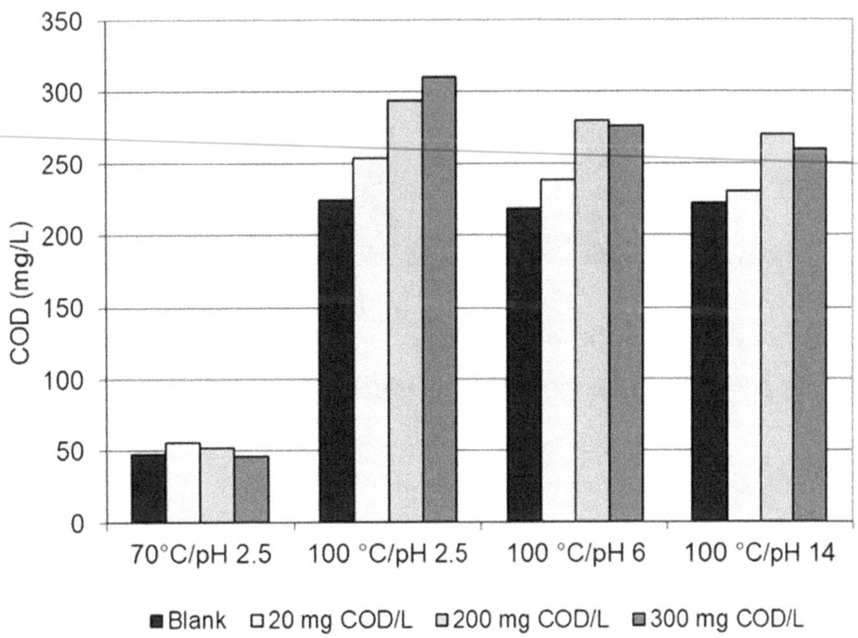

Ce(IV) is a very strong oxidant and is used in 5% sulfuric acid for the determination of medical drugs (Cerimetry in Hansen et al. 2012). It was also reported for the determination of COD in flow injection analyses. However, using this technique for a mixture of glutamic acid and lactose, Korenga et al. (1993) measured oxidation rates similar to those obtained, when $KMnO_4$ was used for the oxidation, and the results were much lower than those determined for oxidation with $K_2Cr_2O_7$. Thus, oxidation with Ce(IV) is not suitable to replace the Cr(VI)-COD.

Mn(III) reaction is based on the comproportionation of MnO_4^- with Mn^{2+} in sulfuric acid after Eq. 5:

$$2\,KMnO_4 + 8\,MnSO_4 + 8\,H_2SO_4 \rightarrow 5\,Mn_2(SO_4)_3 + K_2SO_4 + 8\,H_2O \quad (5)$$

This reaction is used in Hach cuvette tests (Manganese III Reactor Digestion Method, Hach 2014) at 0.5 mL volumes for the rapid determination of COD. Scaling up volumetric amounts for daily practice is somehow not easy. Also, the representative sampling of just 0.5 mL from a vast amount of WW rises problems, in particular, if SS is involved. Therefore, the Mn(III)-cuvette test is good and valid as a rapid test, but for the wet chemical, volumetric determination of COD in WWTP operation laboratories it is less suitable. However, in an experiment with Mn(III)-cuvette test compared with the Cr(VI)-method (DIN 38409-41 (1980)) different reference substances had similar oxidation rates which indicated the potential of Mn(III) as an alternative oxidant for Cr(VI) (Figure 4).

Figure 4. Comparison of COD-values of selected reference substances determined with Cr(VI)-COD according to DIN 38409-H41 (1980) and with Mn(III)-COD Hach cuvette test.

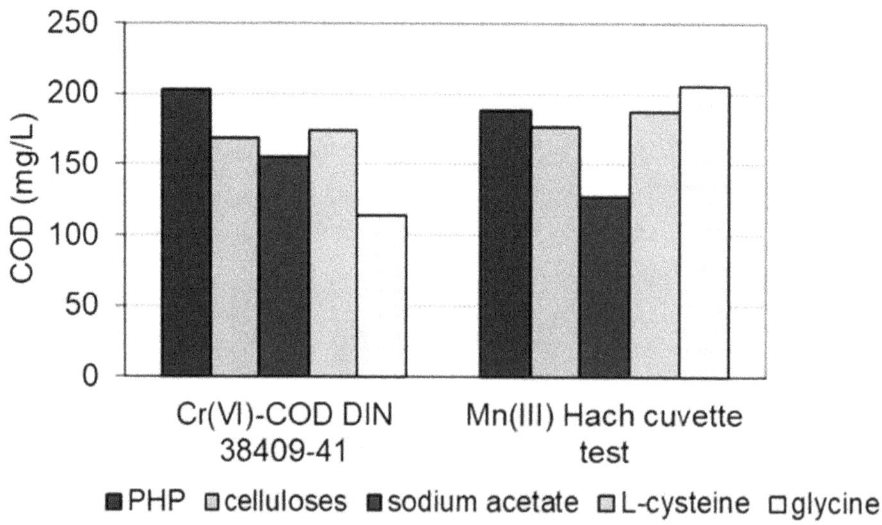

Furthermore, no negative environmental effects have been reported in connection with Mn(III). For humans, chronic damage to the lungs (manganese pneumonia) has been reported after inhalation of Mn(III) oxide dust during the mining and processing of brownite and manganite (IFA 2022). However, such dusts do not occur when using a Mn(III)-reagent.

Manganese (III) as Oxidizing Agent in COD Determination

As a result of the evaluation of the literature and after preliminary experiments, Mn(III) was found to be the best candidate for the substitution of Cr(VI) in the COD determination. In order to obtain a sufficient oxidation yield, temperatures well above 100 °C proved to be necessary. The required increase in boiling points is achieved by sulfuric acid concentrations above 40%-wt. (4.9 mol/L) (Gmelin Institute 1963). In the DIN 38409-H41 (1980), for example, a temperature of 148 °C is achieved at a Cr(VI) concentration of 3.3 mmol/L in the digestion solution by a sulfuric acid concentration of 9 mol/L under reflux.

Based on the work of Miller et al. (2001) and Domini et al. (2009) a reagent solution was developed. The Mn(III)-reagent solution was prepared by mixing a solution of $KMnO_4$ and $MnSO_4/H_2SO_4$. The comproportionation of Mn(VII) of $KMnO_4$ and Mn(II) of $MnSO_4$ resulted in Mn(III) according to Eg. 5. The reaction of Mn(II) with Mn(VII) to Mn(III) was documented by UV-VIS measurements. The spectra of a pure $MnSO_4$ or a pure $KMnO_4$ solution in the concentrations of the final reagent were obviously different from the spectrum of the prepared Mn(III) reagent (Figure 5, pink line). The spectrum of the Mn(III)-reagent has a broad maximum at 514 nm in the VIS region and a shoulder at 290 nm in the UV region. Miller et al. (2001) published the VIS spectrum of Mn(III) from 400 - 700 nm with the same shape indicating that the maximum was at $\lambda = 510$ nm.

Figure 5. UV/VIS spectra of Mn(III)- reagent (0.009 mol/L), $KMnO_4$ (0.0017 mol/L), and $MnSO_4$ (0.038 mol/L) concentrations based on final volume of Mn(III)-reagent, and $KMnO_4$ (0.0017 mol/L) mixed with H_2SO_4 without addition of $MnSO_4$, concentration as for Mn(III) reagent, measured in 1 mm quartz cuvettes

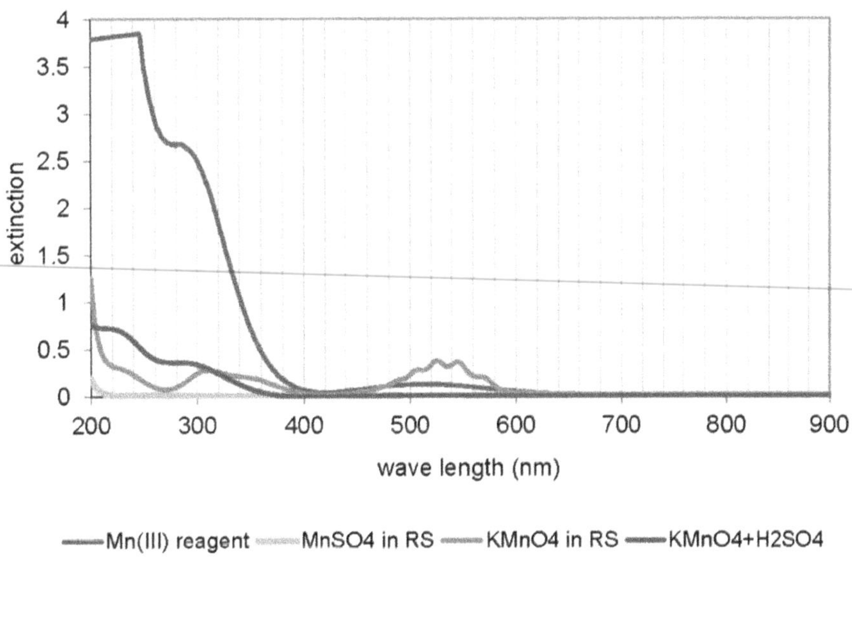

Furthermore, $KMnO_4$ was mixed only with sulfuric acid ($KMnO_4 + H_2SO_4$) at the same concentration as for reagent preparation but without addition of $MnSO_4$ and heated in the same way as the Mn(III)-reagent for COD determination, in order to show that the comproportionation of Mn(VII) and Mn(II)

really occurred. The clear change in the spectrum compared to the water solution of $KMnO_4$ showed decomposition of the $KMnO_4$ (Figure 5, blue line). The spectrum of the reaction product was similar to that of the Mn(III) reagent (Figure 5, red-brown line), but with maxima of much lower intensity. The correspondence of the spectra indicates that a small concentration of Mn(III) was formed by disproportionation of the $KMnO_4$ during decomposition. This indicates that the $KMnO_4$ solution was not stable under the harsh reaction conditions required for oxidation, making $KMnO_4$ in sulfuric acid unsuitable as an alternative reagent. The spectrum of $MnSO_4$ did not show any absorption bands in the measured wavelength range (Figure 5, rose line). It can be concluded from this that the characteristic bands of the reagent spectrum did not result by overlaying the spectra of $KMnO_4$ and $MnSO_4$, but that a reaction actually took place between $KMnO_4$ and $MnSO_4$.

The composition and preparation of the reagent were optimized by varying the concentrations of $KMnO_4$, $MnSO_4$ and H_2SO_4 in different directions (Table 2). The changes were controlled by the COD recovery rates of a PHP standard. Another aspect of the optimization was the stability of the reagent and the reproducibility of the measurement. The oxidizing power did not increase continuously with increasing H_2SO_4 concentration and the associated increase in boiling temperature, but a maximum was obtained at an H_2SO_4 concentration of 8.6 mol/L and thus a boiling point of 135 °C, and Mn(III) concentration of 0.0084 mol/L (variant B2) with recovery rates for PHP of 92% on average (Table 2). At higher or lower concentrations of H_2SO_4, the recovery rates of PHP were reduced.

Table 2. Boiling point of Mn(III) reagent and recovery rates of PHP with a COD of 200 mg/L at different concentrations of Mn and H_2SO_4.

Reagent variants	**c (H_2SO_4) mol/L**	**Temp. (°C)**	**COD PHP (mg/L)**	**Recovery (%)**
c(Mn) = 0.007 mol/L n(Mn(II)/Mn(III))=5/1 n=2				
A1	9,0	138	148	74
A2	10,8	157	170	85
A3	12,6	180	158	79
A4	14,4	184	125	63
c(Mn) = 0.0084 mol/L n(Mn(II)/Mn(III))=3,7/1 n=3				
B1	8,3	130	172	86
B2	8,6	135	189	92
B3	10,8	157	148	74

The concentration data do not take into account the volume contraction that occurred due to the addition of concentrated sulfuric acid to the reagent. The real concentration of Mn(III) was determined for reagent variant B2 by titration with Fe(II). It was 0.009 mol/L. This concentration was near the solubility limit of Mn(III) of 0.01 mol/L given by Miller et al. (2001).

A mean recovery rate of 87.2% ± 4.0% was achieved with the optimized Mn(III)-reagent (0.009 mol/L) in a total of 73 determinations of the Mn(III)-COD of PHP with a theoretical COD of 200 mg/L. In 15 replicates of COD determination with Cr(VI) according to DIN 38409-H41 (1980), a mean recovery of 99.6% ± 2.4% of the PHP was obtained. In order to explain the different recovery rates of PHP with Mn(III) and Cr(VI), a post-oxidation of PHP-samples was performed with Cr(VI), which were first oxidized with Mn(III). However, this did not result in any consumption of Cr(VI)-reagent. This suggests that not parts of PHP or further oxidizable degradation products were responsible for the lower recovery rates. One plausible explanation might be that during the reaction with Mn(III), oxidation stopped in part at CO instead at CO_2. However, this could not be proven experimentally. Prolongation of the digestion time, digestion with microwave, reduction of the sample volume and addition of different catalysts like Ag^+, Cu^{2+} and TiO_2 did not further improve the recovery rates of PHP with the Mn(III)-reagent.

The oxidation of PHP with the optimized Mn(III)-reagent was reproducible and linear ($R^2 = 0.9996$) in a COD range from 20 to 500 mg/L (Figure 6).

Figure 6. Linearity of PHP oxidation with Mn(III)-reagent

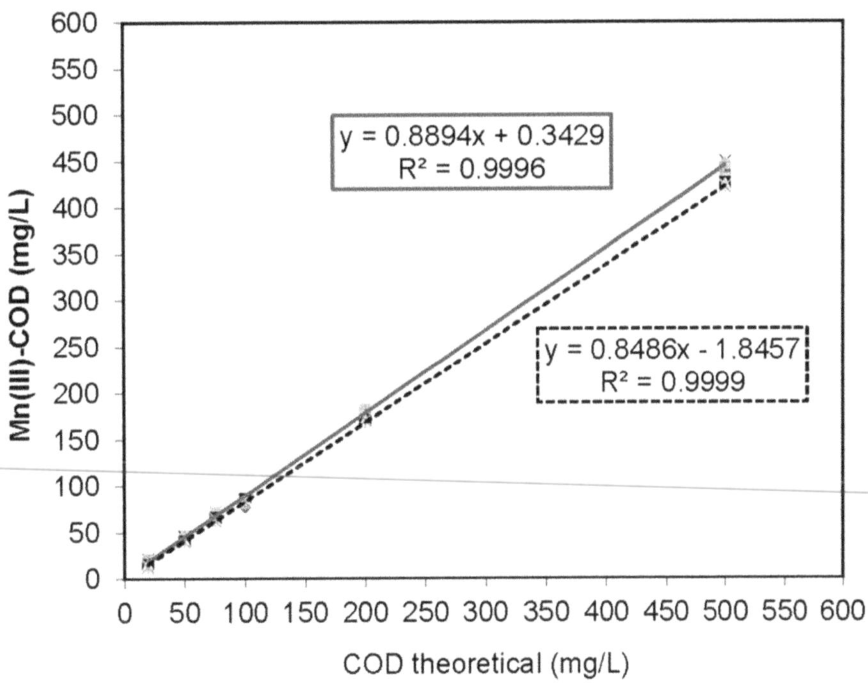

Solid line: Mn(III) oxidation of PHP standard solutions (20, 50, 75, 100, 200, 500 mg COD/L), n= 9 without sample preparation.

Broken line: Mn(III) oxidation of PHP standard solutions (20, 50, 75, 100, 200, 500 mg COD/L) spiked with 1000 mg Cl^-/L with sample preparation by precipitation of AgCl, n= 5.

The relative standard deviation (RSD) in the concentration range of 50 to 500 mg COD/L was ≤ 3.5% (n=9) with a mean recovery rate of 88.7%. For the lowest concentration (20 mg COD/L), the RSD was 8.4%.

PHP standard solutions were also spiked with a high chloride concentration of 1000 mg Cl^-/L in order to investigate the influence of the sample preparation steps for chloride masking (see chapter 4.3). Even under these conditions, a very good linear correlation ($R^2 = 0.9999$) and good reproducibility of the data were obtained. The RSD in the concentration range from 50 to 500 mg/L was ≤ 2.1% (n=5, Figure 6), with a slightly lower mean recovery of 84.1% due to the additional sample preparation step. The RSD for the PHP standard at 20 mg COD/L was 7.7%.

These results were satisfactory, but, recoveries and RSDs for the oxidation of a PHP standard with Cr(VI)-reagent under standard conditions published by Almeida et al. (2013) were slightly higher than those we determined with the Mn(III)-reagent. Almeida et al. (2013) determined in the concentration range of 50 to 500 mg COD/L mean recoveries of 101.9% and relative standard deviations of < 1.45% (n=5). At a concentration of 20 mg/L, their recovery rate was 110% and the coefficient of variation was 2.1%.

Manganese(III)-COD Free of Chromium(VI) and Mercury

In addition to the development of an Cr(VI)-free oxidation method, a Hg-free method for the elimination of interfering chloride ions from water samples was the second target of the method development. Precipitation of chloride with $AgNO_3$ was used as a substitute for masking chloride ions with $HgSO_4$. Other methods using ion exchangers, complexing with Sn(IV) or oxidation of chloride by means of Bi(V) or co-oxidation of chloride and subtraction of the resulting Cl-COD value from the total one were tested, but did not prove to be successful.

However, it was known from the literature (Cameron and More, 1957; Wagner 1976) and confirmed in our own experiments that precipitation of chloride ions as AgCl in the samples reduces the COD content by co-precipitation of suspended solids (SS) and colloids. In order to include these COD fractions in the analysis, a dual filtration was introduced for chloride elimination (cf. Figure 7).

Step 1: If the samples contained suspended solids SS or colloids, they were separated after coagulation and filtration after adding $Al_2(SO_4)_3$ and NaOH to the water samples and forming $Al(OH)_3$. The filter residue FR was digested separately for COD-determination.

Step 2: In the filtrate FL, the dissolved chloride ions were precipitated with $AgNO_3$, the AgCl precipitate was removed with a glass fiber filter in a filter apparatus made of ultra-pure perfluoroalkoxy polymer (PFA). This ultra-pure polymer was used because of its extremely smooth and hydrophobic surface, and thus avoiding any losses of sample contents through physical-chemical interactions like adsorption.

Step 3: During the first two steps, the sample volume was increased from 10 mL to 12 mL. After the separation of the AgCl-precipitate, an aliquot of 5 mL was taken from the filtrate FL and subjected to oxidation with the optimized Mn(III)-reagent in the third step. In this process, the COD_{FL} was obtained. Because of the use of an aliquot of 5 mL for oxidation from 12 ml of the total filtrate, the COD_{FL} had to be multiplied by a factor of 2.4. The total COD is the sum of the two fractions ($COD_{turbid} = COD_{FR} + 2.4 \times COD_{FL}$). In the case of a clear water sample without colloidal contents, filtration was performed only once after the addition of 1 mL $AgNO_3$. In this case, only this filtrate was used for COD-determination. From this filtrate, a 5 mL aliquot was oxidized in step 3. By increasing the sample volume to 11 mL, the COD_{FL} in this case had to be multiplied by a factor of 2.2 ($COD_{clear} = 2.2 \times COD_{FL}$). The details of the sample preparation and digestion with Mn(III) were published in Kolb et al. (2017).

Figure 7. Flow chart of the Mn(III)-COD determination including three working steps

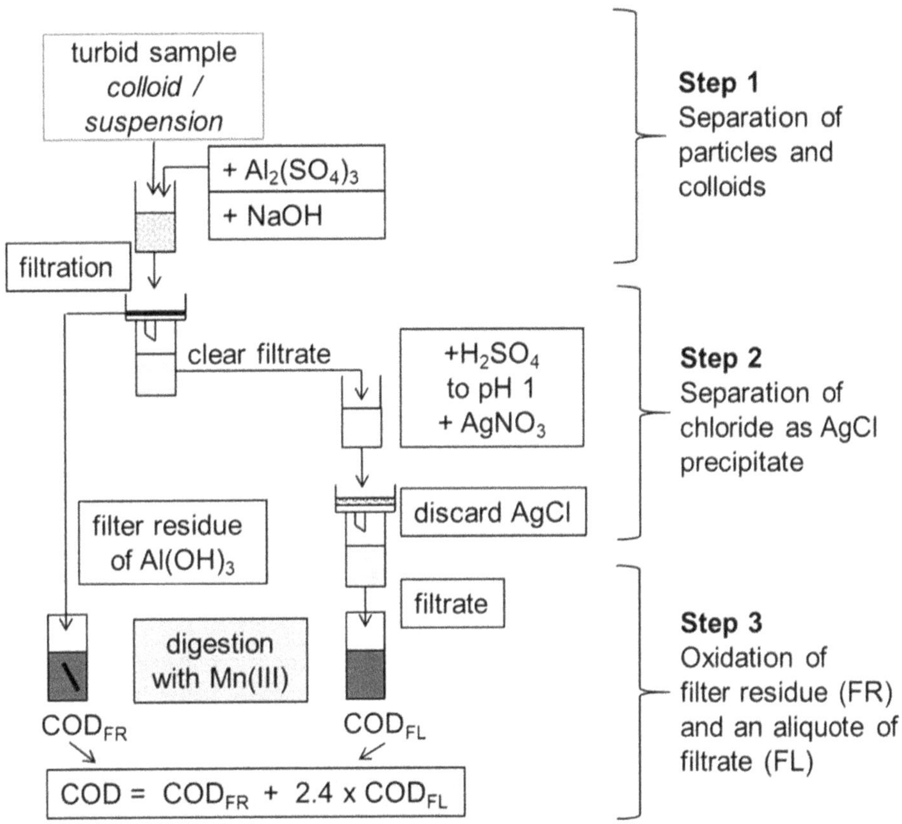

Separate determinations of the two fractions COD_{FR} and COD_{FL} by oxidation with Mn(III) and also with Cr (VI)/Hg from a municipal WW sample and an industrial WW were compared (Table 3). In the case of the oxidation with Mn(III), the COD recovery rates of each fraction were lower than with Cr(VI)/Hg according to the DIN 38409-41 (1980). A recovery rate of 77% was obtained for the municipal WW sample and 82% for the industrial one. These recovery rates were slightly lower than the mean recovery rate of 84.1% for the PHP standard with 1000 mg Cl/L (cf. chapter 3.2). When comparing the two oxidation methods Mn(III) and Cr(VI), the results for the distribution of COD between the two fractions COD_{FR} and COD_{FL} were the same within the analytical range of variation. For the municipal WWs, about 75% of the COD was in the filter residue (COD_{FR}) and 25% of the COD was in the filtrate (COD_{FL}) for each of the oxidation methods. For industrial WW, the distribution was opposite, with 90% of the COD determined in the filtrate (COD_{FL}) and 10% in the filter residue (COD_{FR}) (Table 3).

In addition, total WW samples were oxidized with Cr(VI)/Hg without further sample pretreatment according to the standard method (DIN 38409-41 (1980)). The results for the oxidation of the total samples (COD_{DIN}) and for the oxidation of the two separate fractions (COD_{FR+FL}) with Cr(VI)/Hg were almost identical (municipal WW 716 and 715 mg/L, industrial WW 2766 and 2710 mg/L) (Table 3). This demonstrated that the sample preparation developed to eliminate the chloride interference did not reduce the COD contents in the samples.

Table 3. COD determination of the two fractions COD_{FR} and COD_{FL} after sample preparation of a municipal and an industrial WW with the developed method. Oxidation with Mn(III) (n=2) and Cr(VI)/Hg (n=1) reagents compared to the COD determination with the standard method DIN 38409-41 (1980).

Municipal WW							
	Cr/Hg			Mn/Ag			
	$\bar{X}$ (mg/L)	D (%)	$Rec._{DIN}$ (%)	$\bar{X}$ (mg/L)	RSD (%)	D (%)	$Rec._{DIN}$ (%)
COD_{FR}	535	**74.7**	74.7	405	*1.1*	**73.6**	
COD_{FL}	181	**25.3**	25.3	146	*8.4*	**26.4**	
COD_{FR+FL}	**716**	100	**100**	**550**	***3.0***	100	**76.9**
COD_{DIN}	**715**		100				100
Industrial WW							
	Cr/Hg			Mn/Ag			
	$\bar{X}$ (mg/L)	D (%)	$Rec._{DIN}$ (%)	$\bar{X}$ (mg/L)	S_R (%)	D (%)	$Rec._{DIN}$
COD_{FR}	256	**9.3**	10.6	224	*2.88*	**10.1**	
COD_{FL}	2510	**90.7**	92.6	1992	*1.17*	**89.9**	
COD_{FR+FL}	**2766**	100	**102**	**2216**	**0.76**	100	**81.8**
COD_{DIN}	**2710**		100				100

D: Percentage distribution of COD_{FR+FL} on the two fractions COD_{FR} und COD_{FL}
$Rec._{DIN}$: Recovery rate of (COD_{FR+FL}) compared to COD_{DIN}

The COD values of different WW samples (4 municipal WWs, 3 industrial WWs and 2 landfill leachates) determined by the Mn(III)-method had a high correlation ($R^2 = 0.9935$) with the COD values determined by the Cr(VI)/Hg method according to DIN 38409-41 (1980) (Figure 8). The mean recovery rate of the Mn(III)-COD of all WW samples was 78.1% (± 5.2%) compared with Cr(VI)-COD. This means that the conversion of the Mn(III)-COD to the standard Cr(VI)-COD with a constant factor is possible, which is important for practical application.

Two WW samples with a high COD content were analyzed at different dilution levels with the Mn(III)-method and the DIN 38409-4 (1980) as well. Linear regressions with high coefficients of determination (R^2) of 0.9992 for the industrial and 0.9851 for the municipal WW sample were obtained (Figure 9). This demonstrated that the Mn(III)-method well correlated with the Cr(VI)-method even at different sample dilutions.

Figure 8. Correlation of Mn(III)-COD (COD_{FR} + COD_{FL}) with Cr(VI)-COD determined according to DIN 38409-41 (1980) from different WW samples (COD contents 34-3100 mg/L, chloride contents 50-1600 mg/L, n=2)

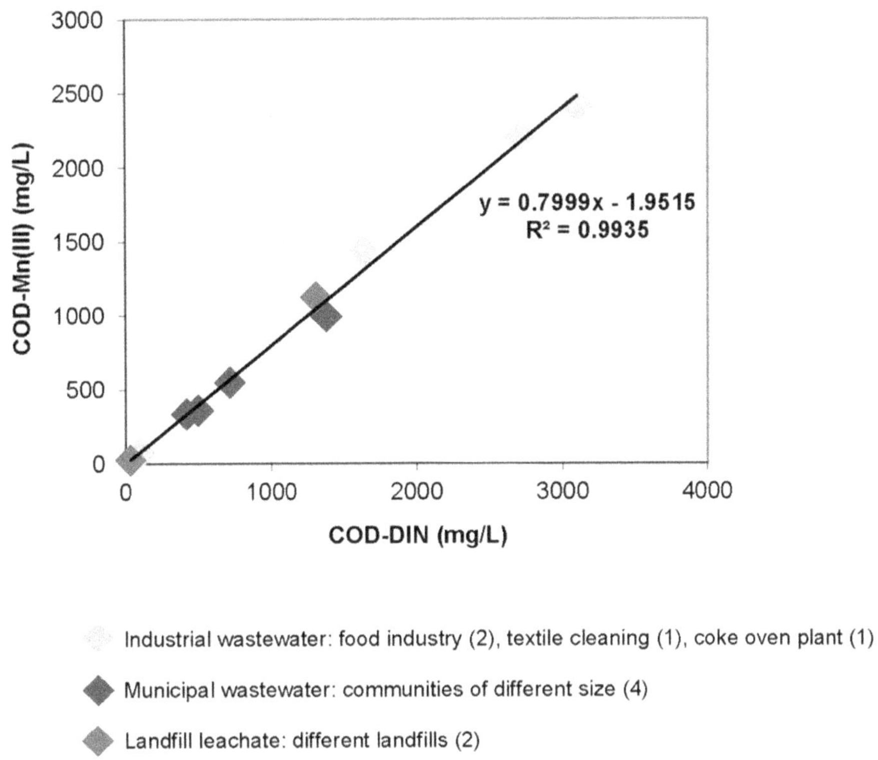

Figure 9. Linear correlation of the COD contents of a dilution series of an industrial and a municipal WW sample after processing with the developed Mn(III)-method

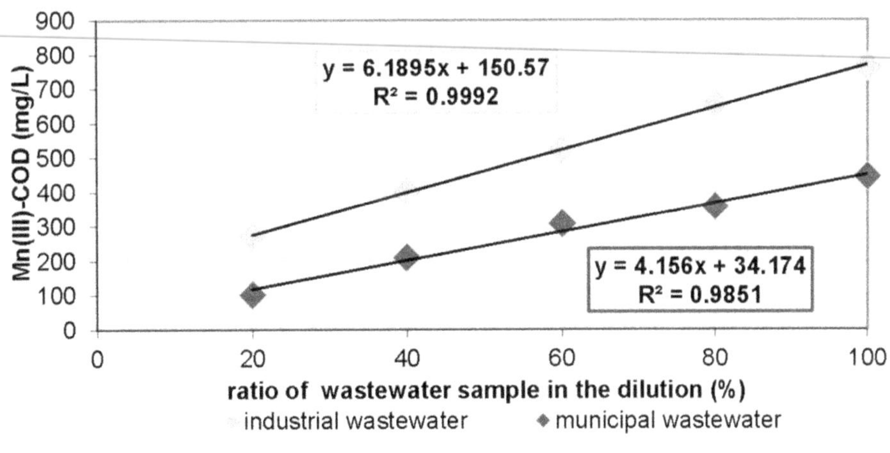

In order to determine the repeatability of the method, 4 to 5 parallel approaches of 3 different WW samples (2 municipal WWs and 1 landfill leachate) were analyzed by Mn(III)-method. For the filter residues (COD_{FR}), the relative standard deviations (RSD) ranged from 4.5% to 21% and thus were significantly higher than the results for the filtrates (COD_{FL}) with RSDs of 1.5% to 2.8%. With regard to the total COD (COD_{FR} + COD_{FL}), the RSD values ranged from 3.6% to 5.1% and thus showed a high repeatability of the Mn(III)-method.

To further qualify the results obtained with the new Mn(III)-method, data from the literature on the analysis of WW samples with the classical Cr(VI)/Hg-method were searched. Almeida et al. (2013) analyzed three different WW samples (domestic WW, dairy WW, and dye WW) five times each using the Cr(VI)/Hg method and determined RSDs ranging from 1.12 to 1.57%. The reproducibility with the Cr(VI)/Hg method was thus somewhat higher in these analyses than in ours with the Mn(III) method.

When digesting a PHP standard, the limit of quantification of the Mn(III)-method was 20 mg COD/L without further sample preparation. At a concentration of 20 mg/L, the volume difference in titration compared to the blank was only about 1 mL, and the value for RSD increased from ≤ 3% to 8%. For COD determinations of real water samples with COD below 50 mg/L, the developed method reached its limits, since the total COD was usually composed of the two fractions COD_{FR} and COD_{FL}. Thus, if the total COD is 50 mg/L, the concentrations for the two fractions CODFR and CODFL are each < 50 mg/L. However, since the limit of quantification for the titration is 20 mg/L, the COD of each fraction must be at least 20 mg/L. This became evident in the determination of Mn(III)-COD of 3 effluent samples from municipal WW treatment plants with COD contents < 35 mg/L. The recovery rates were appreciably over 100%.

CONCLUSION

In the analysis of WW samples with the Mn(III)-method presented here, recovery rates with an average of just 78% in comparison to the standard method with Cr(VI)/Hg were obtained. However, the results showed a high correlation between the Mn(III)-method and with Cr(VI)-method. In comparison to literature data for COD determination of WWs with the Cr(VI)/Hg-method, the reproducibility was somewhat lower but can be described as good overall.

It is thus possible to convert the results obtained with the newly developed Mn(III)-method to the results of the Cr(VI)-standard method using a conversion factor within a certain range of variation.

The requirement for the method to include suspended solids was also met. The separate analyses of suspended solids and colloids in the COD_{FR} can even be of advantage for the operation of WW treatment plants, since both values, that for the particles and that for the dissolved substances, are of importance and interest for the WWTP operators. In this context, the application of the Mn(III) wet-chemical method leads to benefits in terms of both occupational health and safety and avoidance of hazardous waste in accordance with the objectives of the EU within the framework of the EU WFD and REACH, as well as the international development of environmental quality standards.

Another advantage of the Mn(III)-method developed is that the standard digestion apparatus for standard COD determination available in the laboratories can continue to be used. Titration can be carried out with the same equipment and materials as for the standard method. Only the filtration apparatus with a PFA-filter holder would have to be purchased additionally. This means that the method remains cost-effective in terms of the equipment to be used. This is particularly advantageous for the implementation of such a method in developing and newly industrializing countries, where the provision of funds and aspects of occupational health and safety as well as waste disposal are much more problematic than in the richer industrialized countries.

However, the application of the method is not entirely straightforward. It requires a very accurately and precisely working both in the preparation of the Mn(III)-reagent and in the sample preparation because of the required fractionation of turbid samples into a particle- and colloid-containing (COD_{FR}) and a water-soluble fraction (COD_{FL}). The mixing of the reagent with the samples for digestion must also be done correctly to ensure reproducible results. Overall, however, the advantage of avoiding neurotoxic and carcinogenic substances in the determination of this important environmental parameter outweighs the disadvantage of a slightly lower reproducibility, more time-consuming training and implementation by far.

REFERENCES

Almeida, C. A., Savio, M., González, P., Martinez, L. D., & Gil, R. A. (2013). Determination of chemical oxygen demand employed manganese as an environmental friendly oxidizing reagent by a low injection method base on microwave digestion an speciation coupled to ICP-OES. *Microchemical Journal*, *106*, 351–356. doi:10.1016/j.microc.2012.09.007

American Public Health Organization. (2011). APHA 5220, Standard methods for the examination of water and wastewater. Chemical oxygen demand. (COD). doi.org/10.2105/SMWW.2882.103

American Society for Testing of Materials. (2012). ASTM D1252-06, Standard test methods for chemical oxygen demand (dichromate oxygen demand) of water, West Conshohocken, PA, USA.

Bilanovic, D., Loewenthal, R. E., Avnimelech, Y., & Green, M. (1997). Potentiometric measurement of chemical oxygen demand. *Water S.A.*, *23*, 301.

Cameron, W. M., & Moore, T. B. (1957). The influence of chloride on the dichromate-value test. *Analyst (London)*, *82*(979), 677–682. doi:10.1039/an9578200677

Decker, G., Hönicke, P., & Hall, D. (2015). COD analysis and REACH authorization. *International Environmental Technology, Water/Wastewater*, 56.

Deutsches Institut für Normung. (1980). *DIN 38409 Teil 41, Deutsche Einheitsverfahren zur Wasser-, Abwasser- und Schlammuntersuchung Summarische Wirkungs- und Stoffkenngrößen (Gruppe H) Bestimmung des Chemischen Sauerstoffbedarfs (CSB) im Bereich über 15 mg/l (H 41)*. Beuth.

Domini, C. E., Vidal, L., & Canals, A. (2009). Trivalent manganese as an environmentally friendly oxidizing reagent for microwave- and ultrasound-assisted chemical oxygen demand determination. *Ultrasonics Sonochemistry*, *008*(5), 686–691. doi:10.1016/j.ultsonch.2009.01.008 PMID:19250852

Drasch, G., Horvat, M., & Stoeppler, M. (2004). Mercury. In M. Anke, M. Ihnat, & M. Stoeppler (Eds.), *Elements and their compounds in the environment* (2nd ed., p. 931). Wiley-VCH. doi:10.1002/9783527619634.ch38

Duca, G., & Vaseashta, A. (Eds.). (2020). *Handbook of Research on Emerging Developments and Environmental Impacts of Ecological Chemistry*. IGI Global., doi:10.4018/978-1-7998-1241-8

European Chemicals Agency. (2010). Member state committee support document for identification of potassium dichromate as a substance of very high concern because of its CMR properties, adopted on 4 June 2010.

European Chemicals Agency. (2016). Annex XVII to REACH. *Substances restricted under REACH.* https://echa.europa.eu/substances-restricted-under-reach

European Union. (2008). *Directive 2008/105/EC of the Europeans Parliament and of the Council of 16 December 2008, on environmental quality standards in the field of water policy*. EU.

European Union (2011). *Directive 2011/65/EU on the restriction of hazardous substances in electrical and electronic equipment (RoHS2)*. EU.

European Union. (2013). Commission Regulation 2013/348/EU amending Annex XIV to Regulation (EC) No 1907/2006 of the European Parliament and of the Council on the Registration, Evaluation, Authorisation and Restriction of Chemicals (REACH).

Gmelin Institut. (1963). *Gmelins Handbook of Inorganic Chemistry, Sulfur* (Vol. 8). Verlag Chemie. (in German)

Hach. (2014). *Oxygen Demand Chemical, Manganese III reactor digestion method 190067 without chloride removal 30 to 1000 mg/L COD Mn.* Hach Company, DOC316.53.01102.

Hansen, S., Pedersen-Bjergaard, S., & Rasmussen, K. (2012). *Introduction to pharmaceutical chemical analysis. Redox Titrations.* John Wiley & Sons., doi:10.1002/9781119953647

International Standard Organization. (1989). ISO 6060, Water quality - Determination of the chemical oxygen demand. Genf, Schweiz.

International Standard Organization. (2002). ISO 15705, Water quality – Determination of the chemical oxygen demand index (ST-COD – Small-scale sealed tube method, Genf

Kilic, M. Y., Yona, T., & Mert, B. K. (2014). Landfill leachate treatment by Fenton and Fenton-like oxidation processes. *Clean (Weinheim)*, *42*(5), 586–593. Advance online publication. doi:10.1002/clen.201200714

Kolb, M., Bahadir, M., & Teichgräber, B. (2017). Determination of chemical oxygen demand (COD) using an alternative wet chemical method free of mercury and dichromate. *Water Research*, *122*, 645–654. doi:10.1016/j.watres.2017.06.034 PMID:28649029

Korenaga, T., Zhou, X., Okada, K., Moriwake, T., & Shinoda, S. (1993). Determination of chemical oxygen demand by a flow-injection method using cerium (IV) sulfphate as oxidizing agent. *Analytica Chimica Acta*, *272*(2), 237–244. doi:10.1016/0003-2670(93)80575-6

Miller, D. G., Brayton, S. V., & Boyles, W. T. (2001). Chemical oxygen demand analysis of wastewater using trivalent manganese oxidant with chloride removal by sodium bismuthate pretreatment. *Water Environment Research*, *73*(1), 63–71. doi:10.2175/106143001X138705 PMID:11558305

Nederlands Normalisatie Instituut. (2006). *NEN 6633, Water en (zuiverings) slib - Bepaling van het chemisch zuurstofverbruik (CZV).* Niederlande.

Rivas, F. J., Beltran, F. J., Gimeno, O., & Frades, J. (2001). Treatment of olive oil mill wastewater by Fenton's Reagent. *Journal of Agricultural and Food Chemistry*, *49*(4), 1873–1880. doi:10.1021/jf001223b PMID:11308339

South African National Standard. (2010). SANS 6048, Water – Chemical oxygen demand. SABS standards division, Pretoria, South Africa.

Schweiz. Janicke, W. (1983). *Chemical oxidability of water constituents.* WaBoLu Berichte 1/1983, Dietrich Reimer Verlag GmbH.

Stoecker, B. (2004). Chromium. In Anke M., Ihnat M., Stoeppler M. (eds.) Elements and their compounds in the environment. 2nd Edition. Wiley-VCH Verlag, Weinheim, pp. 709.

Wagner, R. (1976). The COD method in Wastewater Charges Act (AbwAG). [in German]. *Vom Wasser*, *76*, 139–159.

KEY TERMS AND DEFINITIONS

Chemical oxygen demand (COD): A sum parameter for the sum of all substances present in water that can be oxidized under certain conditions usually with potassium dichromate as the oxidant. It indicates the amount of oxygen (in mg/L) that would be required for their oxidation if oxygen were the oxidizing agent.

Chloride: interference: The interference of the COD value by the co-oxidation of chloride ions in water samples by the oxidant.

Cuvette test: Analytical method in which chemical parameters of solutions are determined in photometer cells (cuvettes). The analysis solution is reacted with reagents specific to the parameter of interest. This reaction causes a change in the color or other optical properties of the solution. These changes can then be measured by means of photometry and used for quantification.

Digestion: Chemical decomposition of organic ingredients under harsh reaction conditions (strong oxidant, high temperature, low pH).

Phasing out: Strategy of the EU to ban chemicals that are harmful for human beings and the environment step by step, unless the use is essential.

Standard method: Analytical methods developed by consensus from results of science and technology, formulated and issued by recognized organizations and their committees, and made available to the public. The aim is a national or international standardization of the applied methods and thus a better comparability of analytical results.

Theoretical COD: Amount of oxygen in mg/L required for the complete oxidation of an organic compound to carbon dioxide and water, which can be calculated stoichiometrically.

Total organic carbon (TOC): A sum parameter for the total amount of organic carbon in a water sample. The determination is carried out via combustion or total chemical oxidation of the carbon content. The amount of carbon dioxide produced is measured, from which the carbon content in mg/L is calculated.

Chapter 2
Redox Processes in Natural Waters

Viorica Gladchi
https://orcid.org/0000-0002-5847-4466
Moldova State University, Moldova

Gheorghe Duca
Institute of Chemistry, Moldova State University, Moldova

Elena Bunduchi
Moldova State University, Moldova

Lidia Romanciuc
https://orcid.org/0000-0002-4555-1194
Institute of Chemistry, Moldova State University, Moldova

ABSTRACT

This chapter presents a synthesis of the typical mechanisms and patterns of catalytic redox processes that take place in the natural aquatic environment with the participation of oxygen, its reactive species, hydrogen peroxide, OH radicals, and various forms of copper and iron transition metals. There are presented the redox subsystems (slow or fast) which form the redox character of the environment; the mechanisms and conditions for activating oxygen for its involvement in redox processes; biotic and abiotic sources and pathways of the hydrogen peroxide degradation, as well as its role in the formation, maintenance, and regulation of the environment redox state; and formation and peculiarities of hydroxyl radicals' participation in the processes inside the water basins.

INTRODUCTION: REDOX COMPONENTS IN THE NATURAL WATERS

From a chemical point of view, natural water is defined as an open catalytic and photochemical oxidation-reduction system (Sychev et al., 1983; Sychev et al., 2002; Duca et al., 2003). Oxygen, hydrogen peroxide, ions and complexes of transition metals, reducing substances of autochthonous and allochthonous nature, and solar radiation lead to the development of catalytic redox processes in natural waters. The concentration of dissolved oxygen in natural waters varies within the range of values between 0 and 14 mg/l and is determined by two processes: photosynthesis and biotic and abiotic oxygen consumption (Alekhin, 1970; Nicanorov, 1989).

DOI: 10.4018/979-8-3693-0512-6.ch002

Following the fluorine, oxygen is the next strong oxidant which, being reduced to water, removes about 491 kJ/mol (2.73 V) of free energy. Although, from a thermodynamic point of view, oxygen is a strong oxidant, from a kinetic point of view, it is quite inert. The inertness of oxygen is explained by the particularities of the electronic structure: under normal conditions, O_2 has a triplet structure. Since, under the same conditions, the organic substances are in the singlet state, the interaction with oxygen cannot take place due to the spin prohibition (Sychev et al. 1983; Sychev et al., 2002; Duca et al., 2003).

Involvement of dissolved oxygen in pollutants oxidation processes requires its activation, which can be achieved through interaction with substances with reducing properties (metal ions in reduced form, free radicals) or as a result of sunlight absorption. The reactivity of oxygen in natural waters is actually determined not so much by its molecular form, as by the intermediate products of its biogeochemical cycle ($^1O_2$, $O_2^{\bullet -}$, $^{\bullet}OH$, H_2O_2) (Fig. 1).

Figure 1. Biogheochemic cycle of Oxigen
(Duca et al., 2003)

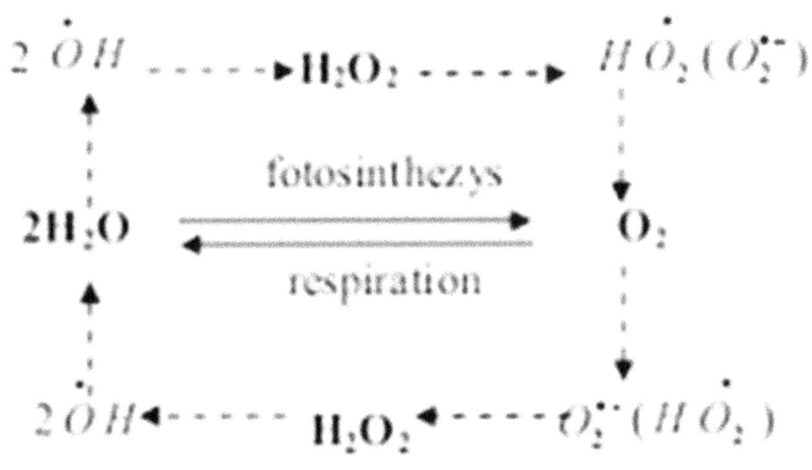

Based on multiple field measurements (Sinelnicov et al., 1974; Zepp et al., 1977; Petterson et al., 1973; Mill et al., 1980; Li et al., 2019; Gladchi, 2018; Wang et al., 2020; Duca et al., 2020; Lis et al., 2021; Garcia et al., 2020; Yon et al., 2021; Ossola et al., 2021; Duca et al., 2002; Bunduchi et al., 2006; Takeda et al., 2004; Vione et al., 2006), the concentrations of these particles were evaluated in the following ranges:

$[^1O_2] = 10\text{-}14 - 10^{-12}$ M, $[H\ ^{\bullet}O_2] = 10^{-8} - 10^{-9}$ M, $[^{\bullet}OH] = 10^{-17} - 10^{-15}$ M, $[H_2O_2] = 10^{-6} - 10^{-5}$ M.

According to the redox and acid-base characteristics of oxygen ($\varphi O_2/H_2O_2 = 0.69$ V, pKa = 11.62), the most stable intermediate product of O_2 reduction is considered hydrogen peroxide system (Sychev et al., 1983; Sychev et al., 2002; Duca et al., 2003; Richard et al., 2007; Wang et al., 2020).

The most important amount of H_2O_2 in natural waters is formed as a result of photochemical processes. When sunlight acts on the dissolved organic substances containing chromophore groups (OSCCG), it results of superoxide anions-radicals formation, which is the precursor of H_2O_2 (Yoon et al., 2021; Yin et al., 2021; Richard et al., 2007):

$$\mathrm{OSCCG} \xrightarrow{h\nu} 2O_2^{\bullet-} \xrightarrow{+2H^+} \mathrm{H_2O_2 + O_2} \tag{1}$$

The process of mechanism is as follows. When OSCCG absorbs solar energy, they are transforming into the excited state (Eq. 1). To return to the initial state, the substances transfer an electron to the dissolved in water oxygen (Eq. 1). As a result, superoxide anion radicals ($H^{\bullet}O_2 / O_2^{\bullet-}$) are formed, which subsequently disproportionate with the formation of H_2O_2 (Travin et al., 2019):

$$\mathrm{OSCCG{+}h\nu} \longrightarrow {}^1\mathrm{OSCCG}^* \longrightarrow {}^3\,\mathrm{OSCCG}^* \tag{2}$$

$${}^3\mathrm{OSCCG}^* + {}^3\mathrm{O_2} \longrightarrow OSCCG^{\bullet+} + {}^{\bullet}OH \tag{3}$$

$$O_2^{\bullet-} + H^+ \rightarrow H^{\bullet}O_2 \tag{4}$$

$$H^{\bullet}O_2 + H^{\bullet}O_2 \longrightarrow \mathrm{H_2O_2 + O_2} \tag{5}$$

Studies of the solar radiation influence on the H_2O_2 formation process have highlighted the fact that, from the UV radiation spectrum that reaches the Earth's surface, the greatest contribution (40%) to the H_2O_2 formation appears in the range of wavelengths 280-315 nm (UV-B). The share of UV-A radiation (315-400 nm) makes 33% and the remaining 27% is coming from PAR radiation (400-700 nm). The photochemical rate of H_2O_2 formation depends both on the wavelengths of the incident radiation and on the season of the year. The amount of H_2O_2 increases from winter to summer (Richard et al., 2007).

A certain amount of H_2O_2 is resulting from the redox-catalytic processes with the participation of organic substances (DH_2). For example, ascorbic and dihydroxyfumaric acids are catalytically oxidized with oxygen by forming hydrogen peroxide:

$$\mathrm{O_2 + DH_2} \xrightarrow{Cu^{2+}/Cu^{+}} \mathrm{H_2O_2 + D} \tag{6}$$

The biological processes represent the other source of H_2O_2 forming in natural waters (Richard et al., 2007; Duca et al., 2002; Yoon et al., 2021; Garcia et al., 2020). It is known that several species of algae (green, diatoms) eliminate hydrogen peroxide as a metabolic product in the external environment. Research carried out with the aim of establishing the influence of solar radiation on the H_2O_2 amount and its formation rate demonstrated that the irradiation of algae with UV rays with wavelengths shorter than 313 nm, resulted in higher H_2O_2 content production, than in case of irradiation with UV rays with wavelengths bigger than 365 nm. At the same time, some microorganisms have fermentation systems arranged on the outer surface, therefore they can catalyze redox processes outside the cells. For example, some algae have such enzymes as deaminases, which catalyze the oxidation process of organic nitrogen compounds to ammonia, simultaneously with the reduction of oxygen to H_2O_2 (Duca et al., 2002):

$$R\text{–}CHNH_2COOH + H_2O + O_2 \longrightarrow R\text{–}CHOCOO^- + NH_4^+ + H_2O_2 \quad (7)$$

Some hydroperoxides can also contribute to supplementing the H_2O_2 content (Sychev et al., 1983; Sychev & Duca, 2002, Duca et al., 2003):

$$ROOH + H_2O \longrightarrow ROH + H_2O_2 \quad (8)$$

H_2O_2 decomposition, as well as formation in natural waters, is determined by biotic and abiotic processes (Sinelnikov et al., 1974; Petterson et al., 1973; Mill et al., 1980; Li et al., 2019; Gladchi, 2018; Wang et al., 2020; Duca et al., 2020, 2002; Lis et al., 2021; Garcia et al., 2020; Yoon et al., 2021; Ossola et al., 2021; Bunduchi et al., 2006; Takeda et al., 2004; Vione et al., 2006).

The biotic decomposition of hydrogen peroxide is determined by the content of microalgae and bacteria, and the reaction order can be zero or first (Duca et al., 2003; Gladchi, 2018). This is particularly related to the mode of action of different enzyme systems of microorganisms. The zero-order reaction represents the contribution of peroxidase oxidation reactions of endogenous substrates (glutathione, ascorbate), which have the role of intramolecular antioxidants. The first-order reaction related to the H_2O_2 concentration is conditioned by the hydrogen peroxide decomposition in the presence of catalase.

Estimating the influence of the biotic community in the decomposition rate of H_2O_2, the authors deduced that the biota with dimensions smaller than 0.22 µm has the largest contribution to the hydrogen peroxide decomposition process.

The processes of abiotic decomposition of hydrogen peroxide in natural waters are described by the processes of catalytic decomposition of H_2O_2, which is presented below (Sychev et al., 2002, 1983; Duca et al., 2003; Travin et al., 2019; Ge et al., 2019; Mopper et al., 1990; Vaughan et al., 1998; Issakhov et al., 2021; Zubaidah et al., 2019; Hopwood et al., 2017; Ndungu, 2019).

Analyzing the thermodynamic characteristics, were deduced that the most reactive particles in the oxygen biogeochemical cycle are OH radicals, which are non-selective and strongly electrophilic particles. The stationary concentration of OH radicals in surface waters varies between $10^{-15} – 10^{-17}$ M (Sychev et al., 2002; Duca et al., 2003; Travin et al., 2019; Ge et al., 2019; Vaughan et al., 1998; Issakhov et al., 2021; Ndungu, 2019).

Currently, the photochemical mechanism of these radicals initiation is considered as one the main one, which was identified during the photolysis of the following components: dissolved organic substances (DOS), hydrogen peroxide, nitrate, and nitrite ions, peroxides, mono-hydroxocomplex of iron (III) (Xu et al., 2020; Lueder et al., 2020; Ge et al., 2019; Mack & Bolton, 1999; Minero et al., 2007; Bu et al., 2020; Mopper & Zhou, 1990; Vaughan & Blough, 1998; Trubetskoi et al., 2019; Sharpless et al., 2003; Vione, 2016; Sychev et al., 2002; Duca et al., 2003):

$$\text{DOS} \xrightarrow{h\nu \geq 300nm} {}^{\bullet}OH \quad (9)$$

$$H_2O_2 \xrightarrow{h\nu = 313nm} 2\,{}^{\bullet}OH \quad (10)$$

$$NO_3^- \xrightarrow{+H^+;h\nu=315nm} NO_2 + {}^{\bullet}OH \quad (11)$$

$$NO_2^- \xrightarrow{+H^+;h\nu=360nm} NO + {}^{\bullet}OH \quad (12)$$

$$Fe(OH)^{2+} \xrightarrow{h\nu=300nm} Fe^{2+} + {}^{\bullet}OH \quad (13)$$

$$ROOH \xrightarrow{h\nu\geq 300nm} R\ {}^{\bullet}O + {}^{\bullet}OH \quad (14)$$

The studies performed for establishing the wavelength influence on the OH radicals generation rate by DOS contained in the arctic waters demonstrated that the most important amount of these particles resulted from exposure to radiation with a wavelength greater than 300 nm. Detailed research based on the application of both solar radiation and monochromatic light from artificial sources, demonstrated that the rate of formation and the amount of OH radicals depends directly on the wavelength of the incident light and is practically not dependent on the location of the pool of water.

Thus, the values of the OH radicals initiation velocities in Arctic waters had the order of 10^{-12} M•s^{-1}, which is comparable to the values obtained in Gulf Stream waters (14,7•10^{-12} M•s^{-1}) and ocean water (2.8-15.9)•10^{-12} M•s^{-1} (Mopper & Zhou, 1990). The same authors, measuring the photochemical production in natural water subjected to solar irradiation, recorded velocities equal to 420•10^{-12} M•s^{-1} in the fresh surface waters. The research carried out on the OH radicals generation by fulvic acids ($\lambda \approx 320$ nm) resulted in the order values of 2•10^{-11} M•s^{-1} (Vaughan & Blough, 1998). Initiation of OH radicals formation velocities of the order of 10^{-11} M•s^{-1} were also obtained during field measurements: the waters of the Volga River (Russia) (Sinelnikov et al., 1974), the waters of the Prut River and the Dniester River (Republic of Moldova) (Duca et al., 2020, 2002; Lis et al., 2021; Bunduchi et al., 2006).

The research carried out on natural waters demonstrated that about 30% of the hydrogen peroxide decomposes with the formation of OH radicals (Duca et al., 2020; Lis et al., 2021). Investigations on the nitrate and nitrite ions' contribution to the OH radicals generation have highlighted the fact that in natural waters, NO_2^- ions account for about 35% of the total production of these radicals, and NO_3^- ions - around 15% (Vione et al., 2006). The same authors mentioned that, although the low nitrite ions concentrations are recorded in natural waters, their contribution to the OH radicals generation in surface waters is comparable or even equal to those of nitrates. For the wavelengths between 340-300 nm, the value of the quantum yield of generating OH radicals during the photolysis of nitrite ions ($\varphi_{NO^{2-}}^{OH}$) is 0.038-0.067, and for nitrate ions, this value is of the 0.01 order.

In addition to photochemical reactions, OH radicals can also form in processes with the participation of metal ions with variable valence and environmental components with redox ligand properties (ascorbic, dihydroxyfumaric acids) (Duca et al., 2020; Lis et al., 2021):

$$CuDH^+ + H_2O_2 \longrightarrow CuOH^+ + {}^{\bullet}OH + D^{\bullet-} \quad (15)$$

$$FeDFH_3^+ + H_2O_2 \longrightarrow FeDFH_2^+ + 2\,{}^{\bullet}OH \quad (16)$$

The catalytic decomposition of H_2O_2 in the presence of Cu(I) ions is accompanied by the formation of OH radicals:

$$Cu^+ + H_2O_2 \longleftrightarrow (CuO)^+H_2O \longrightarrow Cu^{2+}(OH^-) + {}^{\cdot}OH \quad (17)$$

Interaction of ozone (O_3) in concentration of 10^{-8} M inherent to surface waters with the monoanion of hydrogen peroxide (HO_2^-) or with reducing substances results to OH radicals formation (Sychev et al., 2002, 1983; Duca et al., 2003):

$$O_{3\cdot aq} + HO_2^- \longrightarrow {}^{\cdot}OH + 2O_2 \quad (18)$$

$$O_3 + DH^- \longrightarrow {}^{\cdot}OH + D^{\bullet -} + O_2 \quad (19)$$

In its turn, the OH radicals lead to generation of other radicals. Thus, the researchers from (Mill et al., 1980), carrying out photochemical studies of natural waters, besides of OH radicals, identified alkylperoxide radicals ($R\,{}^{\cdot}O_2$) in concentrations of the order of 3×10^{-9} M:

$$RH + {}^{\cdot}OH \longrightarrow {}^{\cdot}R + H_2O \quad (20)$$

$${}^{\cdot}R + O_2 \longrightarrow R\,{}^{\cdot}O_2 \quad (21)$$

The interaction constants of $R\,{}^{\cdot}O_2$ radicals with many of the organic substances, as a rule, have values lower than 1.0 $M^{-1}\cdot s^{-1}$. The lifetime of chemical compounds such as phenols, aromatic amines, hydroxylamines, polyphenols is several days due to the oxidation by peroxide radicals (Duca et al., 2003). According to the data presented by (Zepp et al, 1977), under sunlight action, the effective generation of singlet oxygen ($^1O_2$) takes place in natural waters. The stationary concentration of these particles is 10^{-14} – 10^{-12} M. As higher the DOS content in the water, as higher this value.

$$S \xleftrightarrow{h\nu} {}^1DOS^* \longrightarrow {}^3DOS^* \quad (22)$$

$${}^3DOS^* + O_2 \longrightarrow {}^1O_2 + DOS \quad (23)$$

Singlet oxygen is an effective oxidant for substances that contain in their composition double bonds with high electron density or functional groups that can easily be oxidized (alkylsulfide, thiol, phenolate ions, etc.). For these substances, the lifetime is due a few hours only. Simultaneously with the singlet oxygen the free electrons are generated also as a result of DOS photolysis, Solvated electrons ($\bar{e}_{aq}$) in natural water conditions interact with dissolved oxygen, forming superoxide anion radicals ($O_2^{\bullet -}$), which are the precursors of hydrogen peroxide:

$$\bar{e} + O_2 \longrightarrow O_2^{\bullet -} \quad (24)$$

An important feature of natural waters is the presence of transition metals in its composition. The research results performed (Duca et al., 2003; Yin et al., 2021; Seraghni et al., 2021; Lueder et al., 2020) in the modeled and natural conditions demonstrated that, practically, all metals with variable valence within pH between 5 and 9 (characteristic of natural waters), were found mainly in the oxide colloids, or hydroxide form $[Me^{n+}(OH)_n]$, or in the form of soluble complexes.

At the pH values appropriate to natural waters, the limit concentration of dissolved form, regardless the total content of metal in the water, depends only on its product solubility and the pH value of the environment. Since natural waters are characterized by an oxidizing state, then the transition metal ions are registered in the oxidized form mainly (M^{n+}) and only a very small part of these metals' content can be also found in the reduced form ($M^{(n-1)+}$), participating in the activation of oxygen and hydrogen peroxide.

Among the transition metals presented in natural waters, the most important for redox transformations is considered the contribution of copper, iron, and manganese. Copper ions are the only ones that exist in a neutral environment in a homogeneous form, both in an oxidized state ($[Cu(II)] = 3\times10^{-7}$ M) and in a reduced state ($[Cu(I)] = 10^{-8}$ M). Iron ions in the aerobic environment are mainly found in an oxidized state in the form of micro colloidal hydroxide particles (FeO(OH)) in concentrations of $10^{-6} - 10^{-5}$ M order. The soluble forms of iron (III) are represented particularly in the form of anionic hydroxofulvic complexes. Manganese ions are found either in the form of free Mn^{2+} ions, or in the form of Mn (IV) hydroxide microcolloids, which are formed under natural conditions during the biological oxidation of Mn^{2+} ions (Lueder et al., 2020; Duca et al., 2003).

Possessing the ligand properties, metals can form different complex compounds with natural water components, participating in O_2 and H_2O_2 activation reactions with free radicals. Upon activation of O_2 and H_2O_2, the complex compounds decompose with the formation of hydroxide radicals and superoxide anion (Sychev et al., 2002, 1983; Duca et al., 2003):

$$Cu^+ + O_2 \longleftrightarrow CuO_2^+ \longleftrightarrow Cu^{2+} + O_2^{\bullet -} \tag{25}$$

$$Cu^+ + H_2O_2 \longrightarrow (CuO)^+H_2O \longrightarrow Cu^{2+}(OH^-) + {}^{\bullet}OH \tag{26}$$

or oxidize the reducing substances according to the reactions:

$$CuO_2^+ + DH_2 \longrightarrow Cu^+ + D + H_2O_2 \tag{27}$$

$$CuO^+ + DH_2 \longrightarrow Cu^+ + D + H_2O \tag{28}$$

Unlike copper ions, the iron ions participate in reactions with the transfer of two electrons only (Sychev et al., 2002, 1983; Duca et al., 2003):

$$-Fe(II) \cdots Fe(II) - + O_2\,(H_2O_2) \xrightarrow{+H_2O} -Fe(III) - O - Fe(III) - + H_2O_2\,(H_2O) \tag{29}$$

$$-Fe(III) - O - Fe(III) + DH_2 \longrightarrow -Fe(II) \cdots Fe(II) - + D + H_2O \tag{30}$$

Additionally to oxidants and catalysts, the various reducing substances participate also in the redox processes in natural waters. These substances are produced directly in the aquatic environment (autochthonous substances, the products of the metabolism and decomposition of hydrobionts) or penetrate into runoff, atmospheric precipitation or wastewater (alochthonous substances). Autochthonous organic substances (organic acids, carbohydrates, amino acids, polypeptides, enzymes, etc.) are present in surface waters in concentrations lower than 0.1 mg/l ($< 10^{-5}$ M) and participate in the metabolic cycle of hydrobionts. Allochthonous substances are mainly represented by xenobiotics, substances extraneous to hydrobionts, of anthropogenic nature (petroleum products, phenols, organochlorine compounds, pesticides, heavy metals, radioactive substances, etc.). Their transformation products can be more toxic than the initial substances or vice versa, harmless.

From the total of reducing substances presented in natural waters, the share of organic acids is considerable, their content vary from 62 mg-equiv/l to 230 mg-equiv/l. The main natural sources of these substances are the metabolic products of phyto- and zooplankton. Substantial amounts of organic acids can be found in natural reservoirs and with wastewater discharges from chemical and food enterprises. The composition of natural waters include, in particular, acids: formic, acetic, some of the dicarboxylic acids (oxalic), oxy- and oxoacids (tartaric, glyoxalic, malic, etc.).

Humic acids (HA) and fulvic acids (FA), named with the general formula humic substances (HS), represent the largest part, about 90%, of the total organic acids and form the largest reserve of organic carbon in natural waters. The concentration of humic acids varies from 0.01 to 30 mg/l, while the content of fulvic acids is higher and constitutes 1-100 mg/l and even more. The presence in HS molecules with a large number of functional groups (hydroxyl, carbonyl, carboxyl, amino, etc.), which have the property of coordination centers, leads to the binding of metals in a stable complex of chelate compounds, thus contributing to their migration in natural waters (Kępczyńsqi et al., 2004; Son et al., 2019; Trubetskoi et al., 2019).

Another group of substances identified in natural waters is phenols, of which hydroquinone is made a part. In natural waters, phenols are presented in two groups: biogenic phenols and technogenic phenols (Sprynsky et al., 2007; Grabner et al., 2005; Kępczyńsqi et al., 2004; Prya et al., 2006; Ksibi et al., 2003). Biogenic phenols are products resulting from metabolic processes and from the microbiological decomposition of plant and animal diversity. Technogenic phenols are products resulting from fuel processing and organic synthesis.

The most important representatives of this class of compounds found in natural waters are volatile monohydroxyl phenols (phenol, cresols, xylenols) and nonvolatile di-, polyhydroxy phenols (pyrocatechin, resorcinol, hydroquinone). Volatile phenols are more toxic. In surface waters, phenols can be found in the form of salts, ions, and undissociated molecules. In unpolluted or slightly polluted waters, their content usually does not exceed the value of 12-20 mg/l. The wastewater from the petrochemical, coke-chemical, pesticide and plastics production plants contains from 6 to 2000 mg/l of volatile phenols (Grabner et al., 2005; Kępczyńsqi et al., 2004).

Therefore, in natural water, biotic and abiotic processes of oxidation and reduction equivalents formation are constantly taking place. They interact with each other forming the chemical composition of natural waters and determining its quality.

SELF-PERIFICATION PROCESSES OF NATURAL WATERS

Both the substances resulting from the vital activity of hydrobionts and those that enter in natural waters with wastewater discharges, seepage, and atmospheric precipitation, are involved in physical-chemical and chemical-biological transformation processes. As a result of these processes, the amount of substances decreases to a concentration that does not present a danger to the functioning of the ecosystems (Duca et al., 2003).

Self-purification of the natural aquatic environment involves the physical processes of mass transfer (physical self-purification), metabolism and co-metabolism processes (biological self-purification) and chemical processes of substance transformation (chemical self-purification) (Duca et al., 2003; Gladchi et al., 2018).

Physical processes (dilution, evaporation, sorption), in fact, lead only to redistribution of polluting substances in natural environment, either among the components of the water basin (suspended particles, organisms, etc.), or by their evacuation into neighboring systems (pools aquatic, bottom deposits, atmosphere).

The living organisms presented in the water basin, especially the autotrophic ones (bacteria, algae), participate in transformation of substances reducing the concentration of the pollutant through metabolic and co-metabolic processes. The contribution of microbiological processes in the degradation of substances, not participating in metabolic processes, is relatively small. This type of self-purification is more important in case of easily assimilable substances such as biogens. It is known that many substances, which do not participate in the biota's biological cycle, can become the result of co-metabolism processes. But the effectiveness of these processes depends mainly on the accompanying factors.

For the majority of substances, the most important transformation processes are the chemical ones. Overall, polluting substances (PS) can participate in a limited number of chemical transformations (Vione, 2016; Packer, 2003; Lam & Mabury, 2005; Duca et al., 2003; Ossola, 2020; Gladchi et al., 2018):

- **Hydrolysis,** where water participates in the reaction not only as a solvent, but also as a reagent;
- **Catalytic oxidation,** with the participation of O_2, H_2O_2, transition metal ions and OH radicals;
- **Direct,** sensitized and induced photolysis, transformations under the action of sunlight.

The speeds of these self-purification processes are influenced by the physical-chemical parameters of the aquatic environment in the following way. When the pH changes, the contribution of hydrolysis in the transformation of substances with hydrolytic properties decreases. Raising the water turbidity decreases the input of photochemical transformation processes for substances possessing photochemical activity, but at the same time, sensitized photolysis processes are intensified, increasing the speed of photochemical processes of initiation of free radicals. Pollution with substances possessing ligand properties leads to the redistribution of metal forms, which, on the one hand, influences mass transfer processes, and on the other hand, modifies the catalytic activity of metals. Increasing temperature intensifies temperature-dependent processes, such as hydrolysis and catalytic oxidation. Radioactive pollution leads to an increase in the initiation speed and stationary concentration of OH radicals. This could have a negative impact on the circuit of biogenic elements, involving them in unwanted oxidation processes. Pollution with substances containing sulfur in their composition will lead to decrease of Cu(II) ions concentration in natural waters, as a result of the stable complex compounds formation with low reaction capacity, which at the same time are biologically inaccessible forms for hydrobionts. The impact of pollution with sulfur compounds lead to inhibition of oxidation processes with free radicals, accompanied by the appearance of the toxicity effect towards hydrobionts that exchange water intensively with the external environment.

Catalytic Self-Purification and Free Radicals

As was mentioned, the natural waters constitute an open type of redox system, in which the substances with oxidizing and reducing properties constantly enter and undergo to transformations.

Based on the data accumulated by different researchers, the chemical composition of the natural waters abiotic component as a reaction system can be presented as follows (Duca et al., 2020, 2002; Lis et al., 2021; Garcia et al., 2020; Travin et al., 2019; Ossola et al., 2021):

$$O_{2,aq} – H_2O_2 – \{R_x\} – S – Me^{n+}/Me^{(n-1)+} – L – Sed$$

where, $O_{2,aq}$ –molecular oxygen; H_2O_2 – hydrogen peroxide; R_x – free radicals; S – substrates; $Me^{n+}/Me^{(n-1)+}$ – transition metal ions in oxidized and reduced state; L – ligands; Sed – suspended particles.

The participation of abiotic components in the self-purification processes of natural waters can be presented by the following diagram (Fig.2).

Figure 2. Scheme of the chemical self-purification redox processes in natural waters
(Gladchi et al., 2018)

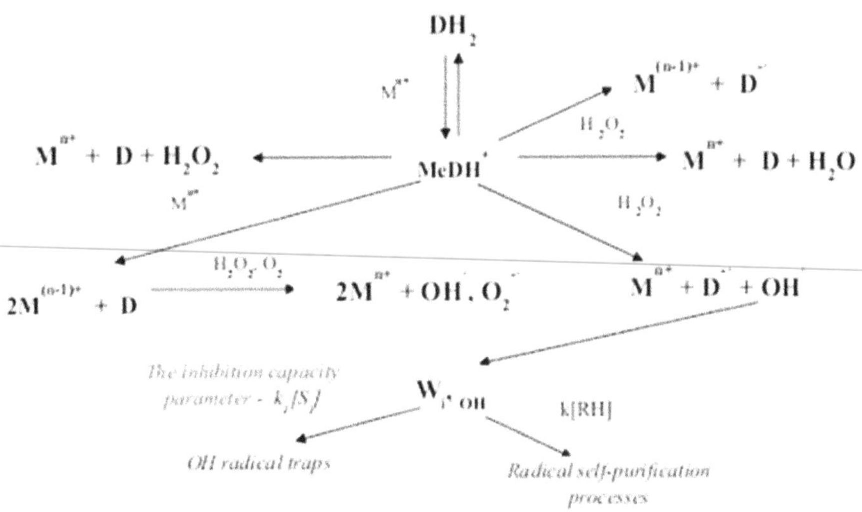

Discovery of metabolic cycles, which regulate the exchange of substances between the external environment and living organisms, as well as the intramolecular metabolic processes, led to extensive studies of oxy- and oxoacids redox transformations.

In the framework of the research carried out by Duca et al. (2020) and Lis et al. (2021) the oxidation processes of tartaric acid and a series of its metabolites with H_2O_2 were studied in an aerobic and anaerobic environment, at pH 3, in the presence of Fe^{2+}, Fe^{3+} and Cu^{2+} ions. Investigations have shown that, in the absence of metals, acids do not undergo oxidation with either O_2 or H_2O_2. The addition of metals, in particular, Fe^{2+} ions, in the oxyacid – O_2/H_2O_2 system leads to the effective activation of O_2 and H_2O_2, so that all acids undergo oxidation. In the presence of Fe^{3+} and Cu^{2+} ions, only acids with pronounced reducing properties are oxidized, such as: tartaric acid, dihydroxyfumaric acid and reductone.

On the example of ascorbic acid, the mechanism of the oxidation process with H_2O_2 catalyzed by Cu(II) ions was established at the pH values characteristic of natural waters. Research has demonstrated that in this system OH radicals are formed at the initiation stage with the participation of metal ions in an oxidized state, although, in specialized literature, it is known that the activation of oxidants with the formation of active particles occurs only in the presence of metal ions in a reduced state. The explanation of the obtained results was as follows. If the reducing substance demonstrates the ligand properties, then upon interaction with the metal in an oxidized state, a compound with partial charge transfer ($CuDH^+$) is formed. In the composition of this compound, the reducing properties of Cu(II) ions are intensified, so that it reduces hydrogen peroxide to OH radicals, at the same time, the metal ion does not change its degree of oxidation:

$$CuOH^- + DH^- \longleftrightarrow CuDH^+ + OH^- \tag{31}$$

$$CuDH^+ + H_2O_2 \longrightarrow Cu^{2+} + D^{\bullet-} + {}^{\bullet}OH + H_2O \tag{32}$$

In case of presence in natural waters the substances that do not possess ligand properties (for example, hydroquinone, glyoxalic acid), the generation or non-generation of OH radicals during the reduction of H_2O_2 will depend on these substrates concentration in natural waters. As a result of Cu^+ ions interaction with H_2O_2, the CuO^+ particle is formed (which can be treated as hydrolyzed Cu^{3+} ion), which is the precursor of the OH radical:

$$Cu^+ + H_2O_2 \longrightarrow CuO^+ + H_2O \tag{33}$$

In the absence of reducing agents, the CuO^+ particle oxidizes water with the OH radicals formation:

$$CuO^+ + H_2O \longrightarrow (CuOH)^+ + OH \tag{34}$$

and in their presence it is consumed during the reducing agents oxidation only, without generating the OH particles:

$$CuO^+ + DH_2 \longrightarrow Cu^+ + D + H_2O \tag{35}$$

So, in case of reducers without ligand properties, the catalytic reduction process of H_2O_2 can take place differently, depending on the reducers concentration in natural waters:

- at low concentrations, reducing agents will not essentially interfere with the OH radicals initiation process, because the CuO^+ particle can oxidize more intensively the water molecules and will be consumed less during the DH_2 oxidation;
- at high concentrations, these substances will act as inhibitors, capturing the CuO^+ particle, the precursor of OH radicals.

Sulfur compounds (glutathione, thiourea, cysteine) are among the substances with pronounced reducing and ligand properties presented in the natural waters composition (*Bunduchi* et al., 2006; Duca et al., 2002). Investigations demonstrated that the oxidation processes of thiols with hydrogen peroxide are not accompanied by the OH radicals generation. The redox process proceeds through the intermediate complex compound of thiol with Cu^+ ions formation:

$$2Cu^{2+} + 2GSH \xrightarrow[-2H^+]{} 2Cu^+ + GSSG \quad (36)$$

$$Cu^+ + GSH \longleftrightarrow Cu(GSH)^+ \xleftrightarrow{+GSH} Cu(GSH)_2^+ \quad (37)$$

which can then react with dissolved oxygen or hydrogen peroxide. Oxidation with oxygen requires participation of two metal ions:

$$Cu(GSH)_2^+ + O_2 \longleftrightarrow (GSH)_2CuO_2^+ \xleftrightarrow{Cu(GSH)_2^+} 2Cu^+ + 2GSSG + 2H_2O \quad (38)$$

and in case of oxidation with hydrogen peroxide, only one metal ion participates in the transformations:

$$Cu(GSH)_2^+ + H_2O_2 \longleftrightarrow Cu^+ + GSSG + 2H_2O \quad (39)$$

For heterogeneous catalysts such as iron microcolloids, both the decomposition of hydrogen peroxide and oxidation of reducing agents are not accompanied by free radicals formation:

$$- Fe(II) \cdots Fe(II) - + H_2O_2 \longrightarrow - Fe(III) - O - Fe(III) - + H_2O \quad (40)$$

$$- Fe(III) - O - Fe(III) + DH_2 \longrightarrow - Fe(II) \cdots Fe(II) - + D + H_2O \quad (41)$$

Comparing the catalytic activity of transition metals, it was observed that at the pH values characteristic of natural waters, the iron (III) ions have a lower catalytic activity than copper (II) ions. The phenomenon of different catalytic activity of these metals is explained by the redox potentials of the pairs $Fe_{het}(III)/Fe_{het}(II)$ ($\varphi^0 = 0{,}41$ V), Cu_{aq}^{2+}/Cu_{aq}^{+} ($\varphi^0 = 0{,}153$ V), O_2/H_2O_2 ($\varphi^0 = 0{,}69$ V), $H_2O_2/2H_2O$ ($\varphi^0 = 1{,}35$ V). The activation of oxygen and hydrogen peroxide takes place more effectively, when the redox potential of the metal has lower values than the redox potential of the pairs O_2/H_2O_2 şi $H_2O_2/2H_2O$ (Sychev et al., 2002, 1983; Duca et al., 2003).

The oxidation processes, which take place in natural waters could involve both peroxidase and OH radicals also. The rate constants of oxidation reactions with these radicals for the most of organic substances are practically equal to those of diffusion ($k_{OH+P} = 10^8 - 10^{10}\ M^{-1}{\cdot}s^{-1}$). In natural waters, OH particles participate mainly in the following reactions (Sychev et al., 2002, 1983):

a) addition of hydrogen proton with formation of water:

$$RH + OH+ \xrightarrow{k \approx 10^7 - 10^9 M^{-1} \cdot s^{-1}} {}^{\bullet}R + H_2O \tag{42}$$

b) addition to double bond (Ar – aromatic ring)

$$Ar + OH+ \xrightarrow{k \approx 10^9 M^{-1} \cdot s^{-1}} {}^{\bullet}ArOH \tag{43}$$

Secondary radicals resulting from interaction of OH particles with reducing agents, depending on their properties, can lead either to maintenance or to destruction of oxidation chains with OH radicals. Results on identifiyng the role of reducing agents in the process of OH free radicals initiation were obtained both on model and real systems (Vione et al., 2006; Duca et al., 2020, 2002; Lis et al., 2021; Bunduchi et al., 2006; Takeda et al., 2004).

Based on studies from (Travin et al., 2019; Duca et al., 2002) were attested that in presence of thiol substances both cases can be achieved. Thus, thioglycolic acid (TGA) leads to completion of radical processes in the aquatic environment because it has the capacity to effectively generate additional amounts of OH radicals. Thiourea (TU) and glutathione (GSH) possess the properties of OH radicals effective acceptors and thereby inhibit radical oxidation of other pollutants. In the simultaneous presence of TGA and Cu(II) ions, the recorded effect is the same as in the absence of metal, but in presence of Fe(III) ions, the process of OH radicals generating is inhibited. In case of thiourea and glutathione, the addition of metal ions further decreased the intensity of radical oxidation processes, because these metals bind the mentioned substrates in the form of $(Tiol)_2Me^{n+}$ complexes, which are more effective radical acceptors than uncomplexed thiols.

In specialized literature, the data about self-purification process with free radicals of various natural waters (Mediterranean Sea, South Atlantic and South Pacific Oceans, Baical lake, Dnieper, Volga, Prut, Dniester) are presented (Vaughan & Blough, 1998; Xu et al., 2020; Takeda et al., 2004; Sinelnikov,1974; Duca et al., 2002; Bunduchi et al., 2006; Hopwood, 2017; Ndungu, 2019).

Based on the modeling and field measurements results, it was deduced that according to the properties of radicals resulting from the interaction with OH particles, the components of natural waters are classified in two groups (Gladchi, 2018; Yoon et al., 2021; Travin et al., 2019):

1. Substances whose radicals react either with O_2 or with transition metal ions, with the final regeneration of ${}^{\bullet}OH$. The oxidation rate of each component in this chain is determined by the ${}^{\bullet}OH$ stationary concentration.
2. Substances whose radicals have low reactivity or, following transformations, pass into non-reactive form. The latter can become traps for OH radicals or substances that, by reacting with this radical, form a weakly reactive peroxide radical with O_2. Then, these radicals disappear following recombination, disproportionation or following reaction with the reduced form of metal. Consequently, the destruction of radical chains takes place. The effective rate constant of the OH radicals interaction with all components of this group determines that the second category of substances belongs to

the parameter characterizing the ability of natural waters to inhibit self-purification processes due to the participation of OH radicals.

Photochemical Self-Purification

Along with molecular oxygen, hydrogen peroxide and transition metals, the solar radiation contributes to redox transformations in natural waters.

The largest part of UV rays of the solar radiation is retained in the upper layers of atmosphere (by ozone layer) and only rays in the near UV (UV-A, UV-B) and visible range of solar radiation spectrum reach the Earth's surface. Under the action of this radiation different photochemical transformations take place in natural waters, which ensure the self-purification capacity of natural waters. These transformations are characteristic for the surface layers of water basins.

In the surface waters affected by solar irradiation two types of photo-degradation reactions take place: direct photolysis and indirect photolysis (sensitized and/or induced) (Sychev et al., 2002, 1983; Duca et al., 2003).

The direct transformation of substance under the solar irradiation action occurs when the region of solar radiation spectrum overlaps with that of substance absorption. This region is called the "action spectrum". The substances become unstable as a result of actinic radiation absorption and subsequently decompose:

$$DH_2 \xrightarrow{+h\nu} DH_2^* \longrightarrow \text{produse} \quad (44)$$

The most of natural waters components absorb UV radiation with l ≤ 350 nm, therefore this part of solar radiation is the most effective for self-purification processes.

In case of sensitized photolysis, the sunlight is absorbed by the sensitizer and the excitation is transmitted to the substance undergoing transformations or to molecular oxygen with the singlet oxygen formation:

$$SOD \xleftrightarrow{h\nu} {}^1SOD^* \longrightarrow {}^3SOD^* \quad (45)$$

$${}^3SOD^* + DH_2 \longrightarrow P \quad (46)$$

$${}^3SOD^* + O_2 \longrightarrow {}^1O_2 + S \quad (47)$$

Induced photolysis is carried out in the presence of reactive intermediates such as hydroxyl radicals ($^{\bullet}OH$), alkyl peroxide radicals ($^{\bullet}RO_2$), solvated electrons ($\bar{e}_{aq}$), which are generated by constituents of natural waters ((SOD, H_2O_2, NO_3^-, NO_2^- etc.):

$$I \xrightarrow{h\nu} {}^{\bullet}R \xrightarrow{+P} \text{prodcts} \quad (48)$$

The studies carried out by researchers Duca et al. (2020) and Lis et al. (2021), concern the investigation of photochemical transformations of a number of oxyacids (tartaric, dihydroxyfumaric, citric, glyoxalic, tartronic, malic) in the acidic environment (pH 3). Most acids do not have pronounced absorption bands at the $\lambda \geq 200$ nm range (except for dihydroxyfumaric acid) and, therefore, their transformations during direct photolysis proceed quite slowly. It has been demonstrated that Fe^{3+} ions addition to these acids solutions leads to their effective photolysis. The photolysis process intensification is due to the fact that Fe^{3+} ions form complex compounds with these acids, moving their absorption bands to 265-360 nm wavelength range.

In Kępczyńsqi et al. (2004) the research results are presented on sensitized photo-oxidation of phenol by humic acids (HA), in aqueous solutions at the neutral and basic pH. The solutions containing phenol and different concentrations of HA were irradiated with monochromatic light (l = 253.7 nm) and polychromatic light (310-420 nm). The results demonstrated that when samples are irradiated with monochromatic light, direct photolysis is the main way of phenol oxidation, while when irradiated with polychromatic light, the process is actually sensitized by HA.

Researchers Lam & Mabury (2005), studying the photochemical processes in surface waters with involvement of pharmaceutical compounds (atorvastatin (ATR), carbamazepine (CAR), levofloxacin (LEVO), sulfamethoxazole (SMX)), have found that direct photolysis is important for decreasing the LEVO and SMX concentrations, while humic substances reduce the degradation rate of these substrates. Indirect photodegradation is the main way to limit the persistence of ATR and CAR, as the destruction speed increases in solutions containing humic acids. Explanation of different behavior of these pharmaceutical products emerges from the absorption spectra of these substances. The LEVO and SMX absorption peaks are located in the near UV range, and for ATR, CAR – in the far UV range.

The research team of Packer et al. (2003) studied the photochemical degradation of other pharmaceutical compounds (naproxen (NA), diclofenac (DC), ibuprofen (Ib) and clofibric acid (AC)), through the direct and indirect photolysis mediated by OH radicals, in model systems and surface waters. NA and DC have been shown to rapidly photodegrade upon sunlight irradiation. Decrease of AC concentration is due to both direct and indirect photolysis, while Ib is transformed mainly through indirect photolysis. Based on research results, it was concluded that transformation of substance through one or another photolytic pathway depends on the location and intensity of its maximum absorption.

In other works (Prya & Giridnar, 2006; Ksibi et al., 2003; Sharpless et al., 2003; Vione, 2016) was discussed that the effectiveness of direct photochemical transformations of substances depends on the benzene ring presence in the composition of chromophore group, which facilitate the absorbtion in the UV-VIZ range, as well as on the type and position of substituent group in the benzene ring.

The researchers Prya & Giridnar (2006) have studied the photocatalytic degradation of seven phenol derivatives containing two substituents, each from the Cl^-, NO_2^-, $^\cdot CH_3$ series irradiated with light at $\lambda =$ 365 nm in the presence of TiO_2. The results demonstrated that chloromethyl derivatives degrade faster than chloronitrophenols.

Ksibi et al. (2003) describe the research of catalytic photodegradation in TiO_2 suspension ($\lambda = 280$ nm) upon UV irradiation with $\lambda = 280$ nm, at the pH of natural water, of two hydroxyphenols (hydroquinone (QH_2) and resorcinol (Rz)) and of three nitrophenols (4-nitrophenol, 4-NF; 2,4-dinitrophenol, 2,4-DNF; 2,4,6-trinitrophenol, 2,4,6-TNF). According to these authors, with regard to the initial destruction rates values, the nitrophenols were placed in the following order: W_f(4-NF) > W_f(2,4 DNF) > W_f(2,4,6 TNF), and the dihydroxyphenols W_f(Rz) > W_f(QH_2). To explain these results the Hammer constant values (σ) were used, which represents the relationship between structure and reactivity in series of aromatic compounds. Based on the obtained results, it was concluded that as lower is the Hammer constant value with the sign plus as higher is the speed of substance photodegradation. So, knowing the value of this constant, it can determine the susceptibility of the substance to photocatalytic transformations.

The indirect photolysis is important for transformation of substances that absorb in UV-C, and direct photolysis for those substances where the absorption maximum is recorded at wavelengths bigger than 300 nm.

KINETIC INDICATORS FOR THE QUALITY OF NATURAL WATERS ESTIMATION

The existence of global water cycle and its ability to dissolve various gases, organic substances and minerals determine its composition. At the same time, being a living environment for vital organisms, water keeps the dynamically balanced state through based on exchange of substances with aquatic biocenosis. Until the anthropogenic action became considerable, the natural self-purification processes were able to oppose changes in water quality. Excessive anthropogenic pollution led to disruption of this balance, generating the eutrophication of water basins, change of biocenoses and appearance of toxicity effects. To prevent these consequences, permanent monitoring of the natural waters quality is carried out.

Taking into account the multitude of substances entering in natural water that form its composition, it is difficult and even useless to carry out the control of each individual component. Therefore, complex indicators have been developed, which provide information on the larger or smaller groups of substances content, called hydrochemical indicators for water quality assessment. The monitoring of aquatic objects quality was carried out based on such indicators as: dissolved oxygen content (DO) and its degree of saturation (DS), chemical (CCO_{Cr}) and biochemical (BCO_5) consumption of oxygen, the content of mineral forms of nitrogen (NH_4^+, NO_2^-, NO_3^-) and phosphorus ((PO_4^{3-}), the content of heavy metals (Hg, Cd, Pb, etc.), organic pollutants (petroleum products, phenols, pesticides, surface active substances, etc.), and so on.

The hydrochemical indicators represent the criteria which determine the quality of water source, its biological value and usefulness for certain purposes (drinking, recreation, fish farms, etc.). The assessment of substances content in aquatic object was made by comparing the water quality parameters results determined during the field studies with the permissible limit concentrations (PLC) or with the background concentrations. Over time, it has been proven that these indicators are not sufficiently informative and cannot explain the reason of the changes occurring in aquatic ecosystems related to biocenoses, the appearance of toxicity effects for hydrobionts and even for humans.

For example, the PLC value established for copper ions in pure water based on its aqua forms content cannot be used with certainty for natural conditions. Due to particularities of these ions (low charge, high capacity of copper ions to form stable compounds, etc.) in natural waters, hydrobionts can feel a deficit of its biologically accessible forms even at much higher concentrations than PLC. The content of polluting organic substances and biogenic elements cannot sufficiently characterize the quality of natural waters. Thus, natural water, depending on the qualitative composition of organic substances determines the value of chemical oxygen consumption (CCO_{Cr}), which can be biologically valuable at high values of CCO_{Cr} and, conversely, toxic at low values of CCO_{Cr}.

The attestation of saturation degree with a sufficient amount of oxygen does not always represent the sure indicator of the ecological well-being of natural waters. In specialized literature, the presence of molecular oxygen in water is considered as an indicator of the oxidizing state, but it is absent and the presence of iron (II) – is the reduced state. Following multiple field measurements and observations, it was established that the state of natural waters can be considered normal (oxidizing) only when, in addition to dissolved oxygen, hydrogen peroxide is present in the waters in physiological concentrations, which is determined as favorable for hydrobionts: 0,3 μg/l < $H_2O_2 \leq$ 100 μg/l (Travin et al., 2019; Gladchi, 2018).

During the evolution process the organisms have adapted in such way when their development can happend in the oxidizing environment, but their perish in the reducing state. For example, daphnia, infusoria and fish caviar perish in the short term when hydrogen peroxide is missing in the aquatic environment. Until the formation of its own oxidation systems, the juvenile fish needs H_2O_2 as an exogenous oxidant. The lack of H_2O_2 is characterized by dominance of reducing substances that can act by blocking the redox centers of the enzymes, violating the redox balance inside the cell. Another impact of the reducing conditions dominance lead to the pathogenic microflora development, which eliminate toxins and even affects humans. The appearance and duration of unstable states can change depending on the season, but the qualitative aspect is preserved.

So, the oxidizing state of natural waters is characterized by hydrogen peroxide dominance, the oxygen as main molecular carrier of oxidation equivalents in natural waters, and by the dominance of substances that can easily be oxidized with H_2O_2 to the reducing state.

The H_2O_2 circuit in water basins is closely linked to the circuit of OH radicals, the content of which in natural waters vary within 10^{-17} M – 10^{-15} M.

Establishment of the low values of OH radicals concentration can be resulted by water pollution with substances that, upon interaction with these particles, form radicals with low reactivity, which are so-called "traps". When the environment is polluted, in particular with nitrates and nitrites (especially during heavy rains), or as a result of radioactive pollution, there is observed an increase of the OH radicals stationary concentration. Increasing the OH radicals concentration provokes the danger of radical processes intensification withn involvement of polluting substances, as well as environmental components that determine the structure and stability of hydrobiocenoses (disruption of biogeochemical circuit of manganese ions).

Based on the field research was demonstrated (Travin et al., 2019; Duca et al., 2003) that the development of an aquatic ecosystem is influenced both by the OH radicals concentration and also by dominance ratio of the OH radical initiation mechanism: photochemical ($W^{h\nu}_{i,{}^\bullet OH}$) and biotic ($W^{bio}_{i,{}^\bullet OH}$). For a normal regime of oxygen and values of hydrochemical indicators, the ratio: $W^{bio}_{i,{}^\bullet OH} : W^{h\nu}_{i,{}^\bullet OH}$ is less than one. The dominance of the biotic initiation channel of OH radicals can be established as a result of the increasing flow of substances with pronounced reducing properties (Red_{ef}) - products of the vital activity of microorganisms (especially blue-green algae). This can lead to the establishment of quasi-reducing state, which is toxic for hydrobionts that exchange intensively the water. The increase in value can be registered in areas with untreated and biologically active water discharges and in the eutrophic water basins.

As a result of field studies, the new indicators were proposed to supplement the group of hydrochemical indicators with: H_2O_2 content, redox state, inhibition capacity, OH radical content, which were defined as „kinetic indicators".

The kinetic indicators of surface water quality characterize the content and dynamics of mono- and bi-electronic activation products of oxygen (hydroxyl radicals, hydrogen peroxide, etc.).

For the normal oxygen regime, depending on the H_2O_2 concentration in water, the reducing agents and substances with strong oxidizing properties and the dynamic redox states can be established in the natural aquatic environment:

- oxidizing, H_2O_2 is present, in concentrations of about 100 μg/l;
- quasi-reducing (Red_{ef}), H_2O_2 is missing, but there are substances with pronounced reducing properties that are "titrated" with H_2O_2 (phenols, dienols, sulfur compounds, etc.);
- super-oxidant (Ox_{ef}) – products and semi-products of technological cycles (Cr(VI), $KMnO_4$, Cl_2, etc.) are detected.

The oxidizing state of aquatic system ensures the high self-purification capacity, as a result of which the polluting substances are transformed and the initial properties of waters are restored.

Water pollution with organic substances, which form radicals with low reactivity upon interaction with OH radicals, was evaluated based on the value of inhibition capacity parameter ($\Sigma k_i[S_i]$). Using the water inhibition capacity criterion regarding the self-purification process with radicals, waters are divided in three types (Gladchi, 2018; Duca et al., 2002):

- pure waters, Σki[Si] < 3·104 s-1;
- moderately polluted waters, Σki[Si] = 3·104 - 3·105 s-1;
- polluted waters, Σki[Si] > 3·105 s-1.

Following the research carried out by Minero & Bolton (1999), the direct link was established between the values of kinetic indicator $\Sigma k_i[S_i]$ and water toxicity. The correlation coefficient between $\Sigma k_i[S_i]$ and the biotesting results was greater than 0.9, in case when the values of the $\Sigma k_i[S_i]$ parameter exceeded $3{\cdot}10^5\ s^{-1}$. Considering the fact that the waste water biotesting with application of test organisms is difficult to achieve, the kinetic indicator $\Sigma k_i[S_i]$ allows the operative determination of both pollution and toxicity of aquatic environment (two integral indicators of waste water quality).

Usually, in the stable aquatic ecosystem the variation range of rate and time formation of redox agents existence does not exceed the average value ± 50%, which was established under the influence of meteorological, hydrophysical, hydrochemical, etc. factors. The excessive or low values of respective indicators accompanied by uncertain dynamics correspond to an ecosystem whose ecological well-being is irreversibly disturbed.

To estimate the human impact on the Nistru River water quality using the classical indicators and the kinetic ones (the content of hydrogen peroxide, redox state, concentration of OH radicals, the inhibition capacity) numerous expeditions have been carried out. The evaluation of waters based on the inhibition capacity indicator gave the results presented below. In October 2005 the inhibition capacity values in the Nistru waters pointed at a normal course of free radical self-purification processes ($2{,}95{\cdot}10^5 – 8{,}02{\cdot}10^5\ s^{-1}$), with the exception of the Cosautsi collection site ($1{,}31{\cdot}10^6\ s^{-1}$) and downstream the Dubasari barrage ($2{,}31{\cdot}10^6\ s^{-1}$). In November the same unfavorable situation was detected in the region of the Cosautsi village ($1{,}24{\cdot}10^6\ s^{-1}$) and downstream the Dubasari barrage ($1{,}40{\cdot}10^6\ s^{-1}$), demonstrating the deviation from the normal course of radical self-purification processes, including the Naslavcea point ($1{,}30{\cdot}10^6\ s^{-1}$). Considering the obtained values of inhibition capacity, the waters within the studied segment can be characterized as being in a normal state (order of $10^5\ s^{-1}$) with the exception of the v. Naslavcea, v. Cosautsi points and downstream the Dubasari barrage (order of $10^6\ s^{-1}$) (Bunduchi, E., et al., 2006).

CONCLUSION

1. Realization of the redox processes in natural waters is possible due to the presence in its composition of oxidative equivalents (O2, H_2O_2, OH, etc.), reducers (DH_2), catalysts ($M^{(n-1)+}/M^{n+}$) and solar radiation.
2. Among the transition metals, the copper and iron input is the most important for catalytic redox transformations in natural waters.
3. The reducing substances involved in oxidation processes may or may not have the ligand properties. The reducers with ligand properties are oxidized with H_2O_2 in the presence of Cu(II) ions with the generation of OH radicals, valuable oxidants in the self-purification processes of natural waters. Hydrogen peroxide will be consumed without formation of OH radicals during the oxidation of reducing agents that do not have ligand properties. This can lead to the reduction of self-purification capacity and, as a result establishment of quasi-reducing state of the waters.
4. Reductants can have different impacts on radical self-purification processes. Secondary radicals resulting from the interaction of reducing agents with OH particles can participate in maintaining or interrupting the self-purification chain with radicals.
5. Oxidation and reduction processes in water basins contribute to the formation of a special water state, defined as redox state.
6. The dynamics of redox agents content in natural water can serve as an indication of the biocenosis state: homeostasis, reversible disturbance and irreversible disturbance.
7. The water quality of an aquatic object must be defined by its biological value besides of the substances content. Waters with adequate biological value determine the normal processes of vital activity and reproduction of hydrobionts at different trophic levels.
8. Redox processes contribute to the natural waters composition formation and chemical properties appropriate to the biological value of habitation.

ACKNOWLEDGMENT

This work was performed under the Moldovan National Research Project Nr. 20.80009.5007.27, "Physico-chemical mechanisms of redox processes with electron transfer involved in vital, technological and environmental systems".

REFERENCES

Alekhin, O. (1970). *Basics of hydrochemistry*. PC Leningrad.

Bu, L., Zhu, N., Li, C., Huang, Y., Kong, M., Duan, X., & Dionysiou, D. D. (2020). Susceptibility of atrazine photo-degradation in the presence of nitrate: Impact of wavelengths and significant role of reactive nitrogen species. *Journal of Hazardous Materials*, *388*, 121760. doi:10.1016/j.jhazmat.2019.121760 PMID:31810806

Bunduchi, E., Duca, Gh., Gladchi, V., Goreaceva, N., & Mardari, I. (2006). *Assessment of kinetic parameters in the water of the Nistru course in the section Naslavcea-Dubasari*. Chemistry Journal of Moldova., doi:10.19261/cjm.2006.01(1).08

Duca, Gh., Blonschi, V., Gladchi, V., Travin, S. (2020). Dynamics of different sulfur forms in natural waters and their influence on the redox state. *American Journal of Physical Chemistry.*. doi:10.11648/j.ajpc.20200903.12

Duca, Gh., Gladchi, V., & Romanciuc, L. (2002). *Processes of pollution and self-purification of natural waters*. CE USM.

Duca, Gh., Skurlatov, Iu., Misiti, A., Macoveanu, M., & Surpăţeanu, M. (2003). *Ecological chemistry*. CE USM.

Garcia, P. E., Gerea, M., & Diéguez, M. (2020). Natural levels of hydrogen peroxide (H_2O_2) in deep clear South temperate lakes: Field and laboratory evidence of photo - and biotic production. *The Science of the Total Environment*, *727*, 138641. Advance online publication. doi:10.1016/j.scitotenv.2020.138641 PMID:32334222

Ge, L., Zhang, P., Halsall, C., Li, Y., Chen, C. E., Li, J., & Yao, Z. (2019). The importance of reactive oxygen species on the aqueous phototransformation of sulfonamide antibiotics: Kinetics, pathways, and comparisons with direct photolysis. *Water (Basel)*, *149*, 243–250. doi:10.1016/j.watres.2018.11.009 PMID:30448736

Gladchi, V. (2018). *Catalytic transformations and the redox state of the environment*. CEP USM.

Grabner, G., & Richard, C. (2005). Mechanisms of direct photolysis of biocides based on halogenated phenols and anilines. *Environmental Photochemistry*, *M*(Part II), 161–192. doi:10.1007/b138183

Hopwood, M. J., Rapp, I., Schlosser, C., & Achterberg, E. P. (2017). Hydrogen peroxide in deep waters from the Mediterranean Sea, South Atlantic and South Pacific Oceans. *Scientific Reports*, *7*(1), 43436. doi:10.1038rep43436 PMID:28266529

Issakhov, A., Alimbek, A., & Zhandaulet, Y. (2021). The assessment of water pollution by chemical reaction products from the activities of industrial facilities: Numerical study. *Journal of Cleaner Production*, *282*, 125239. doi:10.1016/j.jclepro.2020.125239

Jankowski, J., Kieber, D., & Mopper, K. (1999). Nitrate and nitrite ultraviolete actinometers. *Photochemistry and Photobiology*, *70*(3), 319–326. doi:10.1111/j.1751-1097.1999.tb08143.x

Kępczyńsqi, M., Czosnyka, A., & Nowakowska, M. (2004). Photooxidation of phenol in aqueous nanodispersion of humic acid. *Journal of Photochemistry and Photobiology*, *79*(3), 259–264.

Kleszczewska, E. (1999). The spectrophotometry determination of chelate complex: L-ascorbic acid with cuprum (II) and mercury (II) in alkaline solution. *Polish Journal of Environmental Studies*, *8*(5), 313–318.

Ksibi, M., Zemzemi, A., & Bouqchina, R. (2003). Photocatalytic degradability of substituted phenols over UV irradiated TiO_2. *Journal of Photochemistry and Photobiology A Chemistry*, *159*(1), 61–70. doi:10.1016/S1010-6030(03)00114-X

Lam, M., & Mabury, S. (2005). Photodegradation of the farmaceuticals atorvastatin, carbamazepine, levofloxacin, and sulfamethoxazole in natural waters. *Aquatic Sciences*, *67*(2), 177–188. doi:10.100700027-004-0768-8

Li, W., Xu, X., Lyu, B., Tang, Y., Zhang, Y., Chen, F., & Korshin, G. (2019). Degradation of typical macrolide antibiotic roxithromycin by hydroxyl radical: Kinetics, products, and toxicity assessment. [online]. *Environmental Science and Pollution Research International*, *26*(14), 14570–14582. doi:10.100711356-019-04713-1 PMID:30877533

Lis, A., Gladchi, V., Duca, G., & Travin, S. (2021). Sensitized photolysis of thioglycolic acid in aquatic environment. *Chemistry Journal of Moldova.*, *16*(1), 46–59. doi:10.19261/cjm.2021.796

Lueder, U., Jorgensen, B. B., Kappler, A., & Schmidt, C. (2020). Photochemistry of iron in aquatic environments. *Environmental Science. Processes & Impacts*, *22*(1), 12–24. Advance online publication. doi:10.1039/C9EM00415G PMID:31904051

Mack, J., & Bolton, R. (1999). Photochemistry and nitrite and nitrate in aqueous solution. *Journal of Photochemistry and Photobiology A Chemistry*, *128*(1-3), 1–13. doi:10.1016/S1010-6030(99)00155-0

Mill, T., Hendry, D., & Richardson, H. (1980). Tree radical oxidants in natural waters. *Science*, *207*(4433), 886–887. doi:10.1126cience.207.4433.886 PMID:17729869

Minero, Cl., Chiron, S., Falletti, G., Maurino, V., Pelizzetti, E., Ajassa, R., Carlotti, M. E., & Vione, D. (2007). Photechemical processes involving nitrite in surface water samples. *Aquatic Sciences*, *69*(1), 71–85. doi:10.100700027-007-0881-6

Modiri-Gharehveran, M., Hain, E., Blaney, L., & Shah, A. (2020). Influence of dissolved organic matter on carbonyl sulfide and carbon disulfide formation from cysteine during sunlight photolysis. *Environmental Science. Processes & Impacts*, *22*(9), 1852–1864. doi:10.1039/D0EM00219D PMID:32966465

Mopper, K., & Zhou, X. (1990). Photochemical formation of hidroxyl radical by constituents of natural waters. *Science*, *25*, 661. doi:10.1126cience.250.4981.661 PMID:17810867

Ndungu, L. K., Steele, J. H., Hancock, T. L., Bartleson, R. D., Milbrandt, E. C., Parsons, M. L., & Urakawa, H. (2019). Hydrogen peroxide measurements in subtropical aquatic systems and their implications for cyanobacterial blooms. *Ecological Engineering*, *138*, 444–453. Advance online publication. doi:10.1016/j.ecoleng.2019.07.011

Nicanorov, A. (1989). *Hydrochemistry*. Hidrometeoizdat.

Ossola, R., Clerc, B., & Mcneill, K. (2020). Mechanistic insights into dissolved organic sulfur photomineralization through the study of cysteine sulfinic acid. *Environmental Science & Technology*, *54*(20), 13066–13076. doi:10.1021/acs.est.0c04340 PMID:32936630

Ossola, R., Jönsson, O. M., Moor, K., & Mcneill, K. (2021). Singlet oxygen quantum yields in environmental waters. *Chemical Reviews*, *121*(7), 4100–4146. Advance online publication. doi:10.1021/acs.chemrev.0c00781 PMID:33683861

Packer, J., Jeffrey, J., Latch, D., McNeill, K., & Arnold, W. (2003). Photochemical fate of pharmaceuticals in the environment: Naproxen, diclofenac, clofibric acid, and ibuprofen. *Aquatic Sciences*, *65*(4), 342–351. doi:10.100700027-003-0671-8

Petterson, C. O., & Meyers, J. (1973). Photosintetic production of hydrogen peroxide by Anacystis nidulans. *Plant Physiology*, *51*(1), 104–109. doi:10.1104/pp.51.1.104 PMID:16658269

Prya, M., & Giridnar, M. (2006). Kinetics of photocatalytic degradation of phenols with multiple substituent groups. *Journal of Photochemistry and Photobiology A Chemistry*, *179*(3), 256–262. doi:10.1016/j.jphotochem.2005.08.022

Richard, L., Peake, B., Rusak, S., Cooper, W., & Burritt, D. (2007). Production and decomposition dynamics of hydrogen peroxide in freshwater. *Environmental Chemistry*, *4*(1), 49–54. doi:10.1071/EN06068

Seraghni, N., Ghoul, I., Dekkiche, B. A., Bouaziz, C., Debbache, N., & Sehili, T. (2021). Effect of Fe(III)-bicarboxylic complexes in removal pollutant under UV and sunlight in aqueous solutions. *Environmental Technology*. doi:10.1080/09593330.2021.1891142 PMID:33583355

Sharpless, Ch., Seibold, D., & Linden, K. (2003). Nitrate photosensitized degradation of atrazine during UV water treatment. *Aquatic Sciences*, *65*(4), 359–366. doi:10.100700027-003-0674-5

Sinelnikov, V., Demina, V. (1974). About the origin of hydrogen peroxide contained in open water bodies. *Hydrochemical Materials, X.*

Son, M. H., Gong, J., Seo, S., Yoon, H., & Chang, Y. S. (2019). Photosensitized diastereoisomer-specific degradation of hexabromocyclododecane (HBCD) in the presence of humic acid in aquatic systems. *Journal of Hazardous Materials*, *369*, 171–179. doi:10.1016/j.jhazmat.2019.02.035 PMID:30776600

Song, C., Liu, H. Y., Guo, S., & Wang, S. G. (2020). Photolysis mechanisms of tetracycline under UV irradiation in simulated aquatic environment surrounding limestone. *Chemosphere*, *244*, 125582. doi:10.1016/j.chemosphere.2019.125582 PMID:32050352

Song, H., Wignall, P. B., Song, H., Dai, X., & Chu, D. (2019). Seawater temperature and dissolved oxygen over the past 500 million years. *Journal of Earth Science*, *30*(2), 236–243. doi:10.100712583-018-1002-2

Sprynsky, M., Lebedynets, M., Namieśnik, J., & Buszewski, B. (2007). Phenolics occurrence in surface water of the Dniester river basin (West Ukraine): Natural background and industrial pollution. *Environmental Geology (Berlin)*, *53*(1), 67–75. doi:10.100700254-006-0619-0

Sychev, A., & Duca, G. G. (2002). *Fundamental and practical aspects of homogeneous catalysis of metal complexes*. Stiinta I.E.P.

Sychev, A., Travin, S. O., Duca, G. G., & Skurlatov, Iu. (1983). *Catalytical reactions and environmental protection*. Stiinta I.E.P.

Takeda, K., Takedoi, H., Yamaji, S., Ohta, K., & Sakugawa, H. (2004). Determination of hydroxil radical photoproduction rates in natural waters. *Analytical Sciences*, *67*(1), 153–158. doi:10.2116/analsci.20.153 PMID:14753275

Travin, S., Duca, Gh., & Gladchi, V. (2019). *Self-purification of aquatic media from hexachlorocyclohexane in a radical process*. Chemistry Journal of Moldova. doi:10.19261/cjm.2018.537

Trubetskoi, O. A., Patsaeva, S. V., & Trubetskaya, O. E. (2019). Photochemical degradation of organic pollutants in solutions of soil humic acids. *Eurasian Soil Science*, *52*(9), 1075–1080. doi:10.1134/S1064229319090102

Vaughan, P., & Blough, N. (1998). Hydroxil radical photoproduction in the sea and its potential impact on marine processes. *Environmental Science & Technology*, *32*, 2947. doi:10.1021/es9710417

Vione, D. (2016). Photochemical reactions in sunlit surface waters. *Applied Photochemistry*, *92*, 343–376. doi:10.1007/978-3-319-31671-0_7

Vione, D., Maurino, V., & Minero, C. (2006). Sources and sinks of hydroxil radicals upon irradiation of natural water samples. *Environmental Science & Technology*, *40*(12), 3775–3781. doi:10.1021/es052206b PMID:16830541

Wang, F. (2020). *Hydrogen peroxide and antioxidant enzymes moderate interaction of the carbon and oxygen cycles at the redoxcline* [Doctoral dissertation, University of South Carolina]. https://scholarcommons.sc.edu/etd/5982

Wang, H., Zhou, H., Ma, J., Nie, J., Yan, S., & Song, W. (2020). Triplet photochemistry of dissolved black carbon and its effects on the photochemical formation of reactive oxygen species. *Environmental Science & Technology*, *54*(8), 4903–4911. doi:10.1021/acs.est.0c00061 PMID:32227921

Wang, M., Xiang, X., Zuo, Y., Peng, J., Lu, K., Dempsey, C., & Gao, S. (2020). Singlet oxygen production abilities of oxidated aromatic compounds in natural water. *Chemosphere*, *258*, 127308. doi:10.1016/j.chemosphere.2020.127308 PMID:32535450

Xu, H., Li, Y., Liu, J., Du, H., Du, Y., Su, Y., & Jiang, H. (2020). Photogeneration and steady-state concentration of hydroxyl radical in river and lake waters along middle-lower Yangtze region, China. *Water Research*, *176*, 115774. . doi:10.1016/j.watres.2020.115774 PMID:32272319

Yin, R., Chen, Y., Hu, J., Lu, G., Zeng, L., Choi, W., & Zhu, M. (2021). Complexes of Fe(III)-organic pollutants that directly activate Fenton-like processes under visible light. *Applied Catalysis B: Environmental*, *283*, 119663. doi:10.1016/j.apcatb.2020.119663

Yoon, H., Kim, H. C., & Kim, S. (2021). Long-term seasonal and temporal changes of hydrogen peroxide from cyanobacterial blooms in fresh waters. *Journal of Environmental Management*, *298*, 113515. doi:10.1016/j.jenvman.2021.113515 PMID:34403920

Zepp, R.G., Cline, D.M. (1977). Rates of direct photolysis in aquatic environments. *Environ. Sci. Technol., V*(11), 359.

Zubaidah, T., Karnaningroem, N., & Slamet, A. (2019). The self-purification ability in the rivers of Banjarmasin, Indonesia. *Journal of Ecological Engineering*, *20*(2), 177–182. doi:10.12911/22998993/97286

Chapter 3
Redox Self-Purification Mechanism of Natural Waters With the Involvement of Thioglycolic Acid and Thiourea

Angela Lis
https://orcid.org/0000-0003-1111-7485
Moldova State University, Moldova

Gheorghe Duca
Institute of Chemistry, Moldova State University, Moldova

Viorica Gladchi
https://orcid.org/0000-0002-5847-4466
Moldova State University, Moldova

Elena Bunduchi
Moldova State University, Moldova

Vladislav Blonschi
https://orcid.org/0000-0001-8628-6756
Moldova State University, Moldova

ABSTRACT

This chapter studied the influence of TGA and TU on the self-purification processes of aquatic systems on model systems, whose composition was modeled to be as close as possible to the natural waters. As a source of irradiation was used the DRT-400 lamp and the Solar Simulator, Oriel model 9119X, the emission spectra of which overlap with the emission spectrum of the Sun. The results of the research show that on irradiation, TGA can regenerate an additional amount of OH radicals, thus contributing to the intensification of water self-purification processes. But in the presence of transition metals, such as Cu (II) and Fe (III), it has been established that the intensity of self-purification processes decreases due to the formation of the mercaptides, whose catalytic activity is low. Unlike TGA, TU leads to decreasing the intensity of the self-purification processes of aquatic systems, through the intense consumption of OH radicals, and at the same time on the interaction with transition metals, such as Cu (II) and Fe (III), form stable complexes with low catalytic activity.

DOI: 10.4018/979-8-3693-0512-6.ch003

INTRODUCTION

Decades of research have demonstrated that natural waters represent an unbalanced open kinetic system, with various redox processes involving pollutants, oxidizing agents, metal ions and their complexes, etc. Research in the field has shown that thiol substances, especially those of allochthonous origin, are hardly oxidized by dissolved oxygen, but easily by hydrogen peroxide. Thiol substances reduce or remove the content of hydrogen peroxide, which plays an important role in regulating the redox state of natural waters, thus contributing to the formation of the quasi-reducing state of waters, which is toxic to hydrobionts. At the same time, thiols, upon interaction with transition metal ions, form stable compounds, whose catalytic activity is low. Some thiols, such as thiourea, inhibit the activity of microorganisms in the biological treatment of wastewater. In this context, the main goal of this research was to evaluate the participation of thioglycolic acid (TGA) and thiourea (TU) in redox processes on model systems, to predict their influence on the self-purification capacity of natural aquatic systems.

BACKGROUND

Water is an essential component of the natural environment and the basis of many productive activities, consequently, all human societies have been developed around it. During recent decades, there has been constantly increasing concern about natural water quality and different problems related to the presence, utilization, and management of water resources (Vaseashta et al, 2021). Aquatic systems may be represented as redox catalytic and photochemical open-type multi-component. This fact is largely due to the redox components, the most important of which are: oxygen, hydrogen peroxide, solar radiation, transition metal ions, and complexes, as well as autochthonous and allochthonous reducers (Gladchi et al., 2008; Duca et al., 2008; Bunduchi et al., 2006; Patel et al., 2019).

Dissolved oxygen is the most prevalent oxidative component in natural waters, between 0 and 14 mg/L, and is controlled by two main processes: photosynthesis and biotic and abiotic consumption (Gladchi et al., 2022a; Duca & Travin, 2020; Song et al., 2019; Schmidtko et al., 2017; Liu et al., 2018). Oxygen was the chemical substance that determined the change in the redox state of natural waters from reducing to oxidizing, with the appearance on Earth of autotrophic organisms that produce and eliminate molecular oxygen in the aquatic environment (Foyer et al., 2003). In addition to its vital function, oxygen is also a strong oxidizer, but O_2 is normally inert due to its triplet structure (Duca & Vaseashta, 2020, Vaseashta & Duca, 2022). Therefore, for the involvement of oxygen in oxidation processes, it must be activated, either with sunlight or substances with reducing properties (Larson & Mccord, 1977; Wang et al., 2020). Some of the special components of the chemical composition of natural waters are transition metal ions (Subramanian & Madras, 2017; Lo Conte & Carroll, 2013; Chu et al., 2017; Sun et al., 2011). In the presence of dissolved oxygen, metal ions are in an oxidized state. They help to activate oxygen and hydrogen peroxide, which subsequently engage with different substrates and help to convert them. The following active species are generated in natural waters as a result of the activation of molecular oxygen: $^1O_2$, $O_2^{\bullet -}$, $^{\bullet}OH$, and H_2O_2 (Morris et al., 2022; Mill et al., 1980). The ranges of these species concentrations in natural waters have been determined by the authors of various papers: $[^1O_2] = 10^{-14}$-10^{-12} M; $[HO_2^{\bullet}] = 10^{-9}$-$10^{-8}$ M; $[^{\bullet}OH] = 10^{-18}$-$10^{-15}$ M; $[H_2O_2] = 10^{-6}$-10^{-5} M (Morris et al., 2022; Patterson & Myers, 1973; Wong et al., 2003; Yuan et al., 2018).

The most stable intermediate product of oxygen reduction in natural waters is hydrogen peroxide (Gladchi et al., 2022b; Duca et al., 2002). According to numerous academic articles, the formation of H_2O_2 inside the body of water during photochemical processes is the primary source of H_2O_2 in natural waters (Garcia et al., 2020; Fujii & Otani, 2017; Yoon et al., 2021). Studies on how radiation affects the photochemical generation of H_2O_2 in natural waters have shown that UVB radiation (wavelength range 280-315 nm), although it reaches the Earth's surface in the smallest amount, has a large contribution - 48%, to the formation of hydrogen peroxide. The contribution of UVA radiation (315-400 nm) is 24%, and the rest of the 28% is accounted for visible radiation (400-700 nm) (Häkkinen et al., 2004; Gerringa et al., 2004; García et al., 2018).

The abiotic decomposition of H_2O_2 includes catalytic decomposition mechanisms. Under neutral pH settings, hydrogen peroxide from the natural aquatic environment undergoes decomposition processes, but only when copper ions or iron complexes are present, which act as catalysts in this process (Duca & Vaseashta, 2020; Shtamm et al., 1991).

Redox processes, with the participation of oxygen, form a system in which the equilibrium is established very slowly, so the oxidation rates of reducing substances with molecular oxygen are lower than with hydrogen peroxide, which is characterized by a system in which the equilibrium is established quickly. But the slow system has some advantages, namely that the concentration of oxygen in the aquatic environment is much higher than that of hydrogen peroxide, and when the waters are highly polluted with organic substances, which are eventual reducing agents of hydrogen peroxide, it is completely consumed, and then the main oxidant is oxygen. In addition to this, in the oxidation process with the participation of oxygen, additional amounts of H_2O_2 are obtained, because it is an intermediate product in the process of reducing molecular oxygen to water (Duca et al., 2002; Duca & Vaseashta, 2020; Gladchi et al., 2020).

$$O_2 + 2H^+ \xrightarrow{+2e} H_2O_2 \xrightarrow{+2H^+ +2e} 2H_2O \tag{1}$$

Based on the molecular configuration and oxidative potential (E_0), the most reactive particles in the biochemical cycle of oxygen are OH radicals. The OH radical concentration in surface waters range from 10^{-17} M to 10^{-15} M. The most important sources of OH radicals in surface waters are their formation during the photolysis of many compounds such as DOM (dissolved organic matter), nitrates, nitrites, hydrogen peroxide, other peroxides, Fe(III) monohydroxocomplex, etc. (Vione et al., 2006; Brezonik & Fulkerson-Brekken, 1998; Gong et al., 2021; Bu et al., 2020; Song et al., 2020; Southworth & Voelker, 2003):

$$H_2O_2 + h\vartheta \rightarrow 2{}^{\bullet}OH \tag{2}$$

$$DOM + h\vartheta \rightarrow DOM \xrightarrow{+H_2O} DOM - H^{\bullet} + {}^{\bullet}OH \tag{3}$$

$$NO_3^- + h\vartheta + H^+ \rightarrow {}^{\bullet}OH + NO_2 \tag{4}$$

$$NO_2^- + h\vartheta + H^+ \rightarrow {}^{\bullet}OH + NO \tag{5}$$

$$DOM + h\vartheta \rightarrow {}^{1}DOM^{*} \rightarrow {}^{3}DOM^{*} \quad (6)$$

$$^{3}DOM^{*} + H_2O \rightarrow \left[DOM + H\right]^{\bullet} + {}^{\bullet}OH \quad (7)$$

$$Fe^{3+} - L + h\vartheta \rightarrow Fe^{2+} + L^{+\bullet} \quad (8)$$

$$Fe^{2+} + H_2O_2 \rightarrow Fe^{3+} + {}^{\bullet}OH + OH^{-}, \qquad k = 76\ L{\cdot}mol^{-1}s^{-1} \quad (9)$$

$$Fe\left(OH\right)^{2+} + h\vartheta \rightarrow Fe^{2+} + {}^{\bullet}OH \quad (10)$$

$$ROOH + h\vartheta \rightarrow RO^{\bullet} + {}^{\bullet}OH \quad (11)$$

The presence of transition metals in natural waters is of special interest. The most widespread transition metal in natural waters is iron. The iron concentration in surface waters is $\approx 10^{-5}$ M, copper ($\approx 10^{-7}$ M), cobalt, manganese, nickel, and chromium ($\approx 10^{-7}$-10^{-6} M) (Munteanu et al., 2011). The chemical speciation of iron (Fe) in natural waters is influenced by pH value and the presence of ligands. It was determined that the trivalent iron ion (Fe(III)), which forms complexes with ligands, predominates in the composition of natural waters. It is known that the photolysis of hydroxo-complexes of Fe(III) is accompanied by the formation of OH radicals in water, but as the pH increases, the effectiveness of this process decreases (Yuan et al., 2018; Duca et al., 2002; Garcia et al., 2020). The Fe(II) ion exists within pH values 6-10, predominantly in the form of $FeHPO_4$. Thus, the balance $Fe^{2+} \leftrightarrows Fe^{3+}$ is widespread in nature and is of great importance in redox processes (Sun et al., 2011; Seraghni et al., 2021; Ouyang et al., 2020). Besides iron ions, copper ions are of particular interest, possessing important catalytic properties in redox aquatic environment transformations. Depending on the pH value, the copper ions are in states Cu^{2+} (pH=6.5) and respectively Cu^{+} (pH=6.5-7.5). Unlike Fe(III) ions, Cu(II) ions participate in free radical, $^{\bullet}$OH and $O_2^{\bullet -}$, formation reactions within the pH value of natural waters (Sychev et al., 1983).

In the redox processes that take place in natural waters, in addition to oxidants and catalysts, various reducing substances also participate. These substances can be autochthonous (products of the metabolism and decomposition of hydrobionts, such as organic acids, enzymes, amino acids, thiols, etc.) or allochthonous (pesticides, heavy metals, petroleum products, heavy metals, etc.), and they reach natural waters through wastewaters, runoff, atmospheric precipitation, etc. A group of substances with pronounced reducing properties that affect aquatic systems' ecological balance is represented by thiol substances (Patai, 1974). According to their place of origin, thiol substances found in natural waters can typically be classified into two main groups. The first group consists of thiol compounds of autochthonous origin that are produced inside of the aquatic basin, such as cysteine (Cys) and glutathione (GSH). Allochthonous substances, such as thiourea (TU) and thioglycolic acid (TGA), are part of the second group and enter natural waters as a consequence of anthropogenic environmental influence.

Thiourea (TU) is a thiol substance of anthropogenic origin that is used in the synthesis of various substances, in the fertilizer and insecticide industries, and in the production of pharmaceuticals, dyes, plastics, etc. Because the range of applications of TU is very wide and the annual production is quite high (over 10000 t/year), its content in the environment is estimated to be quite high. Based on the physicochemical properties of TU and its use pattern, the hydrosphere is expected to be the main target compartment for this compound. Because TU is not biodegradable and is hydrolytically stable, the transformation of TU into aquatic systems will take place through chemical and photochemical processes (Lis, 2016). TU is a dangerous compound for the aquatic environment because it influences the hydrobionts' activity, but its oxidation products (such as urea) are non-toxic and even beneficial for natural habitats (Sahu et al., 2011).

The main goal of water protection is to maintain and improve the physicochemical and biological characteristics of natural waters. Pollution and self-purification processes can change or maintain these physicochemical and biological characteristics. There is a reciprocal cause-and-effect relationship between the processes of pollution and self-purification. If the effect of pollution exceeds the self-purification capacity, then the water quality is getting worse. Therefore, to avoid disturbing the pollution-self-purification balance, special attention must be paid to the wastewater treatment before its discharge into natural waters (Gladchi et al., 2008; Bunduchi et al., 2019, 2006).

The aquatic environment's self-purification represents the totality of the physical, biological, and chemical processes that take place inside the aquatic basins, and lead to reducing the number of polluting substances in the water to a level that does not pose a danger to the ecosystem's functioning (Gladchi et al., 2008; Duca et al., 2008; Bunduchi et al., 2006).

Monitoring of the self-purification capacity of natural aquatic systems is based on different parameters such as the chemical oxygen demand, biochemical oxygen demand, dissolved oxygen concentration, speed of water decontamination by some pollutants, etc. (Nugraha et al., 2020).

MATERIALS AND METHODS

Chemicals and Standards

Thioglycolic acid, thiourea, *p*-nitrosodimethylaniline (PNDMA), oxygen peroxide, potassium oxalate monohydrate, and iron(III) chloride monohydrate were purchased from Sigma-Aldrich (Germany); iron(III) chloride hexahydrate and copper(II) sulfate pentahydrate from Stanchem (Poland), phosphate buffer solution (KH_2PO_4, $Na_2HPO_4 \cdot 2H_2O$) from Chem-Lab (Belgium). All solutions were prepared with double-distilled water. The absorption was determined using a double-beam UV–Vis spectrophotometer (T80+ PG Instruments). The structural formulas of the analyzed thiols are presented in Table 1.

Table 1. Structural formulas of the analyzed thiol compounds

Thiol compounds	Thioglycolic acid (TGA)	Thiourea (TU)
Structural formula	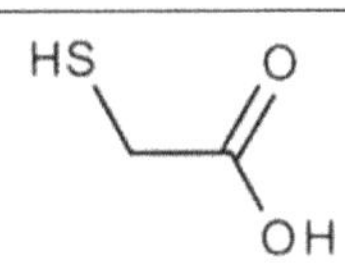	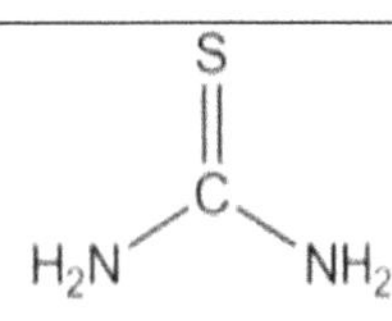

Irradiation Sources

In this study, a DRT-400 lamp (400W, Lisma, Russia) and a Solar Simulator (SS) Oriel Model 9119X (Newport, USA), were used as irradiation sources, some technical parameters are presented in Table 2.

Table 2. The used irradiation sources and their general characteristics

Irradiation source	Filters	Emitted wavelengths, nm	Comments
DRT-400 lamp		220-600	characterized by three intense lines at 312 nm, 365 nm, and 436 nm.
Solar Simulator, Oriel Model 9119X	Air Mass 0 filter (AM0)	250-2800	This filter corrects the illuminator output to better match the solar spectrum in extraterrestrial space. This filter is coated to remove excess IR radiation from the beam.
	Air Mass 1.5 Direct filter (AM1.5D)	250-2800	This filter set simulates the direct solar spectrum with the sun at a zenith angle of 48.2°.
	UVC blocking filter (UVC-bl.)	290-2800	This filter blocks the passage of UV rays with wavelengths less than 290 nm.

Determination of the Hydrogen Peroxide Initial Concentration

The spectrophotometric method was used for the determination of the hydrogen peroxide initial concentration. The mixture consisting of 0.5 mL of concentrated hydrogen peroxide and 9.5 mL of double-distilled water was prepared, and the absorbance of the solution was measured at 300 nm wavelength in a high-performance quartz glass cuvette, spectral range 200-2500 nm, pathlength 10x10 mm, chamber volume 3500 μL (Sigma-Aldrich, Germany). As a blank solution, double-distilled water was used. The hydrogen peroxide initial concentration ($C_{H_2O_2,\ init.}$, mol·L^{-1}) is determined according to Eq. (12) (Duca et al., 2002):

$$C_{H_2O_2,\ init.} = 20 \cdot A^{300} \tag{12}$$

where: A^{300} - the absorbance of the H_2O_2 solution at the wavelength $\lambda = 300$ nm.

Determination of the Initial Concentration of P-Nitroso-N,N-Dimethylaniline Solution (PNDMA)

The initial concentration of the dye solution $[PNDMA]_0$ (M) is determined spectrophotometrically, at the wavelength of 440 nm in a cuvette with a thickness of 1 cm and is calculated according to Eq. (13) (Duca et al., 2002; Travin et al., 2019):

$$[PNDMA]_0 = \frac{n \cdot A^{440}}{l \cdot 3.4 \cdot 10^4} \tag{13}$$

where: n - the dilution coefficient of the initial solution; A^{440} - the absorbance of the colored solution at the wavelength $\lambda = 440$ nm; l - the optical path length (thickness of cuvette), cm; $3.4 \cdot 10^4$- PNDMA molar extinction coefficient.

Determination of the Inhibition Capacity

The determination of the inhibition capacity concerning redox agents is based on the comparison of the rate of 4-nitroso-N,N-dimethylaniline (PNDMA) discoloration, and the interaction with OH radicals in distilled water in the absence and presence of the sample additions, according to Eq. (14) (Duca et al., 2002; Travin et al., 2019):

$$PNDMA\ (yellow\ solution) + OH^{\cdot} \rightarrow product\ (colorless\ solution) \quad (14)$$

OH radicals are formed as a result of hydrogen peroxide photolysis upon irradiation by the sources described in Table 2. Schematically, this reaction can be shown according to Eq. (15):

$$H_2O_2 \xrightarrow{h\vartheta} 2\dot{O}H \quad \begin{matrix} \xrightarrow{kPNDMA} [PNDMA] \\ \xrightarrow{kH_2O_2} [H_2O_2] \\ \xrightarrow{k_i} [S_i] \end{matrix} \quad (15)$$

where: [PNDMA] is the concentration of PNDMA, mol·L^{-1}; [H_2O_2] is the concentration of the hydrogen peroxide, mol·L^{-1}; [Si] is the concentration of radical scavengers, mol·L^{-1}; k_{PNDMA} is the bimolecular constant of the reaction rate of the dye with OH radicals, mol^{-1}·L·s^{-1}; kH_2O_2 is the rate constant of the hydrogen peroxide decomposition; k_i is the reaction rate constant of various substances with OH radicals.

In distilled water, OH radicals interact with the dye and hydrogen peroxide. As a result, the dye discolors and the hydrogen peroxide decomposes into water and oxygen. When $k_{PNDMA}[PNDMA] > k_{H2O2}[H_2O_2]$ the rate of dye discoloration in distilled water (W_{PNDMA}) coincides with the rate of hydroxyl radical formation.

When the sample irradiation takes place, except for H_2O_2, the source of redox agents can be photosensitive compounds such as iron compounds, dissolved organic substances, etc. To estimate the influence of these sources, the sample is simultaneously irradiated in the absence and presence of hydrogen peroxide.

The value of the inhibition capacity ($\sum k_{iOH}[S_{iOH}]$, s^{-1}), is calculated according to Eq. (16):

$$\sum k_{iOH}[S_{iOH} = \frac{V_{tot}}{V_{s.}} \bullet k_{PNDMA} \bullet [PNDMA]_0 \bullet [\frac{W_{d.w.}}{W_{s.}} - 1] \quad (16)$$

where: $W_{d.w.}$($W_{s.}$) – initial rate of PNDMA discoloration upon photolysis of hydrogen peroxide in distilled water (sample); $k_{PNDMA} = 1.25 \cdot 10^{10}$ - the bimolecular constant of the reaction rate of the dye with OH radicals, mol^{-1}·L·s^{-1}; $[PNDMA]_0$ – PNDMA initial concentration, mol·L^{-1}; V_{tot} – the total volume of the mixture, mL; $V_{s.}$ – the volume of the analyzed sample, mL.

If $(\sum k_{iOH}[S_{iOH}] < 10^4\ s^{-1}$ – the water should be classified as very clean;
If $(\sum k_{iOH}[S_{iOH}] < 10^6\ s^{-1}$ – the water should be classified as highly polluted;
If $(\sum k_{iOH}[S_{iOH}] < 10^5\ s^{-1}$ – the water is in its normal state with respect to the flow of free radical self-purification processes.

Method for Determination of ˙OH Radical Steady State Concentration

The method is based on the interaction of ˙OH with PNDMA, resulting in solution discoloration, thus allowing indirect determination of the free radicals using UV-VIS spectroscopy, according to the method previously described (Travin et al., 2019). PNDMA was chosen as an indicator since it acts as a trap for ˙OH formed during the photolysis of hydrogen peroxide. The model system consisted of phosphate buffer–PNDMA–H_2O_2, artificially exposed to UV radiation using the sources mentioned above. After a certain amount of time, the absorbance of the system was measured periodically at 440 nm on a T80+ UV/VIS spectrometer. Oxidation of PNDMA occurs as a result of its interaction with ˙OH generated by H_2O_2 under the action of light. When the rate of radical initiation W_i and the rate constant of destruction are known, the stationary concentration of ˙OH may be determined (Eq.(17)) (Duca et al., 2002; Travin et al., 2019):

$$\left[\dot{O}H\right] = \frac{W_{iOH}}{\sum K_{iOH}[S_{iOH}]} \qquad (17)$$

where, W_{iOH} - rate of radical initiation, M•s^{-1}; $\Sigma k_{iOH}[S_{iOH}]$- inhibitory capacity, s^{-1}.

Experimental Procedure

The solutions were prepared using double-distilledater, and the reaction was carried out at pH 7, which was set by the 0.0013 M phosphate buffer (it was used to keep the pH constant and bringing the modeling conditions closer to the natural ones, since the rates of thiols transformation were found to be highly pH-dependent). Additionally, the pH was periodically monitored (before and after the photolytic session) using the WTW Ionlab 720 pH meter (WTW, Germany). All measurements were carried out under thermostatic conditions at a temperature of 25 °C. The solutions were contained in aerobic conditions, bringing the modeling conditions closer to the natural ones. Hydrogen peroxide was used as a source of HO˙ radicals. Such important parameters as the rate of photochemical discoloration of the PNDMA in the absence and presence of thiols, the rate of initiation of free radicals, the "inhibitory capacity" of the medium, as well as the steady-state concentration of HO˙ radicals, were calculated during this research. The photolytic reactions were induced by irradiating the solution samples using the sources mentioned in Table 2, for 30 minutes, with a time step of 5 minutes, and the absorbance was measured at 440 nm. The constancy of the light flux and its intensity were checked with a standard ferrioxalate actinometer. The samples (50 mL) were placed in a thin-walled quartz glass with a volume of 100 mL and exposed to irradiation. Given the absorption spectrum of hydrogen peroxide and the low concentration of this reagent, the illumination throughout the volume could be considered uniform. However, solutions were mixed with a magnetic stirrer during irradiation, for additional averaging. The initial concentration of hydrogen peroxide in the model systems was determined using the iodometric method ($1{\cdot}10^{-2}$ M). The goal of the work was to study the influence of TGA and TU on the processes of radical self-purification in water systems. PNDMA was selected as an indicator because it is a trap for HO radicals produced during the photolysis of hydrogen peroxide.

FOCUS OF THE STUDY

This study examines the role of thioglycolic acid (TGA) and thiourea (TU) in redox processes by predicting their impact on the ability of natural water systems to self-purification using model systems. In this work, the kinetic method was used by applying the parameters: inhibition capacity and steady-state concentrations of OH radicals. In the present study, the principle of "from simple to complex" was used, which implies a step-by-step complication of the investigated model systems. In all model systems, step-by-step, the concentration of one component varied while the concentrations of the other components were kept constant. To determine the influence of TGA and TU on the self-purification process of aquatic systems, several systems were modeled using the PNDMA trap, which allows the determination of the inhibition capacity and the concentration of OH radicals in the modeled systems. The following systems have been modeled for this purpose: phosphate *buffer-TGA(TU)-*H_2O_2*-PNDMA-hυ;* phosphate *buffer-TGA(TU)-*H_2O_2*-PNDMA-Cu(II)-hυ;* phosphate *buffer-TGA(TU)-*H_2O_2*-PNDMA-Fe(III)-hυ.* All model systems were subjected to irradiation with artificial light sources, as described in Table 2, according to the methodology described above.

Influence of TGA on the Radical Self-Purification Processes of Aquatic Systems

Figure 1 shows the practical results for the inhibition capacity and the concentrations of OH radicals, obtained for the simplest system (phosphate *buffer-TGA-*H_2O_2*-PNDMA-hυ*), at the variation of the TGA concentration. The findings showed that an increase in TGA concentration resulted in a decrease in inhibition capacity and an increase in the steady-state concentration of OH radicals, upon irradiation with all sources. The decrease in inhibition capacity of the studied system is related to the formation of OH radicals, which interact with the PNDMA dye, inducing its oxidation. This fact demonstrates that TGA upon irradiation leads to the regeneration of an additional amount of OH radicals in the system, according to Eq. (18), which is supported by the literature data (Duca et al., 2002).

$$SH - CH_2 - COOH + H_2O_2 \xrightarrow{h\vartheta} SH - CH_2 - CO\grave{U} + 3\grave{O}H \tag{18}$$

This is due to the fact that TGA contains a -OH group in its structure, which breaks upon irradiation in the presence of hydrogen peroxide and subsequently regenerates an additional amount of OH radicals. Figure 1 shows that, in the presence of TGA, the inhibition capacity values are 10^4 s^{-1} for the system exposed to radiation with the DRT-400 lamp (values typical for very pure water) and 10^5-10^6 s^{-1} for the systems exposed to SS, in the presence of all three filters (values characteristic for polluted waters). Also, Figure 1 has shown that inhibition capacity values are dependent on the irradiation source, which is because OH radicals are forcibly generated during the photolytic degradation of hydrogen peroxide (Eq. (1.10)) and their generation rate depends on the emission spectrum of the used lamp (Gladchi et al.,

2014, 2013). The lowest values of inhibition capacity, for TGA concentrations in the range (1.7-6.7)·10^{-4} M, were found for the system irradiated with the DRT-400 lamp (5.4-3.0)•10^{4} s^{-1}. This is explained by the fact that the lamp's emission spectrum and hydrogen peroxide's absorption spectrum best overlap. The highest values of the inhibition capacity were determined for the systems irradiated upon the Solar Simulator, and they are between ((7.3-0.3)•10^{6} s^{-1}) and are two orders of magnitude higher than the values of the system irradiated by the DRT-400 lamp. This is explained by the fact that the Solar Simulator emits spectral radiation approximating that of natural sunlight. The used filters block or allow the passage of a very small amount of radiation with wavelengths shorter than 290 nm. So, the emission spectrum of Solar Simulator, in the presence of the three filters least overlaps with the absorption spectrum of hydrogen peroxide.

The steady-state concentrations of OH radicals are also shown in Figure 1. It was found that with the increase of TGA concentrations in the system, the concentrations of OH radicals increase and are of the 10^{-16} M order for the system irradiated with the DRT-400 lamp and of the 10^{-17}-10^{-18} M order for the system irradiated at SS, in the presence of all three filters. The OH radicals concentration was also discovered to be dependent on the source of irradiation, as shown by their values. Figure 1 shows that the steady-state concentration of OH radicals decreases depending on the irradiation source, in the following order: DRT-400 lamp ((1.8-3.3)•10^{-16} M) > SS, AM0 filter ((0.3-3.3)•10^{-17} M) > SS, UVC-bl. filter ((0.2-2.5)•10^{-17} M) > SS, AM1.5D filter ((0.1-2.0)•10^{-17} M).

Figure 1. The variation of the inhibition capacity and the steady-state concentrations of OH radicals depending on the TGA concentration at different irradiation sources; $[PNDMA]_0=1.1\cdot10^{-5}$ M, $[H_2O_2]_0=1\cdot10^{-2}$ M, pH=7, t=25 °C

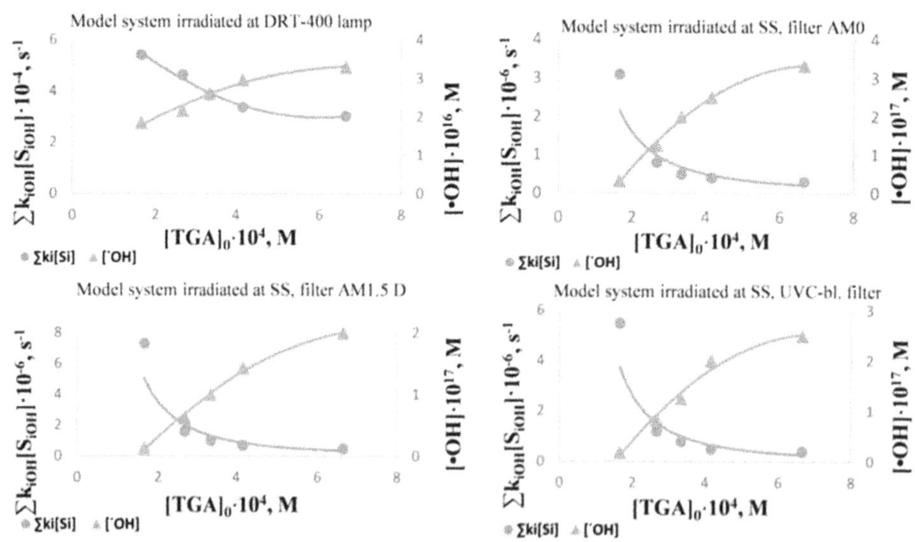

According to the inhibition capacity parameter and steady-state concentrations of OH radicals, it was determined that aquatic systems fell under the categories of very pure and polluted waters at TGA concentrations of the order of 10^{-4} M. It was established that in the presence of TGA in aquatic systems, the intensity of self-purification processes increases because TGA regenerates an additional amount of OH radicals. It has been proven that the inhibition capacity and steady-state concentration of OH radicals depend on the TGA concentration as well as the emission spectrum and radiation intensity of the irradiation source.

The Influence of TGA on the Radical Self-Purification Processes of Aquatic Systems in the Presence of Cu(II) and Fe(III) Ions

The inhibition capacity values $(\sum k_{iOH}[S_{iOH}])$, on the variation of the TGA concentration in the presence of Cu(II) and Fe(III) ions, are presented in Figure 2, and steady-state concentrations of OH radicals are presented in Tables 3-4. Figure 2 show that in the presence of Cu(II) and Fe(III) ions, for all used irradiation sources, the inhibition capacity values decrease. The decrease in the inhibitory capacity of the studied system is related to the formation of OH radicals, which interact with the PNDMA dye, inducing its oxidation. Thus, an increase in the TGA content in the system contributes to the self-purification processes of aquatic systems. This is explained by the fact that TGA in the presence of H_2O_2 can reduce Cu(II) ions to Cu(I) and Fe(III) ions to Fe(II), even at a neutral pH values (pH=6.5-7.0), which is characteristic for natural waters. Respective to irradiation, active particles are generated, such as OH radicals, according to Eqs. (19-24) (Joo et al., 1990):

$$SH - CH_2 - COOH + Cu^{2+} \leftrightarrow [Cu(HOOC - CH_2 - S)]^{+} + H^{+} \tag{19}$$

$$2[Cu(HOOC - CH_2 - S)]^{+} \rightarrow HOOC - CH_2 - S - S - CH_2 - COOH + 2Cu^{+} \tag{20}$$

$$Cu^{+} + H_2O_2 \xrightarrow{h\vartheta} CuOH^{+} + \dot{O}H \tag{21}$$

$$SH - CH_2 - COOH + Fe^{3+} \leftrightarrow [Fe(HOOC - CH_2 - S)]^{2+} + H^{+} \tag{22}$$

$$2[Fe(HOOC - CH_2 - S)]^{2+} \rightarrow HOOC - CH_2 - S - S - CH_2 - COOH + 2Fe^{2+} \tag{23}$$

$$Fe^{2+} + H_2O_2 \xrightarrow{h\vartheta} FeOH^{2+} + \dot{O}H \tag{24}$$

However, it is also clear from Figure 1 that when Cu(II) and, more obviously, Fe(III) ions are added to the system, the situation becomes worse than it would be in their absence. The values of the inhibition capacity in the presence of Cu(II) ions increase by 1.5-3.6 times and depend on the TGA concentration added to the system as well as on the used irradiation source. In the presence of Fe(III) ions, the inhibition capacity values increase by 2.7-6.5 times and also depend on the TGA concentration and used irradiation source. So it can be confirmed that TGA complexes with Cu(II) and Fe(III) ions possess a much higher reactivity towards OH radicals compared to essential TGA, according to the mechanisms described in the literature (Eqs. (25, 26)) (Duca et al., 2002; Joo et al., 1990; Lis et al., 2021):

$$2SH - CH_2 - COOH + Cu^+ \rightarrow (SH - CH_2 - COOH)_2 Cu^+ \quad (25)$$

$$2SH - CH_2 - COOH + Fe^{2+} \rightarrow (SH - CH_2 - COOH)_2 Fe^{2+} \quad (26)$$

The increase in the inhibition capacity values in the presence of Cu(II) and Fe(III) ions demonstrates that the interaction reactions of TGA with Cu(I) and Fe(II) ions take place mainly, and the formed compounds $(TGA)_2Cu^+$ and $(TGA)_2Fe^{2+}$ possess a much higher reactivity towards OH radicals according to the Eqs. (27, 28):

$$(SH - CH_2 - COOH)_2 Cu^+ + 2\dot{O}H \rightarrow HOOC - CH_2 - S - S - CH_2 - COOH + Cu^+ + 2H_2O \quad (27)$$

$$(SH - CH_2 - COOH)_2 Fe^{2+} + 2\dot{O}H \rightarrow HOOC - CH_2 - S - S - CH_2 - COOH + Fe^{2+} + 2H_2O \quad (28)$$

According to the inhibition capacity values for different irradiation sources (Figure 2), the same legality is attested for all sources, but the inhibition capacity values greatly differ from one source to another. The inhibition capacity values decrease with increasing TGA concentration in the system and are of the 10^5-10^6 s^{-1} order, values characteristic of polluted and highly polluted waters. The lowest values of the inhibition capacity $(\sum k_{iOH}[S_{iOH}])$ were determined for the systems irradiated with the DRT-400 lamp, and are between $(1.9\text{-}1.1)\bullet10^5$ s^{-1} – for the system in which Cu(II) ions were present and $(3.5\text{-}1.9)\bullet10^5$ s^{-1} – for the system in which Fe(III) ions were present. This is because the DRT-400 lamp has an emission spectrum closest to the absorption spectrum of H_2O_2. Respectively, the DRT-400 lamp has the largest spectrum of action, and upon irradiation with this lamp, the largest amount of OH radicals are generated. The highest values of the inhibition capacity $(\sum k_{iOH}[S_{iOH}])$ were attested for the systems irradiated with SS, in the presence of the AM1.5 D filter and are between $1.1\bullet10^7\text{-}7.0\bullet10^5$ s^{-1} – for the system in which Cu(II) ions were present, and $2.7\bullet10^7\text{-}2.4\bullet10^6$ s^{-1} – for the system in which Fe(III) ions were present. The same explanation applies here: that the emission spectrum of this filter at least overlaps with the absorption spectrum of H_2O_2, respectively, upon the system's irradiation with it, a smaller amount of OH radicals is formed, which is consumed to oxidize the substrate.

The OH radicals' steady-state concentration increases with increasing TGA concentration in the system (Tables 3-4). In the presence of Cu(II) ions, it was found that the concentrations of OH radicals are of the 10^{-18} M order, for all systems, except for the system irradiated with the DRT-400 lamp. For DRT-400 lamp the values are of the 10^{-18}-10^{-17} M order. But also when comparing the systems with and without the addition of Cu(II) ions, it was established that the OH radical concentrations in the presence of Cu(II) ions decreased and were 1.5-3.6 times lower in their presence. This fact once again confirms that thiols form complexes with transition metal ions and that their complexes have a higher reactivity towards OH radicals. At the same time, it was discovered that the source of irradiation also affects the concentration of OH radicals. The highest OH radical concentrations are generated upon irradiation with the DRT-400 lamp, and the lowest in the systems irradiated with SS, in the presence of AM1.5D filter. Thus, the OH radicals concentration in the presence of Cu(II) ions decreases depending on the irradiation source in the following order: DRT-400 lamp $((5.2\text{-}9.1)\bullet10^{-17}$ M) > SS, AM0 filter $((0.1\text{-}1.7)\bullet10^{-17}$ M) > SS, UVC-bl. filter $((0.1\text{-}1.4)\bullet10^{-17}$ M) > SS, AM1.5D filter $((0.1\text{-}1.2)\bullet10^{-17}$ M).

Figure 2. The variation of the inhibition capacity and the steady-state concentrations of OH radicals depending on the TGA concentration, at different irradiation sources; $[PNDMA]_0=1.1\cdot10^{-5}$ M, $[H_2O_2]_0=1\cdot10^{-2}$ M, $[Cu(II)]_0=5\cdot10^{-6}$ M, $[Fe(III)]_0=5\cdot10^{-6}$ M, pH=7, t=25 °C

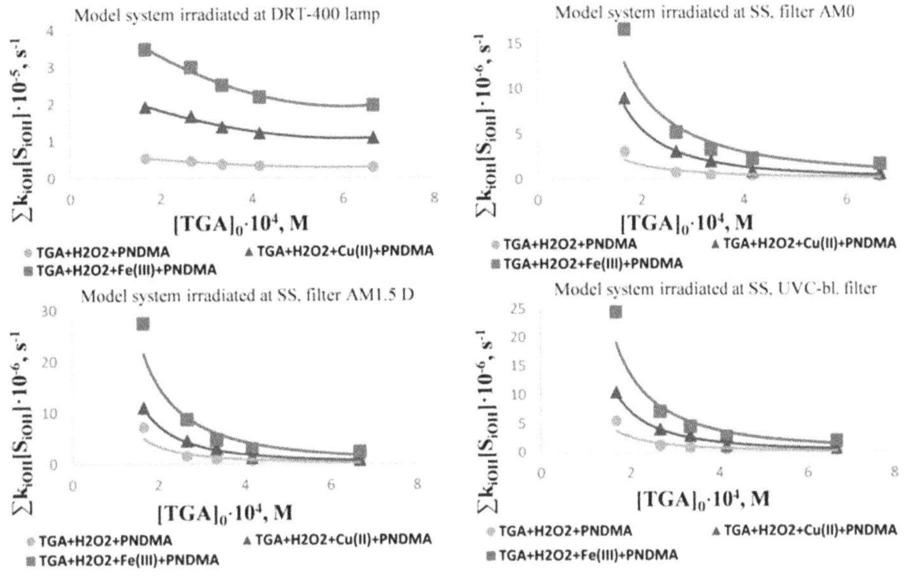

Table 3. Concentrations of OH radicals, generated in the system at different TGA concentrations in the presence of Cu(II) ions; $[PNDMA]_0=1.1\cdot10^{-5}$ M, $[H_2O_2]_0=1\cdot10^{-2}$ M, $[Cu(II)]_0=5\cdot10^{-6}$ M, pH=7, t=25 °C.

	$[{}^{\cdot}OH]_0\cdot10^{17}$, M			
Irradiation source / $[TGA]_0\cdot10^4$, M	*DRT-400*	*AM0*	*AM1.5 D*	*UVC-bloc.*
1.66	5.17	0.11	0.09	0.09
2.66	6.00	0.32	0.21	0.25
3.33	7.19	0.50	0.32	0.34
4.16	8.22	1.10	0.77	0.83
6.66	9.14	1.67	1.25	1.43

The findings in Table 4 demonstrate that when Fe(III) ions were added to the systems, laws corresponding to those in the presence of Cu(II) ions were discovered. However, compared to the presence of Cu(II) ions, the concentrations of OH radicals are 1.7–2.2 times smaller in the presence of Fe(III) ions. This further shows that, in comparison to Fe(III) ions, which were added to the model systems in similar concentrations, Cu(II) ions manifest higher catalytic activity under natural water conditions.

Table 4. Concentrations of OH radicals, generated in the system, at different TGA concentrations, in the presence of Fe(III) ions; $[PNDMA]_0=1.1\cdot10^{-5}$ M, $[H_2O_2]_0=1\cdot10^{-2}$ M, $[Fe(III)]_0=5\cdot10^{-6}$ M, pH=7, t=25 °C.

Irradiation source / $[TGA]_0\cdot10^4$, M	$[^{\bullet}OH]_0\cdot10^{17}$, M			
	DRT-400	AM0	AM1.5 D	UVC-bloc.
1.66	2.87	0.06	0.04	0.04
2.66	3.33	0.19	0.11	0.14
3.33	3.99	0.30	0.21	0.22
4.16	4.57	0.43	0.33	0.37
6.66	5.08	0.59	0.42	0.50

Thus, the OH radical concentration in the presence of Fe(III) ions decreases depending on the irradiation sources, in the following order: DRT-400 lamp ((2.9-5.1)•10^{-17} M) > ((0.6-1.1)•10^{-17} M) > SS, AM0 filter ((0.6-5.9)•10^{-18} M) > SS, UVC-bl. filter ((0.4-5.0)•10^{-18} M) > SS, AM1.5D filter ((0.4-4.2)•10^{-18} M). The obtained findings thus demonstrate that TGA, on the one hand, induces the intensification of the processes of aquatic systems' self-purification, as a result of the regeneration of an additional quantity of OH radicals, which are the most active oxidative particles in aquatic systems. On the other hand, it was established that TGA forms complexes with Cu(II) and Fe(III) ions. The resulting compounds are significantly more reactive to OH radicals. As a result, hydrogen peroxide may be consumed completely and the "redox-toxicity" condition of natural waters may occur. Contrarily, the complexation of transition metals reduces their catalytic activity, which is crucial for the processes of radical self-purification of aquatic systems.

Influence of TU on the Radical Self-Purification Processes of Aquatic Systems

To determine the influence of TU on the self-purification capacity of aquatic systems, the TU initial concentration was varied while the concentrations of the other components were kept constant. The practical results were used to determine the inhibition capacity and steady-state concentrations of OH radicals. From the results shown in Figure 3, it is obvious that increasing the TU concentration in the system leads to increasing the inhibition capacity, regardless of the irradiation source used in the study. The inhibition capacity values are of the 10^6 s^{-1} order, and according to this parameter, they are classified as highly polluted waters for all irradiation sources used in the study. The DRT-400 lamp system, which has values of the 10^5 s^{-1} order and is listed as polluted waters, falls outside of this norm.

Hence, TU has a negative influence on aquatic systems and accordingly leads to the reduction of their self-purification capacity because it consumes an amount of OH radicals for its oxidation, according to Eqs. (29, 30), which is in agreement with previous studies (Duca et al., 2002; Lis, 2016; Sahu et al., 2011):

$$NH_2(NH)C-SH+{}^{\bullet}OH\rightarrow NH_2(NH)CS^{\bullet}+H_2O \quad (29)$$

Figure 3. The variation of the inhibition capacity and the steady-state concentrations of OH radicals depending on the TU concentration, at different irradiation sources; $[PNDMA]_0=1.1\cdot10^{-5}$ *M,* $[H_2O_2]_0=1\cdot10^{-2}$ *M, pH=7, t=25 °C.*

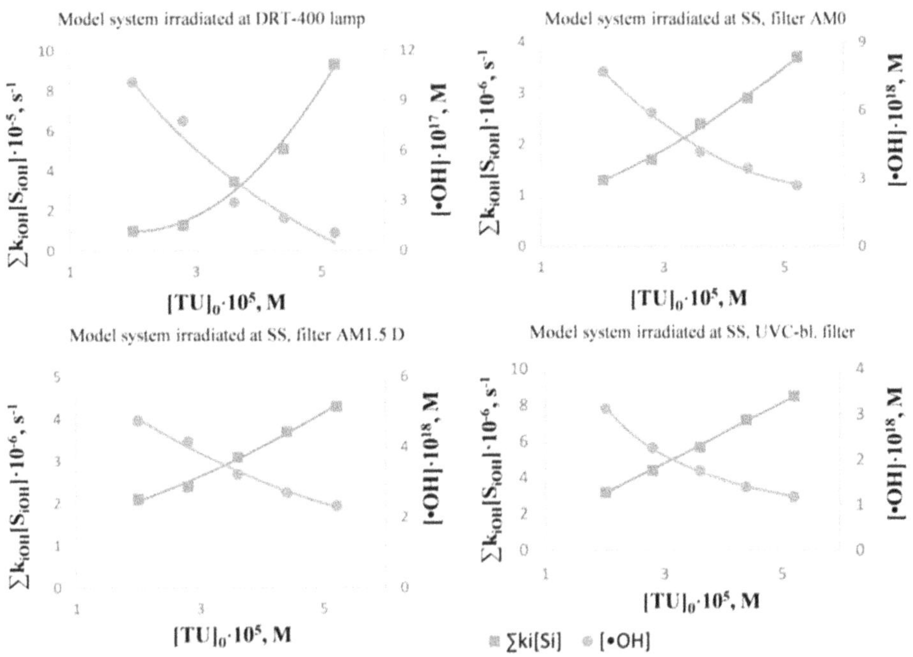

$$NH_2(NH)CS^{\bullet} + NH_2(NH)CS^{\bullet} \rightarrow NH_2(NH)CS - SC(NH)NH_2 \tag{30}$$

Although hydrogen peroxide has a broad absorption spectrum (190-400 nm), its decomposition rate with the generation of OH radicals is higher at lower wavelengths. The lowest values of the inhibition capacity were determined for the system irradiated with the DRT-400 lamp ((0.9-9.3)•10^5 s^{-1}). This is explained by the fact that this lamp emits the highest amount of radiation with shorter wavelengths and in a fairly wide range (240-600 nm). So, the emission spectrum of this lamp best overlaps with the absorption spectrum of hydrogen peroxide. The highest values of the inhibition capacity were recorded for the systems irradiated upon the Solar Simulator, in the presence of all three filters. The values range between ((1.3-8.5)•10^6 s^{-1}) and are practically an order higher compared to the system irradiated with the DRT-400 lamp. This is explained by the fact that the Solar Simulator emits artificial sunlight and provides a very close approximation of natural solar radiation, the used filters block or allow the passage of a very small amount of radiation with wavelengths shorter than 290 nm. So, the emission spectra of SS, in the presence of all three filters, have less overlap with the absorption spectrum of hydrogen peroxide and, accordingly generate the smallest amounts of OH radicals upon irradiation (Gladchi et al. 2008; Duca et al. 2008; Bunduchi et al., 2006).

Another very important parameter used for determining the chemical self-purification capacity of aquatic systems is the steady-state concentration of OH radicals, which is inversely proportional to the inhibition capacity. As a result of the research, it was found that an increase in the TU concentration leads to a decrease in the OH radical concentrations and is of the 10^{-18} M order, for all systems, except for the system irradiated with the DRT-400 lamp, for which the values are of the 10^{-17} M order. The OH radical concentrations decrease depending on the irradiation source, in the following order: DRT-400 lamp ((10.1-1.1)•10^{-17} M) > SS, AM0 filter ((7.7-2.7)•10^{-18} M) > SS, AM1.5D filter ((4.8-2.3)•10^{-18} M) > SS, UVC-bl. filter ((3.1-1.2)•10^{-18} M). Based on the obtained results, it was thus concluded that at TU concentrations of the 10^{-5} M order, according to the inhibition capacity parameter and the OH radical steady-state concentrations, the aquatic systems were included in the category of polluted and highly polluted waters.

The Influence of TU on the Radical Self-Purification Processes of Aquatic Systems in the Presence of Cu(II) and Fe(III) Ions

Starting from the fact that Fe(III) and Cu(II) ions play a very important role in the generation of OH radicals, and respectively in the self-purification processes of aquatic systems (Gladchi et al., 2008; Duca et al., 2008, 2002; Bunduchi et al., 2006), were studied the influence of TU in their presence on the processes of chemical self-purification of waters. Figure 4 shows the values of the inhibition capacity ($\Sigma k_{iOH}[S_{iOH}]$) in the presence of Cu(II) and Fe(III) ions, and Tables 5 and 6 show the steady-state concentration of OH radicals. From Figure 4, it was established that with the addition of Cu(II) and more obviously Fe(III) ions to the system, the self-purification capacity of aquatic systems suddenly worsens, increasing the inhibition capacity values. With the addition of Cu(II) ions, the inhibition capacity values increase by 1.6-5.7 times, and with the addition of Fe(III) ions, they increase by 1.9-7.3 times.

Figure 4. The variation of the inhibition capacity and the steady-state concentrations of OH radicals depending on the TU concentration, at different irradiation sources; $[PNDMA]_0=1.1\cdot10^{-5}$ M, $[H_2O_2]_0=1\cdot10^{-2}$ M, $[Cu(II)]_0=5\cdot10^{-6}$ M, $[Fe(III)]_0=5\cdot10^{-6}$ M, pH=7, t=25 °C

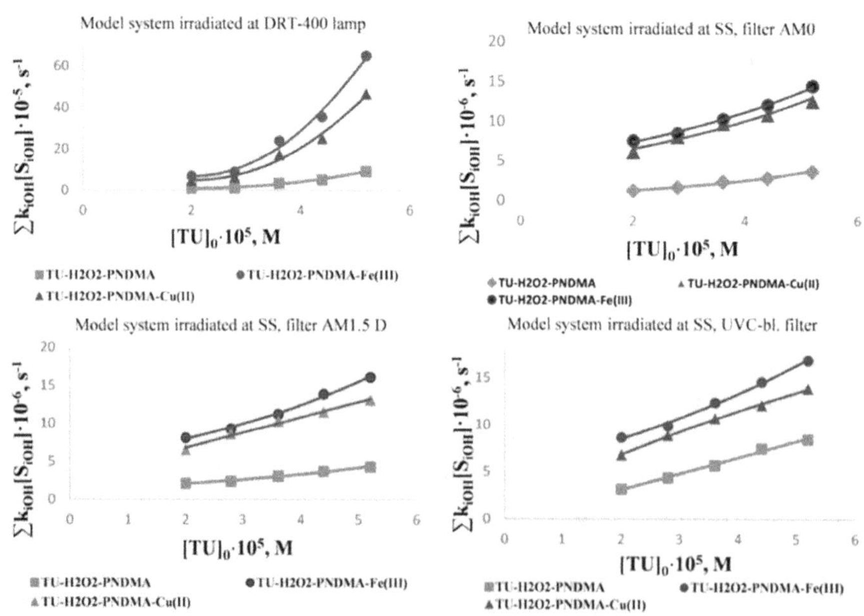

It can be confirmed that TU binds Cu(II) and Fe(III) ions in complex compounds, the most probable ones being $(TU)_2Cu^+$ and $(TU)_2Fe^{2+}$, according to the mechanism described in Eqs. (31-34) (Duca et al., 2002):

$$2NH_2(NH)C-SH+2Cu^{2+}\rightarrow 2Cu^{+}+NH_2(NH)CS-SC(NH)NH_2+2H^{+} \tag{31}$$

$$Cu^{+}+2NH_2(NH)C-SH\rightarrow (NH_2(NH)C-SH)_2Cu^{+} \tag{32}$$

$$2NH_2(NH)C-SH+2Fe^{3+}\rightarrow 2Fe^{2+}+NH_2(NH)CS-SC(NH)NH_2+2H^{+} \tag{33}$$

$$Fe^{2+}+2NH_2(NH)C-SH\rightarrow (NH_2(NH)C-SH)_2Fe^{2+} \tag{34}$$

The formed compounds $(TU)_2Cu^+$ and $(TU)_2Fe^{2+}$ possess a much higher reactivity towards OH radicals compared to TU and lead to their consumption according to Eqs. (35, 36) described in the literature (Duca et al., 2002):

$$(NH_2(NH)C-SH)_2Cu^{+}+2^{\bullet}OH\rightarrow Cu^{+}+NH_2(NH)CS-SC(NH)NH_2+2H_2O \tag{35}$$

$$(NH_2(NH)C-SH)_2Fe^{2+}+2^{\bullet}OH\rightarrow Fe^{2+}+NH_2(NH)CS-SC(NH)NH_2+2H_2O \tag{36}$$

So, it was concluded that TU in real conditions in aquatic systems has a negative influence on radical self-purification processes. On the one hand, it consumes oxidative agents from the aquatic systems, on the other hand, it binds Cu(II) and Fe(III) ions in complexes, thus neutralizing their capacity as catalysts in the processes of generation of OH radicals. On analysis of the inhibition capacity values obtained as a result of irradiation with all sources, similar legalities are attested, but the values of the inhibition capacity differ greatly. It was established that the inhibition capacity values increase with increasing TU concentration in the system and are of the 10^6 s^{-1} order, values characteristic of polluted and highly polluted waters. The lowest values of the inhibition capacity *($\sum k_{iOH}[S_{iOH}]$)* were established for the systems irradiated with the DRT-400 lamp, which fell within the range $6.9\bullet10^5$-$6.5\bullet10^6$ s^{-1} – for the system with Fe(III) ions addition and $4.9\bullet10^5$-$4.7\bullet10^6$ s^{-1} – for the system with Cu(II) ions addition. This is because this lamp has the emission spectrum closest to the H_2O_2 absorption spectrum. Respectively, upon irradiation with this lamp, the largest amount of OH radicals is formed, and at the same time, the emission spectrum overlaps with the TU absorption spectrum, which is predominant in its thiol form (since this lamp emits radiation with longer wavelengths of 300 nm) (Lis, 2016).

The highest values of the inhibition capacity *($\sum k_{iOH}[S_{iOH}]$)* were determined for the systems irradiated with SS, in the presence of the UVC-bl. filter, which took values between $8.7\bullet10^6$-$1.7\bullet10^7$ s^{-1} – the system in which Fe(III) ions were present and $6.8\bullet10^6$-$1.4\bullet10^7$ s^{-1} – the system in which Cu(II) ions were present. This can be explained by the fact that this filter blocks radiation with wavelengths shorter than 290 nm, so its emission spectrum less closely overlaps with the H_2O_2 absorption spectrum, and upon irradiation with this filter, fewer OH radicals are formed, which are then consumed to oxidize the substrate. At the same time, the emission spectrum of the UVC-bl. filter is closer to the emission spectrum of the Sun, so the processes occurring during irradiation with this filter better characterize the processes occurring in real aquatic systems. So, according to the inhibition capacity parameter in the presence of Cu(II) and Fe(III) ions, the values increase depending on the irradiation source, in the following order: DRT-400 lamp < SS, AM0 filter < SS, AM1.5D filter < SS, UVC-bl. filter.

It has been shown that the inhibition capacity values increase with increasing substrate concentration in the system and are of the order of 10^6 s^{-1}, values characteristic of polluted and highly polluted waters. At the same time, it was demonstrated that the inhibition capacity values depend very much on the quality of the emitted radiation. Extrapolating the results obtained on the model systems to natural aquatic systems, it is found that TU slows down the self-purification processes and, at concentrations $(2.0\text{-}5.2)\bullet10^{-5}$ M, it led to the H_2O_2 total consumption, creating premises for a situation of "redox-toxicity".

The steady-state OH radical concentrations, in the presence of Cu(II) and Fe(III) ions upon TU irradiation are shown in Tables 5-6. The steady-state concentration of OH radicals is inversely proportional to the inhibition capacity and was found to decrease with increasing TU concentration in the system. In the presence of Cu(II) ions, the OH radical concentrations are of the 10^{-18} M order, for all systems, except for the system irradiated with the DRT-400 lamp, for which the values were of the 10^{-18}-10^{-17} M order. When analyzing the results presented in Table 5 and Figure 3, it was found that the OH radical concentrations in the presence of Cu(II) ions decreased sharply and were 2-6 times lower in their presence. This once again confirmed that TU complexes with Cu(II) ions have a much higher reactivity concerning OH radicals and consume a larger amount of them. It was established that the steady state OH radical concentrations generated in the analyzed systems also depended on the irradiation source. The highest concentrations of OH radicals are formed in the systems irradiated with the DRT-400 lamp, and the lowest concentrations are generated in the systems irradiated with SS, in the presence of the UVC-bl. filter. The OH radical concentrations, in the presence of Cu(II) ions, decreases depending on the irradiation source, in the following order: DRT-400 lamp ($2.0\bullet10^{-17}$-$2.1\bullet10^{-18}$ M) > SS, AM0 filter ($1.6\bullet10^{-18}$-$7.9\bullet10^{-19}$ M) > SS, AM1.5D filter ($1.5\bullet10^{-18}$-$7.6\bullet10^{-19}$ M) > SS, UVC-bl. filter ($1.5\bullet10^{-18}$-$7.6\bullet10^{-19}$ M).

Table 5. Concentrations of OH radicals, generated in the system at different TU concentrations in the presence of Cu(II) ions; $[PNDMA]_0=1.1\cdot10^{-5}$ M, $[H_2O_2]_0=1\cdot10^{-2}$ M, $[Cu(II)]_0=1\cdot10^{-6}$ M, pH=7, t=25 °C

Irradiation source / $[TU]_0\cdot10^5$, M	$[^{\cdot}OH]_0\cdot10^{18}$, M			
	DRT-400	**AM0**	**AM1.5 D**	**UVC-bloc.**
2.00	20.40	1.61	1.54	1.47
2.80	15.60	1.23	1.16	1.12
3.60	5.81	1.02	0.97	0.93
4.40	3.98	0.92	0.87	0.83
5.20	2.14	0.79	0.76	0.76

Based on the results presented in Table 6, it was found that for the systems in which Fe(III) ions were added, laws similar to those for Cu(II) ions were valid. However, the concentrations of OH radicals in the presence of Fe(III) ions are 1.3-1.5 times lower compared to those in the presence of Cu(II) ions. Analogous to Cu(II) complexes, Fe(III) complexes with TU have a higher reactivity towards OH radicals, and they consume a larger amount than the essential thiol.

Table 6. Concentrations of OH radical generated in the system at different TU concentrations, in the presence of Fe(III) ions; $[PNDMA]_0=1.1\cdot10^{-5}$ M, $[H_2O_2]_0=1\cdot10^{-2}$ M, $[Fe(III)]_0=1\cdot10^{-6}$ M, pH=7, t=25 °C

Irradiation source / $[TU]_0\cdot10^5$, M	$[^{\cdot}OH]_0\cdot10^{18}$, M			
	DRT-400	AM0	AM1.5 D	UVC-bloc.
2.00	20.40	1.61	1.54	1.47
2.80	15.60	1.23	1.16	1.12
3.60	5.81	1.02	0.97	0.93
4.40	3.98	0.92	0.87	0.83
5.20	2.14	0.79	0.76	0.76

Thus, the OH radicals concentration in the presence of Fe(III) ions decreases depending on the irradiation source, in the following order: DRT-400 lamp ($1.4\bullet10^{-17}$-$1.5\bullet10^{-18}$ M) > SS, AM0 filter ($1.3\bullet10^{-18}$-$6.9\bullet10^{-19}$ M) > SS, AM1.5D filter ($1.2\bullet10^{-18}$-$6.2\bullet10^{-19}$ M) > SS, UVC-bl. filter ($1.1\bullet10^{-18}$-$5.9\bullet10^{-19}$ M).

SOLUTIONS AND RECOMANDATIONS

The obtained data reveals the fact that thiols of anthropogenic origin, depending on their structure, can lead to a reduction in the intensity of radical self-purification processes in aquatic systems (for example TU). At the same time, thiols bind transition metals in complexes with low catalytic activity in the process of OH radicals' generation, thus disrupting the redox state of aquatic systems and having a negative impact on their self-purification processes. Based on the obtained results, it is recommended to systematically monitor the thiol concentrations in natural waters to prevent disruption of the redox state and their self-purification capacity. It is recommended to prohibit the discharge into surface waters of wastewaters with thiourea contents over $2\bullet10^{-5}$ M to prevent their negative impact on the self-purification processes of natural waters. It is recommended to use the results obtained by the environmental control bodies to determine the self-purification capacity of natural waters.

FUTURE RESEARCH DIRECTIONS

Resulting from the fact that thiol substances have a strong reducing potential, they actively participate in the redox processes of natural waters, and to establish the biological value of hydrobionts' habitation, it becomes important to prevent water pollution with thiols in concentrations higher than their self-purification capacity. For this, it is necessary to continue studies on model systems using water from natural aquatic objects, because the self-purification capacity is also influenced by other factors, such as the presence and concentrations of other organic pollutants. It is also necessary to study the processes of wastewater treatment containing thiol, from the territories of enterprises, to prevent their penetration into natural aquatic systems. Identifying the effective methods and establishing the optimal conditions for their degradation to non-toxic compounds, so that they can be discharged into natural aquatic systems.

It is known that TGA in the presence of hydrogen peroxide reduces Fe(III) ions to Fe(II), even at a pH close to neutral, which would allow the use of a smaller amount of Fe(II) ions in the wastewaters treatment by the oxidation with the Fenton reagent. At the same time, this would allow the elimination of the water neutralization step in the treatment technology. So, another direction of research in the future would be the use of TGA in wastewater treatment, thus considerably reducing the required amounts of Fe(II) ions and H_2O_2, even at a neutral pH value. In this way, the costs of wastewater treatment can be reduced, and the treated water will be harmless for discharge into natural water bodies.

CONCLUSION

Based on the obtained results, it was concluded that TGA, on the one hand, leads to the intensification of the self-purification processes of aquatic systems because it conducts to the regeneration of an additional amount of OH radicals, which are the most active oxidative particles in aquatic systems. And on the other hand, it was confirmed that TGA interacts with Cu(II) and Fe(III) ions during the formation of complexes. Additionally, the newly formed compounds are much more reactive to OH radicals, which can result in the complete consumption of hydrogen peroxide and, correspondingly, the establishment of the "quasi-reduction" state of natural waters. At the same time, the binding of transition metals in complexes decreases their catalytic activity, which is of great importance in the processes of radical self-purification of aquatic systems.

Unlike TGA, TU in real aquatic systems has a negative influence on radical self-purification processes. On the one hand, it consumes oxidative agents from aquatic systems, on the other hand, like TGA, it binds Cu(II) and Fe(III) ions in complexes, thus neutralizing their capacity as catalysts in the processes of OH radicals generation. At the same time, it was demonstrated that the inhibition capacity values depend very much on the quality of the emitted radiation, and if the results obtained on the model systems are extrapolated to the natural aquatic systems, it is found that TU slows down the self-purification processes. The TU concentrations of $(2.0\text{-}5.2)\bullet10^{-5}$ M, it are driven to the total consumption of H_2O_2 and create the premises for a "redox-toxicity" situation.

It has been shown that thiol substances, according to their behavior in the self-purification processes of aquatic systems, are divided into two groups: TU, which is an effective acceptor of OH radicals, respectively conducts to a decrease in the intensity of the self-purification processes of aquatic systems, and TGA, which regenerates an additional amount of OH radicals upon irradiation in the presence of hydrogen peroxide, respectively directs to an increase in the intensity of the self-purification processes.

This particularity is related to the structure of the studied thiols, TGA has the -OH group in its structure, which breaks upon irradiation in the presence of hydrogen peroxide and thus regenerates an additional amount of OH radicals. Unlike TGA, TU does not have this group, and due to its reducing nature concerning oxidants in aquatic systems (such as OH radicals and H_2O_2), it leads to their consumption and creates the premises for establishing a redox-toxicity state.

ACKNOWLEDGMENT

This research was supported by Moldovan National Research Project No. 20.80009.5007.27 ——Physical-Chemical Mechanisms of the Redox Processes with Electron Transfer in Vital, Technological and Environmental Systems.

REFERENCES

Brezonik, P. L., & Fulkerson-Brekken, J. (1998). Nitrate-induced photolysis in natural waters: Controls on concentrations of hydroxyl radical photo-intermediates by natural scavenging agents. *Environmental Science & Technology, 32*(19), 3004–3010. doi:10.1021/es9802908

Bu, L., Zhu, N., Li, C., Huang, Y., Kong, M., Duan, X., & Dionysiou, D. D. (2020). Dionysios D. Dionysiou: Funding acquisition; Writing - review & editing. Susceptibility of atrazine photodegradation in the presence of nitrate: Impact of wavelengths and significant role of reactive nitrogen species. *Journal of Hazardous Materials, 121760.* doi:10.1016/j.jhazmat.2019.121760 PMID:31810806

Bunduchi, E., Duca, G., Gladchi, V., Goreaceva, N., & Mardari, I. (2006). Assessment of kinetic parameters in the waters of the Nistru course in the section Naslavcea – Dubasari. *Chemistry Journal of Moldova, 1*(1), 56–59. doi:10.19261/cjm.2006.01(1).08

Bunduchi, E., Gladchi, V., & Duca, Gh. (2019). Natural water auto purification capacity through the redox process. *Achievements and perspectives of modern chemistry.* Chisinau: ASM.

Chu, C., Stamatelatos, D., & Mcneill, K. (2017). Aquatic indirect photochemical transformations of natural peptidic thiols: Impact of thiol properties, solution pH, solution salinity and metal ions. *Environmental Science and Pollution Research International, 19*(12), 1518–1527. doi:10.1039/C7EM00324B PMID:29090717

Duca, G., Blonschi, V., Gladchi, V., & Travin, S. (2020). Dynamics of different sulfur forms in natural waters and their influence on the redox state. *American Journal of Physical Chemistry, 9*(3), 52–61. doi:10.11648/j.ajpc.20200903.12

Duca, G., Bunduchi, E., Gladchi, V., Romanciuc, L., & Goreaceva, N. (2008). Estimation of the natural water self-purification capacity from the kinetic standpoint. *Chemistry Journal of Moldova, 3*(1), 10–12. doi:10.19261/cjm.2008.03(1).16

Duca, Gh., Gladchi, V., & Romanciuc, L. (2002). Processes of pollution and self-purification of natural waters. Chisinau: EC MSU, 145.

Duca, Gh., Skurlatov, Yu., & Sychev, A. (2002). *Redox catalysis and ecological chemistry*. PC MSU.

Duca, G., & Travin, S. (2020). Reactions' mechanisms and applications of hydrogen peroxide. *American Journal of Physical Chemistry, 9*(2), 36–44. doi:10.11648/j.ajpc.20200902.13

Duca, Gh., & Vaseashta, A. (2020). *Handbook of research on emerging developments and environmental impacts of ecological chemistry*. IGI Global. doi:10.4018/978-1-7998-1241-8

Foyer, C. H., & Allen, J. F. (2003). Lessons from redox signaling in plants. *antioxidants & redox signaling, 5*(1), 3–5. doi:10.1089/152308603321223487

Fujii, M., & Otani, E. (2017). Photochemical generation and decay kinetics of superoxide and hydrogen peroxide in the presence of standard humic and fulvic acids. *Water Research, 123*, 642–654. doi:10.1016/j.watres.2017.07.015 PMID:28710981

Garcia, P. E., Gerea, M., & Diéguez, M. C. (2020). Natural levels of hydrogen peroxide (H_2O_2) in deep clear South temperate lakes: Field and laboratory evidence of photo- and biotic production. *The Science of the Total Environment, 138641*, 138641. doi:10.1016/j.scitotenv.2020.138641 PMID:32334222

García, P. E., Queimaliños, C., & Diéguez, M. C. (2018). Natural levels and photo-production rates of hydrogen peroxide (H2O2) in Andean Patagonian aquatic systems: Influence of the dissolved organic matter pool. *Chemosphere*. doi:10.1016/j.chemosphere.2018.10.179 PMID:30445400

Gerringa, L. J. A., Rijkenberg, M. J. A., Timmermans, R., & Buma, A. G. J. (2004). The influence of solar ultraviolet radiation on the photochemical production of H2O2 in the equatorial Atlantic Ocean. *Journal of Sea Research, 51*(1), 3–10. doi:10.1016/j.seares.2003.03.002

Gladchi, V., Bunduchi, E., Blonschi, V., & Romanciuc, L. (2022a). Ecological chemistry of natural waters and self-purification processes of aquatic systems. History and development at the State University of Moldova. In: V. Gladchi, A. Aricu (Eds.), Ecological Chemistry: History and Achievements: Academician Gheorghe Duca, 70th Anniversary: Monograph (pp. 37-85). Publishing Center of Moldova State University.

Gladchi, V., Bunduchi, E., Blonschi, V., & Romanciuc, L. (2022b). Hydrochemistry of polluted surface water: case study of Moldova. In A. Vaseashta, G. Duca, & S. Travin (Eds.), *Handbook of Research on Water Sciences and Society* (pp. 501–529). IGI Global. doi:10.4018/978-1-7998-7356-3.ch022

Gladchi, V., Bunduchi, E., & Romanciuc, L. (2020). Ecological Chemistry of the natural water. In I. G. I. Global (Ed.), *Handbook of Research on Emerging Developments and Environmental Impacts of Ecological Chemistry* (pp. 196–211). doi:10.4018/978-1-7998-1241-8.ch009

Gladchi, V., Duca, Gh., Goreaceva, N., Bunduchi, E., & Lis, A. (2013). Chemical composition from the Dniester river tributaries. *Chemistry Journal of Moldova, General. Industrial and Ecological Chemistry, 8*(1), 23–32.

Gladchi, V., Goreaceva, N., Duca, Gh., Bunduchi, E., Borodaev, R., Şurîghina, O., & Lis, A. (2014). Chemical compozition of right bank tributaries of Nistru river and their general impact. Management of Water Quality in Moldova. Springer.

Gladchi, V., Goreaceva, N., Duca, G., Bunduchi, E., Romanciuc, L., Mardari, I., & Borodaev, R. (2008). The study of redox conditions in the Dniester river. *Chemistry Journal of Moldova*, *3*(1), 70–76. doi:10.19261/cjm.2008.03(1).14

Gong, S., Ding, Ch., Liu, J., Fu, K., Pan, Y., Shi, J., & Deng, H. (2021). Degradation of naproxen by UV-irradiation in the presence of nitrate: Efficiency, mechanism, products, and toxicity change. *Chemical Engineering Journal*, *133016*. doi:10.1016/j.cej.2021.133016

Häkkinen, P. J., Anesio, A. M., & Granéli, W. (2004). Hydrogen peroxide distribution, production, and decay in boreal lakes. *Canadian Journal of Fisheries and Aquatic Sciences*, *61*(8), 1520–1527. doi:10.1139/f04-098

Joo, M. S., Tourillon, G., Sayers, D. E., & Theil, E. C. (1990). Rapid reduction of iron in horse spleen ferritin by thioglycolic acid measured by dispersive X-ray absorption spectroscopy. *Biology of Metals*, *3*(3-4), 171–175. doi:10.1007/BF01140575 PMID:2073457

Larson, R. A., & Mccord, J. M. (1977). Environmental chemistry of reactive oxygen species. *CRC Critical Reviews in Environmental Control*, *8*(1-4), 197–246. doi:10.1080/10643387709381663

Lis, A. (2016). Direct photolysis of cysteine and thiourea in aqueous solutions. *Studia Universitatis Moldaviae*, *6*(96), 133–141.

Lis, A., Gladchi, V., Duca, G., & Travin, S. (2021). Sensitized photolysis of thioglycolic acid in aquatic environment. *Chemistry Journal of Moldova*, *16*(1), 46–59. doi:10.19261/cjm.2021.796

Liu, Y., Guo, H., Zhang, Y., Tang, W., Cheng, X., & Li, W. (2018). Heterogeneous activation of peroxymonosulfate by sillenite Bi 25 FeO 40: Singlet oxygen generation and degradation for aquatic levofloxacin. *Chemical Engineering Journal*, *343*, 128–137. doi:10.1016/j.cej.2018.02.125

Lo Conte, M., & Carroll, K.S. (2013). The chemistry of thiol oxidation and detection. *Oxidative Stress and Redox Regulation*, 1-42. . doi:10.1007/978-94-007-5787-5_1

Mill, T., Hendry, D. G., & Richardson, H. (1980). Free-radical oxidants in natural waters. *Science*, *207*(4433), 886–887. doi:10.1126cience.207.4433.886 PMID:17729869

Morris, J., Rose, A., & Lu, Z. (2022). Reactive oxygen species in the world ocean and their impacts on marine ecosystems. Elsevier. *Redox Biology*, *52*, 102285. doi:10.1016/j.redox.2022.102285 PMID:35364435

Munteanu, C., Dumitrascu, M., & Iliuta, R. A. (2011). *Ecology and environmental quality protection.*

Naik, R. M., Agarwal, A., Prasad, S., & Verma, A. K. (2009). Trace determination of thiosulphate and thioglycolic acid using novel inhibitory kinetic spectrophotometric method. *Microchemical Journal*, *93*(1), 43–48. doi:10.1016/j.microc.2009.04.006

Nugraha, W. D., Sarminingsih, A., & Alfisya, B. (2020). The study of self purification capacity based on Biological Oxygen Demand (BOD) and Dissolved Oxygen (DO) parameters. *IOP Conference Series. Earth and Environmental Science*, *448*(1), 012105. doi:10.1088/1755-1315/448/1/012105

Ouyang, Z., Yang, C., He, J., Yao, Q., Zhang, B., Wang, H., & Dang, Z. (2020). Homogeneous photocatalytic degradation of sulfamethazine induced by Fe(III)-carboxylate complexes: Kinetics, mechanism and products. *Chemical Engineering Journal, 126122*, 126122. doi:10.1016/j.cej.2020.126122

Patai, S. (Ed.). (1974). *The chemistry of the thiol group* (Vol. 2). Wiley. doi:10.1002/9780470771327

Patel, M., Kumar, R., Kishor, K., Mlsna, T., Pittman, C. U. Jr, & Mohan, D. (2019). Pharmaceuticals of emerging concern in aquatic systems: Chemistry, occurrence, effects, and removal methods. *Chemical Reviews, 119*(6), 3510–3673. doi:10.1021/acs.chemrev.8b00299 PMID:30830758

Patterson, C. O. P., & Myers, J. (1973). Photosynthetic production of hydrogen peroxide by anacystis nidulans. *Plant Physiology, 51*(1), 104–109. doi:10.1104/pp.51.1.104 PMID:16658269

Sahu, S., Rani Sahoo, P., Patel, S., & Mishra, B. K. (2011). Oxidation of thiourea and substituted thioureas: A review. *Journal of Sulfur Chemistry, 32*(2), 171–197. doi:10.1080/17415993.2010.550294

Schmidtko, S., Stramma, L., & Visbeck, M. (2017). Decline in global oceanic oxygen content during the past five decades. *Nature, 542*(7641), 335–339. doi:10.1038/nature21399 PMID:28202958

Seraghni, N., Ghoul, I., Dekkiche, B. A., Bouaziz, C., Debbache, N., & Sehili, T. (2021). Effect of Fe(III)-bicarboxylic complexes in removal pollutant under UV and sunlight in aqueous solutions. *Environmental Technology*, 1–8. doi:10.1080/09593330.2021.1891142 PMID:33583355

Shtamm, E. V., Purmal, A. P., & Skurlatov, Y. I. (1991). The role of hydrogen peroxide in natural aquatic media. *Russian Chemical Reviews, 60*(11), 1228–1248. http://mi.mathnet.ru/eng/rcr982. doi:10.1070/RC1991v060n11ABEH001141

Song, C., Liu, H.Y., Guo, S., & Wang, S.G. (2020). Photolysis mechanisms of tetracycline under UV irradiation in simulated aquatic environment surrounding limestone. *Chemosphere*, 125582. doi:. 2019.125582. doi:10.1016/j.chemosphere

Song, H., Wignall, P. B., Song, H., Dai, X., & Chu, D. (2019). Seawater temperature and dissolved oxygen over the past 500 million years. *Journal of Earth Science, 30*(2), 236–243. doi:10.100712583-018-1002-2

Southworth, B. A., & Voelker, B. M. (2003). Hydroxyl radical production via the photo-fenton reaction in the presence of fulvic acid. *Environmental Science & Technology, 37*(6), 1130–1136. doi:10.1021/es0207571 PMID:12680665

Subramanian, G., & Madras, G. (2017). Remarkable enhancement of Fenton degradation at wide pH range promoted by thioglycolic acid. *Chemical Communications, 53*(6), 1136–1139. doi:10.1039/C6CC09962A PMID:28054055

Sun, C., Chen, C., Ma, W., & Zhao, J. (2011). Photodegradation of organic pollutants catalyzed by iron species under visible light irradiation. *Physical Chemistry Chemical Physics, 13*(6), 1957–1969. doi:10.1039/C0CP01203C PMID:21082142

Sychev, A., Travin, S., Duca, Gh., & Skurlatov, Yu. (1983). *Catalytical reactions and environmental protection*. Stiinta.

Travin, S., Duca, G., & Gladchi, V. (2019). Self-purification of aquatic media from hexachlorocyclohexane in a radical process. *Chemistry Journal of Moldova*, *14*(1), 47–53. doi:10.19261/cjm.2018.537

Vaseashta, A., Duca, G., & Travin, S. (Eds.). (2022). *Handbook of Research on Water Sciences and Society* (Vols. 1–2). IGI Global. doi:10.4018/978-1-7998-7356-3

Vaseashta, A., Gevorgyan, G., Kavaz, D., Ivanov, O., Jawaid, M., & Vasović, D. (2021). Exposome, Biomonitoring, Assessment and Data Analytics to Quantify Universal Water Quality. In A. Vaseashta & C. Maftei (Eds.), *Water Safety, Security and Sustainability. Advanced Sciences and Technologies for Security Applications*. Springer. doi:10.1007/978-3-030-76008-3_4

Vincoli, J. W. (1997). *Risk Management for Hazardous Chemicals, 2*. CRC Press.

Vione, D., Falletti, G., Maurino, V., Minero, C., Pelizzetti, E., Malandrino, M., & Arsene, C. (2006). Sources and sinks of hydroxyl radicals upon irradiation of natural water samples. *Environmental Science & Technology*, *40*(12), 3775–3781. doi:10.1021/es052206b PMID:16830541

Wang, H., Zhou, H., Ma, J., Nie, J., Yan, S., & Song, W. (2020). Triplet photochemistry of dissolved black carbon and its effects on the photochemical formation of reactive oxygen species. *Environmental Science & Technology*, *54*(8), 4903–4911. doi:10.1021/acs.est.0c00061 PMID:32227921

Wong, G. T. F., Dunstan, W. M., & Kim, D. B. (2003). The decomposition of hydrogen peroxide by marine phytoplankton. *Oceanologica Acta*, *26*(2), 191–198. doi:10.1016/S0399-1784(02)00006-3

Yoon, H., Kim, H., & Kim, S. (2021). Long-term seasonal and temporal changes of hydrogen peroxide from cyanobacterial blooms in fresh waters. *Journal of Environmental Management*, *298*, 113515. doi:10.1016/j.jenvman.2021.113515 PMID:34403920

Yuan, S., Liu, X., Liao, W., Zhang, P., Wang, X., & Tong, M. (2018). Mechanisms of electron transfer from structrual Fe(II) in reduced nontronite to oxygen for production of hydroxyl radicals. *Geochimica et Cosmochimica Acta*, *223*, 422–436. doi:10.1016/j.gca.2017.12.025

KEY TERMS AND DEFINITIONS

Inhibition Capacity: This is a kinetic parameter that involves the indirect determination of the OH radical's steady-state concentrations by using the "trap" of radicals, the dye N,N-dimethyl-4-nitrosodimethylaniline (PNDMA). This is calculated based on the photochemical destruction rates of the PNDMA dye in distilled and natural water. The destruction is achieved as a result of the dye's interaction with OH radicals, which are generated during the hydrogen peroxide photolysis added to the system.

Redox-Toxicity State: A state in which, in the presence of the pollutant, the hydrogen peroxide in the aquatic systems is considerably reduced or completely consumed.

Self-Purification of Surface Waters: This is a totality of biological, physical, and chemical processes within the water body that lead to the diminution of pollutants' concentrations to levels harmless for ecosystem functionality.

Chapter 4
Influence of Thiol Compounds on Redox State of Natural Waters in the Republic of Moldova

Vladislav Blonschi
https://orcid.org/0000-0001-8628-6756
Moldova State University, Moldova

Gheorghe Duca
Institute of Chemistry, Moldova State University, Moldova

Viorica Gladchi
https://orcid.org/0000-0002-5847-4466
Moldova State University, Moldova

Elena Bunduchi
Moldova State University, Moldova

Angela Lis
https://orcid.org/0000-0003-1111-7485
Moldova State University, Moldova

ABSTRACT

The chemical composition of water is determined by the environment with which it comes into contact, with the establishment of a dynamic balance, conditioned by energy and substance flows, expressed by redox equivalents. The valuable biological state of natural waters as a living environment is oxidizing, determined by the presence of hydrogen peroxide. For the characterization of the redox state of the waters, it is not enough to monitor hydrogen peroxide, but it is also necessary to monitor reducing equivalents, such as thiols. The purpose of this chapter is to monitor the content of thiol compounds in some lentic and lotic systems on the territory of the Republic of Moldova for a period of five years. The comparative analysis of the results obtained for rivers and lakes allowed the complex elucidation of the redox state of these waters by characterizing the intensity of redox processes with the participation of thiols. The results indicate that redox processes occur with different intensities in rivers and lakes due to their different chemical composition.

DOI: 10.4018/979-8-3693-0512-6.ch004

INTRODUCTION

In the Republic of Moldova, the most important water resources, for socio-economic activity are mainly surface waters. The country can be divided into three main river basins, the Dniester River's being the biggest of them. (Goreacheva et al., 2004). Raut River is the largest tributary of the Dniester, on the territory of the Republic of Moldova, with a length of 286 km and a water accumulation basin surface – of 7760 km^2. The Ichel River is another significant tributary on the right bank of the Dniester basin, with a length of 101 km, which flows into the Dniester close to Chisinau. Most of the watershed is used for agriculture. In its basin, over 80 minor rivers have a total length of 294 km, of which 75 are less than 10 km lengthy and 7 are between 11 and 20 km. The Dniester River is estimated to supply 54% of the country's economic water requirements. (Blonschi et al., 2022a). Over 90% of the municipality of Chisinau and over 70% of the people of the Republic of Moldova receive their drinking water from the Dniester River. Total water consumption decreased by 4.5 times between 1990 and 2017. The important decrease in water consumption of the Dniester River basin was caused by a slowdown in economic activity overall, especially the abandonment of irrigation (Corobov et al., 2014). Therefore, the rational use and protection of aquatic objects as a living environment are essential. Moreover, the Republic of Moldova does not have any significant rivers that start on its territory, The Dniester River and the Prut River are cross-border rivers. Based on the above, the purpose of this chapter is to determine the amount of thiol compounds in a variety of aquatic objects that are located within the Dniester River basin and to determine their potential impact on the redox state of these waters.

BACKGROUND

Monitoring the ecological status of natural waters provides the necessary information about their well-being. In addition, due to the complexity and even more diversity of the sources of pollution, it is essential to evaluate water quality in a complex manner using a wide range of physical, chemical, and biological parameters. Natural waters have the capacity for self-purification (Gladchi, 2018), which represents a complex series of biological, chemical, and physical processes that partially or entirely restore their initial chemical composition (Figure 1). However, the ecological state of water bodies on the country's territory suggests that the intensity of these processes is low.

The biological self-purification processes are the most active and effective at reestablishing the aquatic object's natural composition among all other self-purification processes, including chemical and physical ones. However, research in the field of ecological chemistry show that the valuable biological state of natural waters as a habitat for hydrobionts is actually formed by the chemical processes of self-purification (Blonschi, 2022b). All self-purification chemical processes come down to a dynamic balance between reducing and oxidative equivalents. This balance cannot be estimated using traditional hydrochemical parameters, so there are specific parameters required, such as the redox state of the waters.

It is well known that hydrogen peroxide is the main oxidative equivalent of dissolved oxygen in the composition of natural waters. Its presence guarantees the waters' oxidizing state, i.e., their valuable biological state. Hydrogen peroxide represents an active intermediate form in the process of divalent reduction of oxygen to the water molecule (Duca et al., 2020a). On the other hand, a wide variety of compounds make up the total amount of reducing equivalents, including hydrobionts metabolic products, substances imported from other ecosystems, organic and mineral substances with a minimum degree of oxidation, etc. All the reducers are divided into two large groups: compounds that are preferentially oxidized by dissolved oxygen and compounds which are preferentially oxidized by hydrogen peroxide. Therefore, the last compounds consume the active oxidative equivalents.

Figure 1. General scheme of self-purification of water basins

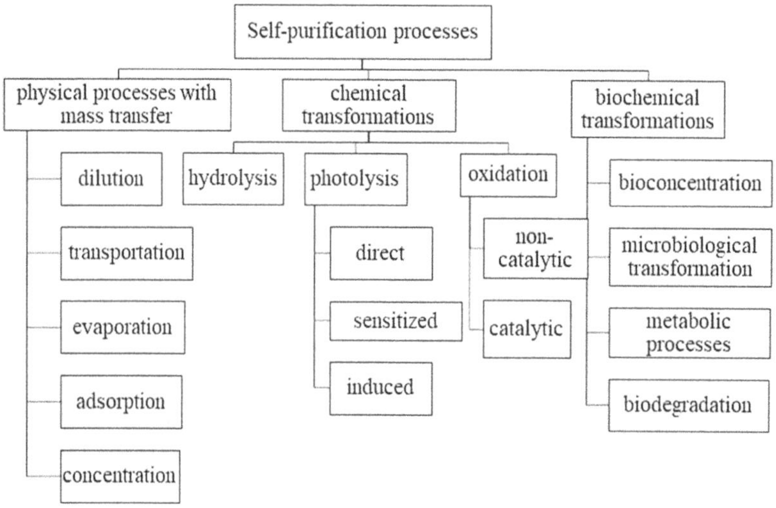

Thiols are organic compounds that contain the -SH group, which gives their reducing abilities. The thiols sources in natural waters are diverse, however, they can be divided into two main groups. The first one includes autochthonous thiols - formed inside the basin as a result of the biological activity of hydrobionts (cysteine, glutathione, arginylcysteine, glutamylcysteine, phytochelatin, Table 1). And the second includes allochthonous thiols - that are introduced into the water's composition of anthropogenic activity (thiourea, thioglycolic acid, etc.) (Chu et al., 2017, Lis et al. 2021).

The assignment of thiols in the category of reducing equivalents does not automatically identify these compounds as pollutants, especially the autochthonous ones. In other words, it is normal for them to be present in the composition of natural waters. Many hydrobionts, including autotrophs, generate large amounts of thiol compounds. According to some studies, thiols are substances used in the defense against active forms of oxygen produced in the "dark stage" of photosynthesis and are synthesized in significant quantities by autotrophic aquatic organisms (Poole, 2015).

The precursors of thiols synthesized *in vivo* are sulfate ions. So, through biochemical processes, it is possible to reduce sulfur from the maximum degree of oxidation +6 (Vione et al., 2019, Tang et al., 2000). The process can be represented by the following schematic equation (Eq. 1):

$$SO_4^{2-} + R{-}H + 6e^- + 8H^+ \rightarrow R{-}SH + 4H_2O \qquad (1)$$

Therefore, based on the legalities previously mentioned, it is reasonable to affirm that thiols are necessarily present in the composition of natural waters, similar to hydrogen peroxide. Therefore, the quality of natural waters as a living habitat is determined by the contents of both equivalents. Their quantitative analysis can provide important information about the redox state or ecological condition of the waters.

Table 1. The abundance of autochthonous thiols and their main biological functions (Chu et al., 2017)

Thiol structure	Biological function	Natural abundance	
		Intracellular, µM	**In the aquatic (marine) environment, nM**
cysteine 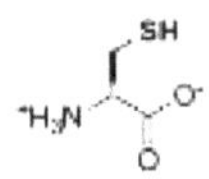	structural function	10- 1100	0,30- 2,11
arginylcysteine 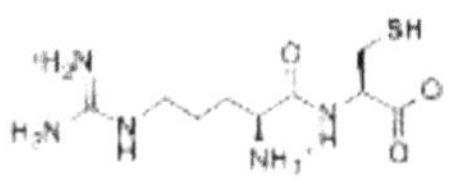	storage of nitrogen sources	≈2900	0,04- 0,50
glutamylcysteine 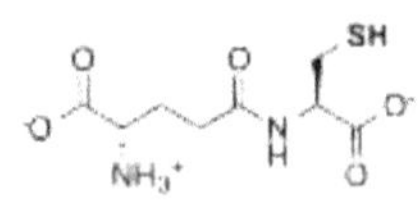	precursor in glutathione synthesis	40-800	0,05- 15,10
glutathione 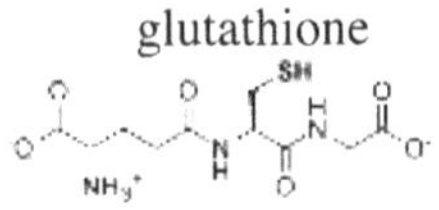	- the main antioxidant in the body, in a free state; - detoxifier, - a precursor in phytochelatin synthesis	780- 3300	0,01- 0,72
phytochelatin 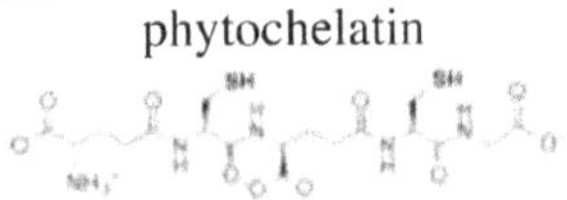	heavy metal detoxifier	2,7- 79,0	0,01- 0,07

MATERIALS AND METHODS

Chemicals and Standards

All chemicals used in this study were chemically pure, purchased from Sigma Aldrich and Acros Organics. The cysteine solution made at the moment of the determination was used as standard. As a source of free OH radicals stabilized hydrogen peroxide (30%) was used. The absorbencies of the analyzed solutions in order to determine the contents of free -SH groups and hydrogen peroxide were recorded on UV-VIS spectrophotometers with double beam T70+ PG Instruments and SP8001, respectively. The pH value of the prepared buffer solutions was checked using the Hanna edge pH meter with glass electrode HI 11310. For each analysis, the working solutions were prepared at the moment of the determination. All solutions were prepared with freshly distilled water (distiller model GFL 2001/4).

Water Sampling

In order to identify the possible transformation laws of thiol compounds in both lotic and lentic systems, samples were taken seasonally. In order to observe the seasonal variation water sampling was performed 5-7 times a year, during the years 2017-2021, from 5 aquatic objects of the hydrographic basin of the Dniester River. As objects of study, 3 lotic systems were selected - the Dniester River, on the course of which 3 catchments were selected (Ustia village, Dubasari district, Criuleni city, Vadul lui Voda city), the Raut River (Ustia village, Dubasari district), the Ichel River (Cosernita village, Criuleni district) and 2 lentic systems - the lakes Ghidighici (Vatra city, Chisinau municipality) and Danceni (Danceni village, Ialoveni district) (Figure 2).

The water samples were collected from the surface of the aquatic objects (about 20 cm deep), until 12:00 am, on days without atmospheric precipitation. The sampling sites were selected so that there isn't an increase in an anthropogenic activity closely. The water samples were analyzed in the laboratory on the day of the sampling.

Figure 2. Water sampling outlets. 1- Dniester, the Dubasari dam, upstream of the Raut River; 2- Raut, the mouth of the Dniester, the Ustia village; 3- Dniester, Criuleni city, downstream of Raut River, upstream of Ichel River; 4- Ichel River, near the discharge into the Dniester River; 5- Dniester, Vadul lui Vodă city, downstream of the Ichel River; 6 – Lake Ghidighici, the region near the dam; 7 – Danceni lake

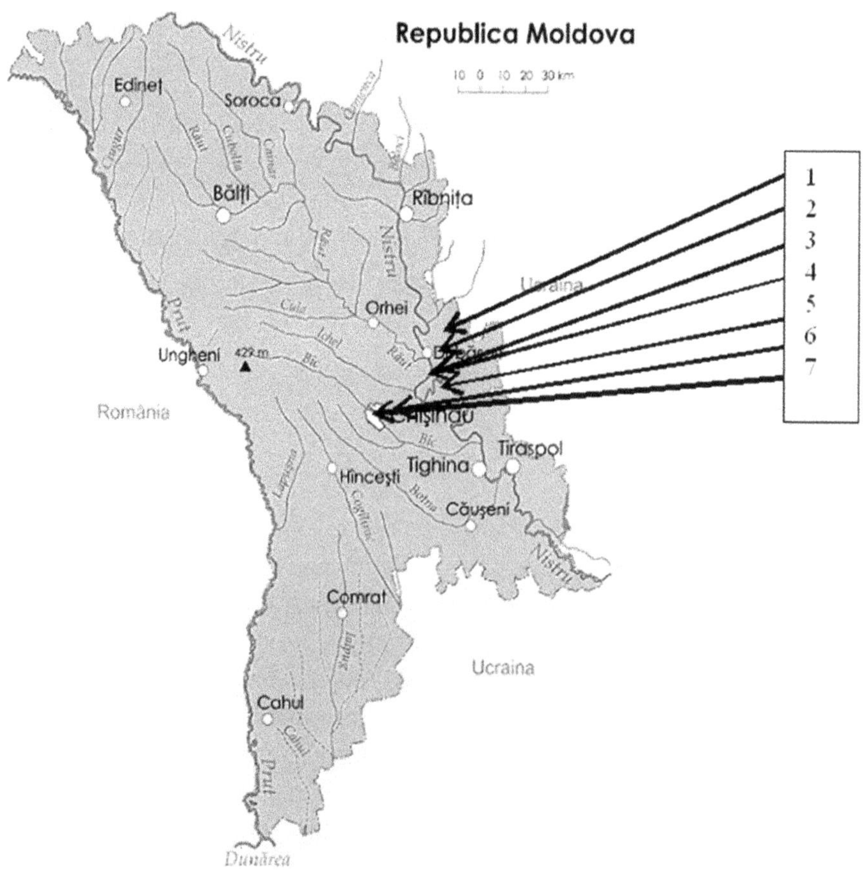

Chemical Analysis

Spectrophotometric determinations were performed using a UV-VIS T70+ spectrophotometer equipped with spectrum recording and data processing software and the UV-VIS spectrophotometer model SP8001.

Determination of the Content of Free –SH Groups

The content of thiol compounds was determined by the derivative spectrophotometric method using 5,5'-dithiobis-(2-nitrobenzoic) acid (DTNB), which at pH 8 interacts with the free -SH groups from the thiol chemical composition with the formation of 2-nitro-5-thiobenzoic acid (TNB). Subsequently, TNB at pH 8, dissociates with the formation of the anion TNB^{2-}, which has an absorption maximum at the wavelength 412 nm (Ellman, 1959) and a high value of the molar extinction coefficient (Riddles et al., 1983). The recorded spectrum of the compound TNB^{2-} formed upon the interaction of cysteine (Cys) and Ellman's reagent confirmed the absorption maximum (Figure 3). The concentration of free –SH groups was calculated according to Eq. 2:

$$C = \frac{A_{412} \cdot n}{\varepsilon \cdot l} \quad (2)$$

where: C – the concentration of free –SH groups from the thiol composition, mol/L; A_{412} – absorbance at the 412 nm wavelength; n – the dilution coefficient of the water sample; ε – the molar extinction coefficient, L/mol·cm; l – the thickness of the layer traversed by the light flux, cm.

Figure 3. UV-VIS absorption spectrum of dissociated 2-nitro-5-thiobenzoic acid. $[Cys]_0=1\cdot10^{-3}$ mol/L, pH=8

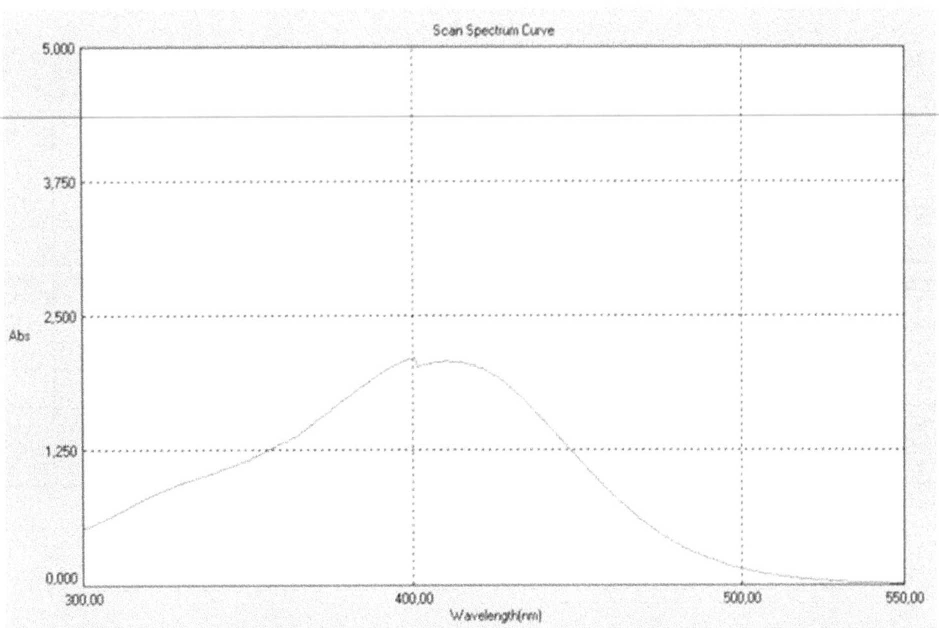

Determination of Hydrogen Peroxide Content

The hydrogen peroxide content was determined by the spectrophotometric method used in the analysis of natural water samples, with malachite green leucobase (Duca et al., 2007). Add aliquot volumes of hydrogen peroxide solution ($1 \cdot 10^{-4}$ mol/L) from 0 to 1 mL, 1 mL phosphate buffer solution pH 4.5, 6 mL water sample, and 0.2 mL each peroxidase solution, to the final volume of 10 mL. In a similar manner, a series of solutions using distilled water was made. The absorbances of the solutions were measured at wavelengths 633 nm and 714 nm. The graphical dependence $\Delta A = f(C(H_2O_2))$ was constructed for the evaluation of the hydrogen peroxide content in the natural water sample. The concentration of hydrogen peroxide in the analyzed samples was calculated using Eq. 3:

$$C_{H_2O_2} = \frac{n \cdot \Delta A \cdot 3,410^7}{\varepsilon_{ef} l} \quad (3)$$

where: $C_{H_2O_2}$ – concentration of hydrogen peroxide, μg/L; n – the dilution coefficient of the water sample; ΔA – the difference in the measured absorbance of natural and distilled water solutions; ε_{ef} – the effective extinction coefficient of the dye, L/mol·cm; l – the thickness of the layer traversed by the light flux, cm.

MAIN FOCUS OF THE CHAPTER

The measurements made between the years of 2017 and 2021 in the 7 sampling sites show that thiol compounds are present in concentrations that are within the range of 10^{-5}-10^{-6} M. For all aquatic systems, the annual and multiyear averages are in the range of 10^{-6} M (Table 2).

Table 2. Annual and multiyear averages of the thiol compounds concentration in aquatic bodies

Sampling points	[R-SH]·10⁶, M					Multiyear average ·10⁶, M
	2017	2018	2019	2020	2021	
Dniester River upstream of Raut River	4.50	3.26	5.15	2.58	1.27	3.35
Raut River	11.01	8.80	11.52	5.40	1.75	7.70
Dniester River upstream of Ichel River, downstream of Raut River	5.88	4.65	3.09	2.38	1.52	3.50
Ichel River	11.11	13.08	14.53	7.32	1.73	9.55
Dniester River downstream of Ichel River	5.19	4.32	6.15	1.73	1.19	3.72
Ghidighici lake	4.45	5.57	5.00	7.49	1.11	4.72
Danceni lake	5.68	5.02	6.12	4.95	0.91	4.54

In all catchments, a comparative analysis of the yearly averages reveals that only the averages for the year 2021 significantly vary from the values for the other years. It can be explained by the fact that 2021 was distinguished by a lot of atmospheric precipitation and cool temperatures. Rains are important sources of oxidative equivalents such as free radicals, especially OH radicals and hydrogen peroxide (Marinoni et al., 2011, Mostofa et al., 2013).

According to the multiannual average that is close to 10^{-5} M, the obtained experimental data show that the concentration of thiols is highest in the waters of the small Rivers, Raut and Ichel, and the lowest values were recorded in the waters of the Dniester River, in all 3 catchments along its course. The average thiol content of the lakes' water is also quite low. In the last two years of monitoring, it was found that the average thiol content in the waters of Danceni Lake had decreased.

Following the discharge of the Raut and Ichel tributaries, the content of thiol compounds along the Dniester River's course was examined, and it was found that in the years 2017, 2018, and 2021, the waters of the Raut River would then contribute to an increase of the thiol content in the waters of the Dniester River. The negative influence of the Ichel River waters was determined only during 2019 when the content of thiol compounds in the waters of the Dniester increased 2 times after the discharge of the tributary.

Through five years of measurements, it was determined that the Raut River had a stronger negative impact on the composition of the Dniester River waters because it was an extra source of thiol compounds. However, the Ichel River waters had a less negative impact on the Dniester River waters' thiol concentration. Based on the multiyear average, the Raut and Ichel tributaries have a negative effect on the Dniester waters, contributing to supplementing them with thiol compounds. But the Dniester waters' self-purification processes are able to transform a load of organic matter supplied by the tributaries, as evidenced by the modest rise in the multiyear average of the content of thiol compounds.

However, the determined thiol compound concentration values do not give the required details about how reducers affect the redox state of these waters. Furthermore, higher concentrations, like those found in small Rivers, do not always indicate anthropogenic or natural pollution. The comparatively low values discovered in lake waters, on the other hand, do not suggest a satisfactory ecological state of these waters. For this reason, the content of oxidative equivalents, expressed by the concentration of hydrogen peroxide, was monitored. This was done to establish some ecochemical legalities based on the detected thiol content (Table 3).

Table 3. Annual and multiyear averages of hydrogen peroxide concentration in water bodies

Sampling points	$[H_2O_2]\cdot 10^7$, M					Multiyear average $\cdot 10^7$, M
	2017	2018	2019	*2020	*2021	
the Dniester River upstream of the Raut River	2.37	4.31	1.28	0	0	1.59
Raut River	2.95	4.59	0	0	0	1.51
the Dniester River upstream of the Ichel River, downstream of the Raut River	1.21	3.38	1.10	0	0	1.14
Ichel River	2.18	3.60	0	0	0	1.16
the Dniester River downstream of the Ichel River	2.10	3.09	1.62	0	0	1.36
Ghidighici lake	3.22	3.69	2.67	0	0	1.92
Danceni lake	2.09	0.82	0.98	0	0	0.78

*The value "0" assumes that the hydrogen peroxide content is below the detection limit of the used analytical method

Thus, in the years 2017-2018, hydrogen peroxide was determined in all aquatic objects. The highest concentrations were in the waters of the Dniester River, and the lowest in the waters of Lake Danceni. During 2019, hydrogen peroxide was not determined in the Raut, and Ichel Rivers. In the last two years of measurements, the content of hydrogen peroxide was not found, which means that its concentration is below the detection limit of the used spectrophotometric method.

Based on the experimental results regarding the content of hydrogen peroxide on the course of the Dniester River, confirms the hypotheses formulated on interpreting the data regarding the content of thiols on the course of this river. As a result, it was found that the Raut River negatively affects the self-purification processes of the Dniester River waters because the amount of hydrogen peroxide in the water slightly decreases within the same order. This means that the waters of the tributary represent a source of reducing equivalents, which leads to the reduction of the self-purification intensity of the Dniester River waters, after the discharge of the Raut. The hypothesis is disproved in the instance of the waters of the Ichel River because the tributary does not change the concentration of this oxidative equivalent, according to the multi-year average of the hydrogen peroxide content.

The decrease in the content of hydrogen peroxide in aquatic objects suggests a decrease in the intensity of chemical self-purification processes, therefore the predominance of the reducing redox state. The put-forward hypothesis is demonstrated by the experimental data of the content of thiol compounds determined in the composition of the analyzed waters. Even if the concentration of thiols decreased in the last years of measurements, however, they remain dominant, while the content of hydrogen peroxide was not detected by the used method.

As a result of the specifics of lotic systems in relation to lentic ones, among which, the energetic exchange of substances between water and soil, the rocks with which it comes into contact, the flow regime, the interaction of water with the atmosphere, etc., the monitoring results require a more detailed interpretation complex, taking into account the biological self-purification processes.

Content of Thiol Compounds in Lotic Systems

The reason for selecting the sampling locations on the monitored River courses was also to observe the impact of the Dniester tributaries, Raut, and Ichel Rivers. Thus, along the course of the Dniester, the three intakes represent intakes upstream and downstream of the tributaries' discharge into the Dniester (Figure 2). Thiol compounds are involved in a variety of biochemical processes, so it stands to reason that their concentration would be inversely proportional to the biological activity of hydrobionts. In other words, there is a seasonal fluctuation in the amount of thiols present in natural waters given that hydrobionts activity varies based on the season, peaking in the summer and declining in the cooler months of autumn and spring. At the same time, the observation of a seasonal fluctuation in thiol content indicates that the processes of self-purification are progressing satisfactorily. The aquatic system's self-purification processes are becoming unregulated if the periodic variation of reducing compounds is not identified.

Based on the seasonal variation in the content of thiol compounds, it is possible to estimate the source of thiols in aquatic objects. The lack of variation may suggest pollution of aquatic systems with biodegradable compounds of a protein nature, such as household water, water from the food industry, etc.

The Dniester River's course in the first catchment (upstream of the Raut River) showed normal seasonal variation throughout the five years of monitoring (Fig. 4a), except for the years 2018 and 2019, when the highest concentrations were observed in spring and autumn, respectively. All values fall within the range of 10^{-6} M.

Figure 4. The multiyear seasonal variation of the content of thiol compounds (a) and of hydrogen peroxide (b) at the sampling point Dniester River (upstream of the Raut River) during 2017- 2021 years

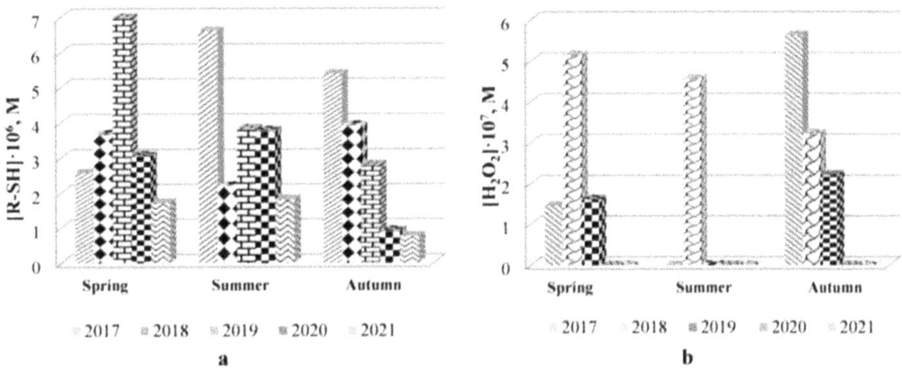

The redox state of the waters is confirmed by comparing the experimental results regarding the concentration of thiol compounds and hydrogen peroxide (Fig. 4b). In the spring of 2017, when the redox state is oxidizing, thiol content is slightly higher than hydrogen peroxide content. During the summer, however, the condition becomes reduced, and the concentration of hydrogen peroxide is less than 10^{-7} M. In the same year, a high concentration of hydrogen peroxide was found in the water basin in the autumn as the intensity of biological processes decreased. Even though thiol compounds are also present, this suggests that the oxidizing condition of these waters has been restored.

Throughout 2018, hydrogen peroxide was found, indicating an oxidizing redox condition of natural waters, while the concentration of thiol compounds did not undergo a natural seasonal variation. This may suggest the dominance of biological self-purification processes with the involvement of hydrobionts which discharge redox equivalents into the aquatic environment. Thus, chemical self-purification processes are not "activated" at high intensities. In 2019, also characterized by the lack of variation of thiols, hydrogen peroxide was found in spring and autumn, periods with the highest concentrations of thiols. The main hypothesis in this situation is the same, the dominance of biological self-purification processes.

The absence of hydrogen peroxide in 2020 and 2021 indicates a dysregulation of the intensity of biological self-purification processes. This in turn causes a reduction in the strength of chemical self-purification processes, resulting in a reducing state. At the same time, it can be concluded about the natural source of thiol compounds in the waters of the Dniester River in the first catchment.

In the second sampling point along the river, downstream of the Raut River and upstream of the Ichel River, the seasonal variation was observed in 2019. In this period the maximum concentration of thiols was attested in the spring, but in the autumn were not found (Fig. 5a). The absence of thiol content in the autumn, however, suggests that the level of self-purification was not significantly deregulated.

The hydrogen peroxide content in relation to the thiol compound content reveals that the redox state was oxidizing in the spring and autumn months of 2017–2019 (Fig. 5b). During the summer, only in 2018, hydrogen peroxide was found. In the summer of the other years, no oxidative equivalents were found within the limit of the detection order. During the last two years of measurements, the hydrogen peroxide content was less than 10^{-7} M. The absence of hydrogen peroxide suggests a reducing state even though the concentration of thiols varied seasonally. Furthermore, rather than redox processes, the reducing equivalents presumably undergo other forms of self-purification. Except for 2019, the number of thiols found in the Dniester course's third catchment, downstream of the Ichel River, varied seasonally as well. The highest concentration during this time was observed in autumn, and in 2021, the springtime concentration of thiols was slightly higher than that of the summertime (Fig. 6).

Figure 5. The multiyear seasonal variation of the content of thiol compounds (a) and of hydrogen peroxide (b) at the sampling point Dniester River (downstream of the Raut River and upstream of the Ichel River) during the years 2017- 2021

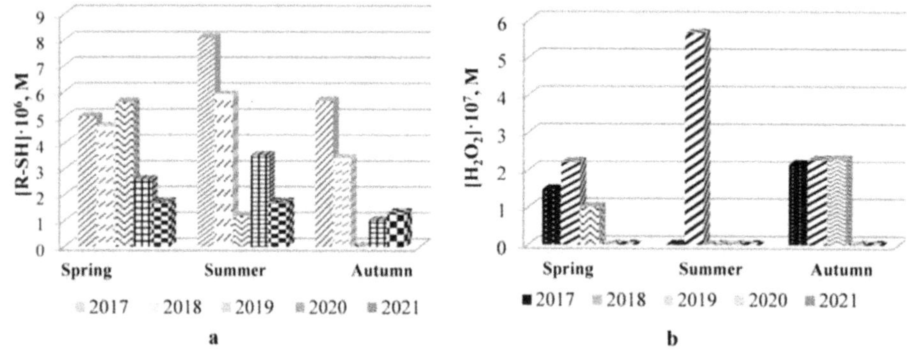

Figure 6. The multiyear seasonal variation of the content of thiol compounds (a) and of hydrogen peroxide (b) at the sampling point Dniester River (downstream of the Ichel River) during the years 2017- 2021

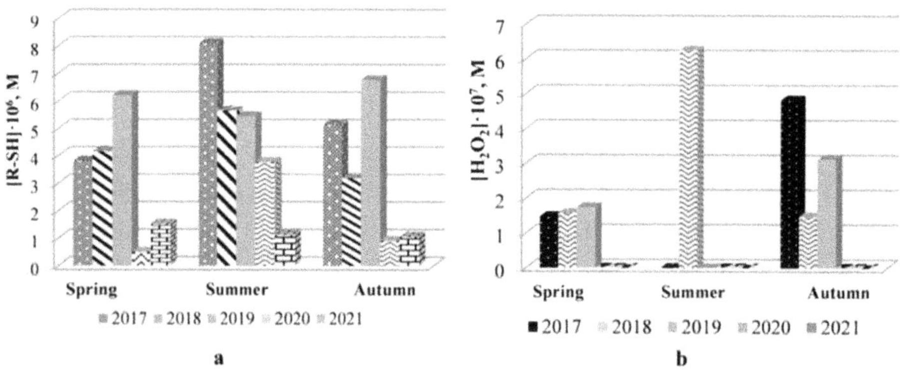

Similar to the sampling sites upstream of the river, the concentration of hydrogen peroxide in the Dniester River catchment after the discharge of the Ichel River suggests an oxidizing state in autumn and spring during the years 2017–2019 (Fig. 6b). But in the summer, except 2019, the redox state is reducing. The lack of seasonal variation in the content of thiol compounds in 2019 may indicate the dominance of biological self-purification processes. So, the content of hydrogen peroxide was also determined, including the autumn of this year. The waters of the Dniester River at this sampling point, during the years 2020 and 2021, are also characterized by a reducing redox state. It is important to note that thiol content naturally varies, indicating their natural genesis.

In the tributary's waters of the Dniester River, the content of thiol compounds detected is much higher, being recorded at levels of the order of 10^{-5} M. Only in the year 2021, out of the five years that the Raut River's waters were being monitored for the presence of thiol compounds, was a natural seasonal fluctuation found (Fig. 7a). At the same time, the highest concentrations of thiols were detected in autumn. This means that the self-purification capacity of the river waters occurs with low intensity.

Figure 7. The multiyear seasonal variation of the content of thiol compounds (a) and of hydrogen peroxide (b) at the sampling point Raut River during the years 2017- 2021

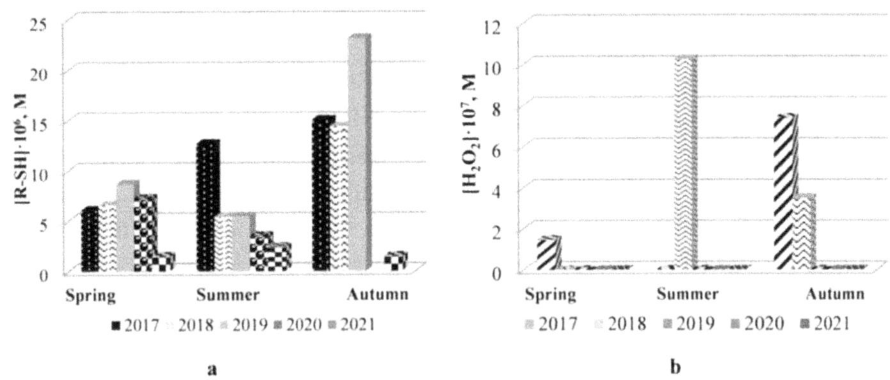

Small Rivers in the Republic of Moldova are distinctive for having high turbidity, which is mainly caused by humic acids (HA). HA represents macromolecular substances, able to absorb in their structure different compounds with low molecular mass, such as thiols (Blonschi et al., 2022). Although the chemical composition of humic acids is not established, it is known that they contain a huge range of functional groups or classes of compounds such as proteins and amino acids (Rodriguez et al., 2009). According to the elemental composition of humic acids, the sulfur content can reach a value of 3.6% (Mahler et al., 2021).

Therefore, high concentrations of thiols in small River waters may be the result of an "apparent self-purification" through absorption by HA. In other words, high concentrations of reducing compounds in the waters of small Rivers is not always indicated a bad ecological condition.

The stated hypothesis is confirmed by the experimental data regarding the content of thiol compounds in the waters of the Ichel River. Thus, the seasonal variation was not detected only during 2017, the highest concentration being recorded in autumn ($15{,}8 \cdot 10^{-6}$ M) (Fig. 8a). Although Ichel's waters have the highest thiol concentrations, there is a clearly marked natural seasonal variation.

Figure 8. The multiyear seasonal variation of the content of thiol compounds (a) and of hydrogen peroxide (b) at the sampling point Ichel River during the years 2017- 2021

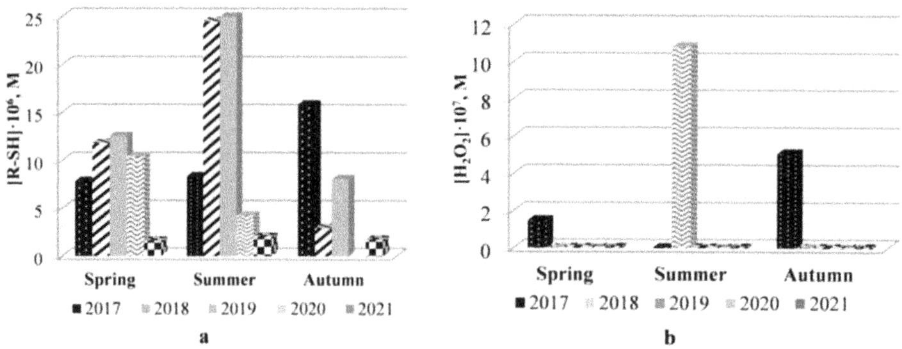

The redox state evaluated by the measured hydrogen peroxide content indicates, in general, a reducing state of these waters (Fig. 8b). The spring and autumn of 2017 and the summer of 2018 denote an oxidizing state. So, the high content of thiol compounds consumes the oxidative equivalents in the waters of the Ichel River. Probably due to the contribution of the other self-purification processes, the natural seasonal variation of thiol content is preserved. At the same time, the variation in the content of thiols indicates that they are the products of the metabolic processes of the hydrobionts.

In all sampling outlets, the seasonal variation of thiol compounds indicates their entrainment in biological self-purification processes, therefore their natural origin. On the other hand, in order to evaluate the impact of thiols on the chemical processes of self-purification, respectively on the redox state and to estimate the mechanisms of thiol transformation, the Pearson linear correlation coefficients were established between the concentration of thiols and hydrogen peroxide.

Thiols are those reducing compounds that almost certainly make the redox state as reducing equivalents by consuming hydrogen peroxide (oxidative equivalent), so there must be an inverse, i.e., a negative correlation between them (Blonschi, 2021, Duca et al., 2020b, Duca et al., 2021, Vaseashta et al., 2022).

Thus, the evaluation of the role played by thiols in creating this dynamic redox equilibrium would be possible through the inference of linear correlations. If a negative correlation cannot be established, it can be assumed that other processes, such as photochemical, biological, and physical (adsorption and mass transport), are used to oxidize thiol compounds. Or it can be concluded that in the composition of natural waters, other compounds with more pronounced reducing properties consume oxidative equivalents.

Since the contents of monitored thiol compounds were interpreted seasonally, the data samples used for calculating the Pearson coefficient reflect multi-year seasonal averages of thiol and hydrogen peroxide contents. According to the type of mechanism, there are two kinds of chemical self-purification processes with a redox character: ion-molecular and radical processes. Hydrogen peroxide participates in both redox processes as an oxidative equivalent and is also the main source of the most reactive free radicals in natural waters, OH• radicals, through redox and/or photochemical processes.

Thus, if the Pearson linear correlation coefficient indicates a strong inverse correlation with the hydrogen peroxide content, it can be stated that these reducing equivalents are oxidized predominantly by an ion-molecular mechanism. The negative value of the Pearson coefficient, for the catchment of the

Dniester River upstream of the Raut River, indicates a strong inverse correlation (Tab. 4). The calculated inverse correlation denotes that thiol compounds are oxidized including the hydrogen peroxide.

The Pearson coefficient for the second catchment outlet along the Dniester River's course is the same as for the first outlet (-0.84), respectively the negative correlation is strong. This suggests that, following the discharge of the latter's waters, the Raut River has little effect on the redox condition of the Dniester waters. However, the input of the dominant mechanism for the oxidation of thiol compounds is unaffected by the tributary waters.

Table 4. Linear correlations between the concentration of thiol compounds and hydrogen peroxide monitored during the years 2017- 2021 in lotic systems

Sampling points	The determined parameter	Spring	Summer	Autumn	Pearson correlation coefficient	Correlation type
the Dniester River upstream of the Raut River	$[R\text{-}SH]\cdot10^6$, M	3.60	3.64	2.75	-0.85	***strong*** *negative correlation*
	$[H_2O_2]\cdot10^7$, M	1.65	0.91	2.12		
Raut River	$[R\text{-}SH]\cdot10^6$, M	6.05	5.90	13.54	0.54	***moderate*** *positive correlation*
	$[H_2O_2]\cdot10^7$, M	0.29	2.05	2.18		
the Dniester River upstream of the Ichel River, downstream of the Raut River	$[R\text{-}SH]\cdot10^6$, M	3.91	4.08	2.28	-0.84	***strong*** *negative correlation*
	$[H_2O_2]\cdot10^7$, M	0.94	1.13	1.34		
Ichel River	$[R\text{-}SH]\cdot10^6$, M	8.83	12.82	7.15	0.77	***strong*** *positive correlation*
	$[H_2O_2]\cdot10^7$, M	0.29	2.16	1.01		
the Dniester River downstream of the Ichel River	$[R\text{-}SH]\cdot10^6$, M	3.24	4.81	3.41	-0.11	***very weak*** *negative correlation*
	$[H_2O_2]\cdot10^7$, M	0.10	0.12	0.19		

Findings from an analysis of the factors in the Raut waters' linear correlation coefficient (0.54) reveal that there is no direct relationship between the oxidative equivalents, represented by H_2O_2, and the reducing equivalents (thiols). The lack of correlation is another argument for the above conclusion. According to this the lack of seasonal variation in the content of thiols and the maximum concentration determined in autumn indicates the deregulation of the self-purification processes of this water's river.

The Pearson coefficient value in the third sampling outlet along the Dniester River's course is significantly different from the values in the first two sampling outlets (Tab. 4). Thus, a very weak negative correlation was determined between the redox equivalents. The Dniester waters' redox state may be influenced by the Ichel River's discharge, or the thiol compounds may be oxidized by radical instead of ion-molecular processes. The second hypothesis is more plausible. This supports the evident seasonal variations in the thiol content of the Dniester waters following the discharge of the Ichel waters. But at the same time, based on multi-year averages, it is kept to have approximately the same content of thiol compounds along the river course (Tab. 2).

The Pearson coefficient value for the Ichel River catchment shows a strong positive correlation. Because seasonal variation was observed during the monitoring period, it is possible that thiols are oxidized by other mechanisms, with the involvement of other substrates, or by biological self-purification processes. The dominance of photochemical reactions with humic acids acting as sensitizers is also possible. Other reactive particles (free radicals) may be produced as a consequence of these processes, which may also cause reducing substances to oxidize.

It was determined that the concentration of these compounds in the Dniester River is not substantially affected by the increased content of thiol compounds in the Raut and Ichel tributaries. The concentration of reducing agents does not significantly vary along the Dniester River course. According to the content of hydrogen peroxide, on the basis of which the redox state is determined, it was found that the concentration of this oxidant decreases, as it is used to neutralize the flows of reducing equivalents from the Dniester tributaries.

As a result, the Dniester River's chemical self-purification processes become slower, and the outflow of its tributaries causes more reducing equivalents to enter the river's waters. Regarding this behavior, the biological value of these fluids as a habitat for hydrobionts will be altered.

So, the content of thiol compounds in lotic systems is strongly influenced by physical-geographical characteristics and climatic conditions. Small Rivers are characterized by high concentrations of thiol compounds, compared to large ones, where the content of reducing compounds is at the limit of detection. For all types of Rivers (small and large) the seasonal variation of thiols is characteristic, which is explained by the high intensity of the biological activity of hydrobionts.

Content of Thiol Compounds in Lentic Systems

During the years 2017 to 2021, the content of thiols was monitored in 2 reservoirs on the courses of two small Rivers that are a part of the Dniester River basin in order to determine the impact of thiol compounds on the formation of the redox state of lentic systems. The Ghidighici lake was built on the course of the Bic River, which is a tributary of the Dniester River. And the Danceni lake, built on the course of the Isnovat River, a tributary of the Bic River.

The multiyear average of the content of thiol compounds in the waters of Ghidighici lake is $4.72 \cdot 10^{-6}$ M (Tab. 2). In only two years of monitoring, 2018 and 2019, was the seasonal variation of thiol concentration confirmed. The content of reducing equivalents is highest in the colder seasons of the other years (spring and autumn). This suggests that reducing equivalents predominates and the rate of chemical self-purification processes has slowed down (Fig. 9a).

Comparing the values of the concentration of thiol compounds found in the water's composition of Ghidighici Lake with the concentration found in the lotic systems reveals that the concentration is relatively low. The low concentration does not, however, indicate an adequate redox condition.

The alkaline lake waters' pH values, which are greater than those of the analyzed river waters (Gladchi et al., 2022a), are a consequence of the low thiol content. (Gladchi et al., 2022b). The pH value of water has a decisive impact on the intensity of thiol oxidation in aqueous systems. The dissociation of the -SH group from the structure of thiols occurs at a pH value of about 9, which greatly speeds up the rate of this group's oxidation with oxygen dissolved in water.

Figure 9. The multiyear seasonal variation of the content of thiol compounds (a) and of hydrogen peroxide (b) at the sampling point Ghidighici lake during the years 2017 – 2021

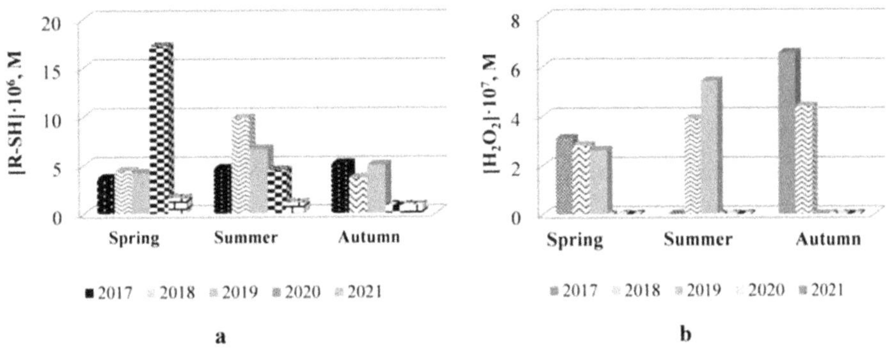

The mechanism of thiol oxidation with molecular oxygen is of the radical type and is present in three conventional steps. The superoxide anion radical ($\dot{O}_2^-$) is a main radical that acts as a scavenger (Oae, 1991, Bindoli et al., 2008):

1) Initiation: dissociation of the –SH group at high pH values:

$$R - SH + R' - O^- \rightarrow R - S^- + R' - OH \tag{4}$$

2) Chain propagation:

$$R - S^- + O_2 \rightarrow R - \dot{S} + \dot{O}_2^- \tag{5}$$

The superoxide radical anion ($\dot{O}_2^-$) oxidizes other thiol molecules:

$$R - SH + \dot{O}_2^- \rightarrow R - \dot{S} + HO_2^- \tag{6}$$

$$R - SH + HO_2^- \rightarrow R - \dot{S} + OH^- + \dot{O}H \tag{7}$$

$$R - SH + \dot{O}H \rightarrow R - \dot{S} + H_2O \tag{8}$$

3) Chain termination:

$$2R - \dot{S} \rightarrow R - S - S - R \tag{9}$$

It is important that the chemical composition of the monitored lakes favors the development of pathogenic organisms (ecological optimum at pH > 9). These pathogenic organisms eliminate large amounts of reducing agents in the environment, thus deregulating the redox state. Just for this reason, additional parameters defining the redox state of the waters must be examined in addition to the thiol content measurement.

Thus, it can be concluded that it is impossible to compare the thiol content of lentic systems with that of lotic systems based on the chemical characteristics of thiol compounds and lakes in the Dniester River basin. It comes out that the concentration of thiol compounds in lake waters should be below the detection limit of the used method in a natural ecological state because one part would have been consumed by hydrobionts and the other part would have oxidized, including by the dissolved oxygen.

However, the measurements' findings show that the Ghidighici lake's waters are not in satisfactory ecological condition. The self-purification processes are slowed down and stimulate the growth of harmful microorganisms that change the redox state into a reducing one.

The data collected on the amount of determined hydrogen peroxide support the hypothesis that was put forth based on the information on the content of thiol compounds. Thus, in 2017, in the spring and autumn, when hydrogen peroxide was found, the content of thiols was lower; however, in the summer, when the oxidant was absent, the content of thiols increased within the same order limit (10^{-6} M) (Fig. 9b).

Since hydrogen peroxide was identified throughout 2018, the oxidizing redox condition of the lake waters could be determined. This year, the natural seasonal variation of thiol compounds was also identified. The year 2019 was characterized by the presence of hydrogen peroxide in spring and summer, with the concentration decreasing below the order of 10^{-7} M in autumn. The presence of the oxidant in 2019 was also accompanied by the seasonal variation of thiol compounds.

In the last two years of measurements (2020 and 2021), the hydrogen peroxide content was below the limit (10^{-7} M), while the seasonal variation of thiol compounds was not determined, even if their concentration decreased compared to previous years. The same explanation can be used to interpret the results regarding the content of thiol compounds in the waters of the Danceni lake (Fig. 10a). The seasonal variation in the content of thiols was not detected only in the last two years of monitoring (2020 and 2021). In this period the highest concentration is attested in the spring, during the other seasons, it decreases.

Figure 10. The multiyear seasonal variation of the content of thiol compounds (a) and of hydrogen peroxide (b) at the sampling point Danceni lake during the years 2017 – 2021

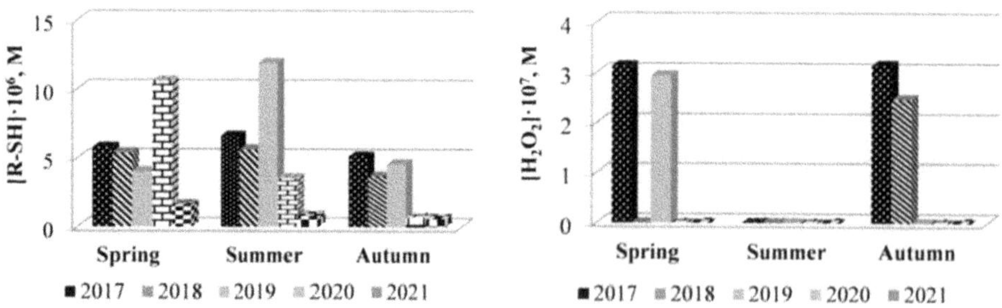

a b

The variation in a small interval of the content of thiols, including in the years when the seasonal variation of these compounds was attested, is another characteristic that indicates a low intensity of the self-purification processes, i.e., on the reducing state of the waters. In other words, the load of reducing compounds in the lake waters is significant throughout the year. The decrease in the content of thiols from spring to autumn suggests disruption of self-purification processes, resulting in the extremely slow oxidation of reducing compounds. It has been discovered that significant amounts of reducing compounds, including thiols, are eliminated in the external environment as a result of the activity of pathogenic organisms, taking into consideration the pH value of the waters, which is favorable for the growth and development of these organisms.

According to the data on the measured hydrogen peroxide content, the redox state of the Lake Danceni waters is predominantly reducing. Thus, the presence of the oxidant was determined in the spring of 2017 and 2019, respectively in the autumn of 2017 and 2018 (Fig. 10b). The absence of hydrogen peroxide during the summertime over the course of the five years of measurements suggests a reducing condition during the warm season when biological activity is at its peak.

The Pearson linear correlation coefficient between the content of thiols and hydrogen peroxide was calculated in order to evaluate the impact of thiol compounds on the redox state of the waters of the investigated lakes. The data samples were obtained similar to the analysis of the data on the investigated Rivers. So, the calculated data represents the multiyear seasonal averages of the content of reducing and oxidative equivalents. Ghidighici lake waters showed a strong negative correlation (-1), and Danceni lake waters showed a moderate negative correlation (-0.50) (Tab. 5).

The strong negative correlation for the water analysis data of Ghidighici lake indicates that thiol compounds are oxidized by hydrogen peroxide through ion-molecular mechanisms. This hypothesis is also supported by the fact that thiol groups are dissociated at the given pH values with the formation of thiolate ions, which manifest stronger reactivity than the undissociated group.

In the waters of Danceni lake, for which a moderate negative correlation (-0.50) was calculated, thiol compounds can be oxidized in other ways as well, such as photochemically or radical oxidation that is driven by OH radicals. So, based on the experimental data on the content of thiol compounds in lotic and lentic systems, it is found that an important specific aspect of lentic systems is the lower intensity of chemical self-purification processes. In a smaller range of them, this also demonstrates the seasonal variation of thiol compounds. In other words, the self-purification capacity of Rivers has a higher intensity than that of lakes. This is mainly because of the distinctive characteristics of these water bodies that were mentioned earlier.

Table 5. Linear correlations between the concentrations of thiol compounds and hydrogen peroxide monitored during the years 2017- 2021 in lentic systems

Sampling points	The determined parameter	Spring	Summer	Autumn	Pearson correlation coefficient	Correlation type
Ghidighici lake	$[R\text{-}SH]\cdot 10^6$, M	6.22	5.33	3.10	-1	***strong*** *negative correlation*
	$[H_2O_2]\cdot 10^7$, M	1.70	1.86	2.19		
Danceni lake	$[R\text{-}SH]\cdot 10^6$, M	5.48	5.70	2.94	-0.50	***moderate*** *negative correlation*
	$[H_2O_2]\cdot 10^7$, M	1.21	0	1.12		

SOLUTIONS AND RECOMANDATIONS

The results of measurements regarding the content of thiol compounds in some surface waters on the territory of the Republic of Moldova reveal that these compounds directly contribute to the formation of the redox state of these waters as reducing equivalents.

The data collected show that lentic systems have a more limited self-purification capacity than lotic systems. As a result, the interdependence between the oxidative equivalents, represented by hydrogen peroxide, and the reducing equivalents, represented by thiol compounds, becomes more evident.

The study's findings led to the conclusion that monitoring the concentrations of thiol compounds and hydrogen peroxide over time enables one to assess the redox state of natural waters and, consequently, to understand the biological state of waters as living habitats.

The research results can be used in the teaching of university subjects in the field of ecological chemistry and environmental protection, as well as serve as a database in studies on the quality of surface waters in the country. The determination of the thiol compounds content, together with the establishment of the redox state, by measuring hydrogen peroxide, can be used as a specific kinetic parameter in the evaluation of the ecological state of water.

FUTURE RESEARCH DIRECTIONS

As a consequence of pollution, monitoring the ecological state of the environment's constituent parts, particularly of natural waters, is a priority direction of research, not only on the territory of the Republic of Moldova. The use of classical hydrochemical parameters does not provide extensive information on the ecological well-being of the water body as a living environment. Thus, new parameters susceptible to the smallest changes in the chemical composition of aquatic systems are needed. However, these parameters must also take into account the dynamics of the given content, which allow for the prediction of certain processes and patterns in addition to the concentration of a particular chemical compound. The data presented in this chapter represent a first attempt to use two specific parameters in the evaluation of the quality of some surface waters. The suggested parameters are complementary to one another and using them separately provides lesser information.

Because monitoring is a continuous activity it is proposed to measure the concentration of thiol compounds and hydrogen peroxide over a longer time period with the following deduction of transformation laws. Taking into account that the reducing equivalents in natural waters are represented by numerous compounds, it is proposed to carry out similar studies to identify them. The determination of some statistical correlations with the oxidative equivalents, expressed by hydrogen peroxide, will allow the adequate assessment of the ecological state of natural waters, by evaluating the kinetic parameter - the redox state. Determination of the chemical nature of the reducing equivalents will enable for the establishment of the exact mechanisms of the chemical processes occurring in the waters—mechanisms about which very little is currently known. Developing analytical methods for the measurement of reducing equivalents that are appropriate for the chemical composition of natural waters is another, equally important aspect.

CONCLUSION

The direct impact of thiol compounds was established on the formation of the redox state of some lotic and lentic systems in the hydrographic basin of the Dniester River on the territory of the Republic of Moldova. This was done by monitoring the content of thiols and hydrogen peroxide over a period of five years. The thiol compounds concentration ranged between 10^{-5} and 10^{-6} M. The highest concentrations are found in the waters of the small Rivers, Raut and Ichel, tributaries of the Dniester River. Based on the chemical parameters, and the concentration of thiol compounds, it was established that Raut and Ichel tributaries have a major impact on the redox state of the Dniester River by supplying its waters with reducing compounds. According to the calculated linear correlations it was established that the discharge of tributaries into the Dniester River affects the intensity and type of redox self-purification chemical processes. Therefore, the ion-molecular mechanisms of these processes in Dniester River are changed to radical ones downstream of the Dubasari dam in the area of Vadul lui Voda. The natural origin of thiols in the monitored aquatic systems was established by detecting the seasonal variations of these compounds. The highest concentrations were recorded in summer when the activity of hydrobionts is highest, and respectively the lowest concentrations in autumn and spring. The lack of seasonal variation of thiol compounds indicates the deregulation of self-purification chemical processes. At the same time, a strong correlation was established between the seasonal variation of thiols and the content of hydrogen peroxide in lentic systems. Therefore, the monitoring of the 3 lotic and 2 lentic systems highlighted the fact that the dynamic balance between oxidative and reducing equivalents in natural surface waters is established with different intensities. Rivers have a higher capacity of self-purification compared to lakes due to the specific characteristics of running water. Because of this, the homogenization of the water layers and the interchange of mass and energy between the layers occur slowly in lakes. The redox state is more readily disturbed in lakes due to these characteristics of the aquatic objects. Also in lakes, oxidation's contribution to self-purification is higher due to the limitations of the other processes. At the same time, the chemical composition of lakes stimulates the growth of pathogenic microorganisms that eliminate more reducing compounds in the water, shifting the chemical balance from the reference value of the redox state. The analysis of the monitoring information regarding the presence of thiol compounds in lentic systems revealed that it is insufficient to determine the redox state only from the measured hydrogen peroxide values. Other chemical factors, such as the pH value and thiol compound content, must be considered when identifying it.

ACKNOWLEDGMENT

This research was supported by Moldova National Research Project Nr. 20.80009.5007.27.

REFERENCES

Bindoli, A., Fukuto, J. M., & Forman, H. J. (2008). Thiol chemistry in peroxidase catalysis and redox signaling. *Antioxidants & Redox Signaling*, *10*(9), 1549–1564. doi:10.1089/ars.2008.2063 PMID:18479206

Blonschi, V. (2021). *Photochemical transformations of some thiol substances and their participation in chemical self-purification processes of natural waters. (No. 57760).* [Doctoral dissertation, Moldova State University]. The National Council on Accreditation and Attestation.

Blonschi, V., Gladchi, V., & Bunduchi, E. (2022a). Monitoring of the thiol substances content in some surface waters. In V. Gladchi & A. Aricu (Eds.), *Ecological Chemistry: History and Achievements: Academician Gheorghe Duca, 70th Anniversary: Monograph* (pp. 141–159). Publishing Center of Moldova State University.

Blonschi, V., Gladchi, V., & Duca, Gh. (2022b). *Participation of thiol substances in chemical self-purification processes of natural waters: Monograph.* Publishing Center of Moldova State University.

Chu, C., Stamatelatos, D., & Mcneill, K. (2017). Aquatic indirect photochemical transformations of natural peptidic thiols: Impact of thiol properties, solution pH, solution salinity and metal ions. *Environmental Science. Processes & Impacts, 19*(12), 1518–1527. doi:10.1039/C7EM00324B PMID:29090717

Corobov, R., Trombitchi, I., Siroedov, G., & Andreev, A. (2014). *Vulnerability to climate change. Moldovan part of the Dniester basin.* Monograph. Elan Poligraf.

Duca, G., Blonschi, V., Gladchi, V., & Travin, S. (2020b). Dynamics of different sulfur forms in natural waters and their influence on the redox state. *American Journal of Physical Chemistry, 9*(3), 52–61. doi:10.11648/j.ajpc.20200903.12

Duca, Gh., Gladchi, V., & Goreacheva, N. (2007). *Practical works on the chemistry of natural waters.* Publishing Center of Moldova State University.

Duca, G., & Travin, S. (2020a). Reaction's mechanisms and applications of hydrogen peroxide. *American Journal of Physical Chemistry, 9*(2), 36–44. doi:10.11648/j.ajpc.20200902.13

Duca, G., & Vaseashta, A. (Eds.). (2020). *Handbook of Research on Emerging Developments and Environmental Impacts of Ecological Chemistry.* IGI Global., doi:10.4018/978-1-7998-1241-8

Ellman, G. L. (1959). Tissue sulfhydryl groups. *Archives of Biochemistry and Biophysics, 82*(1), 70–77. doi:10.1016/0003-9861(59)90090-6 PMID:13650640

Gladchi, V. (2018). *Catalytic transformations and the redox state of the environment.* Monograph. Publishing Center of Moldova State University.

Gladchi, V., Bunduchi, E., Blonschi, V., & Romanciuc, L. (2022a). Hydrochemistry of Polluted Surface Water: Case Study of Moldova. In A. Vaseashta, G. Duca, & S. Travin (Eds.), *Handbook of Research on Water Sciences and Society* (pp. 501–529). IGI Global. doi:10.4018/978-1-7998-7356-3.ch022

Gladchi, V., Bunduchi, E., Blonschi, V., & Romanciuc, L. (2022b). Ecological chemistry of natural waters and self-purification processes of aquatic systems. History and development at the State University of Moldova. In V. Gladchi & A. Aricu (Eds.), *Ecological Chemistry: History and Achievements: Academician Gheorghe Duca, 70th Anniversary: Monograph* (pp. 37–85). Publishing Center of Moldova State University.

Goreacheva, N., & Duca, Gh. (2004). *Hydrochemistry of small Rivers of the Republic of Moldova.* Publishing Center of Moldova State University.

Lis, A., Gladchi, V., Duca, G., & Travin, S. (2021). Sensitized photolysis of thioglycolic acid in aquatic environment. *Chemistry Journal of Moldova*, *16*(1), 46–59. doi:10.19261/cjm.2021.796

Mahler, C. F., Svierzoski, N. D. S., & Bernardino, C. A. R. (2021). Chemical characteristics of humic substances in nature. In A. Makan (Ed.), *Humic Substances* (pp. 1–13). IntechOpen.

Marinoni, A., Parazols, M., Brigante, M., Deguillaume, L., Amato, P., Delort, A. M., Laj, P., & Mailhot, G. (2011). Hydrogen peroxide in natural cloud water: Sources and photoreactivity. *Atmospheric Research*, *101*(1-2), 256–263. doi:10.1016/j.atmosres.2011.02.013

Mostofa, K. M. G., Liu, C. Q., Sakugawa, H., Vione, D., Minakata, D., Saquib, M., & Mottaleb, M. A. (2013). Photoinduced generation of hydroxyl radical in natural waters. In K. Mostofa, T. Yoshioka, A. Mottaleb, & D. Vione (Eds.), *Photobiogeochemistry of Organic Matter. Environmental Science and Engineering* (pp. 209–272). Springer. doi:10.1007/978-3-642-32223-5_3

Oae, S. (1991). *Organic sulfur chemistry*. CRC Press.

Poole, L. B. (2015). The basics of thiols and cysteines in redox biology and chemistry. *Free Radical Biology & Medicine*, *80*, 148–157. doi:10.1016/j.freeradbiomed.2014.11.013 PMID:25433365

Riddles, P. W., Blakeley, R. L., & Zerner, B. (1983). Reassessment of Ellman's reagent. *Methods in Enzymology*, *91*, 49–60. doi:10.1016/S0076-6879(83)91010-8 PMID:6855597

Tang, D., Hung, C.-C., Warnken, K. W., & Santschi, P. H. (2000). The distribution of biogenic thiols in surface waters of Galveston Bay. *Limnology and Oceanography*, *45*(6), 1289–1297. doi:10.4319/lo.2000.45.6.1289

Vaseashta, A., Duca, G., & Travin, S. (Eds.). (2022). *Handbook of Research on Water Sciences and Society* (Vols. 1–2). IGI Global., doi:10.4018/978-1-7998-7356-3

Vione, D., & Scozzaro, A. (2019). Photochemistry of surface fresh waters in the framework of climate change. *Environmental Science & Technology*, *53*(14), 7945–7963. doi:10.1021/acs.est.9b00968 PMID:31241909

Chapter 5
Biodiagnostic Methods in Environmental Chemistry

Vyacheslav Olegovich Shvydkiy
https://orcid.org/0000-0001-7875-218X
Institute of Biochemical Physics, N.M. Emanuel Russian Academy of Sciences, Russia

Sergey Olegovich Travin
N.N. Semenov Federal Research Center for Chemical Physics, Russian Academy of Sciences, Russia

Elena Valentinovna Shtamm
N.N. Semenov Federal Research Center for Chemical Physics, Russian Academy of Sciences, Russia

Elena Grigorievna Cheremnykh
FSBI of Science Scientific Center for Mental Health, Russia

Ludmila Vasilievna Semenyak
Russian Federal Research Institute of Fisheries and Oceanography, Russia

ABSTRACT

As human civilization develops, environmental problems arise more and more often. All forms of the environment are consistently exposed to pollution: water, air, and soil. In the natural aquatic environment, redox processes occur, in which both endogenous and exogenous substances are involved. The processes of antagonism and synergy are also realized, which leads to a decrease or increase in the toxicity of the environment. All this affects intra-aquatic processes, which, as a result, lead to the formation of the quality of the aquatic environment. Historically, there have been three main approaches to assessing water quality using biological objects. These are biodiagnostic methods: bioindication, bio marking, and biotesting. The review reveals the features of biodiagnostic methods and their application in assessing the quality of natural waters.

DOI: 10.4018/979-8-3693-0512-6.ch005

INTRODUCTION

The development of world civilization also implies technological development, which entails the creation of new chemical compounds. New technologies give rise to new possibilities for the synthesis and use of completely new substances, including xenobiotics. Day after day, new organic substances are created, the role of which in the environment is not clear, and sometimes even dangerous. Analytical control of potentially dangerous compounds in real-time is practically impossible since their number is in the millions. In particular, in 2020, more than 159 million substances were entered into the Chemical Analytical Service (CAS) database. The process is close to the exponential law. The entry of new pollutants into the environment is a matter of time. All forms of the environment are consistently exposed to pollution: water, air, and soil.

Particularly, if we consider the natural aquatic environment as an environment for the vital activity of aquatic organisms, and not as a technological solution in industry, then, the first questions of its biological usefulness arise. The answer can only be the response of aquatic organisms, for which water is a vital habitat. This approach also extends to the air and soil environment, where there are other living biological objects adapted to their biologically clean conditions. In the natural aquatic environment, natural redox processes occur, which involve both endogenous and exogenous substances. The processes of antagonism and synergy are also realized, which lead to a decrease or increase in the level of toxicity of the environment. All this affects the intra-aquatic processes and reaction mechanisms, which ultimately lead to the formation of the quality of the aquatic environment. Historically, there have been three main approaches to assessing water quality using biological objects. These are biodiagnostic methods: bioindication, biomarking, and biotesting. These methods can be used separately, but their combined application is the most informative.

BIOINDICATION

Bioindication is the determination of biologically significant loads based on the reactions of living organisms and their communities to them. This method has been used since the reliable identification of living beings of animal and plant origin. Biologically significant loads, as a rule, refer to all types of anthropogenic pollution. The development of methods and criteria that could adequately reflect the level of anthropogenic impacts, considering the complex nature of pollution based on bioindication, is the main task of environmental monitoring.

Bioindication involves the analysis of visible (or imperceptible damage) or abnormal signs of stress exposure in free-living organisms in their natural environment (Parmar et al., 2016). Bioindicators include biological processes, species, or communities and are used to assess the quality of the environment and its changes over time. Organisms and communities of organisms, whose vital functions are closely correlated with certain environmental factors and can be used to assess them, are called bioindicators. All species (or groups of species) tolerate a limited range of chemical, physical and biological conditions that we can use to assess environmental quality. Consistent monitoring makes it possible to diagnose early disturbances in the most sensitive components of biotic communities. The first application of bioindication in geological research is known. Schemes of plants - indicators of rocks were presented at the end of the 19th-century Russian scientist A.P. Karpinsky.

The development of bioindication methods in the field of ecology has made it possible to diagnose various environmental objects. Basically, these are water and soil resources, and bottom sediments, which are subject to pollution both from the air and the direct flow of wastewater of various degrees of purification. Environmental changes are often associated with anthropogenic forcing or natural factors of stressors (e g, drought, late spring frosts), although anthropogenic factors of stressors are the main focus of bioindicator research. Recently, the range of bioindicators has expanded significantly, which helps environmental researchers in the study of all types of environments (i.e., aquatic and terrestrial). In this case, all major taxonomic groups of bioindicators are used, including animals and plants.

However, it should be kept in mind that not all biological processes, species, or communities can serve as successful bioindicators. Physical, chemical, and biological factors, such as the quality of water or air, light, temperature, and competition differ depending on the environment. Over time, populations adapt with a priority on maximizing growth and reproduction within a given range of environmental factors, i.e., achieve maximum adaptability. Beyond the optimum environmental conditions for bioindicators, their physiology, and behavior can be negatively altered, leading to a reduction in overall fitness. A decrease in fitness can subsequently disrupt the dynamics of population development and change the community as a whole. As bioindicators, it is customary to use such types of biocenosis that effectively indicate the state of the environment due to their moderate resistance to environmental variability. At the same time, there are also rare species (or communities of species) that are too sensitive to environmental changes or are too rare to reflect the general reaction of the biota. Similarly, there are ubiquitous species (or communities of species) with very broad persistence. They are less sensitive to environmental changes that disturb the rest of the community. Therefore, the use of bioindicators should not be limited to one species, especially those with limited environmental resistance. Entire communities covering a wide range of environmental conditions can serve as bioindicators and provide multiple data sources for assessing the state of the environment using a “biotic index”.

An example is the use of cutthroat trout as a bioindicator of thermal pollution of water bodies. Cutthroat trout are found in cold water streams in the western United States. The immediate reaction of thugs to heat pollution occurs at the cellular level. In particular, heat shock protein (HSP) synthesis is increased to protect vital cellular functions from heat stress. HSP levels can be analytically quantified to measure heat stress in thugs and evaluate how the environment has changed. Heat stress in cutthroat trout is manifested in reduced growth and development. In extreme cases, persistent thermal changes can reduce populations and even lead to local extinctions (Holt & Miller, 2010). Considering the peculiarities of the influence of thermal pollution on the cellular functions of cutthroat trout, it is possible to separate the influence of thermal pollution from chemical control protein (HSP).

Improved bioindication technology using the functional state of mollusks can also serve as an indicator of excessive pollution of coastal waters by facilities discharging insufficiently treated wastewater. This work of ecological monitoring of ecosystems of coastal marine and freshwater recreational areas (Kholodkevich et al., 2020) was carried out on the example of a number of water bodies in Sevastopol, as well as several freshwater areas in the resort area of St. Petersburg. Given the high expressiveness and ease of use of the technology, the functional state of mollusks can be considered the best for biomonitoring of water quality in recreational reservoirs, in particular, in Russia.

BIOMARKING

Biomarking, as one of the methods of biodiagnostics, assesses the degree of impact of natural and anthropogenic factors on living organisms by changing a number of morphological and functional indicators (biomarkers) that are recorded at the sub-organismal and organismal levels of biological organization (Lima et al., 2018). In fact, biomarkers are enzymes involved in metabolic processes in a living cell or its specific products. Their activity plays a decisive role in the life of the organism, and a decrease in activity to a critical value leads to irreversible, lethal consequences. Changing the activity of enzymes of biomarkers of pollution of aquatic and soil organisms has been used for a long time. And recently, biomarking has found mass application in medicine. At the same time, the choice of enzymes - biomarkers of certain types of pollution and toxicants - is important. Among the factors of paramount importance are a sufficiently high level of their activity in the cell, sensitivity to various external toxic effects, as well as the simplicity and reliability of methods for determining their activity. In particular, the assessment of the degree of pollution of aquatic ecosystems is carried out by the activity of phosphatases in endemic mollusks. In works (Edori et al., 2014; Istomina et al., 2019; Duca et al, 2021; Vaseashta et al., 2022; Vaseashta et al 2021), the pollution of aquatic ecosystems is assessed by the activity of phosphatases in the mollusks *Tympanotonus fuscatus* and the scallop *Mizuhopecten yessoensis*. It has been shown that gasoline and diesel fuel have an inhibitory effect on the enzymes of the snail *Tympanotonus fuscatus*, and copper inhibits the activity of acid and alkaline phosphatases of the scallop *Mizuhopecten yessoensis*. Changes in the activity of acid and alkaline phosphatases in the mussel *Crenomytilus grayanus* made it possible to assess the total pollution of the natural marine ecosystem (Seitkalieva et al., 2016). Digestive enzymes are also actively used as pollution biomarkers (Li et al., 2010; Caruso et al., 2016).

It is known that prolonged exposure to the organotin compound tributyltin leads to significant inhibition of the activities of amylase, lipase, and trypsin in the fry of the common carp *Cyprinus carpio* (Li et al., 2014). And the impact of cadmium, one of the most toxic metals, on the three-spined stickleback *Gasterosteus aculeatus* (Hani et al., 2018) led to a decrease in the activities of trypsin, amylase, and intestinal alkaline phosphatase. Other fish species, for example, the piranha, salmon, cod, and prochilodont fish families have a high sensitivity of cholinesterases and transaminases to organochlorine and organophosphorus pesticides (Narra et al., 2011; Lopes et al., 2014; Greer et al., 2019). In the work (Khare et al., 2019), the synergistic effect of pesticides (carbaryl and methyl parathion) on the tissues of fish of the carp family Catla catla was studied. Synergy occurs when a combination of chemicals has a more pronounced effect on the bioassay as opposed to their individual action. The study used biomarkers of oxidative stress, namely catalase (CAT), glutathione reductase (GSSG-R), and various enzymes such as lactate dehydrogenase (LDH), succinate dehydrogenase (SDH) and acetylcholinesterase (AChE). LDH activity increased significantly (on 748%) in the synergistic group in Catla catla muscle, compared to studies with the same but separate pesticides. There was also a significant increase in the activity of CAT and GSSG-R in all fish tissues compared to the control, which allows the use of CAT and GSSG-R as potential biomarkers of oxidative stress.

The paper (Esimbekova, 2021) compares biochemical markers (protein content, activity of cholinesterase (ChE), catalase (CAT), and glutathione-S-transferase (GST)) with the survival and mortality of *Daphnia magna* after prolonged exposure to Cr (VI) and Cd(II). There was no effect on reproduction and survival up to the maximum tested concentration of Cr(VI) (52.5 μg/l), while the protein content,

ChE, and CAT activity decreased, and GST activity increased. Cd(II) ions affected the reproduction of daphnia above 0.656 μg/l, but the protein content and ChE activity changed at 0.328 μg/l and 0.082 μg/l Cd, respectively. Thus, there is a greater sensitivity of biomarkers in comparison with the Daphnia magna bioassay. However, it should be noted that the *Daphnia magna* as a bio test itself is indeed less sensitive to some chemical contaminants, although it is used in different countries in the study of the toxicity of the aquatic environment. At the same time, there are a number of limitations and disadvantages when using enzymatic tests in situ: the instability of the measured control biochemical characteristics of living organisms, the need to isolate enzymes from the test organism to measure its activity, which increases the time and complicates the analysis procedure. In addition, the number of enzymes whose activity can be determined in a living organism is limited. And there is also a prerequisite - the availability of expensive analytical equipment and qualified personnel.

Nevertheless, the use of enzymes as biomarkers demonstrated their high sensitivity and specificity to the action of a number of pollutants and confirmed that changes in general biological parameters (survival, growth, reproduction, behavior, etc.) are based on biochemical changes in living cells.

In recent years, fruitful work has been carried out to improve and simplify the use of biomarkers. Undoubtedly advantages over enzymatic tests in situ have enzymatic tests in vitro. Working with isolated enzymes makes it possible to study the mechanisms of action of toxicants on the functioning of individual links of the metabolic chain. Working with individual enzymes allows you to simplify and speed up the analysis procedure. It is possible to vary the sensitivity of the methods by changing the conditions of the analysis, for example, by changing the ratio of the components of the reaction mixture, to increase the reproducibility of the analysis due to the transition from a living organism to a reagent. At present, on the basis of in vitro enzymes, it has become possible to develop fundamentally new express adjustable biosensors for the ecological monitoring of environmental objects. The use of enzymes in vitro greatly facilitates the monitoring of the state of the environment. But, as shown above, enzymes are subject to the influence of pollutants of different compositions in comparable concentrations (heavy metals of variable valence, pesticides of various natures, and oil products) (Jemec, 2008). Thus, a detailed analysis by class of pollutants is still required, but knowledge of the mechanisms for reducing enzyme activity allows us to quickly determine the nature of the pollutant.

However, biochemical markers are not always more sensitive than whole-body reactions. Their sensitivity depends on the mode of action, duration of exposure, and the species tested. To adequately identify an environmental hazard, it is required to study the combination of both a series of biomarkers with different levels of biological organization and an array of biomarkers within the same level (Shtamm, 2017).

BIOTESTING

From the point of view of the state of the aquatic ecosystem, the most important indicators are biodiversity, the balance of production and destruction processes (lower links in the trophic chain), fish productivity (fish as the highest link in the trophic chain). The aquatic environment is considered biologically complete if the normal development and reproduction of the main links of the ecological system of the water body are ensured. The most adequate integral characteristic of the state of pollution of wastewater and the natural aquatic environment is toxicity to aquatic organisms of various trophic levels, assessed by biotesting methods.

Since the water of most surface water bodies serves both as a habitat for aquatic organisms and as a source of drinking water supply for the population, the problem of the chemical safety of water from surface sources of drinking water supply is closely related to the problem of the biological usefulness of the aquatic environment as a habitat for aquatic organisms (Centre for Inland Waters, 1988).

Biotesting is a type of biological control of environmental objects, the essence of which is to fix the reaction of a test organism to a toxic factor and includes laboratory studies with living organisms. In addition to the choice of a biological test object, an important role is played by the choice of a controlled (test-) reaction - an organism parameter that changes under the influence of a toxic factor.

For populations, the integral parameters are abundance and biomass, and for ecosystems, the characteristics of the species composition, activity of production, and destruction of organic matter.

In the practice of biotesting, more than two hundred methods are known using various test objects and test reactions (Shtamm, 2011), however, no more than 20 are most widely used. In the USA and Canada, at least three test objects of various trophic levels are used for mandatory toxicological control of water (for example, algae, ciliates, and fish), in European countries - 2 test objects.

Currently, in the practice of water protection and water management, bio-testing methods on cultures of crustaceans, ciliates, bacteria, and microalgae are used, as a rule, to assess the toxicity of wastewater, individual chemicals and their mixtures, water extracts from bottom sediments, waste, etc. (Shtamm, 2017). At the same time, the integral indicator "water toxicity" makes it possible to take into account the presence in natural or wastewater of newly formed compounds, unaccounted for (uncontrolled) toxic compounds, as well as various kinds of interactions of chemical pollutants with each other - synergism, antagonism, additivity, neutralization, etc.

Using a set of two or three bioassays with test objects from the main trophic links of the aquatic ecosystem, one can also obtain sufficiently reliable data to characterize the water quality and the state of the ecosystem of a water body as a whole (Skurlatov, 2020; Zakharov, 1993).

In this case, a sample of water or an aqueous extract from solid materials will be considered toxic, according to the test object that showed the highest sensitivity, i.e., the greatest value of the toxic effect at the lowest concentration of the sample. However, the rapidity of the applied biotesting technique is also of great importance, since with a long-term analysis of the sample, the results of biotesting can characterize not so much the quality of the source water as the toxic properties of the transformation products of the substances present in the source water (Shtamm, 2002).

It follows from the table that the most sensitive test object in relation to an aqueous solution of black liquor was ceriodaphnia (test function - fecundity of ceriodaphnia), while the most express (analysis time 5 and 30 min.) were luminous bacteria (test function - change enzymatic activity of luminous bacteria). Of particular importance is the use of bioassay methods under the detoxification of difficult-to-decompose compounds, for example, surface-active substances (surfactants). In work (Shtamm, 2004) for the detoxification and treatment of wastewater containing surfactants, photo ozonation technology was used in combination with subsequent microbiological treatment with immobilized adapted bacterial strains.

When analyzing the toxicity of surfactant transformation products, test objects from different parts of the trophic chain were used: *ceriodaphnia*, ciliates, green microalgae (*scenedesmus quadricauda*). Ceriodaphnia (*Ceriodaphnia dubia*) and ciliates (*Tetrahymena pyriformis*) turned out to be the most sensitive and informative. Surfactant solutions subjected to 25% transformation as a result of photo-ozonization were fed into bioreactors (see Fig. 1), in which bacterial strains that efficiently decompose surfactants were absorbed on a special carrier, photo 1.

Table 1. Comparative analysis of bioassays on an example of dilute solutions of black liquor

Test system and test function	Minimum concentrations of black liquor (%), in which toxic effect is observed	Exposure time
Reproduction of *Ceriodaphnia*	0.001...0.002	7...10 d
Intensity of LPO processes in native water samples	0.002 (stimulation)	1 h
(rate of MDA accumulation)	0.05 (inhibition)	1 h
Mice SHK: TBA-AP in blood plasma	0.005	12 d
Mice SHK: TBA-AP in liver lipids	0.015	12 d
Activation of fish liver oxygenases (MFO-induction)*	0.02	7 d
Population growth of ciliata *(Tetrahymena pyriformis)*	0.05	24 h
Changes in luciferase activity of luminous bacteria *(Benekea harveyi)*	0.5	5 min
	0.5	30 min
Survival of *Daphnia magna*	0.5	4 d

* These data were obtained in collaboration with P. Hodson at the Institute of Inland Waters (Burlington, Canada).

Figure 1. System of photoreactors and bioreactors for the degradation of surfactant solutions with subsequent biotesting of degradation products

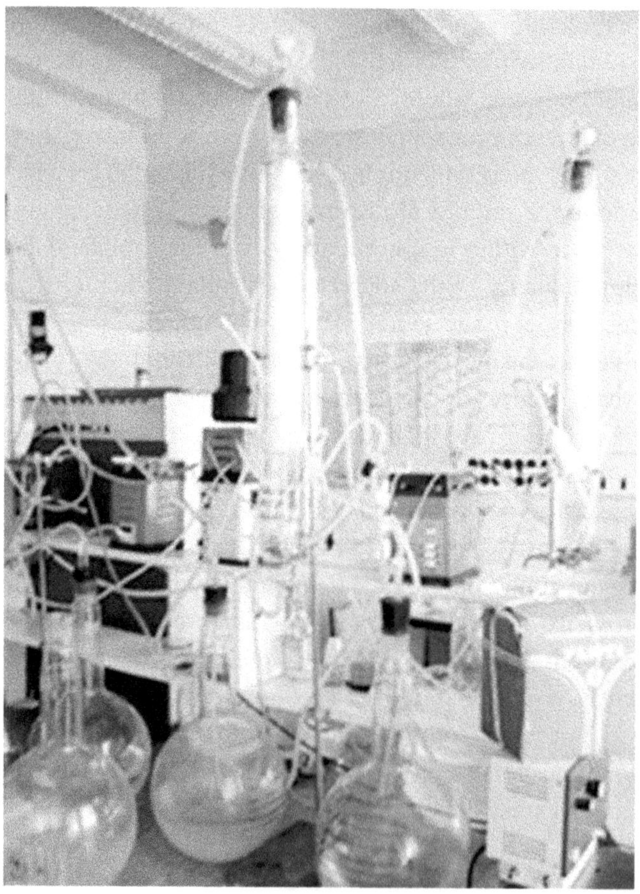

It turned out that with microbiological degradation of photo-ozonation products, the toxicity of anionic surfactant solutions decreases by 2–3 times according to both tests, while in the case of nonionic surfactant, it decreases up to complete detoxification. The influence of photolysis of hydrogen peroxide and nitrite ion on the detoxication efficiency was studied by us in more detail for p-NDA (*para* -nitroso-dimethylaniline) solutions. As can be seen from Fig. 2, the toxicity of our model system decreases symbiotically with the p-NDA concentration, being linearly dependent on the p-NDA concentration but independent of the nature of the photo-initiator (Zaitseva, 2009). The toxicity of the system is determined only by the p-NDA concentration and the sensitized photolysis produces no new intermediate and final highly toxic products.

Figure 2. Kinetic curves of p-NDA decomposition (1,2) and toxicity T decrease (3,4) during photolysis of its solution in the presence of photoinitiators: (1, 3) H_2O_2, (2, 4) nitrite ion. The test function is the change of the enzymic activity of luminescent bacteria.

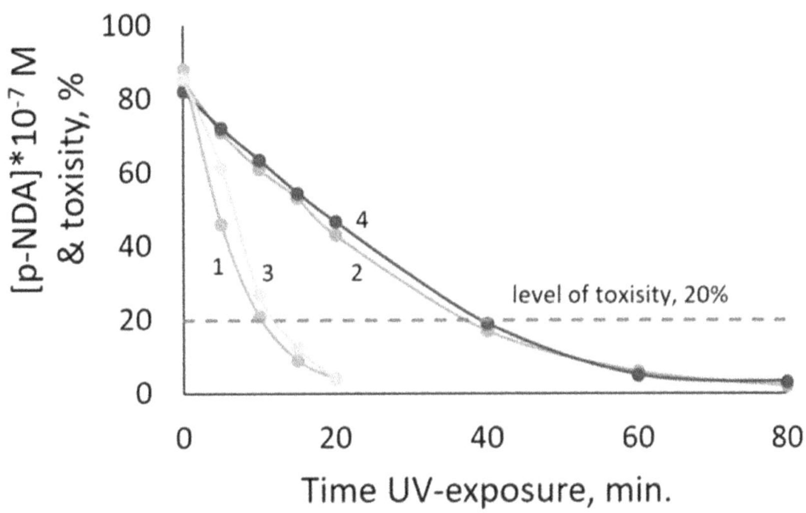

The methodology for the integral assessment of the health of ecosystems and individual species was developed by the International Committee "Biotest" in the early 90s (Ministry, 2002). As biotests, hydrobionts of various links in the trophic chain are used: bacteria, microalgae, ciliates, lower crustaceans (*daphnia* and *ceriodaphnia*), fish, and model systems (liposome lipid peroxidation).

BACTERIA

Bacteria, in particular, luminescent bacteria, are widely used in toxicology as the most suitable tests for express analysis of the toxicity of natural and waste waters (Gil, 1988; Zaitseva, 2009), as well as in quantitative (kinetic) toximetry, which made it possible to carry out, in parallel with the kinetic control of the consumption of starting substances, assessment of the toxicity of analytically undetectable transformation intermediates. This method was used to study the mechanism of FeS oxidation and photochemical transformations of nitrate and nitrite ions in the presence of model toxicants (Ulitzur, 2002).

The method for determining toxicity is based on recording changes in the intensity of bacterial bioluminescence under the total effect of the totality of pollutants present in the analyzed water sample. Under the influence of pollutants, the physiological and biochemical functions of the bacterium change, the activity of bacterial luciferase, the enzyme responsible for the intensity of bioluminescence, changes. Any changes in the metabolism of bacteria affect the luminescence intensity.

A decrease (or increase) in the intensity of bioluminescence indicates the presence of a toxic factor in the analyzed water sample. Quantitative evaluation of the results of biotesting with the use of luminous bacteria is expressed as a dimensionless value - the toxicity index $T(\%)=100(I_k - Ie)/ I_k$, where I_k and Ie are the luminescence intensity in the experimental and control samples, respectively. At $T<20$, the degree of toxicity is considered acceptable, $20<T<50$ - the sample is toxic, $T>50$ - the sample is highly toxic.

Bioassays on luminescent bacteria *Photobacterium* and *Benekea* are highly sensitive, rapid, and sufficiently reliable, which made it possible to test them in natural waters. An analysis of publications concerning biotests on bacteria shows that their advantages include the possibility of creating instrumental biosensors based on them, which is implemented in such facilities as Microtox (on luminescent bacteria), Toxi-Chromotest (on *Escherichia coli* - an enzymatic biotest) and a number of installations used both in Russia and in other countries. In Russia, in 2010, the "Ecolum" test system was included in the Federal Register of Measurement Methods (MMI) for environmental monitoring, based on the test reaction "Change in the intensity of bioluminescence of luminous bacteria based on genetically engineered *Escherichia coli* K12 TG1 with cloned in them lux operon of soil luminous bacteria *Photorhabdus luminescens* ZM1" (Vakhterova, 2020). This method is characterized by speed, accuracy, sensitivity, and simplicity; it allows one to simultaneously control a significant number of toxicants, and also allows one to form an integral idea of the general level of danger of exposure to a toxicant with the possibility of direct extrapolation of the obtained data to human health (Skurlatov, 1987).

MICROALGAE

Algae form the basis of food chains in all aquatic ecosystems. As a result of toxic effects, the primary production of aquatic ecosystems is reduced, therefore, monitoring the state of algocenoses is an obligatory component of the system of environmental monitoring of surface waters. At the same time, microalgae cultures serve as convenient test objects for assessing the extent of the effect of environmental chemical pollution on primary production (Sandmann, 1985; Skurlatov, 1991; Shtamm, 2003; International, 1987).

Representatives of various taxa of algae, various test reactions and toxicity criteria are used as test objects. The most commonly used green algae are *Chlorella vulgaris* and *Scenedesmus quadricauda (Chlorophycota)*. Controlled parameters are the intensity of photosynthesis, the increase in biomass, measured by the content of chlorophyll. Analysis in the acute experiment lasts 72 hours. In addition to direct calculation of the number of cells, an instrumental method for determining the biomass by fluorescence of algae has been developed. Both methods are included in the Federal Register of Russia.

LOWER CRUSTACEANS (*DAPHNIA* AND *CERIODAPHNIA*)

A widely used method is the method for determining acute (by survival) and chronic (by reproductive ability) toxicity to lower crustaceans (*daphnia* and *ceriodaphnia*) ((Shtamm, 2004). Analysis of acute toxicity for ceriodaphnia - 48 hours, and chronic - the number of swept juveniles in three generations - 7 - 10 days. This method is very sensitive to many chemical pollutants and due to its simplicity, it is the most widely used in analytical laboratories of various industries. In addition, ceriodaphnia in combination with ciliates *Tetrahymena pyriformis* turned out to be extremely sensitive and informative test objects in evaluating the effectiveness of new technologies for detoxifying difficult-to-decompose chemical contaminants.

CILIATES

A special place in the biotesting of natural waters is occupied by biotests on ciliates. Infusoria as a test object is suitable for biotesting a wide class of substances (heavy metals, organic compounds), wastewater from various industries, and polluted natural waters. In the literature, there are references to the use of several dozen species of ciliates as test objects, of which two species are most often used in Russia: *Paramecium caudatum* and *Tetrahymena pyriformis*. These bioassays are the most standardized and, in addition, are expressed (the growth of Tetrahymena pyriformis cells in 24 hours is 6-8 times, which makes it possible to trace the effect of the toxicant on several generations of ciliates). The ciliates Tetrahymena pyriformis are representatives of biocenoses of aquatic ecosystems and take an active part in the cycle of substances. They also have sufficient sensitivity to a wide range of toxic substances of a chemical nature. They are characterized by a number of features that bring them closer to higher animals: an acid-base type of digestion, the presence of similar enzymes, well-developed mitochondria that provide energy metabolism, an electron transport system, a universal code of nucleic acids similar to the code of higher animals.

In the analysis of acute toxicity using tetrachymenes as a test object, the coefficient of growth of ciliate cells in the test and control water is compared 24 hours after incubation. E.G. Cheremnykh developed the BioLaT device (Cheremnykh, 2008), which makes it possible to count living, motile cells of ciliate. The device includes two video cameras, devices for lighting and moving the plate with two rows of wells, and a microprocessor control unit. The instrument is controlled and counted by the AutoCiliata software. The calculation is based on the sequential fixation of two frames with further software image processing.

This method of counting living cells was also used in the development of a new affordable and fast method for determining the activity of the blood complement system, based on the cytolytic effect of this system for ciliates *Tetrahymena pyriformis* (Cheremnykh, 2015). Since the complement system is an important component of innate immunity that provides primary protection against pathogens penetrating the body, the use of the developed method for semi-automatic determination of the functional activity of serum complement in everyday clinical practice, including online, can hardly be overestimated.

This test object was also used as the most sensitive indicator of imbalance in intra-water redox processes involving naturally occurring hydrogen peroxide, accompanied by the death of aquatic organisms with intensive water exchange with the aquatic environment, in particular, fish in the early stages of their development (Skurlatov, 1984; Shtamm, 2011). Thanks to toxicological studies of water quality in the river, using the ciliate Tetrahymena pyriformis as a test object, it was possible to predict the onset of a toxic redox state of the Volga water in a day, leading to the mass death of stellate sturgeon larvae (Shtamm, 1989).

FISH

The value of biotesting on fish is determined by their role in aquatic ecosystems as the highest trophic link, which carries integral information about the state of pollution of the aquatic environment. Biotesting on fish is attractive because of the variety of test reactions that can be considered indicators of toxicity. Of particular interest are various biochemical, primarily enzymatic, indicators. The most distinct changes under the influence of toxicants are observed in the activity of cholinesterases, the inhibitors of which are organophosphorus compounds. In the USA and Canada, the state system for monitoring the toxicity of pulp and paper mill effluents located on the Great Lakes is based on determining changes in the monooxygenase activity of rainbow trout liver enzymes (MFO). MFO induction causes a change in the reproductive function of fish living in the wastewater discharge zone and is an early indicator of potential sublethal toxicity in the aquatic environment (Servos, 1992). In toxicological studies using rainbow trout as a test object, it was found that MFO induction is caused mainly by black liquor formed during pulping, it was also shown that treatment of black liquor with hydrogen peroxide leads to detoxification of wastewater (Shtamm, 1993).

LIPID PEROXIDATION OF LIPOSOMES (IN VITRO)

Biotesting methods can be based not only on the use of living organisms, but also on the use of various kinds of model test systems. The toxic effect of many chemical and physical factors is based on damage to cell membranes. Since membrane stability is largely determined by the processes of peroxidation of unsaturated fatty acid residues, components of biomembranes (phospholipids), one of the most informative test systems (in vitro) is based on modeling the processes of lipid peroxidation of liposomes (LPO) (Russian Federation State Committee, 1988; Shtamm, 2005).

The LPO rate depends on temperature, fatty acid composition, and other chemicals present in the incubation medium. Liposomes are the simplest model of cell membranes. Natural water may contain substances that accelerate or slow down lipid peroxidation. Depending on the activity and ratio of these substances, water has pro-oxidant or antioxidant properties. Thus, LPO is a sensitive test system for assessing the biological usefulness of water.

Lecithin (a chicken yolk phospholipid) is used as a model of the cell membrane. The product of the LPO reaction is malondialdehyde (MDA). The MDA analysis method is based on the formation of a colored trimethine complex containing one MDA molecule and two thiobarbituric acid molecules. The complex absorbs light with a wavelength of 532 nm (extinction coefficient $\varepsilon=1.56\cdot10^{5}$ $M^{-1}cm^{-1}$). The reaction is carried out at a temperature of 100°C, pH 6.3-6.4 (Tris-maleate buffer). A catalytic system consisting of ascorbic acid ($2\cdot10^{-4}$ M) and ferrous iron (10^{-4} M) is used as a pro-oxidant; distilled water is used as a control. Accordingly, the LPO rate in the aquatic environment is determined by the amount of formed MDA (C_{MDA}) in a sample of natural water. The criterion of toxicity is the ratio of the concentration of MDA in the sample to the concentration of MDA in the control, in percent. Water is considered non-toxic at C_{MDA} = 80-100%; toxic at $C_{MDA} < 60\%$; highly toxic at $C_{MDA} < 40\%$. The results obtained using the LPO model system correlate well with in vivo experiments on mice and with data obtained on river waters using Daphnia magna as a test object.

AMES TEST

The Ames test for mutagenic activity, also applicable for testing inherently multi-component natural or wastewaters, has become widespread. This test uses strains specially designed by Ames (1975) with TA97, TA98 and TA100 of Salmonella typhymurium bacteria. Mutant bacteria lack a rigid membrane shell, a "tail" for movement, and therefore are not capable of exhibiting pathogenic properties.

A significant drawback of these test systems turned out to be the almost complete absence of enzymes in microorganisms that carry out the metabolic activation of xenobiotics. On cultures of microorganisms, only the action of direct mutagens is recorded - compounds that directly interact with the genetic material of the cell. The majority of fat-soluble xenobiotics acquire mutagenic properties only after metabolic activation in the system of monooxygenases.

In the study of several hundred chemical compounds, it was shown that about 90% of substances that cause carcinogenesis in mammals are mutagenic in the Ames test. On the other hand, more than 90% of the compounds that do not cause mutations in this test are not carcinogenic. Thus, if a substance exhibits mutagenic properties in the Ames test, it has a 90% chance of being carcinogenic and dangerous to animals and humans. In Russia, the Ames test has been adopted at drinking water treatment plants to assess long-term negative effects on human health and subsequent generations.

CONCLUSION

In world practice, the existing systems for monitoring and controlling anthropogenic impact associated with pollution of the natural aquatic environment are usually limited to measuring the content of priority pollutants in environmental objects, compared with standard requirements. At the same time, an adequate assessment of the impact of pollutants on humans and the environment can only be obtained by assessing the impact of the total amount of controlled and uncontrolled toxic substances present in the environment. Among the integral methods for assessing the state of the environment, the most informative are the methods of bioindication, biomarking, and biotesting.

In this work, when choosing biotests, the authors focused on tests that characterize the effect of the aquatic environment on the photosynthetic activity of microalgae, on the reproductive function of aquatic organisms, on intra-water redox processes, and on disturbances in the cell membranes of aquatic organisms, in particular, at the level of lipid peroxidation. Our long-term research experience (Duka, 2002) at natural sites in Russia, Moldova, Ukraine, the USA, and Canada, especially in critical environmental situations, has shown that microalgae, *branchous crustaceans* (daphnia and ceriodaphnia, acute and chronic toxicity), *Tetrahymena pyriformis* infusoria and LPO model system are the most informative and widely available test organisms.

REFERENCES

Ames,, B.N., VcCann, J. & Jamasaki, E. (1975). Methods for detecting carmes mutagenesity tests. *Mutation Research*, *31*(3), 347–364. doi:10.1016/0165-1161(75)90046-1 PMID:768755

Anita, J., Tatjana, T., Damjana, D., Kristina, S., Polona, J., & Milenko, R. (2008). Biochemical biomarkers in chronically metal-stressed daphnids. *Comp Biochem Physiol C Toxicol Pharmacol. Jan., 147*(1), 61. doi:1. doi:0.1016/j.cbpc.2007.07.006

Caruso, G., de Pasquale, F., Mita, D. G., & Micale, V. (2016). Digestive enzymatic patterns as possible biomarkers of endocrine disruption in the red mullet (Mullus barbatus): A preliminary investigation. *Marine Pollution Bulletin*, *105*(1), 37–42. doi:10.1016/j.marpolbul.2016.03.005 PMID:26971230

Centre for Inland Waters. (1988). In B. J. Dutka, & the Dept of Environment Rivers Research Branch National Waters Research Institute Canada (Ed.), Method for microbiological and toxicological analysis of waters, wastewaters and sediments (p. 334). Dept of Environment Rivers Research Branch National Waters Research Institute Canada.

Cheremnykh, E.G., Ivanov P.A., Faktor M.I., Karpova N.S., Vasiljeva E.F., Gusev K.V. & Brusov O.S. (2015). A new method to assess functional activity of serum complement system. *Medical Immunology*, *17*(5), 479-488. . doi:10.15789/1563-0625-2015-5-479-488

Cheremnykh, E. G., Pokataev, A. S., & Gridunova, V. N. (2008). *Device for biological research.* Patent RU 2361913 Publ.27.04.2008 Bull. No. 12. doi:10.17816/phf46438/2713-153X-2020-3-2-64-73

Duca, G., & Vaseashta, A. (Eds.). (2020). *Handbook of Research on Emerging Developments and Environmental Impacts of Ecological Chemistry*. IGI Global. doi:10.4018/978-1-7998-1241-8

Duka, G. & Skurlatov Yu. (2002). *Ecological Chemistry*, 289. Publishing Center M.S.U. Chisinau.

Edori, O. S., Festus, C., & Edori, E. S. (2014). Comparative effects of petrol and diesel on enzyme activity in Tympanotonus fuscatus after sublethal exposure. *Pakistan Journal of Biological Sciences*, *17*(4), 545–549. doi:10.3923/pjbs.2014.545.549 PMID:25911844

Esimbekova, E. N., Torgashina, I. G., Kalyabina, V. P., & Kratasyuk, V. A. (2021). Enzymatic biotesting: Scientific basis and application. [in Russian]. *Siberian Journal of Ecology*, *3*(3), 364–382. doi:10.15372/SEJ20210308

Gil, T. A., Balayan A. E. & Stom D. I. (1988). *Method of biotesting for quenching the luminescence of luminous bacteria. Methods of biotesting of waters*. Chernogolovka: Publishing house of OIKhFI, 17-18.

Greer, J. B., Magnuson, J. T., Hester, K., Giroux, M., Pope, C., Anderson, T., Liu, J., Dang, V., Denslow, N. D., & Schlenk, D. (2019). Effects of chlorpyrifos on cholinesterase and serine li pase activities and li pid metabolism in brains of rainbow trout (Oncorhynchus mykiss). *Toxicological Sciences*, *172*(1), 146–154. doi:10.1093/toxsci/kfz167 PMID:31359069

Hani, Y. M. I., Turies, C., Palluel, O., Delahaut, L., Gaillet, V., Bado-nilles, A., Porcher, J.-M., Geffard, A., & Dedourgegeffard, O. (2018). Effects of chronic exposure to cadmium and temperature, alone or combined, on the threespine stickleback (Gasterosteus aculeatus): Interest of digestive enzymes as biomarkers. *Aquatic Toxicology (Amsterdam, Netherlands)*, *199*, 252–262. doi:10.1016/j.aquatox.2018.04.006 PMID:29677587

Holt, E. A., & Miller, S. W. (2010). Bioindicators: Using Organisms to Measure Environmental Impacts. *Nature Education Knowledge*, *3*(10), 8.

International Organization for Standards Group. (1987). ISO 6341-82. *Water quality. Determination of inhibition of Daphnia magna Straus (Cladocera, Crustacea) mobility, 58,* 13. - M.: Publishing house of standards.

Istomina, A., Chelomin, V., Kukla, S., Zvyagintsev, A., Karpenko, A., Slinko, E., Dovzhenko, N., Slobodskova, V., & Kolosova, L. (2019). Copper effect on the biomarker state of the Mizuhopecten yessoensis tissues in the prespawning period. *Environmental Toxicology and Pharmacology*, *70*, 103189. doi:10.1016/j.etap.2019.04.010 PMID:31103490

Khare, A., Chhawani, N., & Kumari, K. (2019). Glutathione reductase and catalase as potential biomarkers for synergistic intoxication of pesticides in fish. *Biomarkers*, *24*(7), 666–676. doi:10.1080/135475 0X.2019.1651902 PMID:31368361

Kholodkevich, S. V., Kuznetsova, T. V., Kirin, M. P., Smirnov, I. S., Rudakova, O. A., Lyubimtsev, V. A., Manvelova, A. B., Susloparova, O. N., Perelygin, V. V., & Sakharova, O. A. (2020). Bioindication of the ecological state (health) of coastal waters based on the use of automated bioelectronic systems. *Pharmacy Formulas*, *2*(3), 64–73. doi:10.17816/phf46438

Li, Z. H., Li, P., & Shi, Z.-C. (2014). Molecular responses in digestive tract of juvenile common carp after chronic exposure to sublethal tributyltin. *Ecotoxicology and Environmental Safety*, *109*, 10–14. doi:10.1016/j.ecoenv.2014.07.031 PMID:25129219

Li, Z. H., Zlabek, V., Grabic, R., Li, P., Machova, J., Velisek, J., & Randak, T. (2010). Effects of exposure to sublethal propicon azole on intestine-related biochemical responses in rainbow trout, Oncorhynchus mykiss. *Chemico-Biological Interactions*, *185*(3), 241–246. doi:10.1016/j.cbi.2010.02.040 PMID:20193671

Lima, L. B. D., de Morais, P. B., de Andrade, R. L. T., Mattos, L. V., & Moron, S. E. (2018). Use of biomarkers to evaluate the ecological risk of xenobiotics associated with agriculture. *Environmental Pollution*, *237*, 611–624. doi:10.1016/j.envpol.2018.02.011 PMID:29525628

Lopes, R. M., Filho, M. V. S., de Salles, J. B., Bastos, V. L. F. C., & Bastos, J. C. (2014). Cholinesterase activity of muscle tissue from freshwater fishes: Characterization and sensitivity analysis to the organophosphate methyl-paraoxon. *Environmental Toxicology and Chemistry*, *33*(6), 1331–1336. doi:10.1002/ etc.2556 PMID:24648156

Ministry of Natural Resources. (2002). *Guidelines for determining the toxicity of water, bottom sediments, pollutants and drilling fluids by biotesting method.* M.: REFIA, NIA Priroda.

Narra, M. R., Begum, G., Rajender, K., & Rao, J. V. (2011). *In vivo* impact of monocrotophos on biochemical parameters of a freshwater fish during subacute toxicity and following cessation of exposure to the insecticide. *Zeitschrift für Naturforschung. C, A Journal of Biosciences, 66*(9-10), 507–514. doi:10.1515/znc-2011-9-1011 PMID:22191217

Parmar, T. K., Rawtani, D., & Agrawal, Y. K. (2016). Bioindicators: The natural indicator of environmental pollution. *Frontiers in Life Science, 9*(2), 110–118. doi:10.1080/21553769.2016.1162753

Russian Federation State Committee on hydrometeorology (1991) *Method of determination of toxicity of natural and waste waters on the reaction of lipid peroxidation of liposomes*. RD 52. 18. Moscow, 14.

Sandmann, G. (1985). Photosynthetic and respiratory electron transport in Cu^{2+}-deficient Dunaliella. *Physiologia Plantarum, 65*(4), 481–486. doi:10.1111/j.1399-3054.1985.tb08677.x

Seitkalieva, A. V., Menzorova, N. I., & Rasskazov, V. A. (2016). Application of different enzyme assays and biomarkers for pollution monitoring of the marine environment. *Environmental Monitoring and Assessment, 188*(1), 70. doi:10.100710661-015-5066-5 PMID:26721566

Servos, M., Carey, J., & Fergusson, M. (1992). Impact of a modern bleached kraft mill with secondary treatment on white suckers. *Water Pollution Research Journal of Canada, 273*, 423–437.

Shtamm, E. V. (1988). Redox state of water fish reproduction Proceedings of the first all-union workshop on ecological chemistry of natural waters, 279-295. Centre for International Projects, GKNT, Moscow.

Shtamm, E. V., Alexandrova, E. V., Kozlova, N. B., Shishkina, L. N., Semenova, I. V., Skurlatov, Yu. I., & Dedenchuk, I. V. (2005). *Lipid peroxidation of liposomes – sensitive and express method of biotesting of the water quality and biologically active substances*. Third Moscow Intern. Congress Biotechnology: State of the Art and Prospects of Development. Congress Proceedings, Part 2, (pp. 68 – 69). Moscow, Russia.

Shtamm, E. V., Frog, B. N., Skurlatov, Yu. I., Kozlova, N. B., Alexandrova, E. V., & Shishkina, L. N. (2002). Possible role of redused sulfur compounds in toxic properties of pulp and paper mill effluents. *Acta Hydrochimica et Hydrobiologica, 30*(5-6), 256–265. doi:10.1002/aheh.200390007

Shtamm, E. V., Kozlova, N. B., Alexandrova, E. V., Razumovsky, S. D., Konstantinova, M. L., Emnéus, J., & Winther-Nielsen, M. (2004). In G. E. Zaikov (Ed.), *Dinamics of surfactant destruction in complex photoozonation + microbial degradation. Biotechnology and the environment including biogeotechnology* (pp. 35–42). Nova Science Publishers Inc.

Shtamm, E. V., Kozlova, N. B., Alexandrova, E. V., Skurlatov, Yu. I., Frog, B. N., & Frog, D. B. (2003). Microalgae for control of water quality of fishery and drinking reservoirs undergoing the influence of industrial enterprises sewage. *Abstr. Biotechnology of microalgae, 5th European Workshop*, (p. 63).

Shtamm, E. V., & Skurlatov, Yu. I. (1993). *Toxicity caused by disbalance of redox processes in natural waters*. 20-th Annual aquatic toxicity workshop. Quebec, Canada.

Shtamm, E. V., Skurlatov, Y. I., Ernestova, L. S., Baikova, I. S., & Dyubanov, M. V. (2017). Chemical safety of aquatic ecosystems: Biological methods of control. *Russian Journal of Physical Chemistry B. Focus on Physics, 11*(4), 635–642. doi:10.1134/S1990793117040236

Shtamm, E. V., Skurlatov, Yu. I., Kozlova, N. B., Zaitseva, N. I., & Alexandrova, E. V. (2011). Bioassaying in the Assessment of Wastewater Treatment Process Efficiency. *Water Resources*, *38*(2), 237–243. Pleiades Publishing, Ltd.

Skurlatov, Yu. I., Ernestova, L. S., Shtamm, E. V., Shpotova, T. V., & Kalinin, V. B. (1984). Redox state and seasonal toxicity of natural waters. [in Russian]. *Doklady of the USSR Academy of Sciences*, *276*(4), 1014–1016.

Skurlatov, Yu. I., Shtamm, E. V., & Semenyak, L. V. (1991). Role of algae in aquatic environment redox-state formation. Interfacial phenomena in the Environment. *Internat. Conf. Switzerland*, (October), 103–104.

Skurlatov, Y. I., Shtamm, E. V., Travin, S. O., Shvydkiy, V. O., & Semenyak, L. V. (2020). Role of Oxidation-Reduction Processes in Formation of Toxic Properties of Natural Aqueous Environment. *Handbook of Research on Emerging Developments and Environmental Impacts of Ecological Chemistry*, 149-163. IGI Global. https://www.igi-global.com/book/emerging-developments-environmental-impacts-ecological/231905

Skurlatov, Yu. I., Zepp, R. G., & Pierce, J. T. (1987). Algal-induced decay and formation of hydrogen peroxide in water. Its possible role in oxidation of anilenes by algae. *Ann. Chem. Soc. Symp. Ser.*, *327*, 215–224.

Ulitzur, S., Lahav, T., & Ulitzur, N. (2002). A novel and sensitive test for rapid determination of water toxicity. *Environmental Toxicology*, *17*(3), 291–296. doi:10.1002/tox.10060 PMID:12112638

Vakhterova, Y. V., Shtamm, E. V., Shvydkiy, V. O., & Saratovskikh, E. A. (2020). Investigation of the toxicity of the products of photochemical destruction of glyphosate by changing the intensity of bioluminescence of the bacterial test system "ecolum". *Vth International conference actual scientific & technical issues of chemical safety (ASTICS-2020)*. KAZAN.

Vaseashta, A., Duca, G., & Travin, S. (Eds.). (2022). *Handbook of Research on Water Sciences and Society* (Vols. 1–2). IGI Global. doi:10.4018/978-1-7998-7356-3

Vaseashta, A., Gevorgyan, G., Kavaz, D., Ivanov, O., Jawaid, M., & Vasović, D. (2021). Exposome, Biomonitoring, Assessment and Data Analytics to Quantify Universal Water Quality. In A. Vaseashta & C. Maftei (Eds.), *Water Safety, Security and Sustainability. Advanced Sciences and Technologies for Security Applications*. Springer., doi:10.1007/978-3-030-76008-3_4

Zaitseva, N. I., Baikova, I. S., Shvydkii, V. O., Borodulin, R. R., Shtamm, E. V., & Skurlatov, Yu. I. (2009). Kinetics of the Oxidation of Model Toxicants during Photolysis of Nitrites and Nitrates in an Aqueous Medium. ISSN 1990-7931, *Russian Journal of Physical Chemistry B, 3*(2), 185–190. doi:10.1134/S1990793109020031

Zakharov V.M. and Clark D.M. (1993). *Biotest: integral assessment of ecosystems and individual species health.* Moscow department of the International Foundation "Biotest."

Chapter 6
Tocols as Antioxidant Molecules in the Agro-Food Field

Giuseppe Ianiri
https://orcid.org/0000-0002-7949-8432
University of Molise, Italy

Alessandra Fratianni
University of Molise, Italy

Pasquale Avino
University of Molise, Italy

Caroline Vitone
University of Molise, Italy

Ashok Vaseashta
International Clean Water Institute, USA

Gianfranco Panfili
University of Molise, Italy

ABSTRACT

The presence of many different antioxidant molecules in foodstuffs and their by-products has attracted the scientific community's attention to their applications in finished foods due to their high bioactive power. Classes of molecules, such as phenolic compounds (phenolic acids, flavonoids, stilbenoids, and lignans), carotenoids, and tocols, provide numerous benefits to human health, reducing the incidence of cardiovascular disease, diabetes, and cancer. These molecules are innumerable and have similar but different molecular structures. Each can have a greater or lesser antioxidant power depending on its structure and its amounts in the finished food. Among these, tocols are a class of different homologues having antioxidant properties and are well represented in food.

DOI: 10.4018/979-8-3693-0512-6.ch006

INTRODUCTION

The intake of antioxidant compounds through the consumption of foods and functional foods plays a central role in promoting good health and the proper physiological processes in the human body. The predominant role of these molecules is to prevent the oxidation of natural organic substrates (phospholipids, proteins, DNA) in the body from being attacked by reactive oxygen species (Gęgotek & Skrzydlewska, 2022). All living organisms generate reactive oxygen species from their metabolism, which are involved in normal physiological activities. However, when ROS are produced in large quantities, they can cause significant damage to the organism. Such compounds or reaction intermediates are highly reactive and generate chemical instability. Specifically, the instability of free radicals is due to the presence of a mismatched electron on the outermost orbital with a tendency to interact with electrons from other molecules. Oxidative stress has been associated with several diseases such as diabetes, atherosclerosis, hypertension, respiratory diseases, arthritis, cataract, cancer, cardiovascular diseases etc., as free radicals are implicated in the pathogenesis of these diseases (Engwa et al.,2022; Vaseashta et al, 2022). For this reason, the intake of antioxidant molecules through diet helps to combat oxidative stress (Eggersdorfer & Wyss, 2018, Vaseashta et al, 2021). Studies have shown that dietary intake of tocols (Vitamin E) helps prevent the occurrence of neoplasms and heart disease (Shahidi & De Camargo, 2016). The main antioxidants that are generally present in food and consumed by consumers are vitamin A, vitamin C, vitamin E, polyphenols, and carotenoids. Vitamin A can be present in foods in different molecular forms among which the main ones are retinol, retinoic acid, and retinaldehyde. Foods that contain large amounts of vitamin A are liver, milk, and derivates. Vitamin C, or ascorbic acid is found essentially in large amounts in fruits and vegetables. Regarding vitamin E, the latter comprises two large different molecular classes, specifically tocopherols and tocotrienols. Essentially, vegetable oils, fruits, vegetables, and cereals are the foods that have within them the highest amounts of vitamin E in their different forms (Lee et al., 2018). Tocols are also heat-sensitive molecules and tend during storage, preparation, and cooking to be degraded by oxidative enzymes. Moreover, the amounts of tocols that are degraded during the preparation of food strongly depend on the type of food, the cooking method, and its duration (Diamante et al., 2021). Other studies, on the other hand, point to the possibility that in certain food matrices subjected to cooking processes, degradation of bioactive compounds does not occur, but rather that they are released from the matrix to which they were bound, thus making them more bioavailable (Ruiz-Rodriguez et al., 2008). The term polyphenols denote molecules that belong to the following four major molecular classes: phenolic acids, flavonoids, stilbenoids, and lignans. Foods such as olive oil, chocolate, coffee, tea, fruit and vegetables have been seen to contain significant amounts of polyphenols (Da Silva et al., 2012). Carotenoids, on the other hand, are divided into carotenes and xanthophylls. Carotenes consist of a long chain of carbon atoms, often terminating in a ring and with the variable presence of C-C double bonds (lycopene, β-carotene, etc.). Xanthophylls, on the other hand, consist of chains containing oxygen atoms, which gives them a particular sensitivity to being oxidized (lutein, zeaxanthin, astaxanthin, etc.). Foods that contain large amounts of carotenoids are fruits and vegetables. Focusing on polyphenols, despite having an identical basic structure, are characterized by extremely wide molecular variability. Each of these molecules possesses particular antioxidant activity (Saverini et al., 2016). Generally, most of the antioxidant molecules in food have an apolar molecular structure that allows them to be present in fatty tissues where most of the damage due to oxidative stress occurs. Only a few are water-soluble, namely ascorbic acid, retinol, and some phenols. Antioxidant molecules are by definition, molecules that tend, under certain physico-chemical conditions, to be oxidized very rapidly. High temperatures, the presence of specific enzymes, such as polyphenol oxidase, large quantities of oxygen, and limited quantities of antioxidants under oxidative stress conditions promote the almost instantaneous oxidation of these compounds (Penicaud et al., 2011). On this basis, food is sometimes stored in an unsuitable manner, leading to the degradation of antioxidants and, consequently, to a lower intake of food. It is, therefore, extremely important to correctly conduct all storage operations in order to limit the loss of beneficial molecules as much as possible. Proper respect for the cold chain during transport, sale, and home storage, protection from solar radiation and high temperatures, and removal of oxygen are some of the practices to be used to limit premature oxidation. However, even during food processing, many of these molecules can be degraded due to multiple factors. One of the solutions could be having a process line working at low temperatures and limiting oxygen concentrations. In recent years, many industries have been developing the technique of encapsulation of antioxidant molecules. This technique consists of creating an envelope that protects and optimizes the functional properties of bioactive molecules from technological processes (Beecher, 2003). Encapsulation can also assist with the creation of desirable sensory attributes (e.g., aroma, texture, color, and taste), thus playing a major role in the design and development of novel foods and beverages. Encapsulation is one of the few technologies experiencing continued growth due to its unique potentialities, high versatility, and extensive range of applications. In conclusion, the addition of these molecules or ingredients containing bioactive compounds, together with the adoption of techniques for their preservation molecules during technological processes can contribute to obtaining new foods rich in bioactive compounds that positively influence the health status of consumers.

MECHANISMS OF ACTION OF FOOD ANTIOXIDANTS

Antioxidants are compounds capable of retarding or inhibiting oxidation processes that occur under the influence of atmospheric oxygen (O_2) or reactive oxygen species (ROS), including peroxides, superoxides, and radicals. Foods possess endogenous antioxidants that are mainly enzymes, such as superoxide dismutase, catalase, glutathione peroxidase, or non-enzymatic compounds, such as uric acid, bilirubin, albumin, and metallothioneins. When endogenous molecules are unable to ensure the body's complete protection against ROS, exogenous antioxidants, which are contained in food, become necessary. A generic free radical-induced oxidation reaction of an organic substrate takes place in three main steps: initiation, propagation, and termination. The main reactions occurring during the three main phases are briefly outlined.

1) *Initiation reaction*; a generic XH substrate, such as a lipid, is attacked by an R• radical with the formation of a highly reactive X• radical:

$$XH + R\cdot \rightarrow X\cdot + RH \quad (1)$$

2) *Propagation reaction*; the X• radical reacts rapidly with oxygen (O_2) to give a peroxyl radical:

$$X\cdot + O_2 \rightarrow XOO\cdot \quad (2)$$

As a result, the XOO• radical can further oxidise the substrate with the formation of hydroperoxides:

$$XOO\cdot + HX \rightarrow X\cdot + XOOH \quad (3)$$

Hydroperoxides in turn give rise to the formation of a wide range of compounds such as alcohols, aldehydes, ketones and radicals including the alkoxide radical XO•.

3) *Termination reaction*; during the termination phase, two radicals bind together to give a stable compound and this can occur in the following three ways:

a) $XOO\cdot + XOO\cdot \rightarrow XOOX + O_2$ (4)

b) $X\cdot + X\cdot \rightarrow X_2$ (5)

c) $XOO\cdot + X\cdot \rightarrow XOOX$ (6)

Depending on their mechanism of action, food antioxidants can be classified into three main groups: preventive, scavenger and chain breaker. The specific functioning of each category will be explained below.

Preventive antioxidants are those that prevent the formation of initiating radicals (X·):

$$XH + R \rightleftharpoons X + RH \quad (7)$$

Some examples of preventive antioxidants are the chelators of transition metals, such as ethylenediaminetetracetic acid (EDTA), and polyphenols, sulphides, and enzymes such as catalase. The main action of chelators is to 'immobilize' metal cations by making them no longer available in facilitating the initiation reaction (Fratianni et al., 2021). The product of the chelation reaction is called complex. Its formation can be represented by the following equation where M is the metal cation (e.g. Fe^{2+}, Fe^{3+}, Cu^{2+}) and Y is the chelating agent (EDTA):

$$M + Y = MY \quad (8)$$

Sulphides, i.e. salts of hydrogen sulphide, act by eliminating hydroperoxides and avoiding the release of free radicals. The general formula for sulphides is M_xS_y where x corresponds to the oxidation number of S (i.e., 2) and y to the oxidation number of M. Their action could be represented through the following reaction:

$$XOOH + M_2S \rightarrow XOH + M_2SO \quad (9)$$

As far as enzymes are concerned, the main is catalase, which breaks down hydrogen peroxide into water and oxygen, according to the following chemical-enzymatic reaction:

$$2\,H_2O_2 \rightarrow 2\,H_2O + O_2 \text{ (10) CAT}$$

Scavenger antioxidants are substances capable of transforming ROS into non-radical compounds, free of reactivity and therefore toxicity. Scavengers can be specific only for certain types of oxygen radicals or, conversely, have a broad spectrum of action, e.g. superoxide dismutase (SOD) is a specific enzyme for superoxide anions, whereas vitamin E has a generic reactivity towards all free radicals. The main naturally occurring scavengers in plant organisms are flavonoids. The latter is endowed with high antioxidant power, making plants a possible valuable ingredient for enriching food with health-promoting molecules. Chain breakers, on the other hand, can block the propagation of radical chain reactions. In fact, these oxidants act by blocking the peroxyl forms (XOO·) that are created exclusively during the propagation reaction. Chain breakers can act by hydrogen donation, i.e., they interrupt the chain of reaction by donating a hydrogen atom to the peroxyl radicals:

$$XOO\cdot + AH \rightarrow XOOH + A\cdot \quad (11)$$

It is essential that the formed radical A· is unable to react with HX and O_2 as shown below:

$$A + XH \not\rightleftharpoons AH + R \quad (12)$$

$$A + O_2 \not\rightleftharpoons AOOR \quad (13)$$

The most common hydrogen donor antioxidants are those with a phenolic structure. The basic structure of a phenolic compound is as follows:

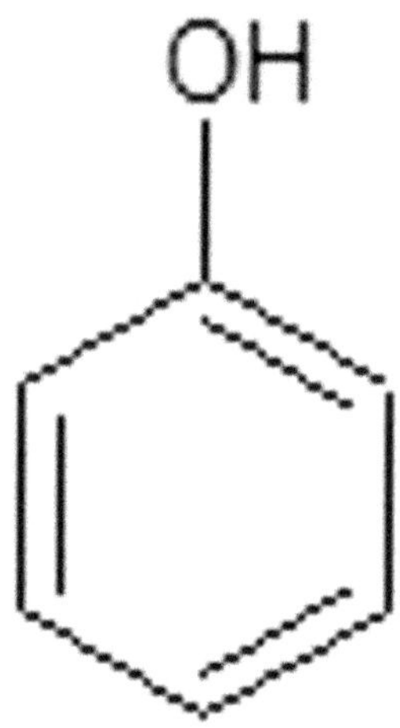

Phenols can simply be represented as AH. In order for the reaction between the hydrogen donor (AH) and the peroxyl radical (XOO·) takes place, effective collisions between the two molecules must occur. Obviously, the effective collisions are those occurring in the proximity of the -OH group of the phenol. For example, if there are bulky substituents in the *ortho* positions of the benzene ring, the percentage of effective collisions decreases, since, as the steric size increases, reactivity as an antioxidant decrease:

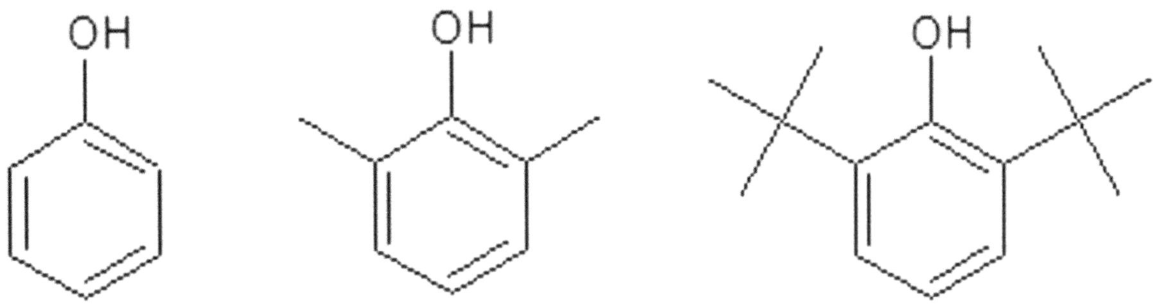

Vitamin E, and in particular α-tocopherol, summarises all the factors that characterize a good antioxidant: a) a reduced steric bulk in the *ortho* position, b) the presence of electron-donor substituents in the *ortho* and *para* position, c) a high lipophilicity. Even polyunsaturated molecules, such as carotenoids, can deactivate peroxyl radicals (XOO·), by donating an allyl hydrogen to form a neutral carotenoid radical, which is strongly stabilized by resonance. In conclusion, peroxidative reactions can be inhibited by two main mechanisms: hydrogen donation and electron transfer. The antioxidant activity of all the described compounds, irrespective of the mechanism by which they work, is explained by the deactivation of ROS or their trapping.

ROLE OF ANTIOXIDANT MOLECULES IN THE DIET

Epidemiological studies carried out on humans indicate that a high energy intake is associated with the risk of developing several chronic diseases, while a diet rich in fruit and vegetables and low in fat exerts a protective role. Cereals, as well as vegetables and pulses, play a major role in regulating various physiological functions of the body. There is growing evidence that antioxidants (mainly vitamins A, C, E, carotenoids, such as lutein and zeaxanthin, and phenols) can help to prevent the onset of numerous diseases (Takur et al., 2022). The causes that can generate free radicals in the body are many: the digestion of food, the metabolism of drugs by which the liver can generate radicals, excessive exposure to ultraviolet rays (sun or tanning lamps), cigarette smoking, or exposure to various kinds of pollutants. The body naturally produces a number of antioxidants referred to as endogenous. These include glutathione, coenzyme Q, and the enzymes superoxide dismutase and catalase. The exogenous antioxidants are introduced daily into the body through food. Hence, the importance of a healthy, balanced diet encourages the introduction of all the necessary antioxidants, many of which are found in abundance in fruit and vegetables. Several pieces of evidence show that cooking and long storage of food can deactivate certain antioxidants. An example is vitamin C, which is very volatile and can be easily degraded during cooking (Burgos et al., 2013; Fratianni et al., 2021). Other antioxidants, such as vitamin E, on the other hand, are more resistant to cooking temperatures. Still, others, such as carotenoids, are degraded by cooking (Fratianni et al., 2021). When the production of free radicals is excessive, or the antioxidant reserve is insufficient, a condition of oxidative stress can develop. This is the case, for example, of a bacterial or viral disease, an unhealthy lifestyle (e.g., cigarette smoking, alcohol or drug abuse), diets rich in sugar and saturated fats and too low in micronutrients, in case of high psychological stresses. An increased dietary intake of antioxidants could be particularly useful in pregnant women, infants, children, the elderly, and athletes. In conclusion, available data show that an increased intake of antioxidants from natural sources, particularly fruit, and vegetables, can be helpful in the prevention of various diseases. On the other hand, data on antioxidant supplementation are still insufficient to recommend the generalized use of antioxidants in quantities exceeding those naturally present in a 'healthy' diet (Zehiroglu C., & Sarikaya S. B, 2019).

ANTIOXIDANTS USED AS ADDITIVE

Antioxidants are used as food additives to increase product 'shelf-life' and prevent spoilage. Antioxidants are an important means of preservation as the oxidation process can occur very quickly even in frozen, refrigerated, and dried foods. Foods par excellence that exhibits a high degree of oxidation are those composed of unsaturated fats, which have an unpleasant taste when attacked. Oxidation is often catalyzed by metals, which is why fatty foods such as butter should never be wrapped in aluminum foil or stored in metal containers. Some foods, such as olive oil are partially protected from oxidation due to their high antioxidant content, although they remain sensitive to photo-oxidation. The table below lists the antioxidants used by the food industry:

Table 1. List of antioxidant molecules added to food

Hydrocaffeic acid	Isoflavones	Sulphurdioxide
Ascorbic acid	Egg and soylecithin	Sodiumsulphite
Nordihydroguaiaretic acid	Nor-conhydrin	Sodiumbisulphite
Anthocyanins	Quercetin	Potassiumsulphite
Catechins	Resveratrol	Potassiumbisulphite
Conidendrine	Sesaminol	Calciumsulphite
Flavones	Sesamol	Calciumbisulphite
Alkylgallates	Tocopherols (vitamin E)	Ascorbic acid salts
Gossypol	Vanillin	Ascorbylstearate
Butylhydroxyanisole (BHA)	Erythorbic acid	
Butylhydroxytoluene (BHT)	Nicotinamide	

TOCOLS: STRUCTURE, ROLE, AND PRESENCE IN FOODS

Vitamin E comprises two different classes of molecules: tocopherols and tocotrienols. The molecular structure of tocopherols consists of a phenolic ring and a saturated isoprene side chain. They exist in nature in multiple stereoisomer forms and in particular as α-tocopherol (α-T), β-tocopherol (β-T), γ-tocopherol (γ-T), and δ-tocopherol (δ-T). Tocotrienols differ in that they have an unsaturated isoprene chain containing three carbon-carbon double bonds. Tocotrienols also exist in different stereoisomer forms, the most common being: α-tocotrienol (α-T3), β-tocotrienol (β-T3), γ-tocotrienol (γ-T3) and δ-tocotrienol (δ-T3). In plants, tocopherols are involved in growth, signal transduction, phytohormonal balance, abscission, and senescence, as well as in many other metabolic processes. They are also known to physically quench and chemically react with O_2 in chloroplasts, thus protecting lipids, other membrane components, and the structure and functions of PSI (Munné-Bosch & Alegre, 2002). In humans, apart from the vitamin E activity of some tocols (with α-T having the highest one), as the major lipid-soluble chain-breaking antioxidants, they have been demonstrated to prevent different chronic diseases (Shahidi & De Camargo, 2016).

Their main sources are vegetable oils, but they are also found in large amounts in different vegetable products with significant nutritional content (Shahidi & De Camargo, 2016; Panfili et al., 2020; Mignogna et al., 2015). A food product can be considered a source of vitamin E as long as its consumption offers a supplement of at least 15% of the vitamin E RDI, as expressly stated in Regulation (EU) No 1169/2011 (European Parliament and Council, 2011). Animal fats also contain tocopherols, however, in a much lower amount than those of vegetable oils (Shahidi & Shukla, 1996). In particular, fats and oils exhibit differences due to natural variation, different refining processes, and their complex profile in vegetable oil blends. The main forms in the diet are α- and γ-tocopherol, due to the highest content in food products. Nevertheless, α-tocopherol is the main form of vitamin E with the highest tissue concentration. α-Tocopherol, which is the most common homologue, is the major tocopherol in many edible oils (peanut, olive, and sunflower oils). The content of γ-tocopherol in some edible oils such as corn oil is similar to or higher than that of α-tocopherol. Palm oil contains three tocotrienol homologues (alpha, gamma, and delta), moreover, also another compound, namely α-tocomonoenol (α-T1) is present. Although α-tocopherol is present in palm oil, it is worth mentioning that palm oil has been reported to be the only edible oil with γ-tocotrienol as the major lipophilic antioxidant among common ones such as coconut, corn, olive, peanut, and sunflower oils (Shahidi & De Camargo, 2016).

Figure 1. Molecular structures of tocols

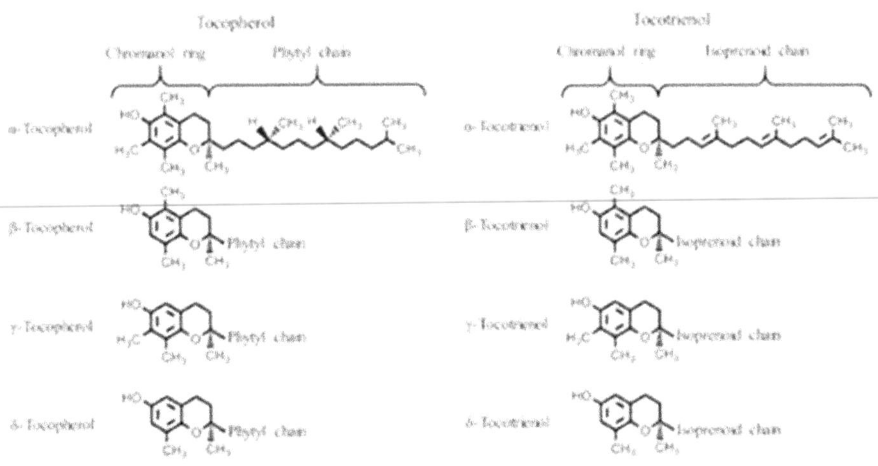

The stability of tocopherols and tocotrienols may be influenced by the system in which they are present and, in foods, by the form in which they are present (free, esterified, or linked to the food matrix), the structure of the food matrix and the chemical nature of the specific compound. Moreover, it could also depend on the fat amount which differs even among different cultivars of the same crop, the species, the cultivar, the agronomic practices, storage and processing conditions, and the cooking procedures employed. The extent of degradation that thermal treatments can cause depends on temperature, light, presence of oxygen, pH, water activity, and interactions with other antioxidants (Fratianni et al, 2020). One example is extra virgin olive oil (EVOO), in which the process of pressing the olive fruit in a solvent-free process, with no heat applied, is beneficial for the retention of its bioactive compounds, including tocols. The number of tocopherol homologues is very low when compared with other antioxidants, such as carotenoids, phenolic acids, and flavonoids, which makes their identification simple. However, because of co-elution, there are several studies in which a complete peak separation is not often achieved. In the last few years, the number of studies employing hyphenated techniques, such as HPLC-MS, has been growing, however, data on these techniques applied to tocols are scarce in the literature.

ANALYTICAL DETERMINATION OF TOCOLS IN FOOD

A suitable analysis for the determination of nonpolar molecular classes with antioxidant activity in food is based on the following steps: sampling, sample pre-treatment, instrumental analysis, and qualitative and quantitative analysis. Sampling is a crucial step in the correct analytical determination of any molecule. With regard to foodstuffs, it is advisable to carry out diffuse sampling, i.e., to take several parts of the foodstuff homogeneously, so the sample is representative. The main purpose of the sample pre-treatment phase is to bring the sample into a suitable form for subsequent instrumental analysis and to prevent compound degradation, by using antioxidants, nitrogen flushing, moderate extraction temperatures, and light protection. Generally, this phase is performed by extracting the molecules of interest through the use of an organic solvent capable of solubilizing the analytes. Obviously, the extraction of the analytes with an organic solvent can only be done for molecules that are apolar and which therefore tend to pass from the aqueous phase to the apolar one. The most used solvent for the extraction of tocols in foods is hexane. Hexane has high extracting power and does not have high purchase costs. A one-step extraction with hexane or other polar solvents or their mixture can be due to extracting free compounds. A preliminary saponification procedure, through an alkaline solution, can be applied to hydrolyze esterified compounds. Some authors (Panfili et al., 2003), use an extraction procedure of tocols from food, consisting of saponification of the sample matrix followed by solvent extraction, in order to release compounds eventually bound to the food matrix, to hydrolyze esterified compounds and to remove interfering molecules for the subsequent instrumental analysis. The sample extraction by *Panfili* et al., (2003) is briefly summarized. At first, samples are placed in a test tube with a hermetic screw cap. Into the test, tubes are added in order, with the amount depending on the sample under treatment: ethanolic pyrogallol (60 g L^{-1}), ethanol (95%), sodium chloride (10 g L^{-1}), potassium hydroxide (600 g L^{-1}) solution and glass beads to prevent tumult boiling. After this step, the test tubes are vortexed and the oxygen inside them is removed by insufflating nitrogen (N_2) for one minute. The saponification reaction takes place by immersing the samples in a bath at 70°C for 45 minutes. At the end of the reaction, the test tubes are cooled rapidly using ice and then 15 mL of a 15% sodium chloride solution and 15 mL of a 9:1 v/v hexane: ethyl acetate solution are added twice. Tocols are extracted by shaking the test tube and recovering the upper solution. The extract is first brought to dryness with a rotary evaporator and then recovered with a known volume of a solution of isopropilialcohol: hexane (1%, v/v). The extraction procedure is summarized in figure 2.

Figure 2. Analytical procedure for the analysis of tocols

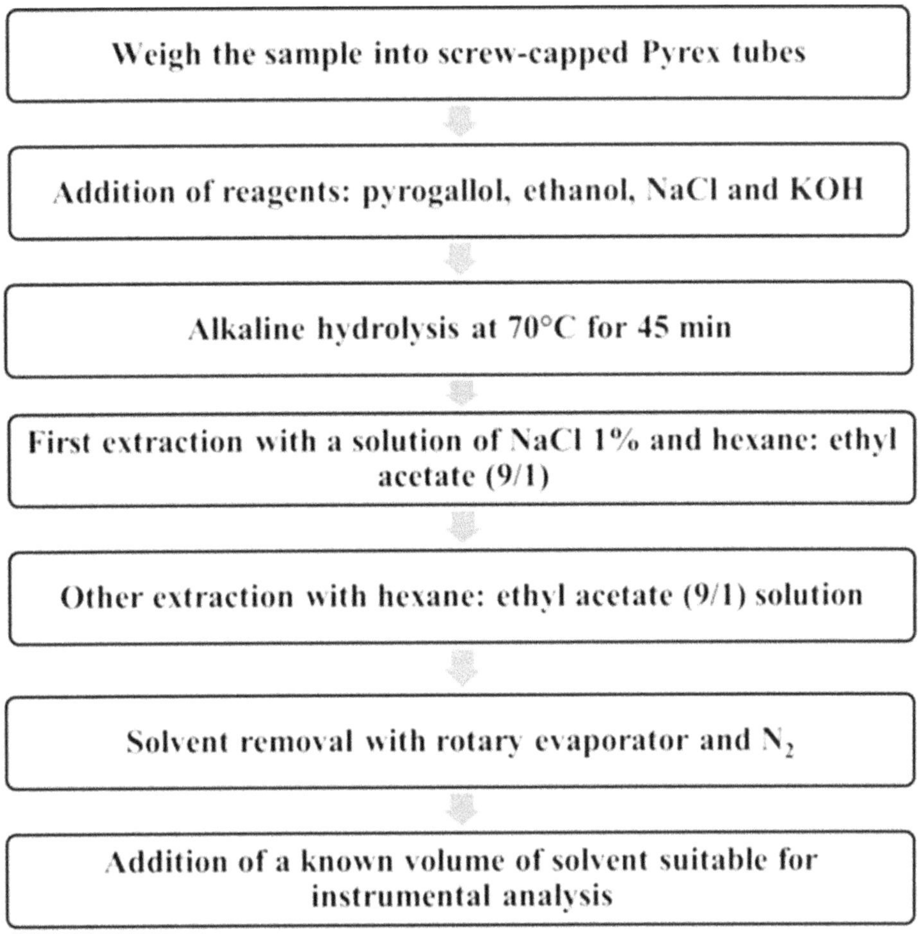

The instrumental analysis involves the separation of tocols through high-performance liquid chromatography (HPLC). The analysis of tocols can be performed using two different chromatography columns with two different fixed phases inside normal and reversed-phase columns. Of course, changing the fixed phase also changes the mobile phases used for eluting the molecules. Although reversed-phase columns are generally known to have the advantages of better stability and longer durability than normal-phase columns, the latter is more efficient in separating β and γ isomers of tocopherols and tocotrienols. A normal phase method for the determination of tocols is that described for cereals and other food matrices by *Panfili* et al., (2003), consisting of a silica column and an isocratic elution with n-hexane/ethyl acetate/acetic acid solution and their detection through a spectrofluorimeter. Tocols can be detected both by spectrophotometric and spectrofluorimetric detection. In fact, tocols have a molecular structure whereby, when hit by radiation, they re-emit other radiation, the wavelength of which depends on that of the exciting radiation. Spectro-fluorimetric detection is more suitable, preventing interfering molecules.

CONCLUSION

In recent years, the work carried out by the authors of this chapter has allowed firstly the development of a robust analytical method suitable for the determination of tocols in foodstuffs, mainly cereals, and vegetables. Secondly, given the high antioxidant power that tocols have, new food formulations have been developed with the addition of ingredients rich in vitamin E and others. In other words, food products (pasta and biscuits) have been made with the addition of ingredients (wild edible plants or WEPs) with high concentrations of tocopherols and tocotrienols. WEPs are usually added in the formulations in the form of lyophilisate, which allows not to use of exaggerated quantities and volumes to reach the desired concentrations of vitamin E. It should be emphasized that the addition of antioxidant molecules in food formulations through the use of food matrices such as WEPs allows greater protection against degradation during heat treatments and unsuitable states of conservation. Furthermore, the bioavailability and absorption by the body can also be favored. In some studies, being published, many foods made with the addition of different plant species, abundantly reach the threshold of 15% of the RDA set by European regulation. Finally, using balanced formulations and appropriate technologies, leafy vegetable pasta and biscuits proved to be a good alternative food, with high nutritional and sensorial quality and high healthy and antioxidant properties.

REFERENCES

Beecher, G. R. (2003). Overview of dietary flavonoids: Nomenclature, occurrence, and intake. *The Journal of Nutrition*, *133*(10), 3248S–3254S. doi:10.1093/jn/133.10.3248S PMID:14519822

Burgos, G., Amoros, W., Muñoa, L., Sosa, P., Cayhualla, E., Sanchez, C., Diaz, C., & Bonierbale, M. (2013). Total phenolic, total anthocyanin and phenolic acid concentrations and antioxidant activity of purple-fleshed potatoes as affected by boiling. *Journal of Food Composition and Analysis*, *30*(1), 6–12. doi:10.1016/j.jfca.2012.12.001

Da Silva, V. B., Barreira, J. C. M., & Oliveira, M. B. P. P. (2016). Naturalphytochemicals and probiotics as bioactive ingredients for functional foods: Extraction, biochemistry and protected-delivery technologies. *Trends in Food Science & Technology*, *50*, 144–158. doi:10.1016/j.tifs.2015.12.007

Diamante, M. S., Vanz Borges, C., Minatel, I. O., Jacomino, A. P., Basílio, L. S. P., Monteiro, G. C., & Pace Pereira Lima, G. (2021). Domestic coking practices influence the carotenoid and tocopherol content in colored cauliflower. *Food Chemistry*, *340*, 127901. doi:10.1016/j.foodchem.2020.127901 PMID:32889204

Eggersdorfer, M., & Wyss, A. (2018). Carotenoids in human nutrition and health. *Archives of Biochemistry and Biophysics*, *652*, 18–26. doi:10.1016/j.abb.2018.06.001 PMID:29885291

Engwa, G. A., Nweke, F. N., & Nkeh-Chungag, B. N. (2022). Free radicals, oxidative stress-related diseases, and antioxidant supplementation. *Alternative Therapies in Health and Medicine*, *28*(1), 144–128. PMID:32827401

Fratianni, A., D'Agostino, A., Niro, S., Bufano, A., Paura, B., & Panfili, G. (2021). Loss or gain of lipophilic bioactive compounds in vegetables after domestic cooking? Effect of steaming and boiling. *Foods*, *10*(5), 960. doi:10.3390/foods10050960 PMID:33924885

Gęgotek, A., & Skrzydlewska, E. (2022). Antioxidative and anti-inflammatory activity of ascorbic acid. *Antioxidants*, *11*(10), 1993. doi:10.3390/antiox11101993 PMID:36290716

Lee, S., Choi, Y., Jeong, H. S., Lee, J., & Sung, J. (2018). Effect of different coking methods on the content of vitamins and true retention in selected vegetables. *Food Science and Biotechnology*, *27*(2), 333–342. PMID:30263756

Mignogna, R., Fratianni, A., Niro, S., & Panfili, G. (2015). Tocopherol and tocotrienol analysis as a tool to discriminate different fat ingredients in bakery products. *Food Control*, *54*, 31–38. doi:10.1016/j.foodcont.2015.01.032

Munné-Bosch, S., & Alegre, L. (2002). The function of tocopherols and tocotrienols in plants. *Critical Reviews in Plant Sciences*, *21*(1), 31–57. doi:10.1080/0735-260291044179

Panfili, G., Fratianni, A., & Irano, M. (2003). Normal Phase High-Performance Liquid Chromatography Method for the Determination of Tocopherols and Tocotrienols in Cereals. *Journal of Agricultural and Food Chemistry*, *51*(14), 3940–3944. doi:10.1021/jf030009v PMID:12822927

Panfili, G., Niro, S., Bufano, A., D'Agostino, A., Fratianni, A., Paura, B., Falasca, L., & Cinquanta, L. (2020). Bioactive compounds in wild Asteraceae edible plants consumed in the Mediterranean diet. *Plant Foods for Human Nutrition (Dordrecht, Netherlands)*, *75*(4), 540–546. doi:10.100711130-020-00842-y PMID:32803719

Penicaud, C., Achir, N., Dhuique-Mayer, C., Dornier, M., & Bohuon, P. (2011). Degradation of β-carotene during fruit and vegetable processing or storage: Reaction mechanisms and kinetic aspects: A review. *Fruits*, *66*(6), 417–440. doi:10.1051/fruits/2011058

Regulation (EU) No 1169/2011 of the European Parliament and of the Council of 25 October 2011 on the provision of food information to consumers, Official Journal of the European Union (L 304/18, Luxembourg) 2011.

Ruiz-Rodriguez, A., Marín, F. R., Ocaña, A., & Soler-Rivas, C. (2008). Effect of domestic processing on bioactive compounds. *Phytochemistry Reviews*, *7*(2), 345–384. doi:10.100711101-007-9073-1

Severini, C., Giuliani, R., De Filippis, A., Derossi, A., & De Pilli, T. (2016). Influence of different blanching methods on colour, ascorbic acid and phenolics content of broccoli. *Journal of Food Science and Technology*, *53*(1), 501–510. doi:10.100713197-015-1878-0 PMID:26787969

Shahidi, F., & De Camargo, A. C. (2016). Tocopherols and tocotrienols in common and emerging dietary sources: Occurrence, applications, and health benefits. *International Journal of Molecular Sciences*, *17*(10), 1745. doi:10.3390/ijms17101745 PMID:27775605

Shahidi, F., & Shukla, V. K. S. (1996). Nontriacylglycerol constituents of fats, oils. INFORM - *International News on Fats. Oils and Related Materials*, *7*(11), 1227–1232.

Thakur, A., Singh, S., Dulta, K., Singh, N., Ali, B., Hafeez, A., & Marc, R. A. (2022). Nutritional evaluation, phytochemical makeup, antibacterial and antioxidant properties of wild plants utilized as food by the gaddis-a tribal tribe in the western himalayas. *Frontiers in Agronomy*, *4*, 1010309. doi:10.3389/fagro.2022.1010309

Vaseashta, A., Duca, G., & Travin, S. (Eds.). (2022). *Handbook of Research on Water Sciences and Society* (Vols. 1–2). IGI Global. doi:10.4018/978-1-7998-7356-3

Vaseashta, A., Gevorgyan, G., Kavaz, D., Ivanov, O., Jawaid, M., & Vasović, D. (2021). Exposome, Biomonitoring, Assessment and Data Analytics to Quantify Universal Water Quality. In A. Vaseashta & C. Maftei (Eds.), *Water Safety, Security and Sustainability. Advanced Sciences and Technologies for Security Applications*. Springer. doi:10.1007/978-3-030-76008-3_4

Zehiroglu, C., & Sarikaya, S. B. (2019). The importance of antioxidants and place in today's scientific and technological studies. *Journal of Food Science and Technology*, *56*(11), 4757–4774. doi:10.100713197-019-03952-x PMID:31741500

Chapter 7
Advanced Physical Analytical Methods for the Assessment of the Environment State

Inga Zinicovscaia
Joint Institute for Nuclear Research, Russia

ABSTRACT

Despite the major efforts that have been made over recent years to clean up the environment, pollution remains a major problem and poses continuing risks to health. Thus, the determination of contaminants in environmental samples is a challenging and critical issue. Highly accurate analytical techniques provide the possibility to assess and map the organic and inorganic pollutants in complex atmospheric, water, and soil matrices qualitatively and quantitatively. The present chapter brings together the wealth of information in the current literature, covering both the fundamentals and applications of these methodologies. The main attention is given to neutron activation analysis, inductively coupled plasma mass spectrometry, and inductively coupled plasma atomic emission spectroscopy. The principles, main strengths, and weaknesses of each technique are discussed. Several examples of techniques applied in air, water, and soil sample analysis are presented.

INTRODUCTION

The interaction between humans and their physical surroundings is thoroughly studied due to the intensive influence of anthropogenic activities on the environment. Pollutants in solid, liquid, or gaseous states enter into the environments, affect the quality of water, soil, and air and as a result are harmful to humans and other living organisms (Manisalidis et al., 2020). For example, air pollution has been estimated to cause 6.7 million premature deaths worldwide annually (Atkinson et al., 2014).

Among pollutants present in the environment, heavy metals are among the most dangerous due to their genotoxic, carcinogenic, mutagenic, and teratogenic properties, persistence, and bioconcentration in living organisms. Heavy metals and metalloids are naturally occurring elements that are found throughout the Earth's crust, part of them being important microelements such as for example, Mo, Cr, Ni, Co,

DOI: 10.4018/979-8-3693-0512-6.ch007

etc. Meanwhile, major environmental contamination and human exposure result from anthropogenic activities, which include mining and smelting operations, industrial production and use, and domestic and agricultural use of metals and metal-containing compounds (Tchounwou et al., 2012).

The tremendous demand in analytical information has created a growing need for techniques capable to measure and quantify a large variety of different analytes in complex matrices, such as environmental samples. Nowadays, multianalyte analysis is a primary goal of analytical laboratories in regulatory, industrial, and academic environments throughout the world (de la Guardia & Armenta, 2011).

Among the analytical methods used for the determination of heavy metals and metalloids inductively coupled plasma atomic emission spectroscopy (ICP-AES), inductively coupled plasma mass spectrometry (ICP-MS), flame atomic absorption spectrometry (FAAS), cold vapor atomic absorption spectroscopy (CV-AAS) and electrothermal atomic absorption spectroscopy (ETAAS) are the most widely used and considered the most sensitive, selective and rapid techniques (Iti, 2020). Besides the abovementioned techniques, neutron activation analysis (NAA) has a special place in the multi-elemental analysis of environmental samples.

The scope of the present chapter is to present the principle of the operation of the advanced analytical techniques, namely NAA, ICP-MS and ICP-AES, to discuss their advantages and limitations as well as to give examples of their application in environmental studies.

NEUTRON ACTIVATION ANALYSIS

Neutron activation analysis (NAA) is an important analytical technique for performing, both qualitative and quantitative, multi-elemental analysis in order to determine the content of major, minor, and trace elements in environmental samples (Gibson, 2004). In 1998, NAA was defined by the Comité Consultatif pour la Quantité de Matière — Métrologie en Chimie as the primary method of measurement, which is a method having the highest metrological properties, whose operation can be completely described and understood, for which a complete uncertainty statement can be written down in terms of SI units (Greenberg et al., 2011).

In general, activation analysis is based on the conversion of stable nuclei to radioactive one *via* nuclear reactions and measurement of the reaction products. In NAA the nuclear reactions occur *via* bombardment of the sample to be analyzed with neutrons. The reaction products to be measured are either the radiation, released nearly instantaneously upon neutron capture; or, if the resulting new nuclei are radioactive, the induced radioactivity by which they decay (Greenberg et al., 2011). The most common occurring reaction is the (n,γ) reaction, but also reactions such as (n,p), (n,α), (n,n′) and (n,2n) are important.

Neutrons can be produced in: (i) isotopic neutron sources, (ii) particle accelerators or neutron generators and (iii) nuclear research reactors. Reactors are the strongest sources of neutrons. The neutron output of research reactors is often quoted as neutron flux in an irradiation facility and varies, depending on reactor design and reactor power, between 10^{11} and 10^{14} n cm^{-2} s^{-1}(Trincherini et al., 2009).

The NAA technique has numerous advantages, among which need to be mentioned: non-destructive nature, possibility of performing bulk analysis, sensitivity and wide applicability, especially for trace elements; the virtual absence of an analytical blank; negligible influence of the matrix; possibility of application in a self-validating manner (Byrne, 1993). Elements such as C, H, N, O and Si, often found as major components of many matrices, do not activate well and contribute little or no gamma-ray activity. The method is suitable for measurement of masses in the order range of 10^{-6}–10^{-9} g and less, depending on the element to be determined (Trincherini et al., 2009). There are about 60 elements, which can be determined by NAA with various detection limits, depending on nuclear characteristics of their isotopes and a neutron fluence rate available for activation (Kučera, 2007). Further advantages of the technique are that minimal sample preparation and possibility of examination of very small samples (Gibson, 2004).

NAA is less widely applied than other analytical techniques due to necessity of having access to a nuclear reactor (Witkowska et al., 2005), also the equipment used for the analysis is rather expensive and requires special laboratories and a highly qualified staff (Helmke, 2018).

Today, NAA is successfully applied in environmental, geological, biomedical, archeological, etc. studies. Beside application for scientific purposes for a long time NAA has been used in many countries for the characterization of candidate reference materials or certification of elements in existing reference materials. Nowadays, there is a trend to use NAA to validate other methods of elemental analysis (Witkowska et al., 2005).

One of the applications of NAA is the analysis of elemental content of samples such as aerosols, ash and particulate matter, organisms biomonitors in order to assess air pollution. Air pollution, which originates from both natural and human activities, the latter dominating in the last few decades, is a critical problem for society. Thus, NAA was used to determine content of Na, Cl, Mn, V, Al, Sm, Mo, W, La, As, Br, Sb, K, Ba, Se, Th, Cr, Rb, Ca, Fe, Ce, and Sc in aerosol samples collected in the city of São Paulo, Brazil. Elemental composition of aerosol particles allowed to determine that soil dust and sea salt are the main sources of air pollution in studied area (Paiva et al., 1990). The method was applied to analyze coarse and fine air particulate samples collected in front of an automotive workshop located at Tasmasipabad on Chaklala Road in Rawalpindi. Soil was defined as the main source of Al, K, Fe, Sr, Na, and Zn in analyzed samples (Siddique et al., 2013). INAA was employed for determination of 34 elements in PM_{10} samples collected at a height of 90 m above ground level in Ostravsko-Karvinská agglomeration, Czech Republic. The contribution of three important industrial sources (iron and steel works, cement works) was identified, and the long-range cross-border transport representing the pollution from the Polish domestic boilers confirmed the most important pollution inflow during the winter season (Pavlíková et al., 2021). NAA is often applied for determination of elemental composition of biological monitors such as mosses, lichens, tree leaves/needles and bark. Moss is one of the well-known biomonitors applied in the framework of International Cooperative Programme on Effects of Air Pollution on Natural Vegetation and Crops (ICP Vegetation) to assess the air quality in many European and Asian countries. For example, NAA was used to assess the atmospheric deposition of heavy metals in Bulgaria (Hristozova et al., 2020), Georgia (Chaligava et al., 2021), Moldova (Zinicovscaia et al., 2021, 2022), Russia (Vergel et al., 2020, 2022), Czech Republic (Krakovská et al., 2020), Romania (Stihi et al., 2017), etc.

Other sample types widely analyzed by NAA, are soils and sediments. The interest in analyzing metal content in sediment and soil samples arises from their property to concentrate these analytes, exhibiting higher concentrations than other adjacent samples, such as air, water or plants (Witkowska et al., 2005). Using NAA content of 35 elements was determined in sediments collected in the Varzob River. The increased values of some pollution indices in the upper part of the Varzob River indicate anthropogenic origin of As, Sb and Mn while the increased contents of Th and U in the lower reaches of the Varzob River could be attributed to the pegmatite field of the Odjuk tributary (Abdusamadzoda et al., 2020). NAA was used to determine content of chromium, manganese, cobalt, zinc, arsenic, molybdenum, cadmium, and barium in 36 soil samples collected from different parts of the Maritza Basin, Edirne, Turkey (Zaim et al., 2016). In many studies NAA is used to determine not only the composition of soil, but also of the plants growing on this soil. Thus, Zinicovscaia and co-authors (2020) applied NAA to determine the elemental composition of fruits and soils collected from four different zones in the Republic of Moldova. In such studies, data obtained by NAA allow calculation of the transfer factor values of risk assessment indices. Thus, in the presented study the values of transfer factor for K and Rb were higher than 1.0, indicating their accumulation in fruits from soil. For elements considered as environmental pollutants transfer factor values were lower than 1.0. The health quotients for all elements, except Sb, in fruits were lower than unit, which implies that all the analyzed varieties of fruits are safe for human consumption. NAA was employed for the determination of the content of inorganic elements in paddy soils and quantified the nutrient value of paddy rice cultivated on various agricultural sites throughout Thailand. In rise from soil were mainly accumulated Zn, Mn and K (Seeprasert et al., 2017).

Water sample analysis by means of NAA is not very popular due to the need of samples pretreatment. At the same time NAA is used to determine composition of water organisms reflecting the state of water objects. Thus, Nekhoroshkov et al. (2021) used NAA to determine chemical elements in mussels collected in 8 polluted and 4 pristine zones, which included Namibia, the west and east coasts of South Africa, and Mozambique. According to factor analysis Co, Ni, Zn, As, Se, Br, I, Sb accumulated in mussels could have anthropogenic origin. Another application of NAA in water chemistry is the assessment of sorption capacity of different sorbents, including biological ones used for metal removal from wastewater (Zinicovscaia et al., 2018, 2020, 2021).

INDUCTIVELY COUPLED PLASMA MASS SPECTROMETRY (ICP-MS)

Today ICP-MS is considered as the main competitor of NAA. The equipment for ICP-MS is becoming more and more common in analytical laboratories. Less expensive than NAA and not requiring nuclear equipment, ICP-MS can be used for more than fifty analyses per day (Pinte et al., 1998) and allow determining elements such as Pb, Cd that cannot be determined using NAA. An ICP-MS instrument uses plasma (ICP) to ionize the elements in a sample and then measures the ions using a mass spectrometer (MS). The ICP mass spectrometer has major components such as an ICP torch, sample introduction unit (nebulizer and spray chamber), interface, mass analyzer (quadrupole), and detector (Nageswaran et al., 2017; Pinte et al., 1998). Below are presented the main steps of ICP-MS measurements, they include:

(i) conversion of a given sample (liquid or solid) into an aerosol or its direct introduction in the gaseous form. Gaseous samples can be directly introduced into the plasma, while liquid sample analysis requires a sample preparation step.
(ii) transport of aerosol/gas into plasma. Argon is the most commonly used plasma gas in ICP-MS, which can ionize or excite most of the elements in the periodic table without forming any stable compound with the analyte (Nageswaran et al., 2017).
(iii) atomization of the introduced compounds, followed by ionization of atoms in the hot plasma environment. Plasma may be classified as direct-current plasma, high-frequency or RF plasma, or micro-wave plasma, according to the electric field used. RF power is commonly used in ICP.
(iv) transport of the generated ions from the plasma operated at atmospheric pressure to the mass separator (analyzer) operating under high vacuum, involving ion extraction through ion optics. There are different types of mass analyzers among them the quadrupole mass analyzer is the most common mass filter used in ICP-MS.

(v) separation of the ions according to their mass to charge ratios (m/z)
(vi) detection of the separated ions by detector, and conversion of ion flux intensity to an electronic signal (cps) (Bulska & Wagner, 2016).

Detailed description of the principle of ICP-MS technique is presented in (Nageswaran et al., 2017) and (Cubadda, 2007) reviews.

The main advantages of technique are: high sensitivity, multi-element capability, wide linear dynamic range, low detection limit, high sample throughput, low sample volume, relatively simple sample preparation, and ability to discriminate between isotopes (Bulska & Wagner, 2016; Cubadda, 2007). Modern ICP-MS systems are capable of analyzing most elements of the periodic table simultaneously at levels of 10 pg/mL with a mass resolution of <1 amu (Trueman, 2004). Limitations of the technique include the relatively high cost of the equipment which requires a high level of staff proficiency, laboratory set up costs, and interferences needed to be controlled (Wilschefski & Baxter, 2019). The ICP-MS is routinely used in diverse research fields such as earth, environmental, life and forensic sciences, food, material, chemical, semiconductor and nuclear industries.

As opposed to NAA which is mainly used for solid samples analysis, ICP-MS is mainly applied to analyze liquid samples. One of the applications of the technique is determination of the level of heavy metals in drinking water. ICP-MS was used to determine 27 elements (Li, Be, B, Na, Mg, Al, Si, K, Ca, V, Cr, Mn, Fe, Ni, Co, Cu, Zn, As, Se, Rb, Sr, Mo, Ag, Cd, Sb, Ba and Pb) in water samples collected in Bholakpur area of Hyderabad, India. Concentrations of some elements (Fe, Cu, Ni, Al and Na) in water exceed the permissible limits established for drinking water by the World Health Organization (Abdul et al., 2012). Søndergaard and co-authors (2015) applied ICP-MS to determine level of V, Mn, Co, Ni, Cu, Zn, Cd and Pb in sea water. The techniques is applicable in wastewater treatment studies (He et al., 2017; Holubová et al., 2012; Jiang et al., 2022), when different type of sorbents are used to elaborate efficient approach of metal removal.

Beside water itself, ICP-MS is applied to assess metal distribution patterns in water organisms, which are used as food and reflect the state of the water objects. The distribution patterns of 18 elements in the different cytosol fractions of mussels, coming from a wild (uncontaminated) coast, an industrialized (contaminated) area in Asturia and commercial mussels from the Galicia coast were evaluated using ICP-MS. Results showed that total content of metals in samples increased following the expected order: wild coast < Galicia coast <industrial area coast (Ferrarello et al., 2000). Köşker et al. (2022) determined using ICP-MS the metal levels in mussels *Mytilus galloprovincialis* sold in five cities in Turkey in order to assess the possible health risks for children and adults. Based on the obtained data Cr, As, and Cd levels posed a carcinogenic risk for almost all locations and consumer groups, while it was found that Pb did not pose any carcinogenic risk.

ICP-MS technique is also widely applied for soil and plant samples analysis in order to evaluate the level of toxic elements as well as to assess their transfer in soil-plant system. Thus, ICP-MS was applied to determine level of metals in rice and paddy soil in a lead-acid battery production area of China. Bioaccumulation factor values for determined elements (Cr, Cd and Pb) were less than unit in all rice samples, suggesting that the bioavailability of these metals in rice was relatively low (Shao et al., 2016). The technique was applied to assess the level of 12 toxic elements in soils and plants from urban area in Iraq. The concentrations of elements were found in the following order—Mg > Al> Ni > Cu > Cr > Pb > Co > As > Se > Cd > Hg > Ti - in leaf samples and in the soil - Mg > Al > Cr > Ni > Cu > Pb> Co > As > Se > Ti > Cd > Hg. Pollution indices showed a moderate level of contamination with Pb,

Cd, and Ni, and a high level of contamination with As and Hg in plant and soil samples (Khurshid et al., 2022). In Gao et al. (2021) study aimed to assess the sources of contamination and pollution levels of urban garden soils, as well as the health risks for adults and children consuming vegetables grown in such environments ICP-MS was used to determine the level of Cd, Pb, As, Cr, and Hg. According to obtained data Cr and As pollution present the greatest concern for human health.

In the chemistry of the atmosphere ICP-MS is used to determine the level of toxic elements in air thought analysis of filter samples, mosses, lichen, etc. ICP-MS has become the most appropriate technique for the determination of metal in workplace air due to their very low concentrations. Seven elements (Cr, Mn, Ni, Cu, Pb, Cd and Sn) were determined in 54 samples of workplace air collected on the factory in China (Lu et al., 2020). ICP-MS is widely applied to obtain the data on heavy metals depositions in moss samples in the framework of ICP-Vegetation programme (Frontasyeva et al., 2020; Lazo et al., 2022; Schröder & Nickel, 2019; Zhou et al., 2017).

INDUCTIVELY COUPLED PLASMA ATOMIC EMISSION SPECTROSCOPY (ICP-AES)

Resulting from the high cost of ICP-MS techniques often ICP-AES is used in routine analysis of environmental samples. ICP-AES is a method of optical emission spectrometry that uses the emission spectra of sample molecules to identify and quantify the elements in samples. The major components of the ICP-AES instrument are sampler, pump, nebulizer, spray chamber, ICP torch, monochromator / polychromator, and detector arranged in a definite order.

The argon gas is supplied to the torch tube and simultaneously an electromagnetic field is generated by high frequency electric current in work coil positioned on the pointed end of the torch tube. The strong magnetic field induces the ionization of argon in torch tube. The ionized gas produced plasma with high electron density, temperature (10.000K) and energy. The newly produced bulk of the energy is used for excitation and emission of molecules in analyzed sample.

Samples in liquid form are introduced with a peristaltic pump into the nebulizer where they are nebulized into tiny droplets like a fine mist. The larger droplets of the sample are going down in the drain but fine droplets are directed into the high plasma. Hot plasma atomizes the molecules and excites them to a higher state. When the excited atoms return to low energy position, emission rays (spectrum rays) are released and the emission rays that correspond to the photon wavelength or emission lines are specific for each metal. All elements present in the radiation source emit their characteristic spectra at the same time. This wavelength tracks the detector ray and the light intensity for each different wavelength is quantified by software, which convert them into concentration units (de la Guardia & Armenta, 2011; Iti, 2020).

ICP-AES can be used to measure almost all the elements of the periodic table with detection limits in the range of micrograms per liter to milligrams per liter, however technique is not suitable for determination of noble gases, halogens, or light elements such as H, C, N, and O (de la Guardia & Armenta, 2011). ICP-AES instruments are limited to the analysis of liquids only that require additional reagents and equipment. At the same time conversion of samples in a liquid form relieve from the necessity to use standards matched to the matrix of the original sample (Cheremisinoff, 1996).

The environment field is probably the main field of activity of ICP-AES. The determination of metal ions in natural waters at low concentration level remains one of the most important tasks in environmental monitoring. Chochorek and co-authors (2010) used ICP-AES to determine ten elements, Zn, Cr, Cd, Cu,

Ni, Pb, Fe, Mn, Mg and Ca, in surface water. ICP-AES is also widely applied for analysis of chemical composition of wastewater. The technique was used to determine heavy metals (Pb, Cu, Cd, Cr, Zn, Al, Fe, Ni, Co, Mn) and major elements (Ca, Mg) in wastewater of Kocabas Stream collected from nine stations (Yayintas et al., 2007). In order to assess the level of pollution Li and co-authors (2014) used ICP-AES to determine the level of Zn, Se, B, Ba, Fe, Mn, Mo, Ni, V, Al, Li, Sr, Be, Cd, Cr, Cu, and Pb in ground and surface water. It was found that for some elements determined concentrations exceeded values established for drinking water by standards. Beside direct water analysis, ICP-AES is widely applied in wastewater treatment experiments to assess the efficiency of applied sorbents. Thus, ICP-AES was used to trace the efficiency of cerium (Zinicovscaia et al., 2021), chromium (Frois et al., 2012), Cd, Cu, Pb and Zn (Alexander et al., 2017) removal from wastewater treated with sorbents of different origin.

ICP-AES showed to be simple, rapid, accurate, and reliable method to determine the level of 24 metal elements and their compounds in the air of workplace (Goullé et al., 2005). Along with ICP-MS and NAA, ICP-AES is one of the main techniques applied in the ICP-Vegetation programme (Barandovski et al., 2012, 2013; Pavlíková et al., 2021; Špirić et al., 2013).

The ICP technique is widely accepted as a means of determining major and minor compositions of soils and plants or heavy metals in soils and sediments. The presence of heavy metals in the samples collected around Kastriot's power plants was proved by ICP-AES. The results showed that the content of some heavy metals (Pb, Zn, Cd, As, Cu) in the area close to the power plants was very high and often exceeded the target and even the intervention limits (Jusufi et al., 2016). The concentrations of seven elements, Cd, Cr, Cu, Fe, Ni, Pb, and Zn in surface sediment samples collected from the Algerian Mediterranean coasts were analyzed using ICP-AES in order to assess the distribution and bioavailability of metals and to study the anthropogenic factors affecting their concentrations. According to enrichment factors values investigated samples were not contaminated with Cr, Cu, and Fe, moderately contaminated with Ni, Pb, and Cd, and contaminated with Cd in some sites (Alomary & Belhadj, 2007).

CONCLUSION

A wide variety of analytical techniques were developed for multi-elemental analysis of the environmental samples. Each of the technique notwithstanding its specific advantages and limitations showed to be applicable for addressing the challenges of environmental chemistry. In the present study the principle of three analytical techniques, NAA, ICP-MS and ICP-AES, was discussed and examples of their application in environmental studies were presented. Since presented techniques have different sensitivity and operating principle the selection of method for elemental analysis depends on the range of elements concentrations in the sample, type of samples and weight of samples.

ACKNOWLEDGEMENT

This research has been performed under the Moldovan National Research Project Nr.20.80009.5007.27 "Physical-chemical redox processes with electron transfer in the vital, technological and environmental processes".

REFERENCES

Abdul, R. M., Mutnuri, L., Dattatreya, P. J., & Mohan, D. A. (2012). Assessment of drinking water quality using ICP-MS and microbiological methods in the Bholakpur area, Hyderabad, India. *Environmental Monitoring and Assessment, 184*(3), 1581–1592. doi:10.100710661-011-2062-2 PMID:21544503

Abdusamadzoda, D., Abdushukurov, D. A., Zinicovscaia, I., Duliu, O. G., & Vergel, K. N. (2020). Assessment of the ecological and geochemical conditions in surface sediments of the Varzob river, Tajikistan. *Microchemical Journal, 158*, 105173. doi:10.1016/j.microc.2020.105173

Alexander, D., Ellerby, R., Hernandez, A., Wu, F., & Amarasiriwardena, D. (2017). Investigation of simultaneous adsorption properties of Cd, Cu, Pb and Zn by pristine rice husks using ICP-AES and LA-ICP-MS analysis. *Microchemical Journal, 135*, 129–139. doi:10.1016/j.microc.2017.08.001

Alomary, A. A., & Belhadj, S. (2007). Determination of heavy metals (Cd, Cr, Cu, Fe, Ni, Pb, Zn) by ICP-OES and their speciation in Algerian Mediterranean Sea sediments after a five-stage sequential extraction procedure. *Environmental Monitoring and Assessment, 135*(1–3), 265–280. doi:10.100710661-007-9648-8 PMID:17342430

Atkinson, R. W., Kang, S., Anderson, H. R., Mills, I. C., & Walton, H. A. (2014). Epidemiological time series studies of PM2.5 and daily mortality and hospital admissions: A systematic review and meta-analysis. *Thorax, 69*(7), 660–665. BMJ Publishing Group. doi:10.1136/thoraxjnl-2013-204492

Barandovski, L., Frontasyeva, M. V., Stafilov, T., Šajn, R., Pavlov, S., & Enimiteva, V. (2012). Trends of atmospheric deposition of trace elements in Macedonia studied by the moss biomonitoring technique. *Journal of Environmental Science and Health. Part A, Toxic/Hazardous Substances & Environmental Engineering, 47*(13), 2000–2015. doi:10.1080/10934529.2012.695267 PMID:22870997

Barandovski, L., Stafilov, T., Šajn, R., Frontasyeva, M. V., & Bačeva, K. (2013). Air pollution study in macedonia using a moss biomonitoring technique, ICP-AES and AAS. *Macedonian Journal of Chemistry and Chemical Engineering, 32*(1), 89–107. doi:10.20450/mjcce.2013.137

Bulska, E., & Wagner, B. (2016). Quantitative aspects of inductively coupled plasma mass spectrometry. *Philosophical Transactions of the Royal Society A: Mathematical, Physical and Engineering Sciences, 374*(2079). Royal Society of London. doi:10.1098/rsta.2015.0369

Byrne, A. R. (1993). Review of neutron activation analysis in the standardization and study of reference materials, including its application to radionuclide reference materials. *Fresenius' Journal of Analytical Chemistry, 345*(2–4), 144–151. doi:10.1007/BF00322575

Chaligava, O., Shetekauri, S., Badawy, W. M., Frontasyeva, M. V., Zinicovscaia, I., Shetekauri, T., Kvlividze, A., Vergel, K., & Yushin, N. (2021). Characterization of trace elements in atmospheric deposition studied by moss biomonitoring in Georgia. *Archives of Environmental Contamination and Toxicology, 80*(2), 350–367. doi:10.100700244-020-00788-x PMID:33236186

Cheremisinoff, N. P. (1996). *Elemental and structural characterization tests. Polymer Characterization, 43–81*. Elsevier. doi:10.1016/B978-081551403-9.50006-6

Chochorek, A., Bobrowski, A., Kiralyova, Z., & Mocak, J. (2010). ICP-OES determination of select metals in surface water - A metrological study. *Polish Journal of Environmental Studies, 19*(1), 59–64. http://www.pjoes.com/ICP-OES-Determination-of-Select-Metals-r-nin-Surface-Water-a-Metrological-Study,88356,0,2.html

Cubadda, F. (2007). Inductively coupled plasma mass spectrometry. In *Food Toxicants Analysis* (pp. 697–751). Elsevier. doi:10.1016/B978-044452843-8/50020-1

de la Guardia, M., & Armenta, S. (2011). Multianalyte determination versus one-at-a-time methodologies. In *Comprehensive Analytical Chemistry, (57)*. Elsevier. doi:10.1016/B978-0-444-53709-6.00006-9

Ferrarello, C. N., Fernández de la Campa, M. D. R., Muñiz, C. S., & Sanz-Medel, A. (2000). Metal distribution patterns in the mussel Mytilus edulis cytosols using size-exclusion chromatography and double focusing ICP-MS detection. *Analyst (London), 125*(12), 2223–2229. doi:10.1039/b008181g PMID:11219056

Frois, S. R., Tadeu Grassi, M., De Campos, M. S., & Abate, G. (2012). Determination of Cr(VI) in water samples by ICP-OES after separation of Cr(III) by montmorillonite. *Analytical Methods, 4*(12), 4389–4394. doi:10.1039/c2ay26125a

Frontasyeva M., Harmens H., Uzhinskiy A., & Chaligava, O. (2020). Mosses as biomonitors of air pollution: 2015/2016 survey on heavy metals, nitrogen and POPs in Europe and beyond. *Report of the ICP Vegetation Moss Survey Coordination Centre.* Joint Institute for Nuclear Research, Dubna, Russian Federation. https://www.researchgate.net/publication/342804243_Mosses_as_biomonitors_of_air_pollution_20152016_survey_on_heavy_metals_nitrogen_and_POPs_in_Europe_and_beyond

Gibson, L. T. (2004). Archaeometry and Antique Analysis - Metallic and Ceramic Objects. In *Encyclopedia of Analytical Science* (2nd ed., pp. 117–123). Elsevier. doi:10.1016/B0-12-369397-7/00020-0

Goullé, J. P., Mahieu, L., Castermant, J., Neveu, N., Bonneau, L., Lainé, G., Bouige, D., & Lacroix, C. (2005). Metal and metalloid multi-elementary ICP-MS validation in whole blood, plasma, urine and hair: Reference values. *Forensic Science International, 153*(1), 39–44. doi:10.1016/j.forsciint.2005.04.020 PMID:15979835

Greenberg, R. R., Bode, P., & De Nadai Fernandes, E. A. (2011). Neutron activation analysis: A primary method of measurement. *Spectrochimica Acta - Part B Atomic Spectroscopy, 66*(3–4), 193–241. Elsevier B.V. doi:10.1016/j.sab.2010.12.011

He, M., Huang, L., Zhao, B., Chen, B., & Hu, B. (2017). Advanced functional materials in solid phase extraction for ICP-MS determination of trace elements and their species - A review. *Analytica Chimica Acta, 973,* 1–24. Elsevier B.V. doi:10.1016/j.aca.2017.03.047

Helmke, P. A. (2018). Neutron activation analysis. In *Methods of Soil Analysis, Part 3: Chemical Methods* (pp. 141–159). Springer US. doi:10.2136ssabookser5.3.c6

Holubová, Z., Moos, M., & Sommer, L. (2012). Simultaneous determination of metal traces by ICP-MS in environmental waters using SPE preconcentration on different polymeric sorbents. *Chemical Papers, 66*(10), 899–906. doi:10.247811696-012-0185-8

Hristozova, G., Marinova, S., Motyka, O., Svozilík, V., & Zinicovscaia, I. (2020). Multivariate assessment of atmospheric deposition studies in Bulgaria based on moss biomonitors: Trends between the 2005/2006 and 2015/2016 surveys. *Environmental Science and Pollution Research International, 27*(31), 39330–39342. doi:10.100711356-020-10005-w PMID:32648219

Iti, S. (2020). ICP-OES: An Advance Tool in Biological Research. *Open Journal of Environmental Biology*, 027–033. doi:10.17352/ojeb.000018

Jiang, H., Wang, Y., Tan, Z., Hu, L., Shi, J., Liu, G., Yin, Y., Cai, Y., & Jiang, G. (2022). Dissolved metal ion removal by online hollow fiber ultrafiltration for enhanced size characterization of metal-containing nanoparticles with single-particle ICP-MS. *Journal of Environmental Sciences (China), 126*, 494–505. doi:10.1016/j.jes.2022.05.034 PMID:36503776

Jusufi, K., Stafilov, T., Vasjari, M., Korca, B., Halili, J., & Berisha, A. (2016). Determination of heavy metals by ICP-AES in the agricultural soils surrounding Kosovo's power plants. *Fresenius Environmental Bulletin, 5*(25), 1313–1321. https://hero.epa.gov/hero/index.cfm/reference/details/reference_id/3462260

Khurshid, C. A., Mahdi, K., Ahmed, O. I., Osman, R., Rahman, M., & Ritsema, C. (2022). Assessment of potentially toxic elements in the urban soil and plants of Kirkuk city in Iraq. *Sustainability (Basel), 14*(9), 5655. doi:10.3390u14095655

Köşker, A. R., Gündoğdu, S., Ayas, D., & Bakan, M. (2022). Metal levels of processed ready-to-eat stuffed mussels sold in Turkey: Health risk estimation. *Journal of Food Composition and Analysis, 106*, 104326. doi:10.1016/j.jfca.2021.104326

Krakovská, A. S., Svozilík, V., Zinicovscaia, I., Vergel, K., & Jančík, P. (2020). Analysis of spatial data from moss biomonitoring in czech-polish border. *Atmosphere (Basel), 11*(11), 1237. doi:10.3390/atmos11111237

Kučera, J., Granja, C., Leroy, C., & Stekl, I. (2007). The use of neutron activation analysis in environmental and biomedical research. *AIP Conference Proceedings, 958*, 28–32. doi:10.1063/1.2825821

Lazo, P., Kika, A., Qarri, F., Bekteshi, L., Allajbeu, S., & Stafilov, T. (2022). Air quality assessment by moss biomonitoring and trace metals atmospheric deposition. *Aerosol and Air Quality Research, 22*(5), 220008. doi:10.4209/aaqr.220008

Li, J., Li, F., Liu, Q., & Zhang, Y. (2014). Trace metal in surface water and groundwater and its transfer in a Yellow River alluvial fan: Evidence from isotopes and hydrochemistry. *The Science of the Total Environment, 472*, 979–988. doi:10.1016/j.scitotenv.2013.11.120 PMID:24355394

Lu, S., Zhang, H., Du, W., & Deng, J. (2020). Determination of metal elements in workplace air by ICP-MS. *MATEC Web of Conferences, 319*, 02005. 10.1051/matecconf/202031902005

Manisalidis, I., Stavropoulou, E., Stavropoulos, A., & Bezirtzoglou, E. (2020). Environmental and health impacts of air pollution: A review. In Frontiers in Public Health, 8, 14. Frontiers Media SA. doi:10.3389/fpubh.2020.00014

Nageswaran, G., Choudhary, Y. S., & Jagannathan, S. (2017). Inductively Coupled Plasma Mass Spectrometry. In *Spectroscopic Methods for Nanomaterials Characterization, 2, 163–194*. Elsevier. doi:10.1016/B978-0-323-46140-5.00008-X

Nekhoroshkov, P., Bezuidenhout, J., Zinicovscaia, I., Yushin, N., Vergel, K., & Frontasyeva, M. (2021). Levels of elements in typical mussels from the southern coast of africa (Namibia, south africa, mozambique): Safety aspect. *Water (Basel), 13*(22), 3238. doi:10.3390/w13223238

Paiva, R. P., Munita, C. J. S., Cunha, I. I. L., Romano, J., & Alonso, C. D. (1990). Determination of trace elements in aerosol samples by instrumental neutron activation analysis. *Biological Trace Element Research, 26–27*(1), 231–242. doi:10.1007/BF02992676 PMID:1704723

Pavlíková, I., Hladký, D., Motyka, O., Vergel, K. N., Strelkova, L. P., & Shvetsova, M. S. (2021). Characterization of pm10 sampled on the top of a former mining tower by the high-volume wind direction-dependent sampler using inna. *Atmosphere (Basel), 12*(1), 1–19. doi:10.3390/atmos12010029

Pavlíková, I., Motyka, O., Plášek, V., & Bitta, J. (2021). Monitoring of heavy metals and nitrogen concentrations in mosses in the vicinity of an integrated iron and steel plant: Case study in Czechia. *Applied Sciences (Basel, Switzerland), 11*(17), 8262. doi:10.3390/app11178262

Pinte, G., Veado, M. A. R., Oliveira, A. H., Khalis, M., Ayrault, S., & Revel, G. (1998). Comparison of neutron activation analysis and ICP-MS used for river water pollution control. *Hydrobiologia, 373–374*(0), 61–73. doi:10.1023/A:1017004329179

Schröder, W., & Nickel, S. (2019). Spatial structures of heavy metals and nitrogen accumulation in moss specimens sampled between 1990 and 2015 throughout Germany. *Environmental Sciences Europe, 31*(1), 33. doi:10.118612302-019-0216-y

Seeprasert, P., Anurakpongsatorn, P., Laoharojanaphand, S., & Busamongkol, A. (2017). Instrumental neutron activation analysis to determine inorganic elements in paddy soil and rice and evaluate bioconcentration factors in rice. *Agriculture and Natural Resources (Bangkok), 51*(3), 154–157. doi:10.1016/j.anres.2016.09.005

Shao, Y., Chen, Y., Liu, J., Wu, P., Ying, Y., & Xie, J. (2016). ICP-MS Determination of potential toxic elements in soil and rice (Oryza sativa L.) and related health risk. *Food Analytical Methods, 9*(12), 3501–3508. doi:10.100712161-016-0536-0

Siddique, N., Jawad, M., & Waheed, S. (2013). Instrumental neutron activation analysis for the study of size-fractionated airborne particulate matter samples. *Radiochimica Acta, 101*(4), 273–277. doi:10.1524/ract.2013.2020

Søndergaard, J., Asmund, G., & Larsen, M. M. (2015). Trace elements determination in seawater by ICP-MS with on-line pre-concentration on a Chelex-100 column using a "standard" instrument setup. *MethodsX, 2*, 323–330. doi:10.1016/j.mex.2015.06.003 PMID:26258050

Špirić, Z., Vučković, I., Stafilov, T., Kušan, V., & Frontasyeva, M. (2013). Air pollution study in croatia using moss biomonitoring and ICP-AES and AAS analytical techniques. *Archives of Environmental Contamination and Toxicology, 65*(1), 33–46. doi:10.100700244-013-9884-6 PMID:23467710

Stihi, C., Popescu, I. V., Frontasyeva, M., Radulescu, C., Ene, A., Culicov, O., Zinicovscaia, I., Dulama, I. D., Cucu-Man, S., Todoran, R., Gheboianu, A. I., Bucurica, A., Bancuta, I., & Dima, G. (2017). Characterization of heavy metal air pollution in Romania using moss biomonitoring, neutron activation analysis, and atomic absorption spectrometry. *Analytical Letters*, *50*(17), 2851–2858. doi:10.1080/00032719.2016.1275661

Tchounwou, P. B., Yedjou, C. G., Patlolla, A. K., & Sutton, D. J. (2012). *Heavy metal toxicity and the environment. EXS, 101, 133–164.* NIH Public. doi:10.1007/978-3-7643-8340-4_6

Trincherini, P. R., Domergue, C., Manteca, I., Nesta, A., & Quarati, P. (2009). The identification of lead ingots from the Roman mines of Cartagena: The role of lead isotope analysis. *Journal of Roman Archaeology*, *22*(1), 123–145. doi:10.1017/S1047759400020626

Trueman, C. (2004). Geochemistry - inorganic. In *Encyclopedia of Analytical Science* (2nd ed., pp. 171–181). Elsevier Inc. doi:10.1016/B0-12-369397-7/00240-5

Vergel, K., Zinicovscaia, I., Yushin, N., Chaligava, O., Nekhoroshkov, P., & Grozdov, D. (2022). Moss biomonitoring of atmospheric pollution with trace elements in the Moscow region, Russia. *Toxics*, *10*(2), 66. Advance online publication. doi:10.3390/toxics10020066 PMID:35202252

Vergel, K., Zinicovscaia, I., Yushin, N., & Gundorina, S. (2020). Assessment of atmospheric deposition in Central Russia using moss biomonitors, neutron activation analysis and GIS technologies. *Journal of Radioanalytical and Nuclear Chemistry*, *325*(3), 807–816. doi:10.100710967-020-07234-1

Wilschefski, S. C., & Baxter, M. R. (2019). Inductively coupled plasma mass spectrometry: Introduction to analytical aspects. *The Clinical Biochemist. Reviews / Australian Association of Clinical Biochemists*, *40*(3), 115–133. doi:10.33176/AACB-19-00024 PMID:31530963

Witkowska, E., Szczepaniak, K., & Biziuk, M. (2005). Some applications of neutron activation analysis: A review. *Journal of Radioanalytical and Nuclear Chemistry*, *265*(1), 141–150. doi:10.100710967-005-0799-1

Yayintas, O. T., Yılmaz, S., Turkoglu, M., & Dilgin, Y. (2007). Determination of heavy metal pollution with environmental physicochemical parameters in waste water of Kocabas Stream (Biga, Canakkale, Turkey) by ICP-AES. *Environmental Monitoring and Assessment*, *127*(1–3), 389–397. doi:10.100710661-006-9288-4 PMID:16897498

Zaim, N., Dogan, C., & Camtakan, Z. (2016). Neutron activation analysis of soil samples from different parts of edirne in Turkey. *Journal of Applied Spectroscopy*, *83*(2), 271–276. doi:10.100710812-016-0280-7

Zhou, X., Chen, Q., Liu, C., & Fang, Y. (2017). Using moss to assess airborne heavy metal pollution in Taizhou, China. *International Journal of Environmental Research and Public Health*, *14*(4), 430. Advance online publication. doi:10.3390/ijerph14040430 PMID:28420186

Zinicovscaia, I., Cepoi, L., Povar, I., Chiriac, T., Rodlovskaya, E., & Culicov, O. A. (2018). Metal Uptake from complex industrial effluent by *Cyanobacteria Arthrospira* platensis. *Water, Air, and Soil Pollution*, *229*(7), 1–10. doi:10.100711270-018-3873-3

Zinicovscaia, I., Chaligava, O., Yushin, N., Grozdov, D., Vergel, K., & Hramco, C. (2022). Moss biomonitoring of atmospheric trace element pollution in the Republic of Moldova. *Archives of Environmental Contamination and Toxicology*, *82*(3), 355–366. doi:10.100700244-022-00918-7 PMID:35266044

Zinicovscaia, I., Hramco, C., Chaligava, O., Yushin, N., Grozdov, D., Vergel, K., & Duca, G. (2021). Accumulation of potentially toxic elements in mosses collected in the Republic of Moldova. *Plants*, *10*(3), 1–13. doi:10.3390/plants10030471 PMID:33801412

Zinicovscaia, I., Sturza, R., Duliu, O., Grozdov, D., Gundorina, S., Ghendov-Mosanu, A., & Duca, G. (2020). Major and trace elements in moldavian orchard soil and fruits: Assessment of anthropogenic contamination. *International Journal of Environmental Research and Public Health*, *17*(19), 1–19. doi:10.3390/ijerph17197112 PMID:32998367

Zinicovscaia, I., Yushin, N., Grozdov, D., Vergel, K., Nekhoroshkov, P., & Rodlovskaya, E. (2021). Treatment of rhenium-containing effluents using environmentally friendly sorbent, saccharomyces cerevisiae biomass. *Materials (Basel)*, *14*(16), 4763. Advance online publication. doi:10.3390/ma14164763 PMID:34443283

Zinicovscaia, I., Yushin, N., Grozdov, D., Vergel, K., Popova, N., Artemiev, G., & Safonov, A. (2020). Metal removal from nickel-containing effluents using mineral–organic hybrid adsorbent. *Materials (Basel)*, *13*(19), 1–21. doi:10.3390/ma13194462 PMID:33050087

Zinicovscaia, I., Yushin, N., Humelnicu, D., Grozdov, D., Ignat, M., Demcak, S., & Humelnicu, I. (2021). Sorption of ce(Iii) by silica sba-15 and titanosilicate ets-10 from aqueous solution. *Water (Basel)*, *13*(22), 3263. doi:10.3390/w13223263

Chapter 8
REDOX Measurements:
Inexpensive and Fast Tool for Preliminary Ecotoxicity Assessment and an Early Warning System for Water Pollution

Igor Cretescu
https://orcid.org/0000-0001-5108-9752
Faculty Chemical Engineering and Environmental Protection, Technical University of Iasi, Romania

Ruslan Borodaev
Faculty of Chemistry and Chemical Technology, Moldova State University, Moldova

Doina Lutic
https://orcid.org/0000-0002-3959-0891
Faculty of Chemistry, Alexandru Ioan Cuza University, Romania

Gabriela Soreanu
https://orcid.org/0000-0002-8341-9039
Faculty of Chemical Engineering and Environmental Protection, Technical University of Iasi, Romania

Gheorghe Duca
https://orcid.org/0000-0001-7265-6293
Institute of Chemistry, Moldova State University, Moldova

Harsha Ratnaweera
https://orcid.org/0000-0003-1456-2541
Faculty of Science and Technology, Norwegian University of Life Sciences, Norway

ABSTRACT

This chapter contains information related to the oxidation-reduction potential (ORP) in aqueous solutions, aiming to extend some aspects from theoretical electrochemistry to practical environmental monitoring and engineering applications. A more comprehensive and precise continuous monitoring of water quality for groundwater, surface water, wastewater, and drinking water could be achieved by deducing relations between ORP, the concentration of the oxidants and reductants in waters, pH, and temperature, delivering valuable information for the better comprehension of the requirements necessary to design an automated water quality monitoring system with real-time transmitted data, assuring better water governance. Such a system is presented as a case study related to an early warning of preliminary ecotoxicity. The case study was conducted by installing water pollution monitoring equipment on a passenger ship sailing on the Sulina Channel in the Danube Delta, Romania.

DOI: 10.4018/979-8-3693-0512-6.ch008

INTRODUCTION

Complex processes occur in the extensive ecosystems of natural protected areas. An example is the Danube Delta, a territory mainly under the jurisdiction of Romanian state authorities. The Danube Delta Biosphere Reserve was the first natural site in Romania to be registered in 1991 on the United Nations Educational, Scientific, and Cultural Organization (UNESCO) World Heritage List. The composition and properties of the surface waters in this ecosystem are not influenced by anthropic factors (Cretescu et al., 2022a). Complex processes (Duca et al., 2003) occur in a vertical direction, from the surface to the bottom, strongly influenced by solar radiation and specific phenomena at the interfaces between soil/water and water/air. Numerous biological and chemical processes must be considered together with physical actions: gas dissolution (oxygen and carbon dioxide) in water, evaporation/transpiration processes, and many others.

The self-cleaning capacity of the waters from complex aquatic ecosystems is well known (Duca et al., 2002) as a natural process of pollutant cleansing. These natural ecosystems are characterized by a unique, smart balance in the habitats between the diverse components of their biota. In waters where such a biological habitat exists, only the normal vital activity and development of hydrobionts at different trophic levels occur. In contrast, anthropogenic pollution from toxic and dangerous compounds threatens aquatic life and the trophic chain. Persistent organic compounds and heavy metal ions have harmful ecological impacts and generate ecotoxicity. The self-purification phenomena can be disturbed or even fully inhibited. If additional measures to manage water pollution are not implemented, the pollution gradually increases until it reaches, in the extreme, complete eutrophication of the water. Eutrophic water drastically reduces water use and has a strong negative ecological impact on the ecosystem, which is costly to reverse.

Numerous specific offline analytical methods and rapid tests to determine the ecotoxicity of water (Gregorin et al., 2021) have been developed by several manufacturers (EBPI, n.d.; Modern Water, n.d.), allowing measurement of the toxicity created by a wide range of harmful pollutants. The procedure for rapid tests comprises the collection of water samples, mixing them with the specific reagents from the kits, and obtaining the results. The results comprise changes in measurable parameters (e.g., color, turbidity, temperature, pH, etc.) after storage under specified conditions (e.g., incubation at a constant temperature, darkness, absence/presence of oxygen, shaking). In addition to the high cost of analytical kits and the high investment costs of offline analyzers, such methods for detecting ecotoxicity are generally difficult and involve considerable time and resource consumption.

This study sought to develop a cheap and rapid detection method to detect the presence/absence of organic compounds or heavy metal ions through online measurement of the oxidation-reduction potential (ORP or REDOX) in an almost continuous regime (at an adjustable sampling frequency of some minutes). The method, however, does involve measuring the concentrations of specific parameters. The toxicity parameter generated by this concept is strongly influenced by the presence of the compounds mentioned above in water. On the other hand, the ecotoxicity could be correlated with the so-called "REDOX toxicity", which is influenced by the presence of some oxidizing species in the aquatic environment, such as different radical species (e.g., $O_2^{\bullet -}$, OH), or hydrogen peroxide (H_2O_2), affecting the activity of hydrobionts, even if the other common water quality indicators are within the normal ranges (Cretescu et al., 2022b; Borodaev et al., 2019). Water with a high concentration of dissolved oxygen (DO) is of good quality and free from organic substances (which usually have a reducing character). Thus, through the correct interpretation of the REDOX potential values, valuable information could be obtained about water oxygenation, even without using a DO sensor. However, the results must be carefully analyzed to avoid possible errors due to reactive oxygen species (which are not beneficial for aquatic life). A corroboration of the obtained results in ORP and those gathered in a DO sensor (highlighting only the presence of molecular oxygen favorable in water) is essential.

BACKGROUND THEORETICAL ASPECTS CONCERNING REDOX POTENTIAL

The REDOX potential of water is a valuable parameter due to its corrosive potential, as well as for determining some critical parameters of its overall quality from the point of view of pollution with several species.

Generally, water could act as a reducing specie (electron donor) and an oxidizing one (electron capturer). Two possible reactions can be formulated in this respect (Tofan et al., 2019):

$$2\,H_2O - 4\,e^- \rightleftharpoons 4\,H^+ + O_2 \quad (1)$$

$$2\,H_2O + 2\,e^- \rightleftharpoons 2\,HO^- + H_2 \quad (2)$$

These two reactions are characterized by the potential values of $E^o_{O2/2H2O} = +1.23$ V and $E^o_{2H2O/2HO^-} = -0.83$ V, respectively, by taking the normal hydrogen electrode as a standard.

The normal hydrogen electrode consists of a platinized platinum plate immersed in a solution of a strong acid of 1 N concentration, in which molecular hydrogen is bubbling at a pressure of 1 atm to ensure saturation; its working function is based on the reaction:

$$2H^+ + 2\,e^- \rightleftharpoons H_2 \quad (3)$$

According to the Nernst Equation, the potential of this reaction is given by:

$$E_{H_2} = E^0{}_{H_2} + \frac{0.059}{2}\lg\frac{[H^+]^2}{p_{H2}} \quad (4)$$

By convention, E^o_{H2} is equal to zero.

The REDOX potentials of the oxygen and hydrogen production according to equations (1) and (2) are calculated using the relations:

$$E_{O_2} = E^0{}_{O_2} + \frac{0.059}{4}\lg\frac{p_{O2}}{[H_2O]^2}[H^+]^4 \quad (5)$$

$$E_{H_2} = E^0{}_{2H_2O/2HO^-} + \frac{0.059}{4}\lg\frac{[H_2O]^2}{[HO^-]^2} \quad (6)$$

Since oxygen and hydrogen production starts when their pressure reaches 1 atm, equations (5) and (6) could be simplified as:

$$E_{O_2} = 1.23 - 0.059\text{pH}\,(\mathbf{V}) \qquad (7)$$

$$E_{H_2} = -0.059\text{pH}\,(\mathbf{V}) \qquad (8)$$

The REDOX potential of water could be expressed by a parameter defined in a manner similar to the pH by using the relation:

$$\text{rH} = -\lg(p_{H2}) \qquad (9)$$

By combining relations (4) and (9), the dependence between rH and pH can be written as:

$$rH = \frac{E_{H2}}{0.0295} + 2pH \qquad (10)$$

or:

$$E_{H2} = 0.0295\ \text{rH} - 2\ \text{pH}\ (\text{V}) \qquad (11)$$

Equations (10) and (11) reveal that the variations in pH and rH are inversely proportional. The high rH values indicate a strong oxidant, while the low rH values are characteristic of a reductant medium.

If the limit values of the pH of 0 and 14 are replaced in Equation (10), the extreme values of 0 and 41.6 result in the rH values by considering the $E_{O2} = 1.23$ (at pH = 0), $E_{O2} = 0.40$ (at pH = 14), $E_{H2} = 0$ (at pH = 0) and $E_{H2} = -0.83$ (at pH = 14), respectively.

Thus, rH values below 28 indicate that the water acts as a reductant, while rH values between 28 and 41.6 reveal that the water acts as an oxidant. These value ranges are typically represented in a Pourbaix diagram (Figure 1).

In real measurement systems, being able to perform potentiometric measurements with pH values, the E_{H2} value could be expressed as:

$$E_h = \text{ORP} + E_{ref} \qquad (12)$$

where ORP is REDOX, and E_{ref} is the reference potential, which usually has a constant value that should be corrected with the temperature changes for more accuracy.

The potentiometric measurement could be performed by joining a reference electrode (E_{ref}) of Ag/AgCl and a platinum disc for the ORP values.

Thus, Equation (11) becomes, after combining it with (12):

$$\text{rH} = (\text{ORP} + E_{ref}) / 29.5 + 2\ \text{pH} \qquad (13)$$

where the values of ORP and E_{ref} are used in mV, and the value of the denominator is transformed accordingly.

Figure 1. Pourbaix diagram of water
(adapted from Tofan et al., 2019)

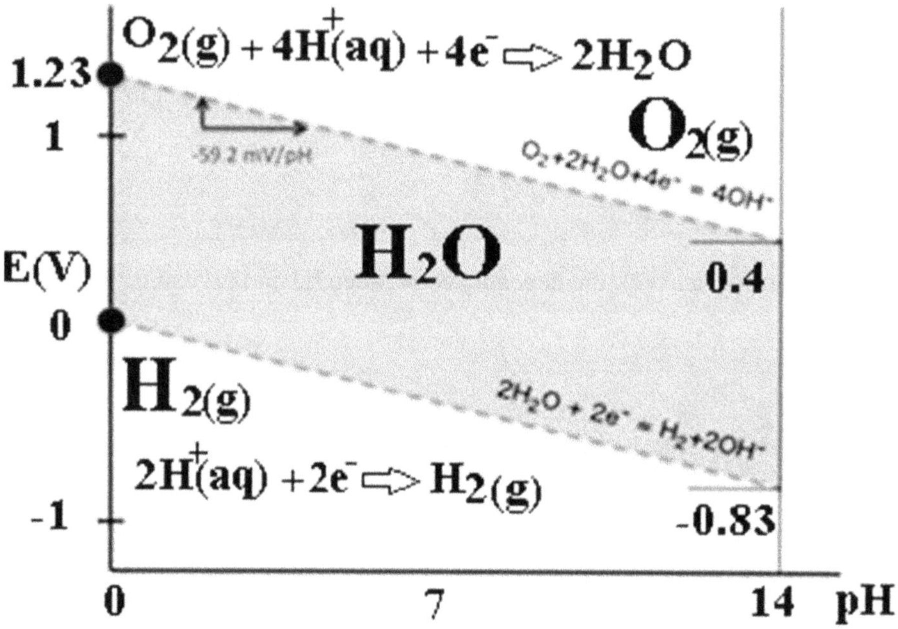

FROM THEORY TO THE PRACTICAL CHARACTERISATION OF SURFACE WATERS

The concentration of DO in clean surface waters varies between 11-14 mg O_2/L in the mountain springs, while in highly polluted waters, much lower values of 1-3 mg O_2/L are common, depending on the water temperature values. This latter range indicates the presence of considerable amounts of substances with reducing character, such as organic matter, hydrogen sulphide derivatives (HS^-, S^{2-}), etc. The DO concentration also varies with pressure; at greater depths, the oxygen concentration decreases compared to the values measured at the water surface at the same location. The DO concentration depends on two primary processes: the biotic and abiotic oxygen consumption, on one part, and the oxygen generation by photosynthesis when solar radiation is available, a process driven by photoautotrophic macro- and microphytes (plants, algae, cyanobacteria, etc.). This kind of biota achieves light energy capture through the participation of pigment-type photoreceptors (mainly, but not limited to, chlorophyll) and transforms it into chemical energy, stored by the formation of new organic molecules accumulated in the composition of the reaction biomass in the meantime with oxygen release (Nicanorov, 2000; Ozcan & Ogun, 2001).

Figure 2. Generation of reactive oxygen species between organisms and the environment (adapted from Duca et al., 2003)

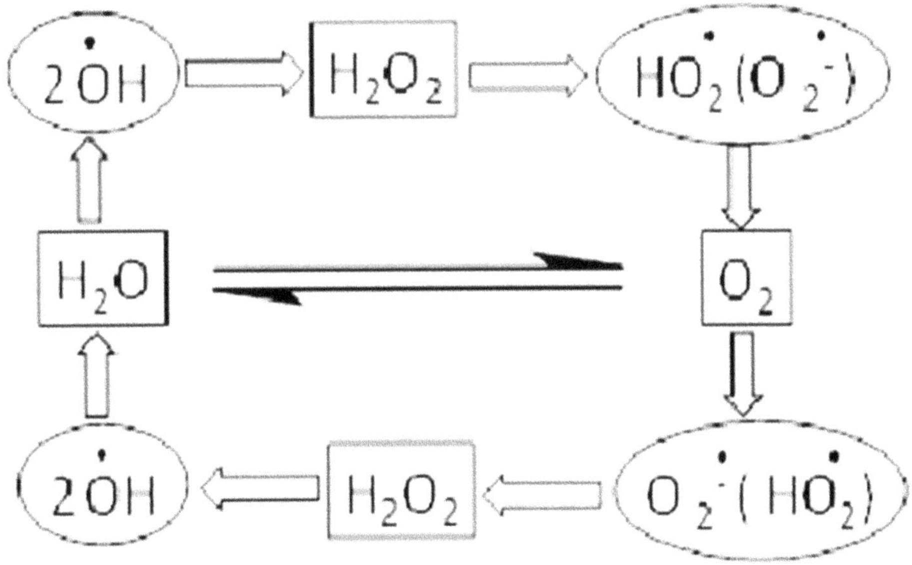

Oxygen is generally considered a strong oxidant. However, oxygen is inert in surface water and has an unexpectedly low reactivity to the reducing substances encountered. This is due to its triplet electronic structure, which prevents interaction with dissolved organic substances owning a singlet electronic structure. Therefore, reactions involving oxygen require an activation procedure, which can be achieved either by interacting with sunlight with the oxygen molecule or by the presence of catalysts, such as transition metal ions (Fe, Cu, etc.). As a result, the oxygen reactivity in surface waters is determined by the intermediate products of its biogeochemical cycle rather than its molecular form (Figure 1).

One can see that the most stable compound from intermediate oxygen-containing reduction products is H_2O_2 (Duca et al., 2003; Duca et al. 2021; Ozean & Ogun, 2014). The most significant amount is obtained from photochemical reactions when sunlight acts on dissolved organic substances with chromophore groups (colored compounds).

After the absorption of solar energy (with different wavelengths and energies), the organic molecules transform into excited states. Upon their return to the fundamental state, an electron is transferred to the oxygen dissolved in water, forming anion radicals of the superoxide type, which subsequently are disproportionate to form H_2O_2 (Travin et al., 2019; Vaseashta et al, 2022).

Another source of H_2O_2 in surface water (Richard et al., 2007; Ozean & Ogun, 2014) is the presence of certain algae species that synthesize it by their normal metabolism. Under the same illumination conditions, the ultraviolet (UV) radiation with a wavelength shorter than 300 nm induced higher yields of H_2O_2 than in the irradiation by visible range wavelength light (longer than 365 nm), meaning that the amount of irradiation energy is essential to the hydrogen peroxide yield. At the same time, some algae can oxidize organic compounds with nitrogen content by enzymatic catalysis, forming ammonia simultaneously with the transformation of oxygen into H_2O_2 (Ozean & Ogun, 2014).

AN INNOVATIVE CONCEPT FOR MONITORING AND EVALUATION

Simple measurements of water pH and ORP, followed by the raw calculation of rH and Eh values, is a promising strategy for monitoring the quality of diverse water types such as spring water, surface water, potable water, flowing rivers, lakes, and even wastewater. Various oxygenated species in the water give precious information about its cleanliness, its safe use for various purposes, and the biota equilibrium. Water pollution from certain chemicals is associated with modifications in hydrobionts. In particular, highlighting the eutrophication process or the toxicity associated with the presence of oxygen-containing compounds can be related to "REDOX toxicity". The concept of online monitoring of the REDOX potential together with easy-to-monitor indicators (e.g., pH, dissolved oxygen and temperature) is of particular importance. It can ensure the diagnosis and forecast of the evolution of water quality based on a large number of measurements made over time and space in a specific area of interest, or when the specific analyses for the evaluation of ecotoxicity are too expensive to be conducted frequently, obtaining valuable real-time data related to the preliminary estimate of water ecotoxicity.

These kinds of measurements are interesting, reliable, and representative when performed within a reasonable time, at a high sampling frequency in areas of high interest for environmental characterization. An innovative, cheap, and convenient strategy to do so was tested by our team by placing a complex water monitoring system on a passenger ship traveling from Sulina to Tulcea in the Danube Delta. This route is essential for economic and tourism interests and practical for establishing environmental strategies and local rules for the long-term protection of this unique area.

A CASE STUDY FOR THE REDOX ASSESSMENT OF DANUBE RIVER WATER IN THE SULINA CHANNEL USING A MOBILE WATER QUALITY MONITORING STATION

This study (Cretescu et al., 2022b) was conducted in September 2022, based on data collected using a mobile water quality monitoring station installed on a passenger ship cruising in the Sulina Channel (colored in red on the map in Figure 3, which was adapted from Google Map) between Sulina and Tulcea. Human communities situated along the Sulina Channel are highlighted in the red ovals in Figure 3.

Water quality monitoring was conducted using the instruments installed on the ship, which sailed from Sulina to Tulcea on even-numbered days in September and from Tulcea to Sulina on odd-numbered days. The cruise duration was almost the same in either direction. The ship's average speed was 25 km/h, with about five-minute stops at various ports. The online measurements continued while the ship was docked at Tulcea and Sulina; there, the ship acted as a fixed monitoring point for water quality. The measurements were conducted every three minutes (sampling frequency) at a water depth of around 1-1.5 meters.

The monitored water parameters for the proposed characterization included temperature, pH, ORP, DO, nitrate (NO_3^-), and nitrite (NO_2^-) ion concentrations. The content of organic compounds was expressed as the chemical oxygen demand (COD).

The overall color-specific pigments from green algae are expressed by the chlorophyll A (Chl-a) and from the blue algae (blue 1 and blue 2, respectively). which are immobilized in the vegetal tissue of the macro/microalgae (therefore, not directly dissolved in water). However, to point out the dissolved organic compounds, other parameters, namely, the chromophore dissolved organic matter (CDOM) (Ji et al., 2019), were monitored using an optical method, employing different light wavelengths to highlight the presence of diverse colored compounds comprising the CDOM with different chemical structures. Therefore, the CDOM includes the colored compounds from water provided by terrestrial inputs and microscopic algae-assisted biological reactions from water. The CDOM could be evaluated by measuring the absorbance values of the analyzed water at two wavelengths: 254 nm for dissolved aromatic compounds and 350 nm for colored compounds (Coelho et al., 2017). The corresponding values were labeled CDOM2 for the 254 nm concentration and CDOM3 for the 350 nm concentration.

Figure 3. Monitored area and the inhabited communities
(adapted from Google Map)

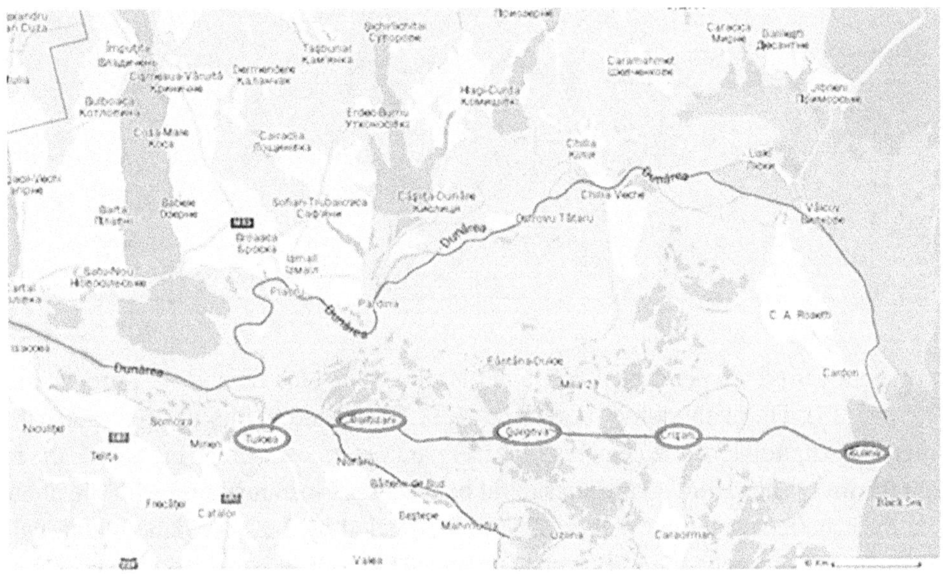

The values for rH and the hydrogen and oxygen potential were calculated with equations (7-13). A combined sensor for pH, Eh (REDOX) and temperature (PONSEL PHEHT from AQUALABO, France) was employed for the pH/ORP measurements. The setup was able to perform simultaneous potentiometric measurements for pH and ORP using two pairs of electrodes: 1) a glass bulb sensitive to H^+ plus a reference electrode (E_{ref}) of Ag/AgCl for pH measurement, and 2) a platinum disc for REDOX (ORP) measurement coupled with the same reference electrode (Ag/AgCl). The reference electrode was Ag/AgCl with a E_{ref} potential value of +205 mV at 25°C, which was corrected with the temperature changes. A Negative Temperature Coefficient NTC thermistor incorporated in this sensor body ensured the temperature measurement and contributed to the necessary temperature compensation by applying specific software during the pH and ORP measurements (Cretescu et al. 2022b).

Two optical multi-parameter probes (for chemical parameters and specific pigments and CDOM) from Trios (Trios, n.d.) were also integrated into the referred innovative water quality monitoring system, equipped with a data storage/transmission module. The water monitoring system was designed following a patent (Cretescu et al., 2022c) and concepts of DOSCON AS, Norway (Nair, et al., 2022).

Using a mobile station for water quality monitoring instead of a classic monitoring station allows for detecting more precisely the time and the place of the pollutant's entry into the surface water body. Thus, additional benefits could be foreseen concerning the better control of water quality indicators (Gladchi et al., 2014), pointing out the changes in the eutrophication and water pollution status in real-time (Ene et al., 2021; Jurminskaia et al., 2021).

The DO concentrations were examined over a period of about four days, as shown in Figure 4 (Cretescu et al., 2022b), in which the day/night cycles correlated with the cruise ship position.

Figure 4. Evolution of DO over time (night-time marked in grey)
(Cretescu et al., 2022b, credited by the source journal).

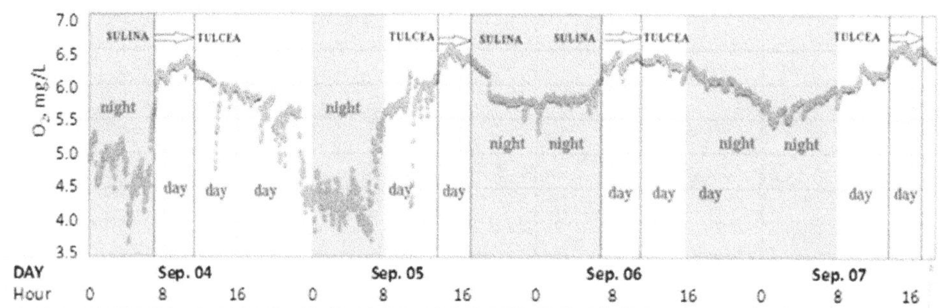

Based on the measured temperature, ORP, and pH, the values for Eh and rH were calculated using equations (12) and (13). The methodology (Cretescu et al., 2022b) and results are presented in Figures 5-7.

For a better understanding of the changes in water parameters over time (Figure 5), the importance of daylight on the water saturation in oxygen should be discussed in connection with the photosynthesis performed by the aquatic vegetation (plants, algae), as revealed by the systematic higher values of DO during the daytime than the night-time. Oxygen is produced through carbon dioxide reduction to biomass.

An excellent correlation can be seen (Figure 5) between the variation in the peaks attributed to the following chemical parameters: DO, ORP, rH, NO_2 and COD. Thus, at t_1 (corresponding to the switch from night to date in Sulina), the peaks for A, B, C, D, and E, corresponding to the above parameters, were simultaneously modified, where the DO, ORP, and rH increased, while NO_2 and COD decreased.

At t_2 (corresponding to the switch from day to night in Sulina), the peaks referred to as A', B', C', D' and E' were also simultaneously changed, but in the opposite direction, compared to the status at t_1. At t_3 (corresponding to the switch from night to date in Sulina), the peaks referred to as A", B", C", D" and E" were also simultaneously changed, but in the opposite direction, compared with t_2. A similar behavior was observed in Tulcea during the first night. The reduced DO associated with peak (a) was correlated with the drop of ORP corresponding to peak (b) and rH corresponding to peak (c). Similar behavior was observed briefly in Tulcea during the second night. The reduced DO associated with the peak A"' was correlated with the temperature increase corresponding to peak T, but also with the slight reduction in the ORP and rH values and a slight increase in COD.

The ship was docked during the night in the two ports, Sulina and Tulcea. The passenger ship was moving during the daytime, and no significant variation in the above parameters was observed. This suggests the lack of additional pollutant sources (other than the existing background chemical composition of Danube water in the Sulina Channel).

In Figure 6, a linear variation between Eh and ORP can be observed in Equation (11).

Figure 5. Comparison between simultaneous changes in the water parameters during the monitoring period

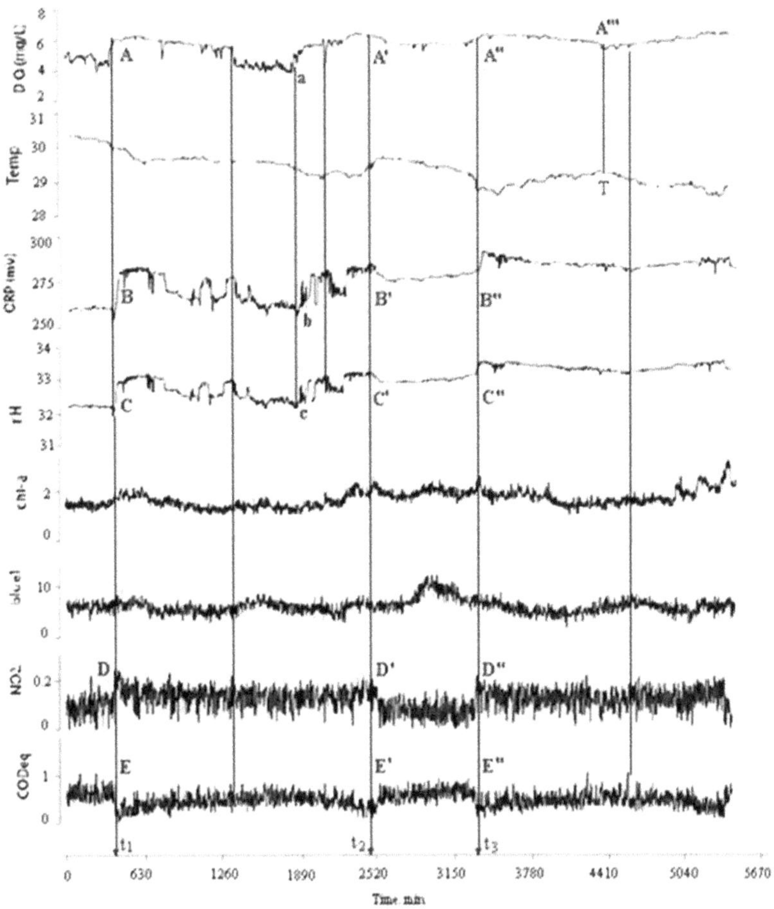

An approximate linear variation between Eh and rH can be observed (Figure 7).

However, as shown in Figure 8, Eh varies with pH (in the presented example, the pH variation limits are 7.82–8.02).

Overall, there was no direct correlation concerning the variation of Eh in the day/night regime. However, an agglomeration of points (mixed points corresponding to the day/night regime) shown in Figure 8 shows a pH interval between 7.88 and 7.98. At the same time, an agglomeration of points corresponding to the diurnal regime is situated in the pH interval of 7.98–8.01.

Figure 9 shows a similar tendency of variations between rH and ORP, as observed in Figure 7, because there is a linear dependence between Eh and ORP (as shown in Figure 6).

Other significant correlations could be observed between general chemical parameters and other parameters, such as the specific pigments (chlorophyll and phycocyanin) associated with the green/blue algae and the CDOM. As shown in Figure 10, a good correlation is observed between the variation of rH and DO, which increases during the daytime due to photosynthesis from sunlight, and the specific pigment of the phytoplankton. Chlorophyll is one of the most critical pigments, the influence of which on rH is shown in Figure 11.

Figure 6. The changes Eh (mV) on ORP during the same monitoring period

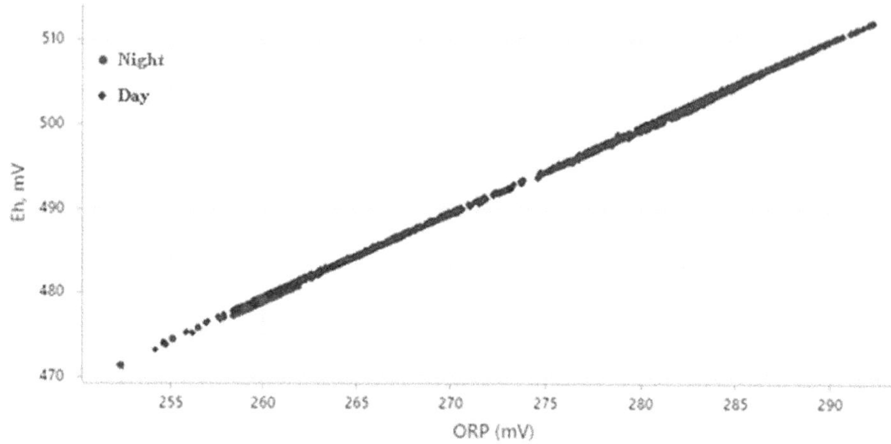

Figure 7. The changes Eh (mV) on rH during the same monitoring period

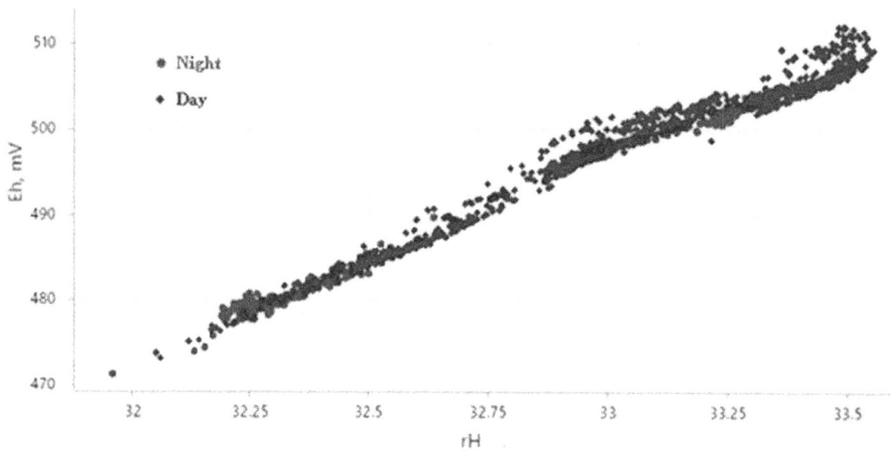

Figure 8. The changes Eh (mV) on pH during the same monitoring period

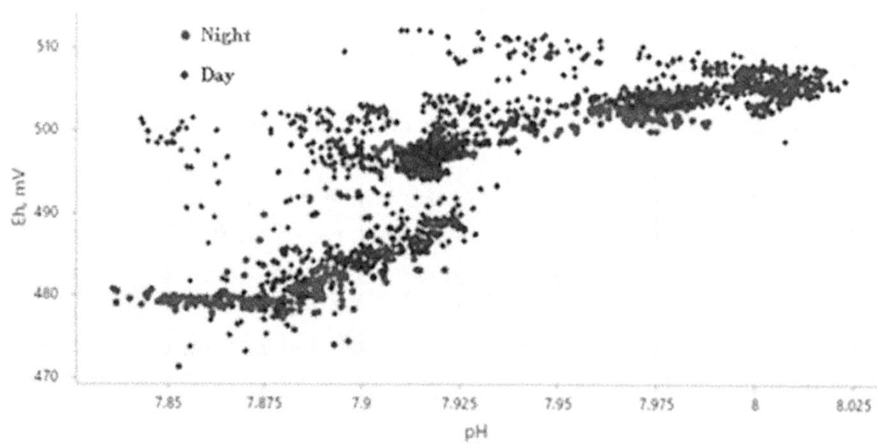

Figure 9. Changes in rH as a function of ORP during the same monitoring

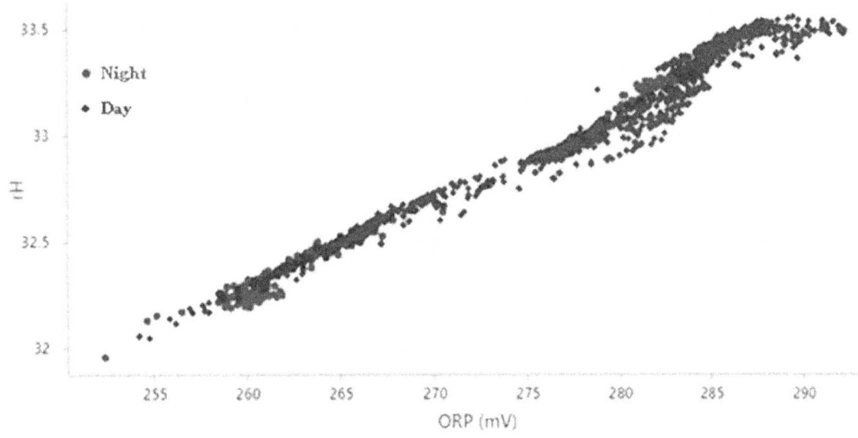

Figure 10. Changes in rH as a function of DO during the same monitoring period

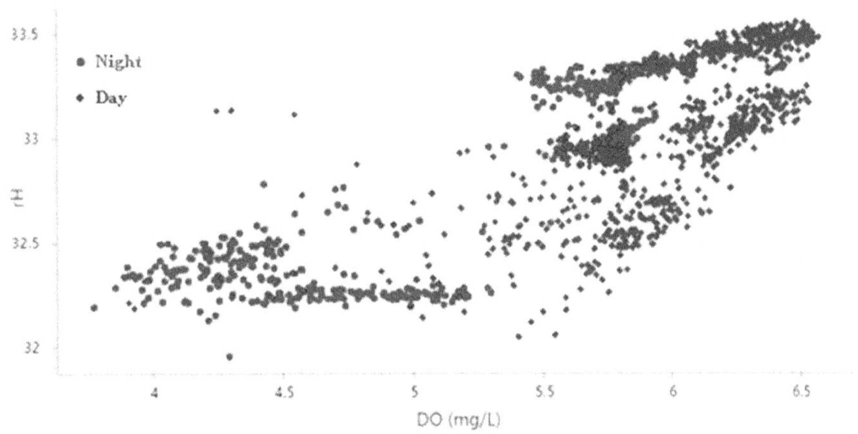

Figure 11. Changes in rH as a function of Chlorophyll a during the same monitoring period

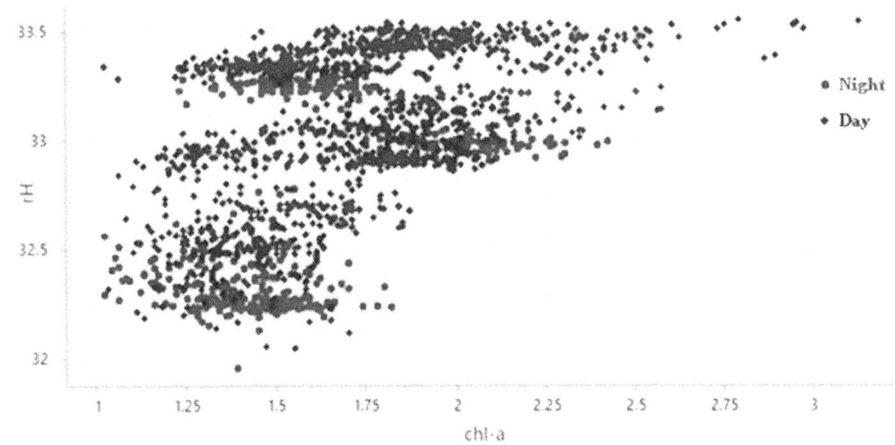

As shown in our previous study (Cretescu et al., 2022b), the Eh and rH values calculated in the daylight experiment on the Sulina Chanel varied in the average ranges, as shown in Table 1.

Table 1. The average values for Eh and rH calculated using the ORP data

ORP, mV*	Eh, mV**	rH**
285 + 1	504 ± 1	33.30 ± 0.2

*Average measured value during the cruise; **Average calculated value

The values measured at different points were exceedingly close to each other, indicating that the changes registered during daylight hours were insignificant. The oxidizing species from natural water prevailed over the reducing ones, as indicated by the value of rH over 28.

Preliminarily, it was assumed that the primary oxidizing agent from the water was the DO. However, a decrease in DO from 5 mg O_2/L during the cruise to 3.8 mg O_2/L in the stationary nighttime (a difference of 1.2 mg O_2/L) noticed in the measurements suggests the reduced photosynthetic activity of the algae and the working functions of the aquatic microorganisms, especially on the benthic ones. Combining this observation with the almost constant values of Eh and rH suggests the occurrence of oxidizing species other than oxygen and peroxide radicals, for instance, possibly emitted by several biological processes, primarily due to hydrobiont metabolism.

The comparison of the actual study results in a somewhat similar case study of the Dniester River (Borodaev, 2008; Borodaev, 2002; Borodaev & Duca, 2002; Gladchi et al., 2008) performed at lower water depths (about 0.2 m), where the ORP was also measured. We noted a similar REDOX status for the Dniester River (Borodaev et al., 2019) as for the Danube River. However, the calculated values of the E_h based on the measured ORP values were significantly higher than the calculated ones, which suggests that the REDOX processes that take place at 0.2 m deepness in the water from the Dniester are different from these corresponding to the Sulina Channel at a depth of around 1-1.5 m. This aspect suggests that other species of peroxide types are supposed to form in natural water by complex biologic mechanisms at different depths, strongly influenced by solar radiation.

The REDOX processes in surface water are due to the presence of several metal ions (Fe, Cu, etc.) acting as catalysts. Interaction with solar radiation generates chemical species with oxidizing characteristics (O_2, H_2O_2, $OH^{\bullet}$). The reducing substances involved in the oxidation processes could also play the role of ligands, generating complex combinations with metal ions. The ligand-bearing species can be oxidized by hydrogen peroxide in the presence of Cu(II) ions, generating $OH^{\bullet}$ radicals, a powerful oxidizing species involved in the self-purification processes of surface waters. If the organic compounds in water have no complexation capacity, the hydrogen peroxide will be consumed without the generation of $OH^{\bullet}$ radicals. Therefore, a quasi-reducing character of surface waters is established, a characteristic of polluted waters. Chemical species with intrinsic reducing characteristics have an adverse impact on the ecotoxicity of water (depending on the process type and involved toxic species, as shown in Table 2). Their interaction with hydroxyl $OH^{\bullet}$ radicals leads to a decrease in the self-purification potential of the surface waters.

Table 2. ORP changes for chemical species involved in different process.

Biochemical process: (YSI, n.d)	**ORP, mV**
Nitrification	+ 100 to + 350
BOD degradation with free molecular oxygen	+ 50 to + 250
Biological phosphorus removal	+ 25 to + 250
Denitrification	+ 50 to - 50
Sulphide (H_2S) formation	- 100 to - 250
Acid formation (fermentation)	- 100 to - 225
Methane production	- 175 to - 400
Chemical process: (Grieve, 2021; Cvetković et al. 2008)	
Normal tap water	+ 400 to + 500
Reduced water	- 100 to - 350
Oxidized water	+ 700 to + 800

SOLUTIONS AND RECOMMANDATIONS

The proposed solution for the automated online monitoring of water quality is related to the above-presented concept, which is still in the early stage of development and requires more inputs for the full validation of the results, taking into consideration that are many additional factors that could influence the functionality of the system. For example, one of the most common such aspects refers to the influence of the temperature on ORP sensors which should be further explored, as it is known that such sensors are based on the Nernst equation. The solution for solving this problem is the automated compensation of the temperature by involving a temperature sensor and using dedicated software, respectively, as it is proposed for the actual system. Anyhow, it is recommended that the REDOX measurements be performed in thermostatic electrochemical cells (where the temperature variation is not more than a few degrees).

Also, applying the periodic cleaning of the electrodes system, followed by its calibration with a standardized solution performed at an average value of the temperature for the immediately following period is highly recommended, as described in EPA recommendation (Grieve, 2021). Even if instrumentation for ORP measuring is simple and cheap, there are some limitations, which should be considered for practical applications such as measuring the dissolved hydrogen gas in water, at different values of pH and temperature. All of these aspects were carefully pointed out in a rigorous study (LeBaron & Sharpe, 2022), followed by practical recommendations regarding the use of ORP, ORP-based hydrogen meters, and accurate methods for measuring dissolved hydrogen gas levels in the water.

FUTURE RESEARCH DIRECTIONS

The obtained preliminary experimental results are promising for the main proposed objective, even if more investigations are needed in order to implement such kind of monitoring system in daily practice and make it widespread application, practical, and cost-effective an automated online monitoring system for water quality assessment in real-time. Future research needs to refer to the study of the developed monitoring system feasibility in rapport with each water quality parameter approached by the system. Taking into consideration that the developed system is based on both optical and electrochemical sensors, should be considered the known drawbacks for each such kind of sensor. Once such aspects are overcome, the automated monitoring system will become indispensable for its use in support of better water governance.

CONCLUSION

1. The monitoring of the dissolved oxygen (DO) concentration, REDOX potential (ORP), and pH values measured in stationary and dynamic conditions at around 1-1.5 m underwater depth, corrected against the temperature, allowed the calculation of Eh and rH for the water flow of the Sulina Channel of the Danube River before reaching the Black Sea. The calculations allowed an evaluation of the ecotoxicity status of the Sulina Channel, revealing the presence of clean water in which oxidizer species were predominant. Together with dissolved oxygen, the presence of other oxidation species naturally produced in the flowing river water (most likely, hydrogen peroxide of biological origin) was suggested by the values of Eh. Confirming this supposition would be possible when combining the ongoing hydro-monitoring with additional studies.
2. The established REDOX state of surface waters is a result of the balance between the oxidizing and the reducing compounds, being an indication to characterize the biocenosis condition from the analyzed biotope, which could be either in an undisturbed state (homeostasis) or disturbed by stress factors (such as chemical or physical pollution, etc.), reversible or irreversible.
3. Even when the water chemical state indicated by individual quality indicator values in a particular water body is situated within limits provided by the environmental legislation, the certainty about its ecological state must also consider several biological aspects connected to specific conditions required for aquatic life, which can simply be quantified through ORP measurements.

ACKNOWLEDGMENT

This study is part of a partnership project funded by the Romanian National Authority for Research and Innovation, CCCDI - UEFISCDI, project number: 107/2019, Cod: COFUND-WW2017 -WATER HARMONY, within PNCDI III and EU project (Horizon 2020): “Closing the Water Cycle Gap – Sustainable Management of Water Resources”.

REFERENCES

Borodaev, R. I. (2002). *Redox conditions of the lower Dniester and the state of dissolved forms of iron and copper. Analele ştiinţifice ale Universităţii de Stat din Moldova*. Chemical-Biological Sciences.

Borodaev, R. I. (2008). *Redox potential of the Dniester water as an indicator of its biological valorization*. In Proceeding of International Conference, Managementul bazinul transfrontalier al fluviului Nistru şi Directiva - cadru a apelor a Uniunii Europene (pp. 38-41). Chisinau, Republic of Moldova.

Borodaev, R. I., Bunduchi, E. G., & Goreacheva, N. V. (2019). *On the mechanism of "Quasi Toxicity" of the photic layer of the waters from the Dniester basin*. In International scientific conference, Achievements in studies of marginal effect in water ecosystems and their practical significance, (p. 14). Odessa, Ukraine.

Borodaev, R. I., & Duca, G. G. (2002). *Characteristics of the redox conditions of the lower Dniester*. In Collection of abstracts of reports of the All-Russian scientific youth symposium, Biosphere Safety–2001/2002 (p. 2). Ekaterinburg, Russia.

Coelho, C., Heim, B., Foerster, S., Brosinsky, A., & De Araújo, J. C. (2017). In situ and satellite observation of CDOM and chlorophyll-a dynamics in small water surface reservoirs in the Brazilian semiarid region. *Water (Basel)*, *9*(12), 913. doi:10.3390/w9120913

Cretescu, I., Borodaev, R. I, Lutic, D., Soreanu, G., Lazar, L., Gitlestad, J., & Zubcov, E. (2022b). Redox state assessment for the Danube water in the Sulina channel using a mobile water quality monitoring station. *Annals of "Dunarea de Jos" University of Galati, Mathematics, Physics, Theoretical Mechanics*, Fascicle II, Year XIV (XLV), 2, 122-125. University of Galati.

Cretescu, I., Kovacs, Z., Lazar, L., Burada, A., Sbarcea, M., Teodorof, L., Padure, D., & Soreanu, G. (2022a). Danube Delta: Water Management on the Sulina Channel in the Frame of Environmental Sustainability. Chapter. In A. J. Manning (Ed.), *River Deltas Research*. IntechOpen Limited. doi:10.5772/intechopen.97877

Cretescu, I., Soreanu, G., & Dragoi, E. (2022c). *Patent RO135089/2022*. Surface water quality monitoring system located on a passenger ship.

Cvetković, S., Purenović, J. M., & Jovićević, J. N. (2008). Change of water redox potential, pH and rH in contact with magnesium-enriched kaolinite–bentonite ceramics. *Applied Clay Science*, *38*(3-4), 268–278. doi:10.1016/j.clay.2007.03.009

Duca, Gh., Gladchi, V., & Romnaciuc, L. (2002). *Processes of pollution and self-purification of natural waters*. Chisinau, Republic of Moldova: CE USM.

Duca, Gh., Skurlatov, Iu., Misiti, A., Macoveanu, M., & Surpăţeanu, M. (2003). *Ecological Chemistry*. Chisinau, Republic of Moldova: CE USM.

Duca, G., & Vaseashta, A. (Eds.). (2020). *Handbook of Research on Emerging Developments and Environmental Impacts of Ecological Chemistry*. IGI Global. doi:10.4018/978-1-7998-1241-8

EBPI. (n.d.). Algae Toxicity Tests. *Biotoxicity*. https://www.biotoxicity.com/index.php/microbiotests/algae-toxicity-tests/fresh-water

Ene, A., Zubcov, E., Bulat, D., Bulat, D., Cristea, V., & Florea, L. Gh. (2021). Pollutants and biodiversity in the Prut River. HydroEcoNex Project Report, Public information seminar, Vladesti, Galati county. *Romania*. Advance online publication. doi:10.13140/RG.2.2.28146.02247

Gladchi, V., Goreaceva, N., Duca, Gh., Bunduchi, E., Borodaev, R., Şurîghina, O., & Lis, A. (2014). Chemical composition of right bank tributaries of Dniester River and their general impact. Chapter in Gh. Duca (Ed.), Management of Water Quality in Moldova (pp. 81-96). Water Sciences and Technology Library, vol. 69, Springer, Cham. https://doi.org/ doi:10.1007/978-3-319-02708-1_6

Gladchi, V., Goreaceva, N., Duca, G., Bunduchi, E., Romanciuc, L., Mardari, I., & Borodaev, R. (2008). The study of redox conditions in the Dniester River. *Chem J Moldova*, *3*(1), 70–76. doi:10.19261/cjm.2008.03(1).14

Gregorin, C., Albarano, L., Somma, E., Costantini, M., & Zupo, V. (2021). Assessing the ecotoxicity of copper and polycyclic aromatic hydrocarbons: Comparison of effects on *Paracentrotus lividus* and *Botryllus schlosseri*, as alternative bioassay methods. *Water (Basel)*, *13*(5), 711. doi:10.3390/w13050711

Grieve, M. (2021). *Field Measurement of Oxidation Reduction Potential*. U.S. Environmental Protection Agency, Laboratory Services and Applied Science Division.

Ji, M., Zhang, J., Li, S., Mu, G., & Di, H. (2019). Evaluating CDOM sources using excitation-emission matrix fluorescence and parallel factor analysis, and their links to water quality in highly polluted rivers in China. *Polish Journal of Environmental Studies*, *28*(3), 1203–1214. doi:10.15244/pjoes/85951

Jurminskaia, O., Ene, A., & Zubcov, E. (2021). *Monitoring of aquatic ecosystems based on biological parameters*. Conference: International Conference, Bender, Republic of Moldova.

LeBaron, T. W., & Sharpe, R. (2022). ORP should not be used to estimate or compare concentrations of aqueous H_2: An in-silico analysis and narrative synopsis. *Front. Food. Sci. Technol.*, *2*, 1007001. doi:10.3389/frfst.2022.1007001

Nair, A., Hykkerud, A., & Ratnaweera, H. (2022). Estimating Phosphorus and COD Concentrations Using a Hybrid Soft Sensor: A Case Study in a Norwegian Municipal Wastewater Treatment Plant. *Water (Basel)*, *14*(3), 332. doi:10.3390/w14030332

Nikanorov, A. M. (2000). *Hydrochemistry*. Gidrometeoizdat.

Ozcan, A., & Ogun, M. (2014), Biochemistry of reactive oxygen and nitrogen species. Chapter in Sivakumar Joghi Thatha Gowder (Ed.). *Basic Principles and Clinical Significance of Oxidative Stress*. IntechOpen, DOI:, https://www.intechopen.com/chapters/48975 doi:10.5772/61193

Richard, L., Peake, B., Rusak, S., Cooper, W., & Burritt, D. (2007). Production and decomposition dynamics of hydrogen peroxide in freshwater. *Environmental Chemistry*, *4*(1), 49–54. doi:10.1071/EN06068

Tofan, L., Paduraru, C., Sluser, B., & Teodosiu, C. (2019). *A small encyclopedia of water*. Ed. EcoZone. (in Romanian)

Travin, S., Duca, G., & Gladchi, V. (2019). Self-purification of Aquatic Media from Hexachlorocyclohexane in a Radical Process. *Chem J Moldova*, *14*(1), 47–53. doi:10.19261/cjm.2018.537

Vaseashta, A., Duca, G., & Travin, S. (Eds.). (2022). *Handbook of Research on Water Sciences and Society* (Vols. 1–2). IGI Global. doi:10.4018/978-1-7998-7356-3

Chapter 9
Modification of Waste Biomass Digestion in the Presence of Additives of Bioactive Substances

Gheorghe Duca
Institute of Chemistry, Moldova State University, Moldova

Ecaterina Covaci
Technical University of Moldova, Moldova

Rodica Sturza
https://orcid.org/0000-0002-2412-5874
Technical University of Moldova, Moldova

Lidia Romanciuc
https://orcid.org/0000-0002-4555-1194
Institute of Chemistry, Moldova State University, Moldova

Olga Covaliova
Institute of Chemistry, Moldova State University, Moldova

Corina Tasca
Institute of Chemistry, Moldova State University, Moldova

ABSTRACT

Waste management in the agro-industrial sector is an important issue as it offers ways to prevent the toxic discharges of harmful pollutants into the environment, and at the same time to obtain a series of value-added products. Liquid organic wastes from the agro-industrial sector are usually in the state of continuing digestion and are capable of rendering toxic effects on plants and living organisms. Conventional technologies require a lot of treatment time and large volumes of reservoirs. The aim of this research was to intensify anaerobic fermentation using the additives of bioactive substances introduced directly into the digested biomass. The results obtained testify that substances of natural origin used as additives demonstrate pronounced effects on the alcoholic fermentation of vinasse under mesophilic conditions. The comparative assessment of different additives' action in the studied processes has demonstrated that dihydrofumaric acid caused the emission of 266 cm3 CO2 during 76 hours, aescinum – 251 cm3 during 55 hours, tomatin – 233 cm3 during 78 hours of fermentation.

DOI: 10.4018/979-8-3693-0512-6.ch009

INTRODUCTION

Waste management from the agro-industrial sector, aimed at the prevention of environmental pollution with toxic components and at the same time, obtaining value-added products, is an important issue requiring a thorough and complex approach. The waste treatment technology should be selected proceeding from the initial wastes' nature, composition, and amounts, which, in their turn, are functions on the main production cycle, raw materials specifics, process technology, and conditions.

This is especially relevant for the agricultural regions having the wine, spirit, beer, and juice-producing industries, where the liquid wastes are generated in the state of continued digestion, which implies the strict prevention of their discharges/dumping directly into the landfills, natural water bodies or other environmental compartments as they can destroy the chemical composition of the soil, violate the natural balance of microorganisms, plants, and other living organisms, including the negative effects on human health (Duca, 2017; Duca et al, 2021, Vaseashta et al, 2022).

Existing methods of solid organic waste management involve their application in agriculture, burning (although the rather costly process associated with air emissions), anaerobic digestion, composting, etc. The most common methods for liquid waste treatment from the agro-industrial sector are sedimentation, decanting in stabilization ponds, anaerobic fermentation, etc.

For ethanol production from carbohydrate-containing raw materials, the digestion process is applied commercially (Busic et al., 2018). Alcohol is obtained by fermentation of the sugars present in different raw materials, which are metabolized by microorganisms belonging to the genera *Saccharomyces, Zymomonas, Kluyveromyces,* and *Zygosaccharomyces* (Moreno-Arribas and Polo, 2005). Structural carbohydrates in biomass components are broken down into sugars with the help of various enzymes. Then, the sugars released are transformed by microorganisms into alcohols, hydrocarbons, or some organic acids during the fermentation process under atmospheric pressure and a temperature of 25-70°C. Intermediate sugars can be also utilized to obtain certain useful chemicals (Binder and Rainer, 2010).

Currently, the operation of numerous processing industries in agriculture, including those dealing with grain processing into alcohol, remains unsafe for the environment and living organisms (Christofoletti et al., 2013; Grossi-Botelho et al., 2012). Meanwhile, closed ecologically safe production cycles will make it possible to resolve fundamental issues such as the rational use of natural raw resources, environmental protection, and improving the quality of the final product (Hidalgo et al., 2019). The alcohol industry affects first the water resources, then air and soil.

Specific consumption of fresh water makes 195-325 m^3/t alcohol (10-17 L water/ 1 L alcohol produced), depending on the raw material type and processing technology, namely, recycled water supply (Wu et al., 2009; Liu et al., 2019). Such high amounts of water used trigger the high volumes of liquid wastes thus formed (Mikucka and Zielińska 2010; Gerbens-Leenes and Hoekstra, 2012; Vaseashta et al, 2021). The admixtures contained in the waste waters contain mineral and organic substances of vegetable origin. Such waste is poorly filtrated, quickly rotting, and releasing unpleasant odors.

Therefore, the indicator of technology efficiency is important, as it determines the degree of transformation of the initial raw material. The amount of substrate that can be transformed in ethanol by microbiological synthesis depends first of all on the type of raw material and selected production technology (Matkovsky et al., 2010). Basically, from 5 to 60% of the substrate, on average 32-36%, depending on its type and ethanol production method, are transformed into production wastes (Pilarski et al., 2021). Thus, the conversion degree of glucose in ethanol makes 95-100%, fructose – 95-100%, xylose – 60-90%, sucrose – 94-100%, hemicellulose – 40-81%, cellulose – 90-100%. The wastes generated from alcohol production include yeasts, vinasse, carbon dioxide, ester-formaldehyde fraction, and fuel oils (0.94 kg/t grain). Distilled grain residue is the main solid waste that contains carbohydrates, proteins, lipids, and some microbial metabolites, which is used as feed, fertilizer, or culture substrate for edible mushrooms (Wang et al., 2013).

Among these residues, the main type is alcohol vinasse (or post-distillery vinasse), which is the biomass separated in the distillation process and represents a liquid with suspended particles, acid pH, and high contents in organic matter. Generally, the vinasse production makes about 10-15 L vinasse/L ethanol (Ferraresi-de Arraujo and Nino-Castillo, 2021). Vinasse is a liquid solution containing suspended particles, with acid pH The liquid wastes from food industries have high amounts of organic matter, and, as a consequence, rather high contents of chemical oxygen demand (COD), biochemical oxygen demand (BOD), and suspended solids (SS) values: wastes from alcohol production from grain have on average 130-160 g O_2/l BOD, 200-220 g O_2/l COD, 340-360 g/l SS and pH=6.5-8.8, whereas the liquid wastes from sugar industry (sugar beet) have around 2370 g O_2/l BOD, 7540 g O_2/l COD, 21320 g/l SS and pH=8.0 (Araujo and Oliveira, 2012).

In some cases, vinasse can be discharged to a sewerage system, reaching the wastewater treatment plant, thus significantly impairing its operation. Vinasse discharges into the environment, as a rule, cause serious environmental problems. As an alternative, it could be used as a liquid food additive for cattle or poultry (López-Campos et al., 2010; Stemme et al., 2005; De Oliveira, 2013), biogas production (Van Lier, 2001; Cruz-Salomón et al., 2017; Parsaee et al., 2019) or for fertilization of agricultural crops (Arsanoglu and Tumen, 2012; Tejada and Gonzalez, 2006; Moran-Salazar et al., 2016). However, the direct suing of vinasse for animals is not always safe or rational and cannot resolve the issue of its complete utilization. A series of standard methods have been developed to evaluate the generation of byproducts and residues in the alcohol production cycle, including the entire supply chain, and minimize negative impacts from crop establishment to consumption of fuel ethanol (Miret et al., 2015; Mikulski and Kłosowski, 2022; Gallejones et al., 2014). High water contents in organic wastes with high fermentable sugar contents and wet biomass imply their obligatory treatment or recovery which is however rather costly (España-Gamboa et al., 2011). Therefore, the development of useful valorization of this waste to obtain other value-added products is a perspective approach.

To break down the waste biomass into sugars that can be further converted into gas or liquid biofuels, or other useful products, biochemical conversion by microorganisms and enzymes are often used (Choeisai, 2014; Ferreira et al., 2010; Ordaz-Diaz and Bailon-Salas, 2020), and the other approaches such as coagulation and oxidation have been also proposed (Guerreiro, 2016). Generally, biochemical digestion is a rather slow process that requires some time for biomass transformation into products. Among these technologies, anaerobic digestion and fermentation are the most popular processes, resulting in producing the biogas which mainly contains biomethane, carbon dioxide, and in smaller amounts biohydrogen, and hydrogen sulfide (Moraes et al., 2014; Adekunle and Okolie, 2-15; Deublein and Steinhauser, 2008).

During anaerobic digestion, bacteria are applied to hydrolyze carbohydrates into the sugars that are digestible by other, methanogenic bacteria, which in their turn, produce biogas from the digestible components of biomass. In dependence on the biochemical digestion conditions, biogas with high contents of methane (Asam et al., 2011; Nevers et al., 2006; Gunaseelan, 1997) or hydrogen (Kamaraj et al., 2020; Karapinar and Kargi, 2006) may be produced under the oxygen-free conditions, or under the elevated pressure (Siciliano et al., 2022; Lindeboom, 2011), or using the other novel approaches such as adding of bioactive compounds accelerating the biochemical digestion and making it possible to obtain much higher amounts of biomethane of biohydrogen per shorter time (Covaliov, 2015; Duca et al., 2020; Covaliov et al, 2015). Biogas can be burnt to produce heat or can be converted to energy, or both energy and heat, using cogeneration devices (Dalpaz, 2020; Lobato et al., 2013).

However, not only the biofuels but also a series of value-added products such as sugar alcohols (xylitol) can be obtained by using combined technologies such as chemical or biochemical processes combined with the thermochemical method (Arcaño et al., 2020; Audemar et al., 2020)]. Production of alcoholic and non-alcoholic beverages generates the wastes and by-products that may be recovered, thus allowing not only to reduce of their disposal costs and minimizing environmental pollution but also to develop the new useful products, in addition to the traditional uses of such wastes as animal feed or soil fertilizers. Dumping in the environment of such residues can provoke problems such as rotting, higher soil acidity, phytotoxicity, methane gas production, etc. (Velásquez-Riaño et al., 2019).

Wine and juice production is a large source of waste in agriculture (Devesa-Rey et al., 2011). Grape marc is generated in the processing of grapes in rather large amounts in Moldova (Musteata et al., 2021). Bioactive compounds are among the potential value-added products that can be extracted from organic wastes. They could be isolated and structurally investigated, and subsequently used for the development of innovative products (Ghendov-Mosanu et al., 2022; Ferri et al., 2020). Thus, the grape marc along with the other solid and liquid wine-production residues is used for feeding cattle and poultry in the agricultural sector, although having a low nutrition value (Yu and Ahmedna, 2013; Moate et al., 2014; Bustamante et al., 2008). According to previous studies, the grape marc is a prospective raw material for obtaining of series of bioactive substances, specifically, fatty acids, pectins, phenols, etc. to be further used in food production, supplements, medical products, cosmetics, colorants (Teixeira et al., 2014; Spinei and Oroian, 2021; Brazinha et al., 2014; Rondeau et al., 2013).

According to numerous works, polyphenols and other components contained in grapes, wine or juice have beneficial effects on human health, such as inhibiting the development of certain cancer cells acting as free radical scavengers (Nerantzis and Tataridis, 2006), antihyperglycemic, cardioprotective, anti-hyperlipidemic, etc. effects (Arvanitoyannis et al., 2006; Chakka and Babu, 2020). Apart from the phenolic compounds, grape marc also contains significant amount of such potentially useful substances, some of them having pronounced antioxidant activity as catechin, epicatechin, hydroxytyrosol, tyrosol, cyaniding glycosides, a series of acids like gallic, caffeic, procatechinic, syrignic, vanillic, *o*-coumaric, *p*-coumaric (Negro et al., 2003). A solid-liquid extraction method (Pinelo et al., 2007; Azmir et al., 2013), hydrothermal treatment (Bassani et al., 2020), and other approaches have been applied to extract the bioactive components from grape pomace, considering the fluctuating composition of its solid fraction for further using in food production.

As waste biomass from the agro-industrial sector can serve as a raw material for not only energy production, but also for food and animal feed production, as a source of bioactive and other substances with useful properties for industry, its conversion efficiency needs to be improved. For this reason, current research is focused on increasing product yields, improving biochemical conversion technologies, increasing the conversion degree of biomass, and improving the entire biochemical conversion system. Biochemical digestion of waste biomass using microorganisms has an important advantage in that no high temperatures are needed (Covaliov et al, 2014). In addition, the wet biomass of organic wastes has high water content, making it necessary to apply the effluent treatment.

Our studies are focused on the investigation of an original approach in enhancing the conversion rate and conversion degree of organic wastes biomass, using the small amounts of biologically active substances introduced into the fermented mixture, to accelerate the process rate and obtain the valuable products and, at the same time, to prevent the discharges of harmful mixtures into the environment. It was of special interest to study the effect of various additives of natural vegetal origin, introduced into the digested mixture of various types of biomasses.

MATERIALS AND METHODS

The scope of the work was to elaborate processes of liquid waste treatment from the agri-food sector, obtaining value-added products and avoiding discharges into the environment of toxic substances which may provoke unfavorable effects on soil, water, air, and living organisms.

The following instruments and equipment were used for the research: thermostat PURA, digital pH-meter PH-3CU, refractometer Brix, titration equipment, and bench-scale biochemical reactor. The studies of the effect of biologically active substances (BAS) on the digestion process have been performed using the vinasse resulting from the grains distillation at the "Garma Grup" company (Hâncești, Republic of Moldova). Also, other types of wastes biomass have been taken for the research, such as waste biomass from corn distillation.

Bench-scale technological experiments have been performed using the laboratory set-up including four parallel closed-up glass reactors in which the nutrition mixture and various types of bioactive additives have been introduced. The nutrition mixture has been prepared using 30 ml fresh grape juice with 10% fermentation yeast/ 10^{10} CFUG), 20 ml inversed sugar (42%), 150 ml vinasse, and 2 ml of BAS (3 g/L).

The biomass mixed with the nutrition mixture was left for alcohol fermentation under mesophilic conditions (20-32°C). The digestion process was followed by the volume of emitted gas (CO_2) which replaced NaOH, determined by standard method (titration with phenolphthalein). The composition and characteristics of the initial vinasse used to study the fermentation process in the presence of bioactive additives is shown in Table 1 and 2.

Table. 1. Composition of initial vinasse (Scull et al., 2012; De Godoi, 2019).*

	Component	*Corn vinasse*	*Grain vinasse*	*Barley vinasse*	*Rye vinasse*
1.	Dry matter, %	8.5	4.2	26.0	8.0
2.	Proteins (% of dry matter)	25.5	34.8	31.3	42.4
3.	Lipids (% of dry matter)	11.7	2.2	10.2	3.5
4.	Fibers (% of dry matter)	10.6	3.4	13.7	5.9
5.	Ash (% of dry matter)	4.7	8.6	2.1	3.5

* P≥0,95

Table. 2. Physico-chemical indices of studied biomass

	Indicator	*Vinasse*	*Grape juice*
1.	Sugar content, g/L	26.27 ± 0.91	124.4 ± 1.6
2.	pH	3.91 ± 0,07	3.77± 0.12
3.	Titratable acidity, g/L sulfuric acid	3.417± 0.241	2.74± 0.322
4.	Contents in amine nitrogen, mg/L	492.2± 0.63	140.8± 1.2

RESULTS AND DISCUSSION

The digestion process was studied with the introduction of biologically active substances of natural origin into the fermented biomass, which included the waste biomass and nutrition mixture. As can be seen from Figure 1, additives of aecinum and betuline suppress the CO_2 emission, whereas using tomatine ensures a higher amount of gas evolved, or a more intensive fermentation process, as compared to the witness test.

Figure 1. Kinetics of carbon dioxide (CO_2) emission during the vinasse alcoholic fermentation in the presence of 0,006 g additives /L

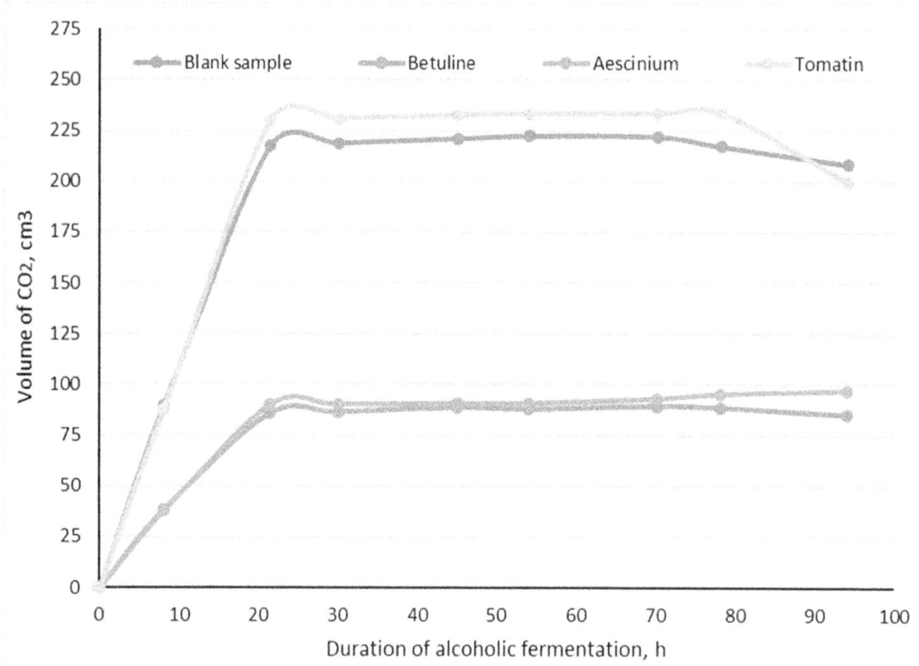

Practical experiences have shown that the introduction of sclareol accelerates biomass digestion, whereas addition of aescinum suppresses this process (Figure 2). Unlike these two bioactive substances, menthol at the first stages of digestion significantly suppresses the CO_2 emission, and after 30 hours of fermentation, it makes this process somewhat more intensive as compared to the witness test.

At the same time, introduction into the alcohol fermentation of sclareol and menthol in different amounts (0.003-0.012 g/L biomass) has shown that the amount of additive introduced has no significant effect on the fermentation rate, although the higher amounts (0.012 g/L) of both additives did not demonstrate higher efficiency; on the opposite, the fermentation rate was lower (Figure 3). As compared to menthol, sclareol provokes more intensive fermentation at a much lower concentration.

In the presence of dihydroxyfumaric acid, the vinasse fermentation proceeds somewhat more easily as compared to the witness test, being approximately similar in the case of the introduction of additive in various concentrations ($0.1013 \div 0.4052 \bullet 10^{-6}$ mol/L), although with the lesser amount of addition ($0.1013 \bullet 10^{-6}$ mol/L) the fermentation was accelerated significantly (Figure 4).

Figure 2. Kinetics of the vinasse fermentation in the presence of additives of aescinum, menthol, and sclareol of 0.006 g /L biomass

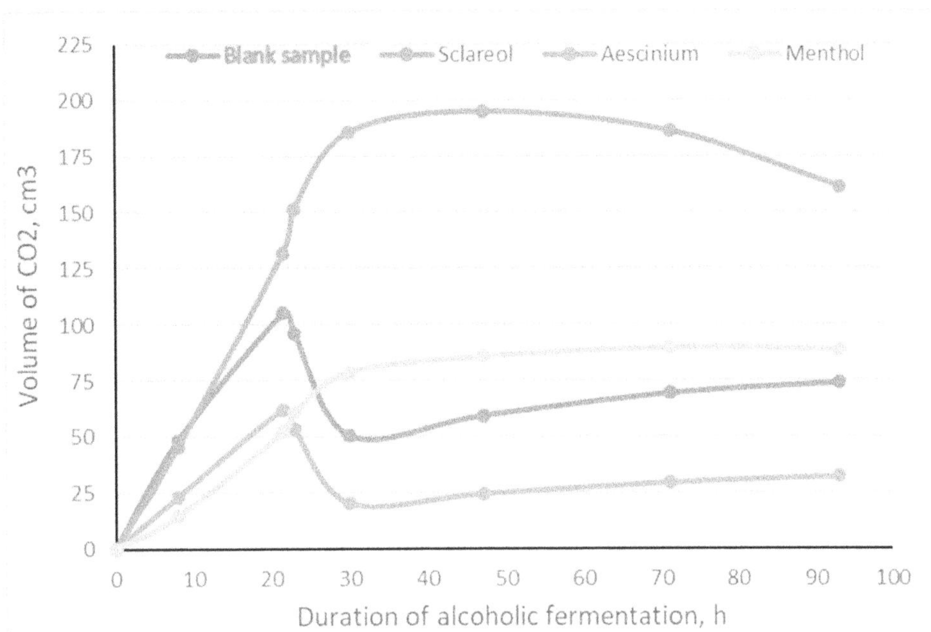

Figure 3. Kinetics of the vinasse fermentation in the presence of sclareol and menthol additions

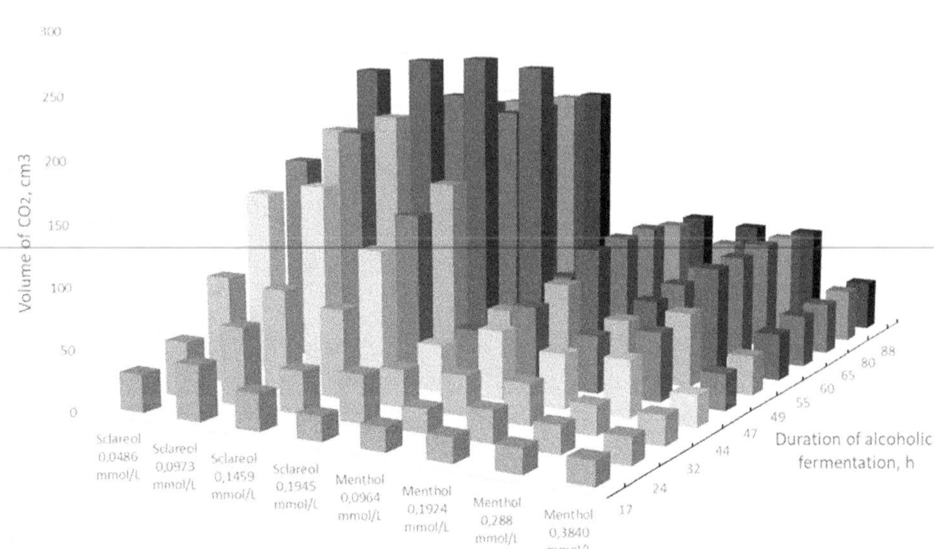

Figure 4. Reaction rate dynamics of the vinasse alcoholic fermentation in the presence of dihydroxyfumaric acid addiction

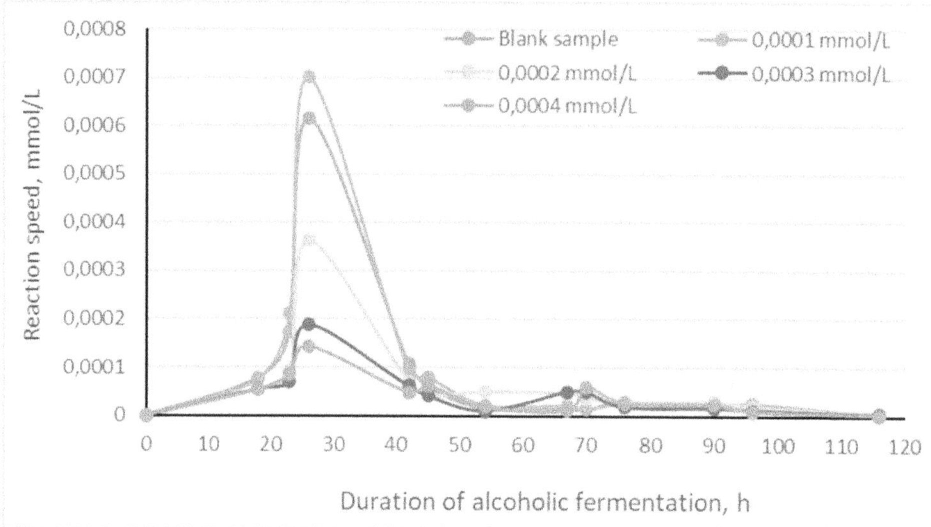

In the case of vanillin introduction into the fermented mixture, likewise, the fermentation process was suppressed with the introduction of a higher amount of additive ($0.19 \div 0.39 \bullet 10^{-3}$ mol/L), whereas the insignificant acceleration of carbon dioxide emission, as compared to the witness test, was observed in case of the using of the lesser amount (0.098 mmol/L) of vanillin addition (Figure 5).

Figure 5. Kinetics of the vinasse fermentation in the presence of vanillin additive

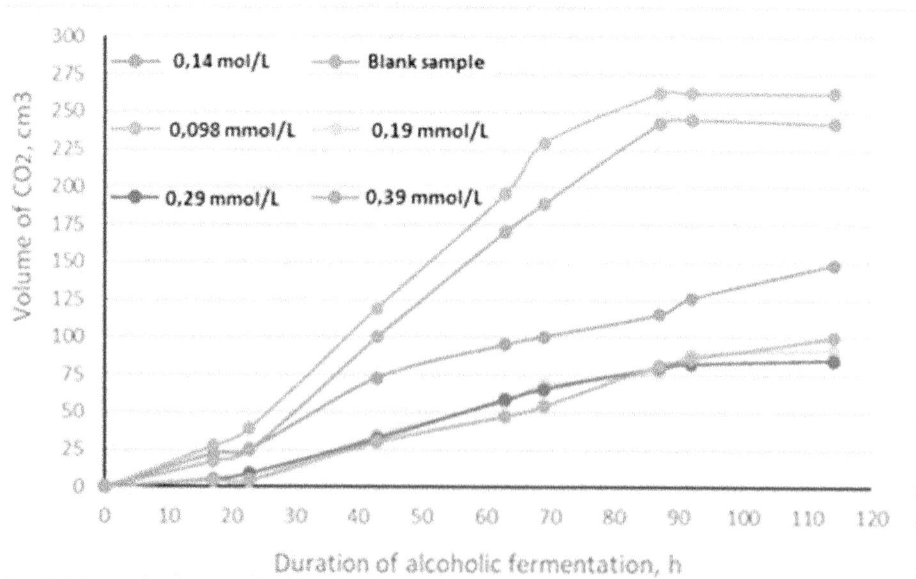

Different amounts of catechin introduced in biomass have also shown a difference in carbon dioxide gas emission: smaller amounts of additive provoked the higher efficiency of fermentation, and the higher concentration made the process less intensive as compared to the witness test (Figure 6).

Figure 6. Kinetics of vinasse fermentation in the presence of catechin additive

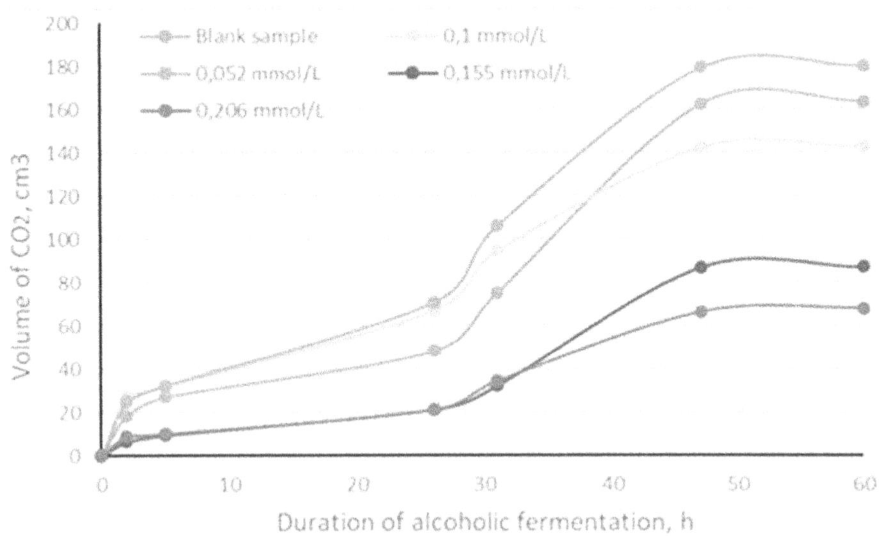

The results obtained demonstrate that the substances of natural origin with bioactive / antioxidant properties show pronounced effects on the alcohol fermentation of vinasse under mesophilic conditions. The most efficient concentrations of studied additives are summarized in Table 3 and Figure 7.

Table 3. Comparative efficiency of different types of additives of bioactive substances in the vinasse fermentation process at a concentration of 0.003 g/L biomass.

Nr. crt.	Bioactive substance used as an additive to the fermented biomass	Total volume of CO_2 emitted gas, cm^3	Fermentation time, h	Average amount of emitted CO_2 gas per hour, cm^3/h
1.	Dihydroxyfumaric acid	266.00	76	3.50
2.	Aescinum	251.01	55	4.56
3.	Tomatin	233.46	78	2.99
4.	Sclareol	232.50	55	4.23
5.	Vanillin	229.00	69	3.32
6.	Catechin	180.00	61	2.95
7.	Betuline	250.00	80	3.13
8.	Menthol	200.00	70	2.86

Figure 7. The average amount of emitted CO_2 gas per hour, cm^3/h independent on the type of bioactive additive introduced into the digested biomass in the amount of 0.003 g/L biomass

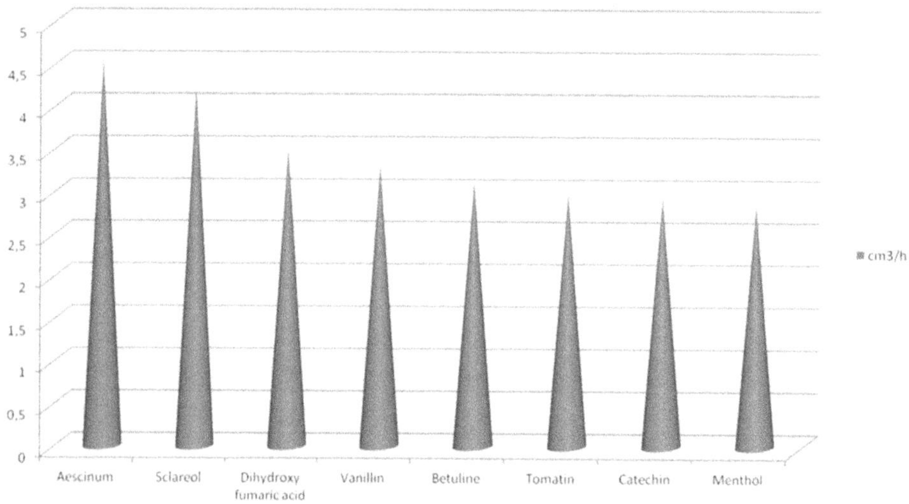

The highest average amount of emitted carbon dioxide gas per hour was reached in the case of aescinum and scareol introduced in the amount of 0.003 g/L biomass, followed by dihydroxyfumaric acid, vanillin, and other studied additives (Figure 7). The results also demonstrate that introduction of higher compounds of additives into the fermented biomass is not rational, as this generally does not accelerate the fermentation process. However, this effect could be useful in case there is a need to suppress the biomass digestion process. It can be suggested that the ability of studied additives to accelerate the fermentation of liquid wastes from the agro-industrial sector relates to their molecular structure and antioxidant, antihypoxant, antimutagen, etc. properties.

CONCLUSION

The study of the influence of additives of biologically active substances on waste biomass fermentation from the agro-industrial sector, along with the revealing of these processes' mechanisms can open a new perspective research direction in this area. This would allow not only to manage of the fermentation processes, accelerating or suppressing them for the specific scopes but also to obtain a series of value-added products under more favorable conditions, as well as to prevent the discharge of harmful wastes into the environment. It was found that various additive types of biologically active substances, being introduced into the biomass in the process of aerobic/alcohol digestion, demonstrate different behavior, accelerating or inhibiting the digestion. At the same time, it became obvious that the introduction of higher amounts of additives in many cases makes no sense as it does not provoke further acceleration of the alcoholic fermentation process, instead, the higher amounts of studied additives were shown to suppress the biomass digestion.

Products of alcohol fermentation can have added value in agriculture and adjacent fields. Thus, carbon dioxide emitted during the biomass digestion can be used, for instance, in the microalgae cultivation basins, which in their turn, could serve as a valuable feed for poultry/cattle in the zootechnical sector, while the water separated from solid fraction can be used for technical scopes at the industries (cleaning machines and apparatuses, cleaning and washing up the production halls and territory, etc.), or watering the plants in the city parks, flower beds, etc. The solid fraction after the fermentation, with the condition of its controlled composition, can be used as fertilizer for technical crops.

ACKNOWLEDGMENT

This research has been performed under the Moldova National Research Project Nr. 20.80009.5007.04 "Physical-chemical redox processes with electron transfer in the vital, technological and environmental processes".

REFERENCES

Adekunle, K., & Okolie, J. (2015). A review of the biochemical process of anaerobic digestion. *Advances in Bioscience and Biotechnology*, *6*(3), 205–212. doi:10.4236/abb.2015.63020

Araujo, G.; Oliveira, S. (2021). Energy and environmental analysis of vinasse processing using internal circulation biodigesters and concentrators for different production ranges of ethanol and its potential impact on Brazil. *Environment, Development and Sustainability: A Multidisciplinary Approach to the Theory and Practice of Sustainable Development, 23*(3), 3130-3163.

Arcaño, Y. D., García, O. D., Mandelli, D., Carvalho, W. A., & Pontes, L. A. (2020). Xylitol: A review on the progress and challenges of its production by chemical route. *Catalysis Today*, *344*, 2–14. doi:10.1016/j.cattod.2018.07.060

Arsanoglu, H., & Tumen, F. (2021). Potassium struvite (slow-release fertilizer) and activated carbon production: Resource recovery from vinasse and grape marc organic waste using thermal processing. *Process Safety and Environmental Protection*, *147*, 1077–1087. doi:10.1016/j.psep.2021.01.025

Arvanitoyannis, I. S., Ladas, D., & Mavromatis, A. (2006). Potential uses and applications of treated wine waste: A review. *International Journal of Food Science & Technology*, *41*(5), 475–487. doi:10.1111/j.1365-2621.2005.01111.x

Asam, Z. U. Z., Poulsen, T. G., Nzami, A.-S., Raxique, R., Kiely, G., & Murphy, J. D. (2011). How can we improve biomethane production per unit of feedstock in a biogas plant? *Applied Energy*, *88*(6), 2013–2018. doi:10.1016/j.apenergy.2010.12.036

Audemar, M., Ramdani, W., Junhui, T., Raluca Ifrim, A., Ungureanu, A., Jérôme, F., Royer, S., & Oliveira Vigier, K. (2020). Selective hydrogenation of xylose to xylitol over Co/SiO2 catalysts. *ChemCatChem*, *12*(7), 1973–1978. doi:10.1002/cctc.201901981

Azmir, J., Zaidul, I. S. M., Rahman, M. M., Sharif, K. M., Mohamed, A., Sahena, F., & Omar, A. K. M. (2013). Techniques for extraction of bioactive compounds from plant materials: A review. *Journal of Food Engineering*, *117*(4), 426–436. doi:10.1016/j.jfoodeng.2013.01.014

Bassani, A., Alberici, N., Fiorentini, C., Giuberti, G., Dordoni, R., & Spigno, G. (2020). Hydrothermal treatment of grape skins for sugars, antioxidants and soluble fibers production. *Chemical Engineering (Albany, N.Y.)*, 79.

Binder, J., & Raines, R. T. (2010). Fermentable sugars by chemical hydrolysis of biomass. *Proceedings of the National Academy of Sciences of the USA (PNAS)*, *107*(10), 4516-4521. 10.1073/pnas.0912073107

Brazinha, C., Cadima, M., & Crespo, J. G. (2014). Optimization of extraction of bioactive compounds from different types of grape pomace produced at wineries and distilleries. *Journal of Food Science*, *79*(6), E1142–E1149. doi:10.1111/1750-3841.12476 PMID:24891032

Bušić, A., Marđetko, N., Kundas, S., Morzak, G., Belskaya, H., Ivančić Šantek, M., Komes, D., Novak, S., & Šantek, B. (2018). Bioethanol production from renewable raw materials and its separation and purification: A review. *Food Technology and Biotechnology*, *56*(3), 289–311. doi:10.17113/ftb.56.03.18.5546 PMID:30510474

Bustamante, M. A., Moral, R., Paredes, C., Pérez-Espinosa, A., Moreno-Caselles, J., & Pérez-Murcia, M. D. (2008). Agrochemical characterization of the solid by-products and residues from the winery and distillery industry. *Waste Management (New York, N.Y.)*, *28*(2), 372–380. doi:10.1016/j.wasman.2007.01.013 PMID:17433659

Chakka, A. K., & Babu, A. S. (2022). Bioactive compounds of winery by-products: Extraction techniques and their potential health benefits. *Applied Food Research*, *2*(1), 100058. doi:10.1016/j.afres.2022.100058

Choeisai, P., Jitkam, N., Silapanoraset, K., Yubolsai, C., Yoochatchava, W., Yamaguchi, T., Onodera, T., & Syutsubo, K. (2014). Sugarcane molasses-based bio-ethanol wastewater treatment by two-phase multi-staged up-flow anaerobic sludge blanket (UASB) combination with up-flow UASB and down-flow hanging sponge. *Water Science and Technology*, *69*(6), 1174–1180. doi:10.2166/wst.2013.794 PMID:24647181

Christofoletti, C. A., Escher, J. P., Correia, J. E., Marinho, J. F. U., & Fontanetti, C. S. (2013). Sugarcane vinasse: Environmental implications of its use. *Waste Management (New York, N.Y.)*, *33*(12), 2752–2761. doi:10.1016/j.wasman.2013.09.005 PMID:24084103

Covaliov, V. (2015). Improving the biogas production systems. *Academos*, *4*(39), 38–44.

Covaliov, V., Malina, J. F., Duca, Gh., Covaliova, O., & Bobeica, V. (2014). Application of Bioactive Substances in the Biogas Technology. In G. Duca (Ed.), *Management of Water Quality in Moldova*. doi:10.1007/978-3-319-02708-1_12

Covaliov, V., Senicovscaia, I., Nenno, V., Bobeica, V., & Covaliova, O. (2015). Effects of amaranths' seeds on dehydrogenase activity and gases emission in methanogenic bioreactors. *Studia Universitatis Moldaviae. Real and Natural Sciences Ser.*, *1*(81), 230–235.

Cruz-Salomón, A., Ríos-Valdovinos, E., Pola, F., Meza Gordillo, R., Lagunas Rivera, S., & Ruíz-Valdiviezo, V. (2017). Anaerobic treatment of agro-industrial wastewaters for COD removal in expanded granular sludge bed bioreactor. *Biofuel Research Journal*, *16*(4), 715–720. doi:10.18331/BRJ2017.4.4.3

Dalpaz, R., Konrad, O., Da Silva Cyrne, C., Barzotto, H. P., Hasan, C., & Guerini Filho, M. (2020). Using biogas for energy cogeneration: An analysis of electric and thermal energy generation from agro-industrial waste. *Sustainable Energy Technologies and Assessments*, *40*, 100774. doi:10.1016/j.seta.2020.100774

De Godoi, L. A. G., Camiloti, P. R., Bernardes, A. N., Sanchez, B. L. S., Torres, A. P. R., da Conceição Gomes, A., & Botta, L. S. (2019). Seasonal variation of the organic and inorganic composition of sugarcane vinasse: Main implications for its environmental uses. *Environmental Science and Pollution Research International*, *26*(28), 29267–29282. doi:10.100711356-019-06019-8 PMID:31396875

De Oliveira, M. C., da Silva, D. M., & de Almeida Faleiros, C. (2013). The effect of including liquid vinasse in the diet of rabbits on growth performance. *Revista Brasileira de Zootecnia*, *42*(4), 259–263. doi:10.1590/S1516-35982013000400005

Deublein, D., & Steinhauser, A. (2008). *Biogas from Waste and Renewable Resources: An Introduction*. Wiley-VCH. doi:10.1002/9783527621705

Devesa-Rey, R. X., Vecino, J. L., Varela-Alende, J. L., Barral, M. T., Cruz, J. M., & Moldes, A. B. (2011). Valorization of winery waste vs. the costs of not recycling. *Waste Management (New York, N.Y.)*, *31*(11), 2327–2335. doi:10.1016/j.wasman.2011.06.001 PMID:21752623

Duca, Gh. (2017). Ecological and environmental chemistry. *Chemistry Journal of Moldova. General. Industrial and Ecological Chemistry.*, *12*(1), 9–19.

Duca, Gh., Covaliov, V., & Covaliova, O. (2020). Intensive biochemical processes of wastewater treatment with high caloricity biogas production. In *Emerging Developments and Environmental Impacts of Ecological Chemistry* (pp. 291–306). IGI Global Publ. House. doi:10.4018/978-1-7998-1241-8.ch013

Duca, G., & Vaseashta, A. (Eds.). (2020). *Handbook of Research on Emerging Developments and Environmental Impacts of Ecological Chemistry*. IGI Global. doi:10.4018/978-1-7998-1241-8

España-Gamboa, E., Mijangos-Cortes, J., Barahona-Perez, L., Dominguez-Maldonado, J., Hernández-Zarate, G., & Alzate-Gaviri, L. (2011). Vinasse: Characterization and treatments. *Waste Management & Research*, *29*(12), 1235–1250. doi:10.1177/0734242X10387313 PMID:21242176

Ferraresi-de Arraujo, J. G., & Nino-Castillo, I. N. (2021). Vinasse: Current concept, challenges, and opportunities for sustainability. *ECORFAN-Journal-Republic of Nicaragua.*, *7*(13), 1–9. doi:10.35429/EJRN.2021.13.7.1.9

Ferreira, L. F., Aguiar, M., Pompeu, G., Messias, T., & Monteiro, R. T. (2010). Selection of vinasse degrading microorganisms. *World Journal of Microbiology & Biotechnology*, *26*(9), 1613–1621. doi:10.100711274-010-0337-3

Ferri, M., Vannini, M., Ehrnell, M., Eliasson, L., Xanthakis, E., Monari, S., & Tassoni, A. (2020). From winery waste to bioactive compounds and new polymeric biocomposites: A contribution to the circular economy concept. *Journal of Advanced Research*, *24*, 1–11. doi:10.1016/j.jare.2020.02.015 PMID:32181012

Gallejones, P., Pardo, G., Aizpurua, A., & Del Prado, A. (2014). Life cycle assessment of first-generation biofuels using a nitrogen crop model. *The Science of the Total Environment*, *505C*, 1191–1201. PMID:25461117

Gerbens-Leenes, W., & Hoekstra, A. Y. (2012). The water footprint of sweeteners and bioethanol. *Environment International*, *40*, 202–211. doi:10.1016/j.envint.2011.06.006 PMID:21802146

Ghendov-Mosanu, A., Cojocari, D., Balan, G., Patras, A., Lung, I., Soran, M.-L., Opris, O., Cristea, E., & Sturza, R. (2022). Chemometric optimization of biologically active compounds extraction from grape marc: Composition and antimicrobial activity. *Molecules (Basel, Switzerland)*, *27*(5), 1610. doi:10.3390/molecules27051610 PMID:35268711

Grossi-Botelho, R., Tornisielo, V. L., Alves de Olinda, R., Maranho, L. A., & Machado-Neto, L. (2012). Acute toxicity of sugarcane vinasse to aquatic organisms before and after pH adjustment. *Toxicological and Environmental Chemistry*, *94*(10), 2035–2045. doi:10.1080/02772248.2012.738516

Guerreiro, L. F., Rodrigues, C. S., Duda, R. M., de Oliveira, R. A., Boaventura, R. A., & Madeira, L. M. (2016). Treatment of sugarcane vinasse by a combination of coagulation/flocculation and Fenton's oxidation. *Journal of Environmental Management*, *181*, 237–248. doi:10.1016/j.jenvman.2016.06.027 PMID:27353374

Gunaseelan, V. N. (1997). Anaerobic digestion of biomass for methane production: A review. *Biomass and Bioenergy*, *13*(1-2), 83–144. doi:10.1016/S0961-9534(97)00020-2

Hidalgo, D., Martin-Marroquin, J. M., & Corona, F. (2019). A multi-waste management concept as a basis for a circular economy model. *Renewable & Sustainable Energy Reviews*, *111*, 481–489. doi:10.1016/j.rser.2019.05.048

Kamaraj, M., Ramachandran, K. K., & Aravind, J. (2020). Biohydrogen production from waste materials: Benefits and challenges. *International Journal of Environmental Science and Technology*, *17*(1), 559–576. doi:10.100713762-019-02577-z

Karapinar, K. I., & Kargi, F. (2006). Bio-hydrogen production from waste materials. *Enzyme and Microbial Technology*, *38*(5), 569–582. doi:10.1016/j.enzmictec.2005.09.015

Lindeboom, R. E. F., Fermoso, F. G., Weijma, J., Zaqt, K., & van Lier, J. B. (2011). Autogenerative high pressure digestion: Anaerobic digestion and biogas upgrading in a single step reactor system. *Water Science and Technology*, *64*(3), 647–653. doi:10.2166/wst.2011.664 PMID:22097043

Liu, H., Ren, L., Zhuo, H., & Fu, S. (2019). Water footprint and water pinch analysis in ethanol industrial production for water management. *Water (Basel)*, *11*(3), 518. doi:10.3390/w11030518

Lobato, L. C. S., Chernicharo, C. A. L., Pujatti, F. J. P., Martins, O. M., Melo, G. C. B., & Recio, A. A. R. (2013). Use of biogas for cogeneration of heat and electricity for local application: Performance evaluation of an engine power generator and a sludge thermal dryer. *Water Science and Technology*, *67*(1), 159–167. doi:10.2166/wst.2012.549 PMID:23128634

López-Campos, O., Bodas, R., Prieto, N., Frutos, P., Andrés, S., & Giráldez, F. (2010). Vinasse added to the concentrate for fattening lambs: Intake, animal performance, and carcass and meat characteristics. *Journal of Animal Science, 89*(4), 1153–1162. doi:10.2527/jas.2010-2977 PMID:21148788

Matkovsky, P.E.; Yarullin, R.S.; Startseva, G.P.; Sedov, I. (2010). Bioethanol: technologies of its production from renewable raw materials and application areas. *Alternative power engineering and ecology, 6*(86), 95-105.

Mikucka, W., & Zielińska, M. (2020). Distillery stillage: Characteristics, treatment, and valorization. *Applied Biochemistry and Biotechnology, 192*(3), 770–793. doi:10.100712010-020-03343-5 PMID:32557233

Mikulski, D., & Kłosowski, G. (2022). Integration of first- and second-generation bioethanol production from beet molasses and distillery stillage after dilute sulfuric acid pretreatment. *BioEnergy Research, 15*(6), 454–465. doi:10.100712155-021-10260-w

Miret, C., Chazara, P., Montastruc, L., Negny, S., & Domenech, S. (2015). Design of bioethanol green supply chain: Comparison between first and second generation biomass concerning economic, environmental and social criteria. *Computers & Chemical Engineering*, 85.

Moate, P. J., Williams, S. R. O., Torok, M. C., Hannah, B. E., Ribaux, M. H., Tavendale, M. H., Eckard, R. J., Jacobs, J. L., Auldist, M. J., & Wales, W. J. (2014). Grape marc reduces methane emissions when fed to dairy cows. *Journal of Dairy Science, 97*(8), 5073–5087. doi:10.3168/jds.2013-7588 PMID:24952778

Moraes, B. S., Junqueira, T. L., Pavanello, L. G., Cavaletti, O., Mantelatto, P. E., & Bonomi, A. (2014). Anaerobic digestion of vinasse from sugarcane biorefineries in Brazil from energy, environmental, and economic perspectives: Profit or expense? *Applied Energy, 113*, 825–835. doi:10.1016/j.apenergy.2013.07.018

Moran-Salazar, R. G., Sanchez-Lizarraga, A. L., Rodriguez-Campos, J., Davila-Vazquez, G., Marino-Marmolejo, E. N., Dendooven, L., & Contreras-Ramos, S. M. (2016). Utilization of vinasses as soil amendment: Consequences and perspectives. *SpringerPlus, 5*(1), 1007. doi:10.118640064-016-2410-3 PMID:27441131

Moreno-Arribas, M. V., & Polo, M. C. (2005). Winemaking biochemistry and microbiology: Current knowledge and future trends. *Critical Reviews in Food Science and Nutrition, 45*(4), 265–286. doi:10.1080/10408690490478118 PMID:16047495

Musteata, G., Balanuta, A., Restica, A., Filimon, R. V., Baetu, M. M., & Patras, A. (2021). Capitalization of secondary wine products – an opportunity for the wine sector of Republic of Moldova and Romania. *Journal of Social Sciences, IV*(2), 117–127. doi:10.52326/jss.utm.2021.4(2).12

Negro, C., Tommasi, L., & Miceli, A. (2003). Phenolic compounds and antioxidant activity from red grape marc extracts. *Bioresource Technology, 87*(1), 41–44. doi:10.1016/S0960-8524(02)00202-X PMID:12733573

Nerantzis, E., & Tataridis, P. (2006). Integrated enology utilization of winery by-products into high added value products. *E-. Journal of Science and Technology, 1*, 79–89.

Nevers, L., Ribeiro, R., Oliveira, R., & Alves, M. M. (2006). Enhancement of Methane Production from Barley Waste. *Biomass and Bioenergy*, *30*(6), 599–560. doi:10.1016/j.biombioe.2005.12.003

Ordaz-Diaz, L. A., & Bailon-Salas, A. M. (2020). Molecular identification of microbial communities in the methane production from vinasse: A review. *BioResources*, *15*(2), 4528–4552. doi:10.15376/biores.15.2.Diaz

Parsaee, M., Kiani, M. K. D., & Karimi, K. (2019). A review of biogas production from sugarcane vinasse. *Biomass and Bioenergy*, *122*, 117–125. doi:10.1016/j.biombioe.2019.01.034

Pilarski, K., Pilarska, A. A., Boniecki, P., Niedbała, G., Witaszek, K., Piekutowska, M., Idzior-Haufa, M., & Wawrzyniak, A. (2021). Degree of biomass conversion in the integrated production of bioethanol and biogas. *Energies*, *14*(22), 7763. doi:10.3390/en14227763

Pinelo, M., & Rodriguez, R. (2007). Supercritical fluid and solid-liquid extraction of phenolic antioxidants from grape pomace: A comparative study. *European Food Research and Technology*, *226*(1), 199–205. doi:10.100700217-006-0526-3

Rondeau, P., Gambier, F., Jolibert, F., & Brosse, N. (2013). Compositions and chemical variability of grape pomaces from French vineyard. *Industrial Crops and Products*, *43*, 251–254. doi:10.1016/j.indcrop.2012.06.053

Scull, I., Savón, L., Gutiérrez, O., Valiño, E., Orta, I., Mora, P. O., Orta, H., Ramos, Y., Molineda, A., Coto, G., & Noda, A. (2012). Physico-chemical composition of concentrated vinasse for their assessment in animal diets. Cuban. *Journal of Agricultural Science*, *46*(4), 385.

Siciliano, A., Limonti, C., & Curcio, C. M. (2022). Performance evaluation of pressurized anaerobic digestion (PDA) of raw compost leachate. *Fermentation (Basel, Switzerland)*, *8*(1), 15. doi:10.3390/fermentation8010015

Spinei, M., & Oroian, M. (2021). The potential of grape pomace varieties as a dietary source of pectic substances. *Foods*, *10*(4), 867. doi:10.3390/foods10040867 PMID:33921097

Stemme, K., Gerdes, B., Harms, A., & Kamphues, J. (2005). Beet-vinasse (condensed molasses solubles) as an ingredient in diets for cattle and pigs - Nutritive value and limitations. *Journal of Animal Physiology and Animal Nutrition*, *89*(3-6), 179–183. doi:10.1111/j.1439-0396.2005.00554.x PMID:15787991

Teixeira, A., Baenas, N., Dominguez-Perles, R., Barros, A., Rosa, E., Moreno, D., & Garcia-Viguera, C. (2014). Natural bioactive compounds from winery by-products as health promoters: A review. *International Journal of Molecular Sciences*, *15*(9), 15638–15678. doi:10.3390/ijms150915638 PMID:25192288

Tejada, M., & Gonzalez, J. L. (2006). Effects of two beet vinasse forms on soil physical properties and soil loss. *Catena*, *68*(1), 41–50. doi:10.1016/j.catena.2006.04.025

Van Lier, J. B., Tilche, A., Ahring, B. K., Macaria, H., Moletta, R., Dohanyos, M., Hulstoff Pol, L. W., Lens, P., & Versaete, W. (2001). New perspectives in anaerobic digestion. *Water Science and Technology*, *43*(1), 1–18. doi:10.2166/wst.2001.0001 PMID:11379079

Vaseashta, A., Duca, G., & Travin, S. (Eds.). (2022). *Handbook of Research on Water Sciences and Society* (Vols. 1–2). IGI Global. doi:10.4018/978-1-7998-7356-3

Vaseashta, A., Gevorgyan, G., Kavaz, D., Ivanov, O., Jawaid, M., & Vasović, D. (2021). Exposome, Biomonitoring, Assessment and Data Analytics to Quantify Universal Water Quality. In A. Vaseashta & C. Maftei (Eds.), *Water Safety, Security and Sustainability. Advanced Sciences and Technologies for Security Applications*. Springer. doi:10.1007/978-3-030-76008-3_4

Velásquez-Riaño, M., Meneses-Sánchez, J. S., & Carvajal Arias, C. E. (2019). Evaluation of acute toxicity of vinasse by means of *Daphnia magna* and *Aliivibrio fischeri*: A comparative study. *Interdisciplinary Toxicology*, *2*(3), 143–148. doi:10.2478/intox-2019-0017 PMID:32210703

Wang, T., Wang, S.-P., Zhong, X.-Z., Sun, Z.-Y., Huang, Y.-L., Tan, L., Tang, Y.-Q., & Kida, K. (2017). Converting digested residue eluted from dry anaerobic digestion of distilled grain waste into value-added fertilizer by aerobic composting. *Journal of Cleaner Production*, *166*, 530–536. doi:10.1016/j.jclepro.2017.08.075

Wu, M., Mintz, M., Wang, M., & Arora, S. (2009). Water consumption in the production of ethanol and petroleum gasoline. *Environmental Management*, *44*(5), 981–997. doi:10.100700267-009-9370-0 PMID:19774326

Yu, J., & Ahmedna, M. (2013). Functional components of grape pomace: Their composition, biological properties and potential applications. *International Journal of Food Science & Technology*, *48*(2), 221–237. doi:10.1111/j.1365-2621.2012.03197.x

Section 2
Technology and Redox Processes

Chapter 10
Palladium-Copper Catalyst Supported on Carbon Fiber Material for Oxidation of Carbon Monoxide by Air Oxygen

Tatyana A. Kiose
I.I. Mechnikov Odessa National University, Ukraine

Tatyana L. Rakitskaya
Odessa I.I. Mechnikov National University, Ukraine

Alim A.-A. Ennan
Physico-Chemical Institute of Environmental and Human Protection, Ukraine

Alla S. Truba
https://orcid.org/0000-0002-3416-2142
Odessa I.I. Mechnikov National University, Ukraine

ABSTRACT

The phase composition, textural, protolytic properties, and activity of catalysts for the carbon monoxide low-temperature oxidation based on two series of non-woven carbon fiber material samples CFM-I and CFM-II, and K2PdCl4, Cu(NO3)2, KBr basic components were studied in this work. Catalysts and CFMs were studied by XRD, SEM, FTIR spectroscopy, nitrogen ad/desorption, pH-metric, and atomic absorption spectrophotometry methods. It has been found that the basic components K2PdCl4 and Cu(NO3)2 deposited on a microporous carbon surface undergo changes and form an X-ray amorphous Pd0 phase and a paratacamite – Cu2(OH)3Cl crystalline phase. The catalysts were tested in the range of carbon monoxide initial concentrations from 50 to 300 mg/m3 with an effective contact time of the catalyst with the gas-air mixture from 0.12 to 0.50 s. Catalysts provide air purification from CO to concentrations below MPCCO (20 mg/m3) and can be used in respiratory devices.

DOI: 10.4018/979-8-3693-0512-6.ch010

INTRODUCTION

It is a well-known fact that carbon monoxide is dangerous for the environment and humans. Relevant is the issue of respiratory protection of workers located around industrial sources of gases containing carbon monoxide is of great relevance. It can be thoroughly made aware that carbon monoxide is oxidized to CO_2 only in the presence of catalysts. In this regard, the universal respiratory protection development is closely linked to progress in the effective low-temperature catalysts development for the oxidation of carbon monoxide, which would provide stable air purification at high humidity and ambient temperature, would meet sanitary-chemical, hygienic, and organoleptic requirements presented to the respiratory protective equipment (RPE). Moreover, there are issues related to the production of the catalyst with renewable characteristics. According to the active component form, the catalysts can be oxide, metal, and metal complex, and according to the temperature criterion, they are divided into low-temperature and high-temperature ones. Catalysts provide air purification from CO to concentrations below MPC_{CO} (20 mg/m^3) and can be used in respiratory devices. This article deals with the results of the development of a low-temperature Wacker-type catalyst for the oxidation of carbon monoxide with atmospheric oxygen, which consists of K_2PdCl_4, $Cu(NO_3)_2$, KBr and a non-woven carbon fiber material carrier. The purpose of this work is to study the effect of textural and physicochemical properties of non-woven carbon fiber materials on the activity and protective properties of palladium-copper catalysts for low-temperature oxidation of carbon monoxide with atmospheric oxygen.

BACKGROUND

The conducted analysis of the research showed that Wacker-type metal-complex catalysts deposited on various carriers demonstrate stable protective properties and may be promising for RPE use (Rakitskaya et al., 2022). The basic components of such catalysts are palladium (II) and copper (II) salts, as well as some additives supported on various carriers. SiO_2 (Rakitskaya et al., 2022), Al_2O_3 (Bruk et al., 2018; Du et al., 2015; Shen et al., 2013; Titov et al., 2012) activated carbons and carbon fiber materials (CFM) (Park et al., 2000; Radkevich et al., 2014, 2008), as well as natural minerals such as tripoli, clinoptilolite, mordenite, and bentonite (Rakitskaya et al., 2009, 2013, 2016, 2017, 2019, 2020) are used as carriers. The carrier is not an inert support and affects the mechanisms of surface complexes formation (Rakitskaya et al., 2015, 2022), as well as the oxidation state of palladium in the case of carbon carriers (Bruk et al., 2018; Shen et al., 2013; Park et al., 2000; Radkevich et al., 2014).

Activated carbon fiber materials have unique properties: high specific surface area, homogeneous microporous structure, and surface functional groups variety; the ratio of which can be varied by chemical and thermal treatment (Bulushev et al., 2002, 2004; Li et al., 2001a; Muniz et al., 1998; Romanenko, 2007) and thus change the activity of catalysts. The advantages of CFM include the absence of external and internal diffusion inhibition of adsorption and catalytic processes, and low values of pressure drop, which is favorable for studying gas-phase reactions kinetics and in the case of using these materials in RPE. The CFM obtaining technology is quite energy-intensive, which causes the high cost of these materials. Nevertheless, the investigation has shown their effectiveness in the development of phosphine oxidation catalysts (Rakitskaya et al., 1987, 2012), ozone decomposition (Rakitskaya et al., 2000, 2001, 2019), carbon monoxide oxidation (Bulushev et al., 2002, 2004; Radkevich et al., 2008, 2014) and as sulfur dioxide oxidation adsorbents and catalysts (Davini, 2003; Li et al., 2001b; Martin et al., 2002). The literature analysis has shown that the study of CFM as catalyst carriers for carbon monoxide oxidation is very limited, which inhibits the development of lightweight RPE for workers in various industries.

METHODS AND METHODOLOGY

Materials. In the given work the samples from two batches of non-woven carbon fiber material Carbopon V-Active-200-65-A (CFM-I and CFM-II), which precursor was hydrated cellulose, were used as a carrier of catalyst components. Table 1 shows the standard characteristics of the CFM (manufactured by Svetlogorsk, Chemvolokno, Belarus). CFM samples were preliminarily dried at 110°C to constant weight (CFM -I and CFM -II) or washed with distilled water and dried at 110°C for 3 hours (H_2O-CFM). The catalyst composition was constant and recalculated for precursors as $C_{Pd(II)} = 1{,}7 \cdot 10^{-4}$; $C_{Cu(II)} = 4{,}68 \cdot 10^{-4}$;

Table 1. Technical characteristics of CFM

Value	CFM -I	CFM -II
Surface density, g/m²	153	145
Bulk density, g/dm³	60	81
Thickness, mm	2.6	1.9
MB* adsorption activity, mg/g	256	370
The limiting volume of the adsorption for benzene vapor, cm³/g	0.41	0.56
Mass fraction of ash, %	3.0	3.4

MB* –methylene blue

$C_{Br^-} = 6{,}0 \cdot 10^{-4}$ mol/g.

The carrier weighing 0.5-2.5 g was impregnated with a water-alcohol solution containing palladium and copper precursors in the specified ratios in the forms of K_2PdCl_4, $Cu(NO_3)_2$ and auxiliary component KBr. Hereinafter, the catalyst was conventionally designated Pd(II)-Cu(II)/CFM (CFM - CFM-I and CFM-II). The wet mass was kept in a closed Petri dish at 20-25 °C for 20-24 hours, then dried in an oven in the air at 110 °C to constant weight. The samples were cooled in a desiccator to room temperature.

Research Methods and Techniques

X-ray phase analysis was carried out using a Siemens D500 (Siemens, Germany) powder diffractometer in copper radiation (Cu K_{α}, λ= 1.54178 Å), with a secondary beam graphite monochromator. To record the diffractograms, the test samples were thoroughly ground in an agate mortar and placed in a glass cuvette with a working volume of 2×1×0.1 cm³. Scanning electron microscopy (SEM) studies of morphology and determination of local composition were performed by electron probe microanalysis on a scanning electron microscope JSM-6390LV (Jeol, Japan). Samples were examined by IR spectroscopy using FT-IR Spectrometer Frontier (400-4000 cm⁻¹, with a resolution of 4 cm⁻¹) (Perkin Elmer, USA). The spectra were registered in KBr tablets obtained at a ratio of 1 mg of substance per 200 mg of KBr.

Nitrogen ad/desorption measurements were performed by a volumetric method on a Sorptomatic adsorption instrument (Thermo Electron Corporation) (Carlo Erba, Israel) at liquid-nitrogen temperature (~196 °C). Before the measurements, samples were outgassed at 200 °C for 5 h. To characterize the protolytic properties of surface functional groups, 0.2 g of samples were suspended in 20 mL of bidistilled water and an equilibrium pH value was measured by a pH-340 (Wissenschaftlich-Technische Werkstatten GmbH (WTW), Germany) instrument equipped with an ESL 43-07 glass electrode and an EVL 1M3 chlorsilver electrode at the continuous stirring of the suspension at 20 °C. A suspension effect, ΔpH_s, was estimated using the following equation:

$$\Delta pH_s = pH_{st} - pH_0$$

where pH_0 and pH_{st} are pH values of a suspension measured in 15 s and after the equilibrium attainment.

The desorption of Pd(II) from the catalyst samples was performed under static conditions by varying the type of the desorbent, temperature, and time of sample treatment. The concentration of the desorbed Pd(II) ions in the solutions was determined by the atomic absorption spectrophotometry using an AAS-1N atomic absorption spectrophotometer (Carl Zeiss Jena, Germany) with propane–butane–air flame. The interfering matrix effect of the solution was solved by the method of standard addition with background correction. In particular, the wavelength of resonance radiation in the palladium(II) concentration measuring was 247.6 nm with a background correction at $\lambda = 246.7$ nm. The error of Pd(II) determination was ± 5%.

Catalyst Testing

The tests of catalyst samples were carried out in a gas-flow thermostat at 293 K, installation, in a reactor type - "allonge", with a fixed layer of catalyst. The dimensions of the reactor and the linear velocity of the gas-air mixture (GAM) correspond to the mode of ideal displacement and the course of the reaction in the kinetic region:

- CO initial concentration, mg/m^3 – 50; 100; 200; 300
- volumetric speed of GAM, L/ min – 1.0
- linear speed of GAM, cm/s – 2.1
- the allonge diameter into the "light" (working area), cm – 3.5
- temperature, °C – 20
- relative humidity, % – 65

GAM with a given concentration of CO was obtained by diluting the concentrated gas (98-99 vol.% CO) with purified air to the required concentration of - 50; 100; 200; 300 mg/m^3. Initial $\left(C_{CO}^{in}\right)$ and final $\left(C_{CO}^{f}\right)$ concentrations of carbon monoxide were determined using a gas analyzer 621EX04 ("Analit-prilad", Ukraine) with a sensitivity of 2 mg/m^3. The reaction rate was calculated by the formula:

$$W = \frac{\omega\left(C_{CO}^{in} - C_{CO}^{f}\right)}{m_{ef}}, \text{ mol/(g·s)} \qquad (1)$$

where $\omega = 1{,}67 \cdot 10^{-2}$ – volume consumption, L/s; C_{CO}^{in}, C_{CO}^{f} – initial and final CO concentrations, mol/L; m_{ef} – the effective mass of the catalyst sample, g per diameter of the allonge working area.

The degree of CO conversion in the steady state (η_{st}) was determined by the formula:

$$\eta_{st} = \frac{\left(C_{CO}^{in} - C_{CO}^{f}\right)}{C_{CO}^{in}} \cdot 100,\ \%. \tag{2}$$

RESULTS AND DISCUSSION

Phase Composition

Figure 1 shows the X-ray diffraction patterns of the CFM-II samples and the Pd(II)-Cu(II)/CFM-II catalyst.

Figure 1. X-ray diffraction patterns of CFM-II samples (a) and catalyst Pd(II)-Cu(II)/CFM-II (b)

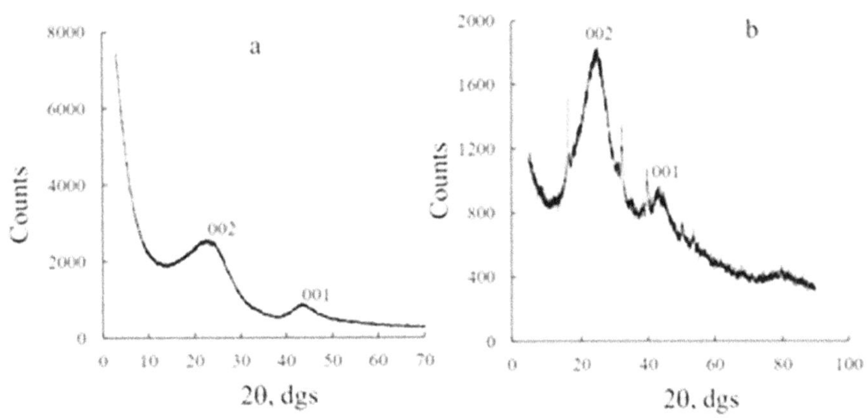

The obtained data indicate that CFM-II has regions of coherent scattering (reflections 002 and 101) of crystals, which indicates fiber graphitization. The space between the ordered regions is filled with amorphous carbon. The criterion for fiber graphitization can be a decrease in the interplanar spacing (002 reflection) up to 3.35 Å, which is typical for a graphite crystal (Bajaj & Dhawan, 1997; Fitzer, 1998; Wang et al., 2006; Zemskova et al., 2004;). The diffraction pattern of the Pd(II)-Cu(II)/CFM-II catalyst indicates the following changes: reflections 002 and 101 become less diffuse, and at $2\theta = 16.2°$, $32.1°$, $39.8°$ and $53.5°$ new reflections appear, which belong to the paratacamite – $Cu_2(OH)_3Cl$ phase [IPCDS 25-1427]. The formation of this phase was discovered by the authors of (Bruk et al., 2018; Radkevich et al., 2014). Based on the XRD data, it is impossible to make a conclusion about the state of the second component of the catalyst, namely, palladium.

Morphology

The research using SEM has shown that the surface morphology of Carbopon, a non-woven CFM, is typical for such materials (Klimuk et al., 2008; Martin-Gullon et al., 2001; Simonova et al., 2008). Fibers with a width of 4.6-8.5 microns are randomly intertwined. On individual fibers, impurities that are difficult to remove are observed, which are formed during decomposition, for example, organosilicon compounds used for structuring fibers. The impurities are distributed unevenly (Figure 2 a). On the surface of a fiber with a minimum content of impurities at a magnification of 10,000 times, cavities of various diameters and depths are clearly visible, which are formed during the activation of the CFM with a mixture of H_2O+CO_2 at high temperature (Fig. 2b). In the case of a catalyst (Figure 2c), fiber erosion increases, and very small agglomerates are visible on the surface, which are formed by crystallites of the pacatacamite $Cu_2(OH)_3Cl$ phase (Figure 1b).

FT-IR Spectroscopy

Figure 3 shows the IR spectra in the range of 4000-400 cm^{-1} for the carrier CFM-II (a) and the catalyst based on Pd (II)-Cu (II)/CFM-II (b).

The spectra were identified considering the generalized data (Romanenko, 2007) for carbon materials, as well as the results of the authors' study (Bekhterev & Zolotarev, 2006; Mikova et al., 2014). According to our results in the IR spectrum of CFM (Figure 3 a, Table 2) there appears a wide, fairly intense absorption band at 3440 cm^{-1}, which is due to the valence vibrations of OH-groups of different nature, namely, in adsorbed water molecules, carboxyl, phenolic, alcohol groups. Deformation vibrations of OH groups correspond to the absorption band at 1632 cm^{-1}. These absorption bands undergo some changes when the catalyst components are applied: they are less blurred and appear at 3369 cm^{-1} and 1651 cm^{-1}, which indicates the excitation of hydrogen bonds.

Valence vibrations at 1426 cm^{-1} and 1384 cm^{-1} are referred to as vibrations of the CO bond in the carboxyl group; the position of these bands does not change in the case of the catalyst, but the intensity of the band at 1384 cm^{-1} increases due to the imposition of valence vibrations of the nitrate group in the composition of $Cu(NO_3)_2$ - a component of the catalyst composition. The same picture is in the case of the quinone group (1517 cm^{-1}) and the aromatic ring (1400 cm^{-1}).

Figure 2. SEM image of CFM-II (Carbopon) (a, b) and catalyst Pd (II)-Cu(II)/CFM-II (c)

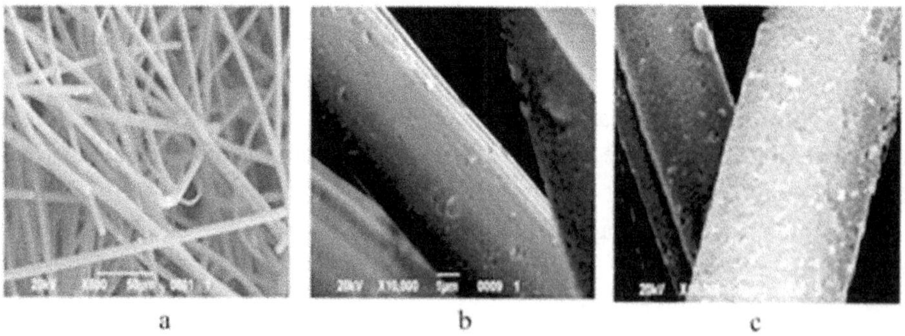

a b c

Figure 3. IR spectra of CFM-II (a) and the catalyst Pd(II)-Cu(II)/CFM-II (b)

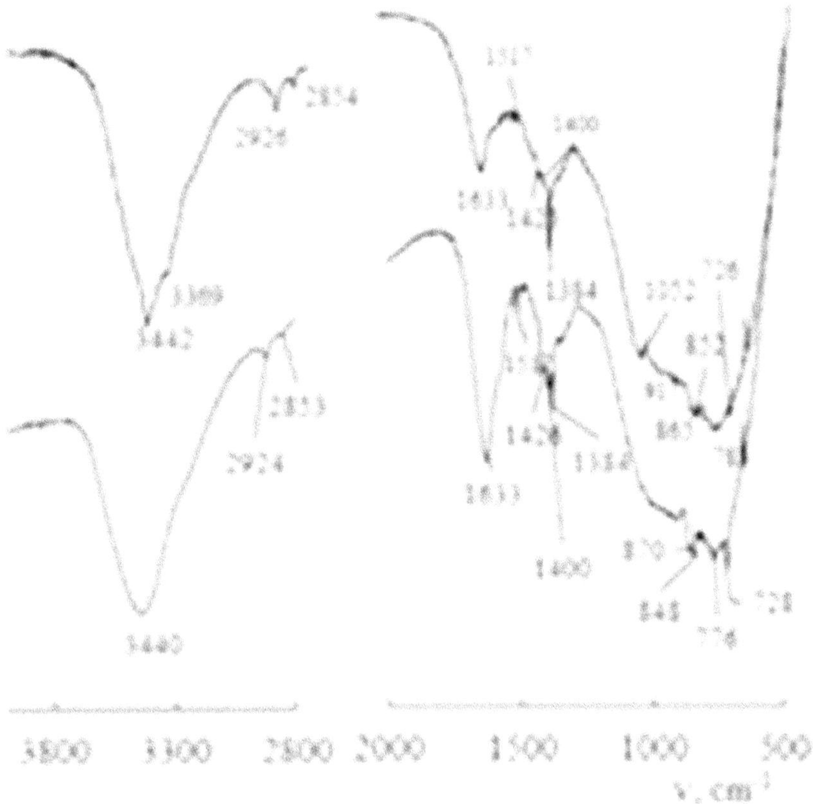

In the catalyst spectrum, in contrast to the CFM carrier, the absorption band at 1052 cm^{-1} is clearly detected, which can be attributed to C–O–C valence vibrations in the lactone or ether and epoxy groups, although C–O–C deformation vibrations groups at 870 cm^{-1} are present in the spectra of both samples. It is known that the content of oxygen-containing surface groups (carboxyl, phenolic, lactone) increases with oxidation of the carbon surface by nitric and sulfuric acids, hydrogen peroxide, ammonium persulfate (Al-Wadhav, 2012; Gimaeva et al., 2011; Mikova et al., 2014) and ions Pd^{2+}, Au^{3+}, Ag^{+} (Simonova et al., 2008). Carbon materials are characterized by a π-electron system, the fragments of which are hexagons of the carbon net, individual bonds in these hexagons, as well as fragments >C=C< and -C≡C- carbon chains (Romanenko, 2007). A new low-intensity absorption band at 917 cm^{-1} is detected in the catalyst spectrum, which is attributed to π (=CH_2) fluctuations (Kolesnik & Sapoletova, 2011) and which can appear in the catalyst spectrum due to the formation of the π-complex with $PdCl_2$ (Simonov et al., 1997), . In the region of 1000-700 cm^{-1} in the spectra of the two samples there is a wide band with individual absorption maxima, the number of which increases in the case of the catalyst. The following absorption bands are common: at ≈ 850 cm^{-1} - interplanar vibrations of carbon atoms (Bekhterev & Zolotarev, 2006); at ≈ 725 cm^{-1} - pendulum vibrations of methylene groups (Polovetskaya et al., 2015); at ≈ 870 cm^{-1} - vibrations in the Si-OC fragment (Bogacheva et al., 2009), which may belong to the residues of organosilicon compounds, which are used as a structuring substance during the carbonization process of cellulose hydrate. The presence of impurities is also indicated by SEM images of CFM (Figure 2).

Table 2. Wave numbers (υ, cm^{-1}) of absorption maxima in the IR spectra of CFM-II and Pd(II)-Cu(II)/ CFM-II catalyst

CFM functional group	Type of vibrations. Chemical bond	υ, cm^{-1}	
		CFM-II	Catalyst
Aliphatic chains ($-CH_3$)	υ (C–H)	2853	2854
Aliphatic chains ($-CH_2-$)	υ (C–H)	2924	2926
Carboxyl: region 3200-3400 cm^{-1}	υ (O–H) δ (O–H)	3440 1633	3442, 3369 sh. 1651 sh., 1633
region 1000-1400 cm^{-1}	υ (C–O)	1426, 1384	1426, 1384
Quinone (ketone) groups	υ (C=O)	1517	1517
Aromatic ring	υ (C=C)	1400	1400
Lactone (ether, epoxy) groups	υ (C–O–C)	–	1052
	δ (C–O–C)	870	870
π-electronic system	π (C=C)	–	917

Texture Characteristics

Figure 4 shows the nitrogen adsorption/ desorption isotherms by CFM samples from two batches – CFM-I, and CFM-II, after the treatment in distilled water for 30 minutes (H_2O-CFM-II) and the catalyst Pd(II) -Cu(II)/CFM-II. According to the IUPAC classification, adsorption isotherms are classified as type I. We see that the adsorption mainly occurs at low values of relative pressure, $P/P_0 < 0.01$, which indicates the microporous structure of the samples. At $P/P_0 > 0.9$, nitrogen adsorption increases for all samples, which may indicate the presence of some mesopores in the samples.

Desorption branches of isotherms are characterized by hysteresis. The hysteresis loop remains almost horizontal in the region of high relative pressure and closes at $P/P_0 \sim 0.45$. According to the IUPAC classification, hysteresis is classified as H4-type (Burtch et al., 2014). The obtained nitrogen adsorption-desorption isotherms are typical for carbon materials (Baek, 2000; Gorina & Cheblakova, 2015; Kowalczyk1 et al., 2002; Lu & Zheng, 2001; Ostafiychuk et al., 2013; Shcherban & Filonenko, 2017; Tamarkina et al., 2008;).

The following methods were used to calculate the porous structure parameters of CFM and catalyst samples: BET, t-plot, Dubinin-Radushkevich (DR) (Gregg & Sing, 1982), Horvat-Kavazoe (HK) (Horváth & Kawazoe, 1983). In Table 3 the porous structure parameters of the studied samples are generalized. The adsorption method isotherms were treated by the BET in the range P/P_0 0.020-0.098 and obtained the values of the monolayer capacity (a_m, mmol/g), specific surface area (S_{BET}, m^2/g) and constant C, which characterizes the affinity of the adsorbate to the adsorbent. Using the t-plot method, data on the specific volume of micropores+supermicropores, $V_{mi+supermi}$, cm^3/g, a specific surface area taking into account the area of mesopores and outer surface, S_{me+ext}, m^2/g, and only the outer surface, S_{sp}, m^2/g. According to the DR method (Dubinin-Radushkevich) a_m, V_{mi} and S_{sp} were calculated. The HK (Horvath and Kawazoe) method for microporous materials makes it possible to characterize the mean (d_m) and maximum (d_{max}) pore diameters, total (cumulative) pore volume, V_c, cm^3/g and total specific surface area, S_c, m^2/g. In addition, the size distribution curves of micropores were calculated by the HK method (Figure 5).

Figure 4. Isotherms of adsorption-desorption of nitrogen samples: 1 – CFM-I; 2 – CFM-II; 3 – H_2O-CFM-II; 4 – Pd(II)-Cu(II)/CFM-II

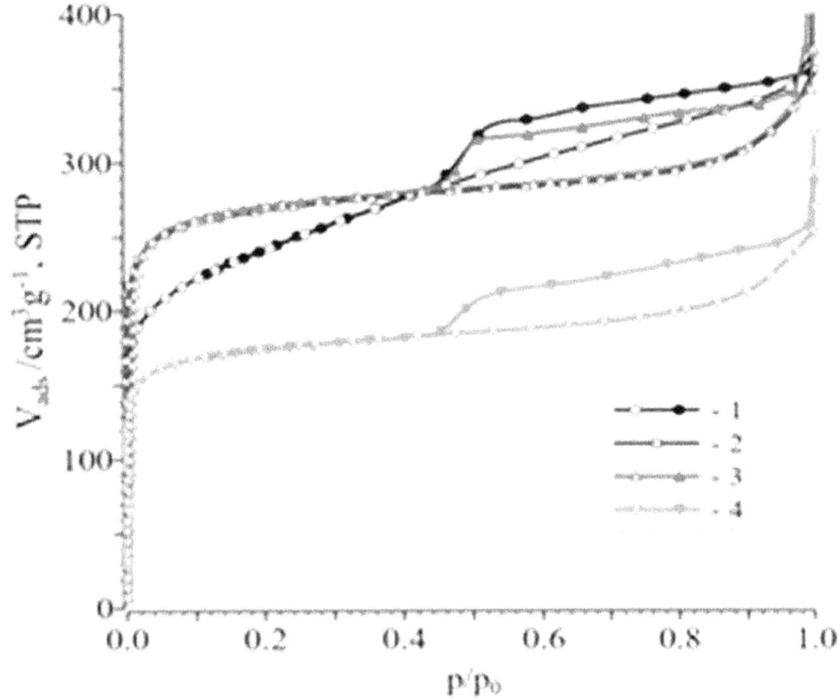

Figure 5. Distribution of pore volume by size in samples: 1 – CFM-I; 2 – CFM-II; 3 – H_2O-CFM-II; 4 – Pd(II)-Cu(II)/CFM-II

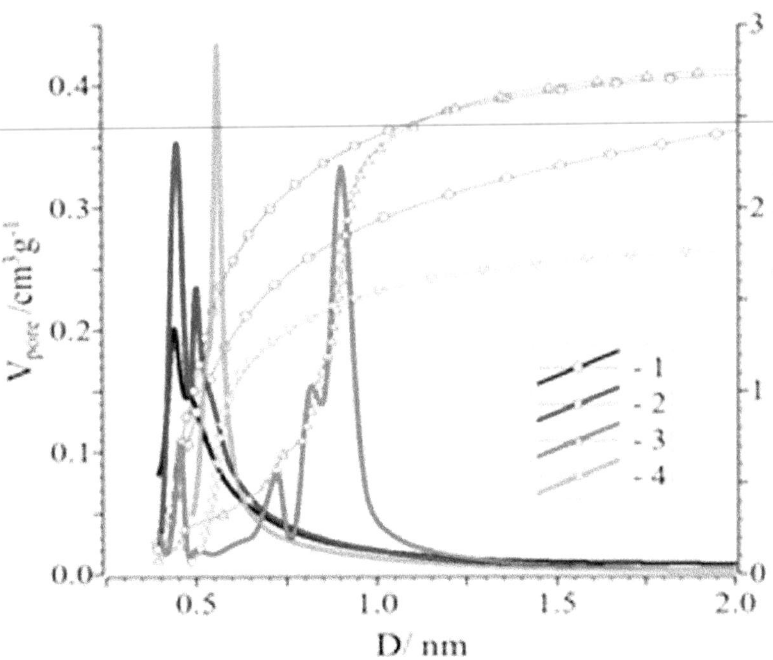

The conclusions from the analysis of the CFM porous structure and the catalyst parameters are as follows. The samples from two batches of Carbopon differ in texture parameters. Thus, the CFM-II sample is characterized by larger values of V_{mi}, S_{sp}, although the average pore diameter (d_m) is smaller. For the H_2O-CFM-II sample, V_{mi}, S_{sp}, d_m and d_{max} increase, which may indicate the expansion of micropores due to the removal of residual technological impurities. The application of catalyst components on the CFM-II carrier leads to a significant reduction in V_{mi}, S_{sp}, d_m and d_{max}. The size distribution data of pores in the studied samples (Figure 5) indicate the following: the structure of the CFM-I microporous sample is homogeneous with a maximum pore volume of 0.4 nm in diameter.

The second batch sample (CFM-II) has two types of pores, the maximum volume of which is 0.4 and 0.5 nm in diameter. After boiling the CFM-II sample in distilled water, the pattern changes. In the structure of the H_2O-CFM-II sample, pores with a diameter of 0.4 nm remain (their volume is much smaller) and larger pores appear, namely 0.75, 0.8, and 0.9 nm. Such structural changes are because CFM-II water treatment opens pores that have been blocked by so-called "technological impurities", which are used to activate and structure carbon fiber materials. As a result of treatment with water in the sample, H_2O-CFM-II increases the volume of micropores and S_{sp}. The structure of the catalyst is homogeneous, only pores are detected, the diameter of which differs from the diameters of the source materials and is 0.55 nm. The obtained data show that as a result of carrier impregnation the catalyst components block the pores with $d \geq 0.75$ nm (they disappear), which leads to a significant reduction in the volume of micropores and the specific surface area. Some increase in pore size up to 0.55 nm may be the result of partial oxidation of the carbon surface by palladium (II) (Simonova et al., 2008).

Table 3. Structural and adsorption characteristics of CFM samples and Pd (II)-Cu (II) / CFM-II catalyst

Sample	BET			t-plot (standart D 4641-87)			Dubinin-Radushkevich			Horvath and Kawazoe		
	a_m, mmol/g	C	S_{sp}, m²/g	$V_{mi+supermi}$, cm³/g	S_{me+ext}, m²/g	S_{sp}, m²/g	a_m, mmol/g	V_{mi}, cm³/g	S_{sp}, m²/g	d_m / d_{max}, nm	V_e, cm³/g	S_e, m²/g
CFM-I	9.0581	605.52	883.7	0.4534	53.2	53.2	9.1617	0.3174	893.8	0.6159 / 0.942	0.4032	1185.0
CFM-II	10.464	-2005	1020.9	0.3991	53.99	53.99	11.348	0.3931	1107.1	0.5431 / 0.4494	0.4223	1408.2
H_2O-CFM-II	10.675	23040	1041.5	0.4030	54.176	54.084	12.336	0.4273	1203.5	0.8884 / 0.9052	0.4287	1031.5
Pd(II)-Cu(II)/ CFM-II	6.9649	1188.8	679.49	0.2491	57.555	57.555	7.4115	0.2567	723.05	0.5851 / 0.5606	0.2774	824.0

Protolytic Properties

Studies of the CFM functional groups' nature indicate the presence of acidic, basic, and neutral groups (Romanenko, 2007). Acid groups (carboxyl, phenolic) are ionized in an aqueous medium, due to which the pH of the medium will decrease. The main groups of CFM (Bransted basic groups) in the aqueous medium will bind hydrogen ions, so the pH of the medium will increase. In the first and second cases, after some time, a protolytic equilibrium is established. The authors (Dzyazko et al., 2020). also point to the possibility of a proton complex formation of with an π-electron system of the carbon surface, as a result of which the surface pH will increase:

$$C_\pi + 2H_2O \rightarrow C_\pi H_3O^+ + OH^-.$$

Neutral groups do not ionize in aqueous medium, so they do not affect the pH change. Figure 6 shows the curves of changes in time of the pH of the samples CFM-I, CFM-II, and H_2O-CFM-II suspensions, which indicates that for three samples the profiles of the curves are the same - the pH decreases and reaches equilibrium. The difference ΔpH_s is negative (table 4).

Figure 6. The pH dependence of the CFM samples aqueous suspension of on time (τ):1 – CFM-I;2 – CFM-II; 3 – H_2O-CFM-II

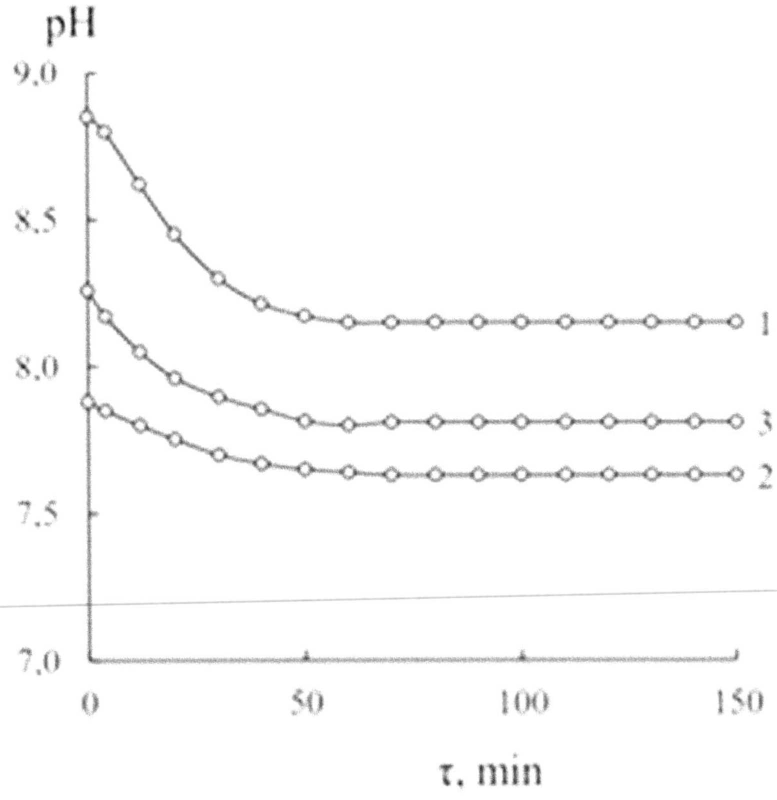

These characteristics specify that the CFM samples are dominated by acidic surface groups. Studies have shown that the textural and physicochemical properties of samples H_2O-CFM-II and CFM-II, as well as the catalysts activity based on them differ slightly, therefore the catalysts were obtained on the CFM-I and CFM-II original samples for further studies.

Table 4. Characteristics of CFM surface acidity

Sample	pH_0	pH_{st}	ΔpH_s
CFM-I	8.85	8.15	-0.7
CFM-II	7.88	7.63	-0.25
H_2O-CFM-II	8.26	7.81	-0.45

Kinetics of Carbon Monoxide Oxidation by Air Oxygen in the Presence of Pd(II)-Cu(II)/CFM catalysts

The State of Palladium-Copper Complexes on a Carbon Surface

Many researchers have found that palladium is in the Pd^{2+}, Pd^{+}, Pd^0 forms (Shen et al., 2013; Bruk et al., 2018; Park et al., 2000; Radkevich et al., 2008, 2014), and copper is in the form of $Cu_2(OH)_3Cl$ heterogeneous phase (Bruk et al., 2018; Du et al., 2015; Titov et al., 2012). The presence of a Pd^0 phase cannot be ruled out on the surface of CFM Busofit (Radkevich et al., 2014). At the same time, attempts were made to find a correlation between the ratio of different forms of palladium, the content of the hydroxochloride form of copper, and the activity of the catalyst. It was shown in (Rakitskaya et al., 2009, 2013, 2015, 2016, 2017, 2019, 2020) that in the case of natural carriers, the active components of the catalyst, especially at low concentrations, are in their stable oxidation states, and the composition of the Pd(II) and Cu(II) surface complexes is determined by the carrier nature.

The XRD results of the Pd(II)-Cu(II)/CFM-II catalyst sample definitely indicated the presence of $Cu_2(OH)_3Cl$ (Figure 1). IR spectral studies of the catalyst sample (Figure 3, Table 2) indicate the appearance of a new absorption band at 1052 cm^{-1}, which is assigned to C–O–C valence combinations in the lactone group and is due to the oxidation of the carbon surface by palladium(II). The XRD data (Figure 1) and SEM data (Figure 2) did not reveal the formation of $[Pd^0]$n agglomerates, that correlate with the data (Simonov et al., 1997; Simonova et al., 2008),. However, additional erosion of the CFM surface in the catalyst may indicate that the CFM surface is oxidized by palladium (II).

The form of palladium in the catalyst composition was judged on the basis of palladium desorption using different desorbents under different conditions of its leaching (Table 5). As a rule, the presence of palladium in the Pd(II) form is confirmed by desorption with water (weak binding to the carrier), HCl or HNO_3 acids at room temperature (medium bond strength with the surface), and boiling (strong binding) (Rakitskaya et al., 2011, 2015). As can be seen, palladium is desorbed only with the help of aqua regia. Consequently, palladium is in the composition of the catalyst in the form of X-ray amorphous Pd^0.

The data obtained indicate that the initial complex forms of palladium(II) are reduced to Pd^0, which is quite strongly retained by the surface (it is not desorbed by water and HCl), when deposited on a carbon surface, but dissolves in aqua regia, oxidizing to palladium(II), which determined by the AAS method. Obviously, the degree of palladium extraction can be increased by increasing the temperature and contact time of the sample with aqua regia.

Table 5. The palladium desorption from the catalyst results

Sample	Desorbent	Contact time, min	Temperature, °C	Desorbed Pd mg/g (%)
K_2PdCl_4-KBr/CFM C_{Pd} = 18 mg/g	H_2O	90	20	Not found
	H_2O	10	100	Not found
	1.5M HCl	90	20	0.076 (0.4)
	aqua regia $HCl+HNO_3$ (3:1)	90	20	10.0 (55.5)
		10	100	10.52 (58.4)

Influence of the Catalyst-Bed Configuration on the Degree of CO Conversion

Figure 7 shows the kinetic curves of the change in the carbon monoxide final concentration $\left(C^f_{CO}\right)$ over time, obtained by varying the number of catalyst beds that were fixed in the reactor of the "allonge" type. Kinetic curves 1-4 characterize the activity of the 1,2,3,4-layer Pd(II)-Cu(II)/CFM-I catalyst, and curve 5 characterizes the activity of the 5-layer Pd(II)-Cu(II)/CFM-II catalyst.

Figure 7. Change in C^f_{CO} over time with varying the number of Pd(II)-Cu(II)/CFM-I catalyst layers from one to four (curves 1-4); five-layer Pd(II)-Cu(II)/CFM-II catalyst (curve 5)

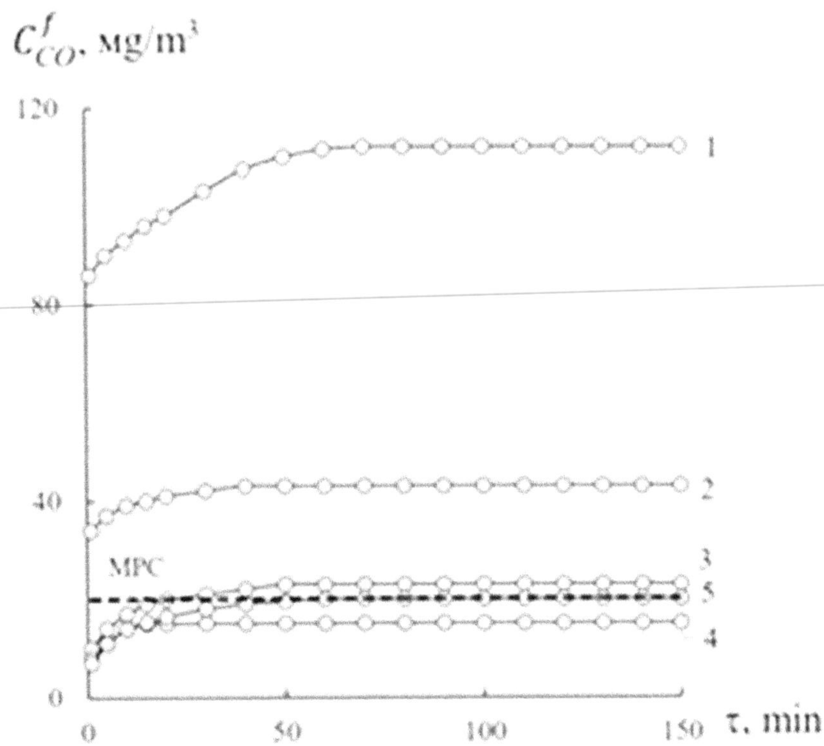

All kinetic curves have the same profile: within 10–50 minutes, the CO final concentration increases, and then a stationary regime is established, in which the CO concentration remains constant for a long time. A change in the configuration of the catalyst layer causes a change in its macro characteristics: the height of the layer (h), the effective contact time (τ_{ef}), and the effective mass of the catalyst (m_{ef}), namely, the catalyst mass, which falls on the allonge area "in the light" (Table 6). On the basis of the data obtained, the following conclusions can be drawn: with an increase in the catalyst layers number, the CO conversion degree in the stationary mode increases, and the condition $C^{f}_{CO} \leq MPC_{CO}$ (20 mg/m^3) is fulfilled for a 4-layer catalyst Pd(II)-Cu(II)/CFM -I and 5-layer Pd(II)-Cu(II)/CFM-II catalyst.

Table 6. Influence of the Pd(II)-Cu(II)/CFM catalyst-bed configuration on its activity in the oxidation of carbon monoxide with air oxygen. C^{in}_{CO} = 300 mg/m^3; U = 2.1 cm/s

Number of layers	h, cm	τ_{ef}, s	m_{ef}, g	C^{f}_{CO}, mg/m^3 stationary mode	η_{st}, %
Pd(II)-Cu(II)/CFM-I					
1	0.26	0.12	0.17	112	63
2	0.52	0.25	0.34	43	86
3	0.78	0.37	0.51	23	92
4	1.04	0.50	0.68	15	95
Pd(II)-Cu(II)/CFM-II					
4	0.76	0.36	0.64	47	84
5	0.95	0.45	0.80	20	93

Thus, although CFM-II is characterized by a more developed texture (Table 3), a more active catalyst was obtained based on CFM-I. As in many other cases, the specific surface area does not determine the activity of the Wacker-type catalyst. An important factor is the nature of the carrier, which affects the composition of the active forms of palladium and copper (Rakitskaya et al., 2022).

Influence of the Carbon Monoxide Initial Concentration on the Degree of its Conversion

The influence of the initial concentration of carbon monoxide in the gas-air mixture on the activity of the Pd(II)-Cu(II)/CFM-II catalyst was studied at two values of the effective contact time of the GAM with the catalyst, namely 0.36 s (4-layer configuration) and 0.45 s (5-layer configuration) (Figure 8).

It can be seen from the presented data that with an increase in C^{in}_{CO} from 50 to 300 mg/m^3, the profiles of kinetic curves do not change, however, the degree of CO conversion in the stationary mode depends on the effective contact time (Table 7).

In two experimental series, with the increase in C^{in}_{CO}, the reaction initial rate (W_n) grows up proportionally, which indicates the first-order reaction in terms of carbon monoxide concentration. In the stationary mode, the first order of the reaction with respect to CO is observed strictly in batch 2 (proportional dependence and almost constant degree of CO conversion). In batch 1, there is a slight deviation from the proportional dependence, which may be due to the influence of macro kinetic factors. For all values of C_{CO}^{in}, with increasing τ_{ef}, the degree of CO conversion rises, but only in the second batch the condition $C_{CO}^{f} \leq MPC_{CO}$ is satisfied in the C_{CO}^{i} entire range.

Figure 8. The change of C_{CO}^{f} in time during the carbon monoxide oxidation with atmospheric oxygen in the presence of a Pd(II)-Cu(II)/CFM-II catalyst with an effective contact time of 0.36 s (a) and 0.45 s (b) and different C_{CO}^{in}, mg/m³: 1 – 50; 2 - 100; 3 - 200; 4 - 300

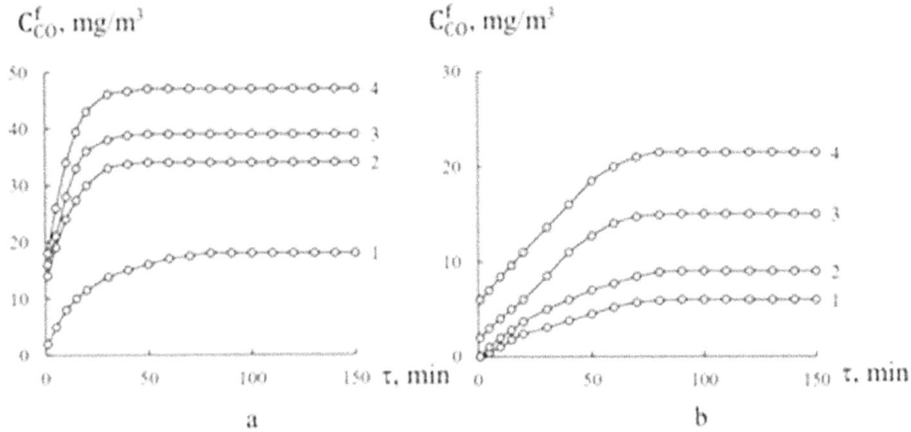

Table 7. The impact of initial CO concentration effect on the carbon monoxide conversion degree in the presence of a Pd(II)-Cu(II)/CFM-II catalyst at different effective contact times

C_{CO}^{in} mg/m³	$W_{st}\cdot10^7$, mol/g·s		C_{CO}^{f} mg/m³	η_{st}, %
	W_{in}	W_{st}		
Batch 1, τ_{ef} = 0.36 s				
50	0.42	0.28	18	64
100	0.76	0.58	34	66
200	1.62	1.42	39	81
300	2.49	2.23	47	84
Batch 2, τ_{ef} = 0,45 s				
50	0.35	0.31	6	88
100	0.70	0.64	9	91
200	1.40	1.31	15	93
300	2.08	1.96	20	93

CONCLUSION

In this paper, two batches of non-woven carbon fiber material samples, CFM-I and CFM-II, that precursor was hydrated cellulose, were used as a carrier of the basic components (K_2PdCl_4, $Cu(NO_3)_2$, KBr) of a catalyst for the low-temperature oxidation of carbon monoxide with atmospheric oxygen. The initial carriers based on their catalysts were studied by XRD, SEM, FT-IR spectroscopy, nitrogen ad/desorption, pH-metric, and atomic absorption spectrophotometry (AAS). XRD studies have shown that, in addition to amorphous carbon, the initial CFM and catalyst contain graphitized fibers, as indicated by the [002] and [001] reflections in the coherent scattering region of crystals. Additional reflections on the diffraction pattern of the catalyst refer to the paratacamite $Cu_2(OH)_3Cl$ phase.

SEM studies confirmed the typical CFM morphology and found, in the catalyst case, very fine agglomerates, which are formed by the paratacamite crystallites phase, and surface erosion caused by palladium(II) oxidation of the carbon surface. The FT-IR spectrum of the catalyst showed a new absorption band at 1052 cm^{-1} assigned to the lactone group. The presence of reduced palladium in the composition of the catalyst was also confirmed by desorption with various desorbents H_2O, HCl or HNO_3 and aqua regia (HCl + HNO_3). With only the help of aqua regia, palladium was desorbed from the carrier surface. It has been found that the K_2PdCl_4 and $Cu(NO_3)_2$ components deposited on the CFM surface undergo changes and form the X-ray amorphous Pd^0 phase and the crystalline paratacamite $Cu_2(OH)_3Cl$ phase.

Based on the parameters of the porous structure of the CFM and the catalyst, the following conclusions can be drawn. Samples CFM-I, CFM-II, and H_2O-CFM-II are microporous. CFM-I is characterized by the presence of the same size 0.4 nm pore diameter, while CFM-II has two types of 0.4 and 0.5 nm pore diameters: for H_2O-CFM-II, the presence of 0.4, 0.75, 0.8 and 0.9 nm pore diameters. The structure of the catalyst based on CFM-II becomes homogeneous with a pore diameter of 0.55 nm. The S_{sp} value for CFM-II and H_2O-CFM-II is greater than for CFM-I, and it significantly decreases when the catalyst components are applied.

The CFM protolytic properties are characterized, for which the suspension pH is in the range of 7.81-8.15, which is favorable for the $Cu_2(OH)_3Cl$ phase formation. Testing samples of the catalyst based on CFM-I and CFM-II in the oxidation of carbon monoxide showed that despite the more developed structure, the catalyst based on CFM-II is less active than the catalyst based on CFM-I; the degree of CO conversion increases with increasing effective contact time. Catalyst based on CFM-II at $\tau_{ef} = 0.45$ s in the range of C_{CO}^{in} from 50 to 300 mg/m^3 provides stable air purification from CO much lower than MPC_{CO} (20 mg/m^3) and can be recommended for application in respiratory devices.

REFERENCES

Al-Wadhav, Kh. A. (2012). Carbon carriers and the synthesis of palladium catalysts based on them. *Bulletin of MITHT*, *7*(1), 3–18.

Baek, I. (2000). Preparation of semi-activated carbon fibers. *Korean Journal of Chemical Engineering*, *17*(5), 553–558. doi:10.1007/BF02707165

Bajaj, P., & Dhawan, A. (1997). PAN-based activated carbon fibres: Production, characterization and applications. *Indian Journal of Fibre and Textile Research*, *22*, 222–235.

Bekhterev, A. N., & Zolotarev, V. M. (2006). IR spectroscopic study of vibrational states in low-dimensional nanocrystalline carbon with a changing structure. *Scientific and technical bulletin of information technologies, mechanics and optics*, (34), 91-96.

Bogacheva, S. Yu., Kapitanov, A. F., & Izmailov, B. A. (2009). Patent N° 2370583. *A method for imparting hydrophilic properties to fibrous materials*. Russia.

Bruk, L., Titov, D., Ustyugov, A., Zubavichus, Y., Chernikova, V., Tkachenko, O., Kustov, L., Murzin, V., Oshanina, I., & Temkin, O. (2018). The Mechanism of Low-Temperature Oxidation of Carbon Monoxide by Oxygen over the $PdCl_2$–CuCl2/γ-Al2O3 Nanocatalyst. *Nanomaterials (Basel, Switzerland)*, *8*(4), 217. doi:10.3390/nano8040217 PMID:29614029

Bulushev, D. A., Kiwi-Minsker, L., Yuranov, I., Suvorova, E. I., Buffat, P. A., & Renken, A. (2002). Structured Au/FeOx/C catalysts for low-temperature CO oxidation. *Journal of Catalysis*, *210*(1), 149–159. doi:10.1006/jcat.2002.3632

Bulushev, D. A., Yuranov, I., Suvorova, E. I., Buffat, P. A., & Kiwi-Minsker, L. (2004). Highly dispersed gold on activated carbon fibers for low-temperature CO oxidation. *Journal of Catalysis*, *224*(1), 8–17. doi:10.1016/j.jcat.2004.02.014

Burtch, N. C., Jasuja, H., & Walton, K. S. (2014). Water stability and adsorption in metal–organic frameworks. *Chemical Reviews*, *114*(20), 10575–10612. doi:10.1021/cr5002589 PMID:25264821

Davini, P. (2003). Flue gas desulphurization by activated carbon fibers obtained from polyacrylonitrile by-product. *Carbon*, *41*(2), 277–284. doi:10.1016/S0008-6223(02)00282-8

Du, X., Li, H. Y., Yu, J., Xiao, X., Shi, Z., Mao, D., & Lu, G. (2015). Realization of a highly effective Pd–Cu–Cl_x/Al_2O_3 catalyst for low-temperature CO oxidation by pre-synthesizing the active copper phase of $Cu_2Cl(OH)_3$. *Catalysis Science & Technology*, *5*(8), 3970–3979. doi:10.1039/C5CY00545K

Dzyazko, Y. S., Volfkovich, Y. M., & Chaban, M. O. (2020). Composites containing inorganic ion exchangers and graphene oxide: hydrophilic–hydrophobic and sorption properties. In *Nanomaterials and Nanocomposites, Nanostructure Surfaces, and Their Applications* (pp. 93–110). Springer.

Fitzer, E. (1988). *Carbon fibers and carbon composites*. Moscow: Mir.

Gimaeva, A. R., Valinurova, E. R., Igdavletova, D. K. & Kudasheva, F. C. (2011). Sorption of heavy metal ions from water by activated carbon adsorbents. *Sorption and chromatographic processes, 11*(3), 350-356.

Gorina, V. A., & Cheblakova, E. G. (2016). Influence of activation modes on the specific surface and development of a microporous structure of viscose-based carbon fibers. *Russian Journal of Non-Ferrous Metals*, *57*(7), 734–738. doi:10.3103/S1067821216070087

Gregg, S., & Sing, K. (1982). *Adsorption, Surface Area, and Porosity* (2nd ed.). Academic.

Horváth, G., & Kawazoe, K. (1983). Method for the calculation of effective pore size distribution in molecular sieve carbon. *Journal of Chemical Engineering of Japan*, *16*(6), 470–475. doi:10.1252/jcej.16.470

Klimuk, A. I., Obvintseva, L. A., Kuchaev, V. L., Shepelev, A. D., Sadovskaya, N. V., Tomashpolsky, Yu. Ya., Kozlova, N. V., & Avetisov, A. K. (2008). Interaction of ozone with microfiber materials. *Journal of the Russian chemical about them D. I. Mendeleev*, *LII*(5), 102–111.

Kolesnik, I. V., & Sapoletova, N. A. (2011). *Infrared spectroscopy*. Moscow state university.

Kowalczyk, P., Terzyk, A. P., Gauden, P. A., & Rychlicki, G. (2002). Numerical Analysis of the Horvath–Kawazoe Equation – The Adsorption of Nitrogen, Argon, Benzene, Carbon Tetrachloride and Sulphur Hexafluoride. *Adsorption Science and Technology*, *20*(3), 295–305. doi:10.1260/026361702760254478

Li, K., Ling, L., Lu, C., Liu, Z., Liu, L., & Mochida, I. (2001a). Influence of CO-evolving groups on the activity of activated carbon fiber for SO_2 removal. *Fuel Processing Technology*, *70*(3), 151–158. doi:10.1016/S0378-3820(01)00175-8

Li, K., Ling, L., Lu, C., Qiao, W., Liu, Z., Liu, L., & Mochida, I. (2001b). Catalytic removal of SO_2 over ammonia-activated carbon fibers. *Carbon*, *39*(12), 1803–1808. doi:10.1016/S0008-6223(00)00320-1

Lu, A. H., & Zheng, J. T. (2001). Study of microstructure of high-surface-area polyacrylonitrile activated carbon fibers. *Journal of Colloid and Interface Science*, *236*(2), 369–374. doi:10.1006/jcis.2000.7425 PMID:11401385

Martin, C., Perrard, A., Joly, J. P., Gaillard, F., & Delecroix, V. (2002). Dynamic adsorption on activated carbons of SO_2 traces in air: I. Adsorption capacities. *Carbon*, *40*(12), 2235–2246. doi:10.1016/S0008-6223(02)00108-2

Martín-Gullón, I., Andrews, R., Jagtoyen, M., & Derbyshire, F. (2001). PAN-based activated carbon fiber composites for sulfur dioxide conversion: Influence of fiber activation method. *Fuel*, *80*(7), 969–977. doi:10.1016/S0016-2361(00)00186-1

Mikova, N., Ivanov I., Chesnokov N., & Kargin V. (2014). Effect of oxidative treatment on the properties of carbon materials from microcrystalline cellulose. *Chemistry of plant raw material, 2*, 51-59.

Muniz, J., Herrero, J. E., & Fuertes, A. B. (1998). Treatments to enhance the SO_2 capture by activated carbon fibres. *Applied Catalysis B: Environmental*, *18*(1-2), 171–179. doi:10.1016/S0926-3373(98)00039-3

Ostafiychuk, B., Budzulyak, I., Ivanichok, N., Lisovsky, R., & Rachiy, B. (2013). Injection of thermal modification on the development of the microporous structure of the carbonaceous material. *Physics of solid body, 14*(4), 891-896.

Park, E. D., Choi, S. H., & Lee, J. S. (2000). Active states of Pd and Cu in carbon-supported Wacker-type catalysts for low-temperature CO oxidation. *The Journal of Physical Chemistry B*, *104*(23), 5586–5594. doi:10.1021/jp000583z

Polovetskaya, O. S., Arsenieva, T. A., & Polovetsky, M. D. (2015). Study of the organic matter of peat by IR-Fourier spectroscopy. *Privolzhsky Nauch. West.*, *7*(47), 12–18.

Radkevich, V.Z., Senko, T.L., & Khaminets, S.G., S. G., Wilson, K., & Egiazarov, & Yu. G. (2008). Catalytic systems based on carbon carriers for low-temperature CO oxidation. *Kinetics and Catalysis*, *49*(4), 570–576. doi:10.1134/S0023158408040149

Radkevich, V. Z., Wilson, K., Khaminets, S. G., & Sen'ko, T. L. (2014). Effect of preparation conditions on the formation of the active phase of carbon fiber catalytic systems for the low-temperature oxidation of carbon monoxide. *Kinetics and Catalysis*, *55*(2), 252–267. doi:10.1134/S0023158414020086

Rakitskaya, T., Dzhyga, G., Kiose, T., & Volkova, V. (2020). Natural Nanobentonites as Supports in Palladium(II)–Copper(II) Catalysts for Carbon Monoxide Oxidation with Air Oxygen. In *Nanooptics and Photonics, Nanochemistry and Nanobiotechnology, and Their Applications* (pp. 141–157). Springer. doi:10.1007/978-3-030-52268-1_11

Rakitskaya, T. L., Bandurko, A. Y., Ennan, A. A., Paina, V. Y., & Litvinskaya, V. V. (2000). Low-temperature catalytic decomposition of ozone microconcentrations by carbon fibrous materials. *Advances in Environmental Research*, *3*(4), 472–487.

Rakitskaya, T. L., Bandurko, A. Y., Ennan, A. A., Paina, V. Y., & Rakitskiy, A. S. (2001). Carbon-fibrous-material-supported base catalysts of ozone decomposition. *Microporous and Mesoporous Materials*, *43*(2), 153–160. doi:10.1016/S1387-1811(00)00358-9

Rakitskaya, T. L., Dzhyga, G. M., Kiose, T. A., Oleksenko, L. P., & Volkova, V. Y. (2019). Pd(II), Cu(II), and pillared clay based nanocatalysts for low-temperature CO oxidation. *SN Applied Sciences*, *1*(4), 1–8. doi:10.100742452-019-0314-x

Rakitskaya, T. L., & Ennan, A. A. (2012). *Phosphine. Physicochemical Properties and Practical Aspects of Elimination*. Astroprint.

Rakitskaya, T. L., Kiose, T. A., Ennan, A. A., Golubchik, K. O., Oleksenko, L. P., & Gerasiova, V. G. (2016). The Influence of Conditions of Acid-Thermal Modification of Clinoptilolite on Catalytic Properties of Palladium-Copper Complexes Anchored on it in the Reaction of Carbon Monoxide Oxidation. *Russian Journal of Physical Chemistry A. Focus on Chemistry*, *90*(6), 1128–1135. doi:10.1134/S0036024416060182

Rakitskaya, T. L., Kiose, T. A., Golubchik, K. O., Ennan, A. A., & Volkova, V. Y. (2017). Acid-modified clinoptilolite as a support for palladium–copper complexes catalyzing carbon monoxide oxidation with air oxygen. *Chemistry Central Journal*, *11*(1), 1–10. doi:10.118613065-017-0256-6 PMID:29086820

Rakitskaya, T. L., Kiose, T. A., Truba, A. S., & Ennan, A. A. A. (2022). Effect of Water on Activity and Protective Properties of Catalysts Used in Respiratory Protective Equipment. In *Handbook of Research on Water Sciences and Society* (pp. 469–499). IGI Global. doi:10.4018/978-1-7998-7356-3.ch021

Rakitskaya, T. L., Kiose, T. A., Vasylechko, V. O., Volkova, V. Y., & Gryshchouk, G. V. (2011). Adsorption-desorption properties of clinoptilolites and the catalytic activity of surface Cu(II)–Pd(II) complexes in the reaction of carbon monoxide oxidation with oxygen. *Chemistry of metals and alloys, 4*(3-4), 213-218.

Rakitskaya, T. L., Kiose, T. A., Voloshchuk, A. G., Oleksenko, L. P., Volkova, V. Y., & Reznik, L. I. (2009). Effect exerted by acid modification of bazalt tuff on catalytic activity of fixed acido complexes of palladium (II) and copper (II) in the reaction of carbon (II) oxide oxidation with air oxygen. *Russian Journal of Applied Chemistry*, *82*(2), 204–208. doi:10.1134/S1070427209020074

Rakitskaya, T. L., Kiose, T. A., Zryutina, A. M., Gladyshevskii, R. E., Truba, A. S., Vasylechko, V. O., Demchenko, P. Y., Gryschouk, G. V., & Volkova, V. Y. (2013). Solid-state catalysts based on bentonites and Pd (II) Cu (II) complexes for low-temperature carbon monoxide oxidation. []. Trans Tech Publications Ltd.]. *Diffusion and Defect Data, Solid State Data. Part B, Solid State Phenomena*, *200*, 299–304. doi:10.4028/www.scientific.net/SSP.200.299

Rakitskaya, T. L., Litvinskaya, V. V., Abramova, N. N., Redko, T. D., & Popova, N. A. (1987). Influence of adsorption-structural characteristics of carbon fiber material on the catalyst activity of phosphine oxidation. *Journal of applied chemistry*, *60*(6), 1415–1417.

Rakitskaya, T. L., Truba, A. S., & Ennan, A. A. (2019). *Ozone. Physical and chemical properties, and catalytic methods of decomposition*. Astroprint.

Rakitskaya, T.L., Truba, A.S., Kiose, T.A., & Raskola, L.A. (2015). Mechanisms of formation of d metal complexes on porous carriers and their catalytic activity in redox reactions. *Odesa National University Herald. Chemistry, 20*(54), 27-48.

Rakitskaya, T. L., Vasylechko, V. O., Kiose, T. A., Gryshchouk, G. V., Dzhiga, A. M., & Volkova, V. Y. (2015). Catalytic activity of Pd(II) and Cu(II) complexes anchored to natural and pre-modified bentonite in the oxidation of carbon monoxide. *Chemistry of metals and alloys, 8*(1-2), 32-38.

Romanenko, A. V., & Simonov, P. A. (2007). Carbon materials and their physical and chemical properties: textbook, 128. M.: Kalvis.

Shcherban, N. D., & Filonenko, S. M. (2017). Synthesis, power, and determination of the functionalization of carbon nanostructures. *Nanosystems, nanomaterials, nanotechnologies, 15*(1), 113-131.

Shen, C., Li, H., Yu, J., Wu, G., Mao, D., & Lu, G. (2013). A First Principles DFT Study on the Active Sites of Pd-Cu-Cl_x/Al_2O_3 Catalyst for Low-Temperature CO Oxidation. *ChemCatChem*, *5*(10), 2813–2817. doi:10.1002/cctc.201300356

Simanova, S. A., Shchukarev, A. V., & Lysenko, A. A. (2008). Adsorption of chloro complexes of palladium, platinum, and gold by carbon fibers of different structures. *Journal "Chemical fibers", 4*, 1-7.

Simonov, P. A., Romanenko, A. V., Prosvirin, I. P., Moroz, E. M., Boronin, A. I., Chuvilin, A. L., & Likholobov, V. A. (1997). On the nature of the interaction of H2PdCl4 with the surface of graphite-like carbon materials. *Carbon*, *35*(1), 73–82. doi:10.1016/S0008-6223(96)00129-7

Tamarkina, Yu.V., Bovan, L. A., Tsyba, N. N., & Kucherenko, V. A. (2008). Development of brown coal porosity during thermolysis with alkalis of different nature. *Chemistry, physics and surface technology, 14*, 224-229.

Titov, D. N., Ustyugov, A. V., Tkachenko, O. P., Kustov, L. M., Zubavichus, Y. V., Veligzhanin, A. A., & Temkin, O. N. (2012). State of active components on the surface of the $PdCl_2$-$CuCl_2$/γ-Al_2O_3 catalyst for the low-temperature oxidation of carbon monoxide. *Kinetics and Catalysis*, *53*(2), 262–273. doi:10.1134/S0023158412020140

Wang, S., Chen, Z. H., Ma, W. J., & Ma, Q. S. (2006). Influence of heat treatment on physical–chemical properties of PAN-based carbon fiber. *Ceramics International*, *32*(3), 291–295. doi:10.1016/j.ceramint.2005.02.014

Zemskova, L. A., & Sheveleva, I. V. (2004). Modified sorption-active carbon fiber materials. *Russian chemical journal*, *48*(5), 53-57.

Chapter 11
Photosensitive Nanopesticides for Environmentally Friendly and Sustainable Agriculture

Natalia Sucman
https://orcid.org/0000-0001-8733-3040
Institute of Chemistry, Moldova State University, Moldova

Fliur Macaev
https://orcid.org/0000-0002-3094-1990
Institute of Chemistry, Moldova State University, Moldova

ABSTRACT

The utilization of nanopesticides is a young and very promising direction in agriculture. Their use can significantly improve many parameters of the used drugs. Nevertheless, there are many disputes and hypotheses about the impact of such pesticides on the environment. There are many examples of nanopesticides in the literature with an inorganic active substance. They are widely used in practice and are sufficiently described in the literature. In this chapter, emphasis will be placed on the use of nanopesticides with photosensitive organic active substances. Methods for their preparation, effect on the target pathogen, and degradation pathways will be described.

INTRODUCTION

Nanotechnologies are becoming extra widespread and repeatedly moving from theoretical to practical environments. In chemistry, there are many publications about using various nanoparticles as reagents and catalysts. There is also a tremendous amount of work on producing macromolecules (fullerenes, rotaxanes, nanotubes, etc.), which can be attributed to nanoparticles in size and exhibit some exciting and unique properties compared with ordinary substances. Nanotechnology is well-represented in the pharmaceutical industry. There are examples of not only obtaining but also using nanoparticles. For instance, nanoencapsulation of the drug active substance is quite widely developed, which makes it

DOI: 10.4018/979-8-3693-0512-6.ch011

possible to improve its solubility and, accordingly, bioavailability, and thus makes it possible to reduce the concentration of the active substance and thereby reduce the overall chemical load on the body. In addition, this approach provides a prolonged action of the drug. For this purpose, natural and modified saccharides (cyclodextrins, dextrans, pectin, chitin, etc.) are frequently employed in the pharmaceutical industry. Although nano preparations have been utilized in pharmaceuticals and their effect has been studied for a long time (Madhusudhan et al., 2017; Macaev et al., 2013; Luo et al., 2014; Bhattarai et al., 2010), they are not so common in agriculture. It is worth noting that today copper hydroxide nano preparations are almost universally used in agriculture to combat fungal plant diseases (Kaur et al., 2014). Despite promising prospects for the benefit of organically supported nanoparticles for plant protection, they are not expansively used in practice. Moreover, there are quite a few patents on this subject (Asrar, 2003, 2010, 2016; Carr, 1993; Doane, 1990; Dellicoll, 1980).

In addition, although agriculture, in general, has never been especially green, and there are many examples of the overuse of plant protection products and fertilizers in pursuit of high yields, the possibility of using nano pesticides has caused a lot of environmental disputes. However, there is a lack of systematic data on the fate of such products in the environment, and the possibility of evaluation of the nano pesticides by the current pesticide regulatory framework is unclear. At the same time, there are a limited number of examples in the literature of studying the pathways of degradation and removal from the environment of nano pesticides with organic active ingredient (AI), as well as their accumulation in soil, plants, and further along the food chain. It should also be comprehended that, despite the recent popularity of organic farming, the need to feed an ever-increasing population of the planet will lead to the continuous use of pesticides in ever-increasing quantities. The purpose of this chapter is to explore the available information on nano pesticides containing some organic active substances. It will consider the influence of nanocarriers on the photodegradation of the active substances in nanoformulations.

BACKGROUND

The annual consumption of active ingredients of pesticides is approximately four million tons worldwide and nearly half-a-million tons in Europe (FAO, 2018). This represents a massive, long-term, and mostly unavoidable input into the soil, water, air, and crops of highly hazardous to human health and ecosystem compounds (Malakof & Stokstad, 2013; Arias-Estevez et al., 2008). However, only a small fraction of the active ingredients reach the target pests, while over 90% of the conventional pesticide formulations used are lost to the environment through runoff, spray migration, degradation, and leaching (Hofmann et al., 2020; Singh et al., 2020). The poor utilization of active ingredients results in the frequent application of traditional pesticide formulations to achieve a satisfactory control effect (Singh et al., 2020; Tleuova et al., 2020). The unreasonable and excessive use of pesticides not only leads to the resistance of pests and waste of resources but also causes the destruction of soil biodiversity and the eutrophication of water (Rai et al., 2018; Stubenrauch et al., 2021). That's why the development of a sustainable strategy for prolonging the effective duration, improving the utilization efficiency, and reducing the nontarget toxicity of pesticides is a very urgent target.

So, over the past decade, new types of plant protection products, called nanopesticides or pesticide nanoformulations (NFs), have been developed, which have employed nanosized elements to gain novel technological, toxicological, or environmental features (Gogos et al. 2012; Kah et al. 2013; Perlatti et al. 2013; Kookana et al. 2014; Campos et al. 2014; Nuruzzaman et al. 2016). Stimuli-responsive pesticide delivery systems based on nanocarriers have attracted widespread attention due to their ability to improve technological properties of AI such as stability, solubility, stiffness, adhesion or miscibility, improve target activity owing to greater persistence, slow/controlled release, targeted/smart delivery, improved bioavailability, and permeability, and minimize the adverse effects on the environment risks by higher degradability, and exclusion of flammable or toxic solvents and surfactants (Liang et al., 2020, 2022; Xu et al., 2021; Zhang et al., 2022). Compared with traditional pesticides, nano-pesticide formulations can provide better leaf wettability, chemical stability, and water dispersibility in line with the development of sustainable agriculture.

The term nanopesticide encompasses a great variety of products covering a wide range of properties and levels of development (Kah et al., 2018). Developments in nanotechnology over the past decade have allowed formulation scientists to gain a better understanding and better control of structures at the nanoscale and allowed an improvement in formulation design to suit a given purpose. Nanopesticides are generally classified into two types: (a) organic molecular active ingredient (AI)-nanocarrier complexes (polymers) (Kah et al., 2016) and (b) inorganic nanoscale AI without carriers, such as $Cu(OH)_2$ nanopesticides (X. Zhang et al., 2019).

Regardless of the nanopesticide formulation type, it is expected to be able to improve pesticide efficacy, enhance AI stability, prolong its effective duration, and reduce environmental loads of pesticides. Currently, nanoagriculture focuses on farming by employing different nanocarrier polymers to form bionanocomposites with superior properties to improve livestock and crop production (Perlati, 2013). The theory of incorporation of bioactive in polymers includes adsorption (Karimi et al., 2016; Otero et al., 2012) dispersion of the active component in the matrix, and encapsulation (Neubauer et al., 2014). Moreover, the covalent bonding between functional groups of the polymer and the active component also serves as a matrix.

Nanocarriers

The nanonature of pesticide formulations alters the environmental fate of AI properties (sorption, degradation, mobility, persistence), exposure (bioavailability, bioaccumulation), (eco)toxicity, and risk of the active ingredient (Kah et al. 2013, 2018; Kah & Hofmann 2014). The physicochemical properties of the nanocarrier, the kinetics of the release of the active ingredient from the nanocarrier, and protection from photochemical degradation will play an essential role in these processes (Kookana et al. 2014; Kah & Hofmann 2014; Walker et al. 2018).

When using natural polymers as carriers, the protection of the active substance from photolysis occurs primarily due to the slow release of the ingredient into the environment. However, the release rate should not be too low to provide a sufficient concentration of pesticide-active substances, nor too high to prevent the excess active ingredient from being photodegraded. The composition, structure of nanocomposites, and the conditions under which the optimal release rate occurs are called controlled release. Several release mechanisms have been proposed and discussed by Neri-Badang et al. (2019). They are largely dependent on the design of the controlled release (CR) polymer matrix. Figure 1 shows the most common ones, such as diffusion from pores, desorption from the surface, and release due to erosion of the carrier. The diffusion-controlled mechanism relies on the differences in the rates between diffusion of the active compound with the relaxation of the polymer chains. Surface desorption of the actives adsorbed on nanoparticles usually exhibits high initial release or burst release which can be modulated by increasing the crosslinking density or using a high molecular weight polymer. Finally, the release of the active compound can also be controlled by the erosion of the matrix. Polymer matrix degradation can be caused by pH, temperature, pressure, enzyme, and other environmental factors. Furthermore, mechanical triggers, such as shaking, sonication, and vortexing, can also release physically bound to the polymer matrix pesticides.

Figure 1. Mechanisms of AI release from polymer matrix

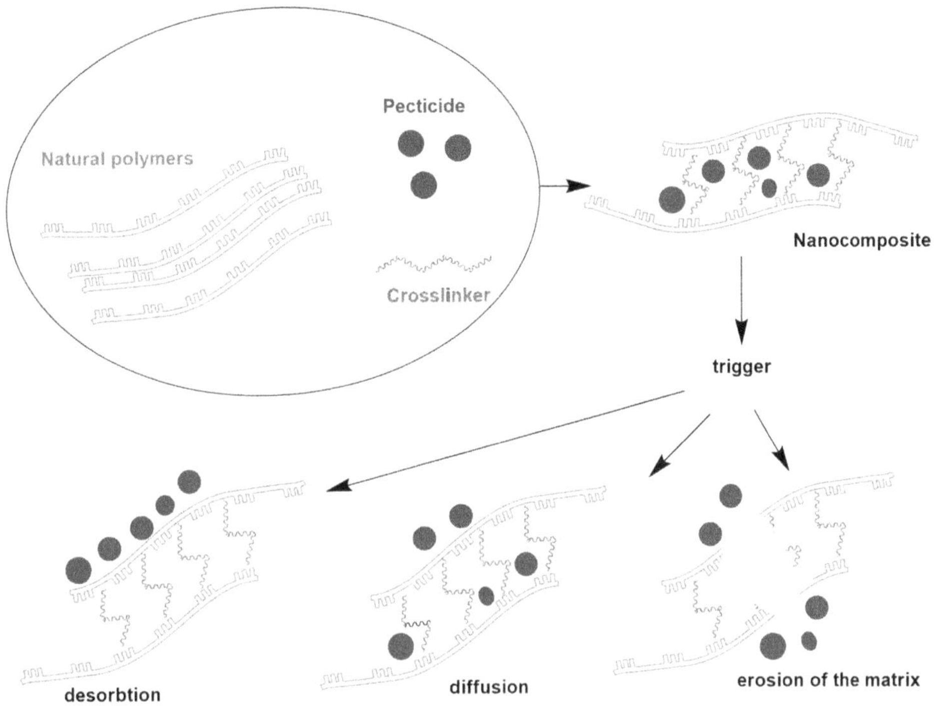

Controlled-release technology addresses problems associated with the excessive use of toxic agricultural chemicals. Nanoformulation, nanoencapsulation, and the development of polymeric nanocomposite systems resolve the main issues of controlled release and photo-protective properties. Depending on the properties of the active substance and the final features of the target composite, various polymer/macromolecular additives for the synthesis of nanoformulations can be utilized. The behavior of these novel nanomaterials depends on the type and content of reinforcement, the structure of the polymer/nanofiller, the conditions of the process, and the final dispersion. Some strategies as incorporating pendant groups, cross-linking, and grafting with other polymers are required to enhance the natural polymer properties as controlled release (CR) nanocarriers. To find out the effect of the carrier on the properties of the active ingredient, and the way in which it can be used in the directed design of nanoformulations, the main characteristics of the various carriers will be briefly presented below.

Polysaccharides

Different polysaccharides – alginate, starch, cellulose, cyclodextrin, pectin, chitosan, and others – are used as vehicles to encapsulate or absorb active ingredients to elaborate CR systems for agrochemicals. The use of each of them has several advantages and some properties that limit their use. Over the years, ***chitosan*** has gathered attention from researchers due to its polycationic nature, biological versatility, and exceptional physicochemical properties. Chitosan is deacetylated derivative of chitin (β-(1–4)-poly-N-acetyl-D-glucosamine) obtained by partial chemical or enzymatic hydrolysis (Knidri et al., 2020). The last one is the most abundant natural polymer. It was found in the exoskeletons of shrimps, crabs, cuticles of insects, and cell walls of fungi (Younes & Rinaudo, 2015). Chitosan is known for its wide range of distinct properties like chelation, viscosity, solubility in different vehicles, film-forming and mucoadhesive, polyelectrolyte behavior, and polyoxysalt formation.

It was established that the unbranched and linear form of chitosan has an excellent viscosity. Furthermore, the viscosity of chitosan modifies due to changes in the deacetylation conditions. The structure of chitosan is of great significance since it affects its physicochemical characteristics. Chitosan consists of three reactive positions: one amino group and two hydroxyl groups in every glycosidic residue. The solubility of chitosan depends on the pH-responsive amino groups protonated at lower pH, permitting chitosan to dissolve (Kumiriska et al., 2011). Also, its solubility depends on the positioning of the acetyl groups along the chain, deacetylation methods, and ionic strength.

Various studies have reported that chitosan is a non-toxic, biocompatible, biodegradable polymer (Kim et al., 2008; Bagheri-Khoulenjani et al., 2009; Shariatinia, 2019). Both chitosan and its derivatives exhibit anti-bacterial (Ardila et al., 2017; Wang et al., 2020; Verlee et al., 2017), and anti-fungal (Ashrafi et al., 2020; Shih et al., 2019) properties. Because chitosan exhibits its own bioactivity, it becomes even more attractive for use as a carrier. A synergistic effect, which leads to an increase in the activity of the active substance in nanoformulations, can also be observed.

However, the application of chitosan is limited by its solubility in many polar solvents and water. Usually, this obstacle is resolved by modifying the chitosan molecule through chemical/enzymatic methods to produce new depolymerized derivatives. The modifications of chitosan are influenced by several factors, such as molecular weight, viscosity, and reaction conditions. The amino and hydroxyl groups in the chitosan structure form the basis of their association with other polymers and biomolecules. There are known several methods to modify chitosan in the literature, for example, physical modification (Eidelman & Simon, 2004; Khor & Lim, 2003; Park et al., 2001), phosphorylation (Amaral et al., 2005), thiolation (Kaur et al., 2020), hydroxyalkylation (Aranaz et al., 2010), sulfation (Jayakumar et al., 2007), carboxyalkylation (Mourya et al., 2010), N-alkylation (Bobu et al., 2011), O-alkylation (Chen et al., 2012), and other (Shukla et al., 2013; Jayakumar et al., 2005; Torres et al., 2012). The hydrophilic (D-glucosamine) and hydrophobic parts (N-acetylated residues) (Figure 2) in chitosan molecules allow it to form emulsion systems.

Bionanocomposites have been identified as the most viable hybrid products made by incorporating nano-sized organic fillers into a biopolymeric matrix under controlled conditions. The formation of chitosan-based bionanocomposites is represented in Figure 3.

Since chitosan is a natural biopolymer, it has become an attractive alternative to synthetic plastic polymers hazardous to the environment (Kumari, 2016). ***Cyclodextrins*** are a family of compounds that consists of three cyclic natural oligosaccharides and their semi-synthetic derivatives. Naturally occurring α-, β- and γ-cyclodextrins are torus-like macro-rings built up from 6, 7, and 8 glucopyranose units, respectively. The internal cavity of the cyclodextrins is lined by hydrogen atoms and the glycosidic oxygen bridges, which produce a high electron density, causing some hydrophobicity (Szejtli, 2004). In an aqueous solution, the cyclodextrin cavity is occupied by water molecules that are energetically unfavored due to polar-non-polar interactions, and therefore can be readily substituted by any less polar guest-molecule added to the solution (Figure 4). The main reasons why the cyclodextrins are so popular as hosts for the inclusion of different substances are: they are produced from a renewable natural material (starch) and are produced in amounts of thousands of tons per year; they are nontoxic in consumable concentrations; they are biodegradable and do not pollute the environment (Morin et al., 2021).

Figure 2. The hydrophilic and hydrophobic parts in chitosan molecules

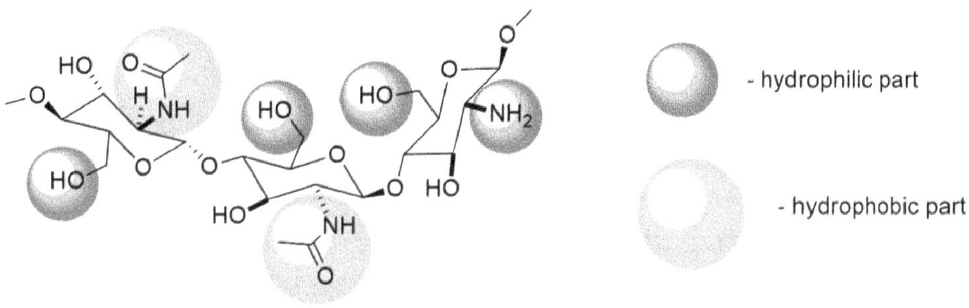

Figure 3. The formation of chitosan-based bionanocomposites

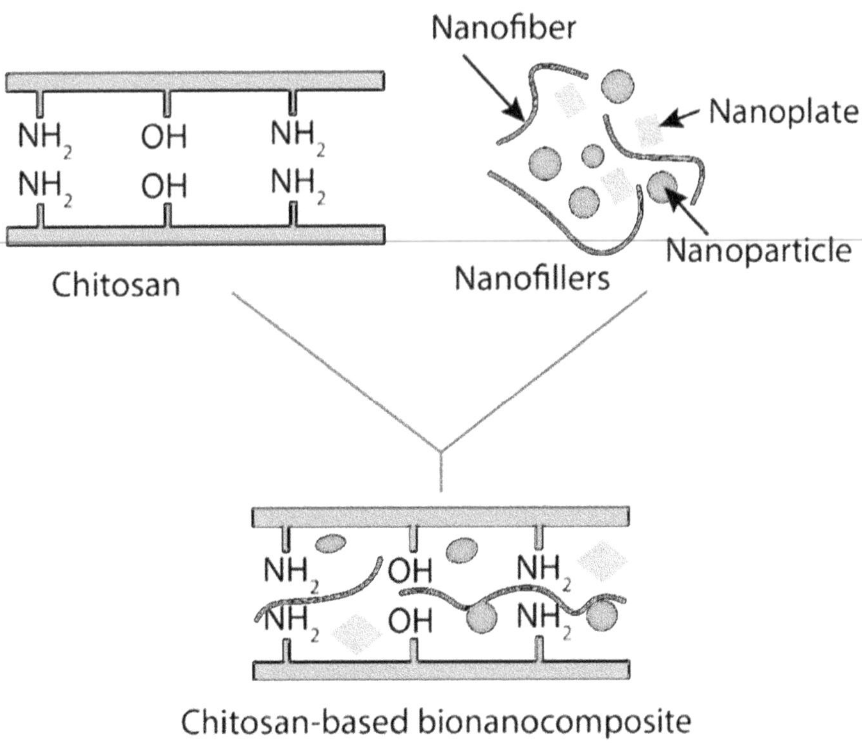

Figure 4. Formation of the inclusion complexes of cyclodextrins with endogenous substances as guest molecule

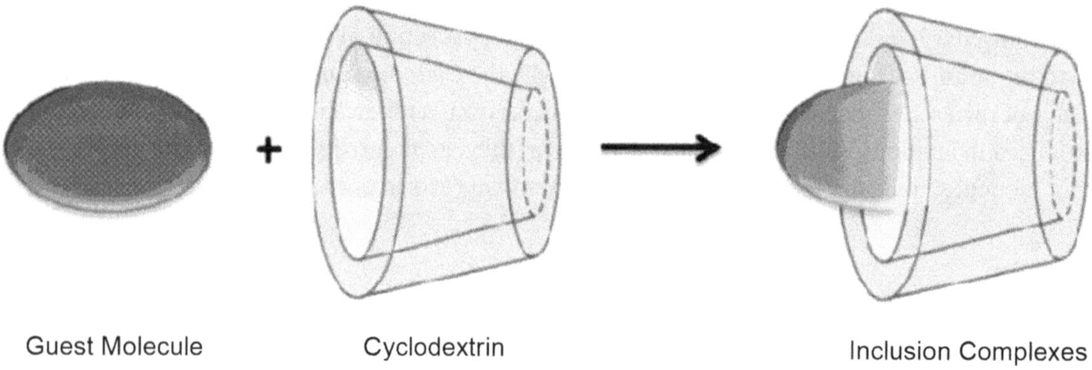

In the agricultural sector, cyclodextrin complexation has assisted pesticide formulation in many ways by enhancing water solubility, and bioactivity, and through the formation of crystalline substances from volatile liquids. An important feature of cyclodextrin-formulated pesticides is the water-triggered release of active compounds. A number of articles concerning the use of cyclodextrins in agrochemistry were filed (Szente & Szejtli 1996; Morillo, 2006). The underivatized or native β-cyclodextrin's low water solubility restricted its advantage. For this reason, a number of derivatives with different functions for various industrial uses such as hydrophilic methylated derivatives, sulfur- and nitrogen-, imidazole or histamine-containing derivatives, alkylated and acyl derivatives, halogenated products, polymers from cyclodextrins, etc. have been synthesized (Sébille, 1987).

Lignocellulosic building blocks (cellulose, lignin, and hemicellulose) are ideal candidates to manufacture eco-friendly nanocarriers given their low cost, abundance, and sustainability (Zhang et al., 2019; Shrestha et al., 2021). Moreover, lignocellulosic material can be obtained from agriculture side products, which is ideal to reduce costs during the production of nanopesticides, which are intended to be applied in large-scale agricultural operations. The application of lignocellulosic in nano-enabled agriculture is currently related to the controlled delivery and release of pesticides and nutrients (Worrall et al., 2018). Also, lignocellulosic nanomaterials could potentially stabilize emulsions (such as those used in oil-borne pesticides) instead of conventional surfactants in multiphase systems (Tardy et al., 2021).

Cellulose, as the most abundant natural material, is regarded as one of the most ideal substitutes for non-degradable supporting materials for CRFs thanks to its good biocompatibility and biodegradability. As raw cellulose materials suffer several problems, such as poor mechanical strength, fast release rate, etc., chemical modifications are commonly performed to improve their properties (Pang et al., 2019). Cellulose is a kind of carbohydrate polymer consisting of D -anhydroglucopyranose units linked by β -(1→4) -glycosidic bonds. Due to the extensive hydroxyl groups on the cellulose backbones, several kinds of chemical modification can be performed. The modified cellulose materials can be rationally designed for CRFs of agrochemicals via esterification, etherification, grafting, and crosslinking are detailly introduced.

Proteins

Protein materials are promising candidates as the building blocks for functional and high-performance bionanocomposites, owing to their unique and well-developed nanoscale structure, rich chemical functionality, excellent mechanical properties, biocompatibility, and biodegradability. Rational integration of protein materials with synthetic organic and inorganic nanomaterials through tailored interfacial interactions leads to synergistic enhancement in the properties compared to the individual components. Due to the complexity of protein composition and structure, proteins possess multiple function properties, such as solubility, gelation, elasticity, emulsification, and cohesion-adhesion. Proteins are amphoteric molecules, and they can migrate spontaneously to an air-water interface or an oil-water interface. Once at the interface, proteins can interact with the neighboring molecules and form a strong cohesive, viscoelastic film that can withstand thermal and mechanical motions. Therefore, proteins are used as adhesives in nanoformulations. It should be mentioned that natural zein as a potential carrier for nanoencapsulation of pesticides has the disadvantages of large amounts of organic solvents and poor dispersibility, but also is widely used as a result of different manipulations.

The protein nanoparticles are formed due to the conformational changes of protein, which depend upon its composition, cross-linking, concentration, and different chemical conditions like pH, ionic strength, type of solvents, etc. (Wang et al., 2014). During the unfolding process, protein expresses various interactive groups like disulfides, thiols, hydroxides, etc. Interestingly, thermal and chemical cross-linking of the groups guide the formation of protein nanoparticles with entrapped drug molecules (Figure 5).

Figure 5. Schematic diagram of obtaining the protein-based nanomaterial

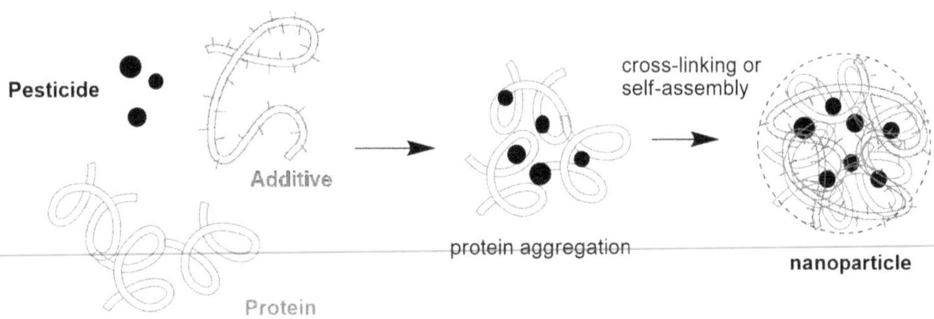

Casein, collagen, gelatin, soy protein, zein protein, and others are low-cost and high availability heteropolymers bionanocomposites conformed by different amino acids and can play a fundamental role like N source after their disintegration. Despite the benefits, proteins are used less as CR systems for agrochemicals (Zhao et al., 2008).

THE INFLUENCE OF NANOCARRIERS ON PHOTOSENSIBILITY OF FORMULATIONS

Developing long-acting and sustained-release systems is a feasible and effective way to increase the efficiency of pesticides, especially for photosensitive pesticides. Pesticide deposits post-treatment and before diffusing inside the plants are exposed to sunlight. The phenomenon of photolysis constitutes one of the natural degradation pathways of a molecule when it is subjected to a particular light radiation. When this radiation can be absorbed by the molecule, it causes the photoexcitation of this molecule. The absorbed energy causes the electron excitation of the electrons constituting the peripheral layers of the atoms and of the bonds between atoms, the vibrations caused can thus cause the rupture of the chemical bonding of a molecule. Photolysis contributes to the formation of radicals by the rupture of carbon-carbon bonds. As follows, many pesticides degrade into a variety of photo products that may be harmful to living beings through accidental ingestion. Therefore, the use of ultraviolet absorbers in pesticides or nanocarriers that would protect the active substance from sunlight is an attractive strategy to prevent the photodegradation of pesticides. Some of the successful attempts to protect active substances from sunlight using nanotechnology will be presented below.

Protection from the degradation process of benzimidazole anthelmintics appears very important because it uses high in tropical nations where temperature and light are particularly elevated. The mechanism of photodegradation was studied by Ragno et al. (2006). The formation of two main photoproducts was described, formed from two sequential reactions, a first demethylation of the estheric group followed by a decarboxylation of the carbamic group. Light was demonstrated to be the effective cause of degradation whereas high temperature seemed not to be influent. To prevent photodegradation, cyclodextrin polymer (CDP)-valved, benzimidazole functionalized, MoS_2-embedded mesoporous silica nanopesticides (MoS_2-MSN-CDP) were obtained by the group of Dong et al. (2021a).

MoS_2 nanosheets are strong photothermal agents in the NIR spectral region and have excellent photothermal conversion efficiency (PCE) due to their narrow direct bandgap (ca. 1.8 eV) (Rice et al., 2013), even better than graphene and gold nanorods. The photothermal effect of MoS_2-MSN-CDP was investigated and these nanoparticles (NPs) dispersed in water showed obvious UV–vis–NIR absorption. PCE of MoS_2-MSN-CDP was calculated to be 15.7% under irradiation of NIR light at 808 nm. The photothermal effect of the MoS_2 cores generated by exposure to near-infrared light (NIR) from sunlight (the light in the NIR region accounts for about 52% of the total solar irradiation energy), weakened the host-guest interactions of the supramolecular nanovalves, resulting in the controlled release of encapsulated pesticides. Furthermore, the splash and bounce of the aqueous droplets containing MoS_2-MSN-CDP nanoparticles on superhydrophobic surfaces were well inhibited and available for excellent wetting to prevent pesticides from losing to the environment. Authors report about the foliage-triggered controlled release of pesticides, owing to the competitive binding of epicuticular wax components to CDP. The controlled-release formulation has demonstrated much stronger fungicidal activity against pathogenic fungi than commercial formulations and proposes a fascinating strategy for reducing environmental pollution.

The same group of scientists, led by Dong (Dong et al., 2021b) used carboxylated porous carbon nanoparticles (PCN) to encapsulate water-insoluble pesticides and subsequently capped with chitosan (CS) to prepare the CS-gated PCN (PCN-CS) nanopesticides for the controlled release of paraquat (PQ), in response to acidic pH and elevated temperature with good fungicidal efficacy. Water-soluble paraquat (1,10-dimethyl-4,40-bipyridinium dichloride) is one of the most widely used herbicides in agriculture in the world. The way of its photodegradation was studied (Florencio et al., 2004). During this process, several intermediate and reaction products are sequentially formed, to which structures are proposed. PQ can cause phytotoxicity rapidly upon contact with the green tissues of plants, dependent on the photosynthetic activity of plants exposed to sunlight and on O2. The controlled release of PQ from the nanoherbicides was realized at elevated temperatures owing to the weakening of the strong π–π interactions, aiming to eliminate weeds via the photothermal effect of PCNs under natural sunlight.

Avermectin (AVM) is a widely used insecticide, acaricide, and nematicide. AVM is a combination of avermectins, containing more than 80% avermectin B1a and less than 20% avermectin B1b. AVM is a group of 16-membered macrocyclic lactone. These naturally occurring compounds are generated as fermentation products by *Streptomyces avermitilis*, a soil actinomycete. However, the structure of the conjugated double bond and 16-membered ring macrolide of AVM can be easily degraded under UV irradiation, resulting in lower efficacy and limited application (Liu et al., 2012; Crouch et al., 1991, Blizzard et al., 1988). Encapsulation of AVM with a carrier can protect avermectin from photodegradation and oxidation by isolating it from the external environment.

It is known about Polyamine-modified zein (AM-zein), which was synthesized by incorporating ethylenediamine into zein to improve its stability as a nanocarrier and enhance the electrostatic force between the carrier, pesticides, and plant foliage. Using AM-zein as carrier AVM-loaded nanoparticles were prepared. Compared to the commercial emulsifiable concentrate (EC), these nanoparticle dispersions exhibited better leaf affinity, and the retention of dispersion increased twice. It was discovered that the UV barrier effect of zein increases while prolonging the UV exposure time. The insecticidal activity of AVM-nanoparticles did not improve compared to EC, but they demonstrated significantly lower toxicity against zebrafish (Chen et al., 2021).

To enhance the hydrophilicity of zein and achieve a nano-scale pesticide system, phosphorylated zein (P-Zein) was incorporated with graft copolymer CMCL-g-PDMDAAC by electrostatic interaction, in which carboxymethyl cellulose (CMCL) and diallyl dimethylammonium chloride (DMDAAC) as monomers. P-Zein/CMCL-g-PDMDAAC was applied to encapsulate model pesticide avermectin via electrostatic interaction to achieve AVM@P-Zein/CMCL-g-PDMDAAC nano-pesticides (Hao et al., 2019). The thickness of CMCL-g-PDMDAAC was increased after encapsulation of AVM-P-Zein, meanwhile, the anti-UV light performance increased by almost 10%.

As a result of the study of Dai et al. (2022), water-soluble zein peptides were obtained by hydrolyzing under alkaline conditions at increased temperatures, and the stable nanopesticides were prepared through crosslinking after self-assembly. After the zein peptides spontaneously assembled into spherical particles in an aqueous solution, the nanopesticides-AVM, with uniform size distribution can be adjusted by the glutaraldehyde. Such nanoparticles have great stability and can be redispersed in water after freeze-drying. Moreover, nanoencapsulation protects avermectin from UV light, with 24% of the encapsulated AVM remaining intact after exposure to UV for 60 min, compared to less than 6% for the bare AVM. In addition, the nanoformulation had similar insecticidal activity to commercial EC and the cross-linking of particles slowed down the release rate.

Preparation and characterization of porous hollow silica nanoparticles (PHSNs), with various shell thicknesses in the range of 5–45 nm and a pore diameter of about 4–5 nm, were investigated (Li et al., 2006). The produced PHSNs were applied as controlled pesticide release carriers to study the effects of the shell thickness on the loading efficiency for AVM, the UV-shielding property for the loaded AVM, and the controlled release of the loaded AVM from the carriers. It was found that the amount of loaded AVM decreases with the increase of shell thickness, while the UV-shielding property of PHSNs for AVM is improved as the shell gets thicker. In addition, shell thickness has a significant impact on avermectin release. Increasing the shell thickness in the range of 5–45 nm leads to a more sustained release by decreasing the release rate of the pesticide from PHSNs, showing that the shell thickness is one of the main controlling factors for the active agent release from such systems.

Due to the selective permeation and protection properties of the semi-permeable membrane, pesticide microcapsules have been widely used. He et al. (2013) developed a novel two-step method for synthesizing highly stable silica–glutaraldehyde–chitosan composite AVM microcapsules. The silica shell was formed through the hydrolysis and polycondensation of tetraethyl orthosilicate (TEOS) at the oil droplet–water interface by using TEOS as the silica precursor and hexadecyl trimethyl ammonium chloride as a surfactant. Then the silica shell was modified with 3-aminopropyltriethoxysilane. Chitosan nanospheres were prepared by adjusting the pH value of the solution and then cross-linking with modified silica at the surface of the silica shell in the presence of glutaraldehyde to form double-shelled AVM microcapsules. The results showed that the resulting microcapsules had a remarkable loading ability for AVM (about 40%) and protect AVM against photo- and thermal degradation effectively. Compared to single-shelled microcapsules, the double-shelled ones had better-controlled release properties under all conditions.

Another example of a novel type of nanopesticide based on avermectin and composite nanocarriers made from the functionalization of boron nitride nanoplatelets with 3-mercaptopropyl trimethoxysilane and poly(ethylene glycol) diacrylate (PEG) was presented (Hao et al., 2020). Such nanocarriers exhibit a high pesticide loading capacity due to their layered nanostructure and their favorable interactions with AVM molecules achieved through hydrophobic effect, π-π stacking, and electrostatic interactions. By changing pH from 7 to 11, the release kinetics could be transformed to a zero-order process from a first-order process, with about a two-to-three-fold increase in the release rate. The pH-responsive behavior was ascribed to the hydrolysis of ester groups of PEG under alkaline conditions. These pesticide nanocarriers reduced the degradation rate of AVM against UV irradiation by about 30%. In summary, novel nanopesticides based on BN composite nanocarriers open new avenues for the delivery of hydrophobic pesticides in the field of agriculture and crop protection.

Oliveira et al. (2018) prepared zein protein nanoparticles loaded with two pesticides, geraniol, and R-citronellal, using an anti-solvent precipitation technique and analyzed them by various material characterization methods. Zein NPs containing plant repellents showed high encapsulation efficiency (>90%), good physicochemical stability, and protection against UV degradation.

α-Amylase-responsive carrier for controlled release of AVM was prepared based on α-cyclodextrin (α-CD) anchored hollow mesoporous silica (HMS) using α-CD as a capping molecule (Kaziem et al., 2018). The release of AVM was studied at different temperatures, pH values, and in the presence or absence of α-amylase. The results revealed that the AVM-encapsulated controlled release formulation (AVM-CRF) has a drastic enzymatic dependence, an excellent encapsulation efficacy reaching 38%, and outstanding UV and thermal shielding ability. The technical AVM was easier photodegradation than AVM in AVM-CRF. The technical AVM exhibited more than 98% decomposition after 3 h of UV radiation. After 24 h of UV radiation, however, less than 9% of AVM in the CRF was photolyzed. The revealed data confirmed that the prepared CRF can protect AVM from photolysis, which demonstrated the extraordinary UV-shielding property of the AVM-CRF. Furthermore, the AVM-CRF biological activity survey shows also excellent toxicological properties against *Plutella xylostella* larvae, which confirms that α-CD caps could be uncapped enzymatically *in vivo* and release AVM, inducing *P. xylostella* larval death. AVM-CRF has a notable capability to keep AVM biologically active until the 14th day with 83.33% mortality of the target insect, which was 40% higher than that of treated with AVM commercial formulation. The study provides a theoretical basis for the application of pesticide reduction.

Peng et al. (2020) reported the preparation of anionic and cationic lignin-based AVM-loaded materials with strong hydrophobicity by using Sodium Lignosulfonate (SL) and cetyltrimethylammonium bromide (CTAB) via self-assembly. The obtained SL/CTAB was used as drug-loaded material and emulsifier to prepare AVM nano-formulation, which can be automatically dispersed into nano drug-loaded nanospheres in water. The cold and hot storage experiments show that the physical and chemical stability of the nano-formulation is good. The nano-formulation exhibits controlled-release performance, and the cumulative release amounts range from 56.27% to 87.33% in 62 h. Meanwhile, the release rates slow down with increasing SL/CTAB dosage. After UV irradiation for 50 h, the retention rates of AVM in the nano-formulation range from 46.67% to 63.41%, which is 2.18-2.96 times higher than commercial EC.

Stimuli-responsive pesticide formulations improve the utilization efficiency of target pests and reduce harm to non-target organisms and the environment. So, multi-stimuli-responsive AVM polyurea microcapsules (AVM@CMCL-SS-PU) have been prepared by interfacial polymerization with modified carboxymethyl cellulose CMCL-SS-NH_2 as the wall material and hexadecane as the temperature-responsive core (Zhao et al., 2023). The encapsulation efficiency of AVM was 88.23%, and the photostability of AVM@CMCL-SS-PU was 5 times higher than that of the AVM solution. The insecticidal effect of AVM solution and AVM EC decreased to 13.3% and 16.6%, respectively, after UV irradiation for 180 min, whereas that of AVM@CMCL-SS-PU remained at 50.0%. AVM@CMCL-SS-PU has better foliar affinity and releases under the stimuli of temperature, glutathione, pH, cellulase, and urease. Hence, it has high insecticidal activity and biosafety. This smart controlled-release pesticide formulation provides a promising solution for green agriculture.

Yan et al. (2013) proposed the use of cyanobacteria as a potential carrier for pesticide avermectin. The efficiency of cyanobacteria, an environmental waste generated in eutrophic water, for the controlled release and photostability of AVM was evaluated. Carbopol was used as a resin to coat the avermectin-loaded cyanobacteria. Utilization of the cyanobacteria could prolong the release of avermectin in a mixture of ethanol/water (4:1 v/v), especially with the addition of carbopol resin. As the cyanobacteria contained a large amount of scytonemin, which can absorb UV light, it could effectively protect AVM against UV radiation.

Abamectin (ABM), is a member of the avermectin family and is a natural fermentation product of soil-dwelling too. ABA is very sensitive to UV light. In general, the photostability of abamectin is improved by encapsulation. To increase both the adhesion to foliage and the effective utilization rate of this pesticide, Yu et al. (2017) developed three types of functionalized abamectin poly(lactic acid) (ABM-PLA) nanoparticles (CH_3CO-PLA-NS, HOOC-PLA-NS, and H_2N-PLA-NS) with different adhesion abilities to cucumber foliage. The prepared ABM-PLA nanoparticles were spherical with diameters of around 450 nm, and their maximum ABM loading rate was around 50%. The nanoparticles exhibited better continuous-release behavior and photostability compared with active ABM. The photodegradation rate of active abamectin is relatively fast, with almost 50% decomposing after 38 h of continuous UV irradiation. In contrast, very little of the ABM loaded in the ABM-PLA nanoparticles decomposed after the same time. These results indicate that ABM-PLA nanoparticles possess higher photostability than that active ABM, resulting from the protection of PLA as a carrier.

Emamectin benzoate (EB), a relatively novel insecticide, was derived from the avermectin family with improved thermal stability, greater water solubility and a broader spectrum of insecticidal activity than avermectin. The results of the investigation indicated that UV photodegradation has a significant effect on the efficacy of emamectin benzoate and the effect is concentration-dependent (Zhu et al., 2011). It was reported (Huang et al., 2019) about an environment-friendly emamectin benzoate-loaded alginate/

chitosan/sodium lignosulfonate (EB-loaded ALG/CS/SL) polyelectrolyte was successfully prepared via layer-by-layer self-assembly. Besides the high encapsulation efficiency (93.3 ± 2.08%), this protective coating strategy exerted an outstanding performance of sustained-release (up to 204 h) and anti-photolysis for EB. As a result, the degradation amount of EB in EB-loaded ALG/CS/SL could be reduced by 5.7 times than that of uncoated EB under UV light.

Song et al. (2019) report a multifunctional nano platform, carboxymethyl chitosan modified carbon nanoparticles (CMC-CNP), as the carrier for emamectin benzoate, and investigate its sustainable antipest activity. EB was loaded on CMC-CNP nanocarrier via a simple physisorption process, with a high loading ratio of 55.56%. The EB-CMC-CNP nanoformulation showed improved solubility and dispersion stability in an aqueous solution, which is of vital importance to its practical application. Different from free EB, EB-CMC-CNP exhibited pH-responsive controlled release performance, leading to sustained and steady EB release and prolonged persistence time. In addition, the significantly enhanced anti-UV property of EB-CMC-CNP further ensured its antipest activity. Therefore, EB@CMC@CNP exhibited superior pest control performance than free EB.

Controlled release formulations are considered an effective way to solve the low bioavailability of traditional pesticides. However, a disadvantage of CRFs prepared by coating or encapsulation is the explosive release of the ingredients. Sustained-release pesticides prepared by coupling with a carrier can overcome this shortcoming. Huang et al. (2022) using sodium lignosulfonate (SL) as a carrier, an emamectin-lignin-sulfonic acid conjugate (EB-SL) was prepared in which emamectin was linked by sulfonamide bonds to lignin. The sustained-release results showed that EB-SL maintained its original structure when released in pure water and soil columns, and the sulfamide bond did not break. The photolysis test displayed that the photolysis half-life $T_{0.5}$ of EB-SL was increased by 1.5 times compared with the EB-SC concentrate. Bioactivity tests in the greenhouse showed that EB-SL not only had similar insecticidal toxicity to emamectin EC against *Ostrinia nubilalis* but also displayed a longer duration. The lethality of EB-SL on *O. nubilalis* was maintained at more than 70% across 19 days, whereas EB-EC as the control was less than 50% after 11 days of application.

Another team (Cui et al., 2019) developed high-speed emulsification and ultrasonic dispersion by electrostatic self-assembly of SL and EB a nanocomposite EB-SL-Nf with a particle size of 150–250 nm. The results of photostability and the controlled release test for EB-SL-Nf showed that the antiphotolysis ability of the nanoformulation is four times higher than that of the commercial EB-emulsifiable concentrate, and the nanoformulation has a pH-responsive controlled release function. Moreover, the duration of the insecticidal effect of EB-SL-Nf was more than one time longer than that of EB-suspension concentrate in the greenhouse. Therefore, EB-SL-Nf has wide application prospects in agricultural production and environmental protection due to its easy preparation, low cost, and environmental friendliness. Thus, the strategy is suitable for large-scale industrial production.

Pyraclostrobin is a strobilurin fungicide and is widely used for the protection of plant cultivation from several high-risk diseases. In Liang's et al. (2022) work, an intelligent redox and pectinase dual stimuli-responsive pesticide delivery system was constructed by bonding pectin with metal-organic frameworks (FeMOF nanoparticles) which were loaded with pyraclostrobin (P) ([FeMOF-P-protected] nanoparticles). The results indicated that the loading capacity of [FeMOF-P-] nanoparticles for pyraclostrobin was approximately 20.6%. The pectin covered on the surface of [FeMOF-P-] nanoparticles protects pyraclostrobin from photolysis effectively.

Fernandes et al. (2014) evaluated the influence of hydroxypropyl-b-cyclodextrin on the photostability of the fungicide pyrimethanil. The formation of the pyrimethanil/HP-β-CD inclusion complex increased the aqueous solubility of this fungicide around five times. To assess the influence of microencapsulation on the environmental photostability of the fungicide, the photochemical degradation of pyrimethanil and pyrimethanil/HP-β-CD inclusion complex has been investigated in different aqueous media under simulated solar irradiation. The studies allow the conclusion that the pyrimethanil/HP-β-CD inclusion complex significantly increases the photostability of the fungicide in aqueous solutions. The half-life of the pyrimethanil/HP-β-CD inclusion complex was increased approximately by a factor of four when compared to the free fungicide.

It is known that pyrethrins are among the safest insecticides on the market due to their rapid degradation in the environment. Synthetic pyrethroids have relatively longer environmental persistence than pyrethrins. Pyrethrins have shorter environmental persistence than synthetic pyrethroids because their chemical structure is more susceptible to the presence of UV light and changes in pH.

Ferna´ndez-Pe´rez et al. (2014) evaluated the effectiveness of lignin and lignosulphonate-based formulations to protect insecticidal pyrethrins against photodegradation and volatilization. They prepared the lignin-based formulation by mixing pine kraft lignin with a natural polymer of alginate via an ionotropic gelation technique. Meanwhile, the lignosulphonate-based formulation was prepared by a spray-drying technique. The lignin-based formulation was more suitable for encapsulating pyrethrins (80% encapsulation efficiency) than the lignosulphonate-based formulation (77% encapsulation efficiency). Both formulations enhanced phytostabilization and reduced volatilization of pyrethrins. However, the lignosulphonate-based formulation exhibited a better protection effect for the pesticide.

The ability of chitosan-coated beeswax solid lipid to protect the insecticide deltamethrin against photodegradation was investigated by Nguyen et al. (2012). The chitosan-coated beeswax solid lipid nanoparticles were synthesized through a combination of hot homogenization and sonication methods. Under optimal conditions, the nanoparticles could encapsulate and load deltamethrin at up to 95 and 12.5%, respectively. The use of nanoparticles resulted in a better-controlled release of deltamethrin and protection against direct and indirect UV radiation as the non-degraded amount of insecticide was nearly 2.0 and 2.6% higher than its free form.

A photoresponsive micellar system based on the conjugation of the herbicide 2,4-dichlorophenoxyacetic acid to polyethyleneglycol with 3-nitro-4-bromomethylbenzoic acid as a crosslinker was successfully developed by Ding et al. (2016). In an aqueous environment, the polymeric micelles self-assembled into core-shell micelles with 2,4-dichlorophenoxyacetic acid acting as hydrophobic moieties and the polyethyleneglycol backbone acting as the hydrophilic moieties. When exposed to UV irradiation, the size of the micelles decreased, and the herbicide was released from the micelle due to photolysis of the polymer chain. With the presence of UV irradiation and simulated sunlight, the release of the herbicide from the polymer micelle was nearly complete in 9 min and 8 h, respectively, which therefore confirmed the slow-release behavior.

Like most traditional pesticides, prochloraz (Pro) is highly degradable under UV light, which greatly reduces its efficiency in the field (Zhao et al., 2018). To reduce the loss caused by photodegradation, encapsulating Pro into sunscreen carriers was carried out by Wu et al. (2022). Herein, a degradable silicon-based iron-doped nanoparticle (Fe-MSN) was prepared as a porous carrier. After loading the pesticide Pro, tannic acid (TA) was used to seal the pores and stabilize the nanocarrier structure. Due to enough pore surface areas, the loading rate of Pro@Fe-MSNs/TA nanoparticles was up to 31%. Nanocarrier could effectively protect Pro against photodegradation. Under UV light only 5.1% of Pro remained undegraded after 24 h of UV irradiation. In contrast, Pro encapsulated by Fe-MSNs/TA nanocarriers was particularly stable, with 56.9% of Pro photodegraded at 24 h and about 75.1% of Pro decomposed by

UV after 72 h. The improvement in photostability was mainly attributed to the absorption or reflection of UV light by Fe-MSNs/TA nanocarriers, which could effectively block UV light from Pro loaded in nanoparticle mesoporous channels.

It is known from the literature about the possibility of using nanocarriers capable of absorbing UV irradiation and thereby protecting active substances from photodegradation. So Cellulose-rosin based thermoplastic elastomers with UV absorption properties were fabricated by a combination of rosin, poly(butyl acrylate), and ethyl cellulose (EC) prepared by a simple esterification reaction between EC, dehydroabietic acid (one of the rosin's resin acids) and 2-bromoisobutylryl bromide (Lu et al., 2018). The introduction of rosin affords these elastomers with UV absorption properties. All EC-rosin grafted copolymers showed an excellent UV absorption performance and maintained outstanding UV absorption capability after continuous UV irradiation or being heated to 115 °C for 1 h. As a result, these EC-rosin grafted copolymers have a potential application in carrier materials with UV absorption properties.

Also, was reported by Narayanan et al. (2017) fabrication of another modified natural polymer with UV absorbance properties based on the poly(lactic acid)-rosin biopolymer. Such composites also offer a negligible percentage of transmittance in the most harmful UV-B region.

Above mentioned hypothesis was realized in practice by Zhao et al. (2021). Environmentally friendly pesticide formulations based on two compounds from natural materials carboxymethyl cellulose and rosin (RS) to form CMCL-g-PRSG carrier have been prepared by Zhao's scientific group. The model pesticide avermectin was encapsulated through hydrophobic interaction, and self-assembled to form nanopesticide AVM-CMCL-gPRSG with an average particle size of 167 nm. The prepared nanopesticide displays enhanced dispersibility and stability of AVM in water and can effectively adhere to the leaves to prevent loss. The release rate of AVM encapsulated in the nanocarrier can be controlled by adjusting pH, and AVM half-life under ultraviolet radiation shows a 3-fold increase allowing control of pests for prolonged periods in practical applications. Biological safety tests showed that AVM-CMCL-g-PRSG effectively reduces the toxicity of AVM to aquatic animals. Therefore, the cheap and degradable carrier CMCL-g-PRSG improves the effect of hydrophobic pesticides.

CONCLUSION

For continued commercial and industrial progress, the development of agriculture is a prerequisite, which can be made possible via the use of nanocomposites based on biopolymers or similar environmentally friendly nanomaterials, such as nanocomposites of fertilizers and pest control agents. Currently, the agricultural sector is in an emerging smart phase, and nanopesticides can be a critical factor in the evolution of this sector by adopting a long-term research agenda. Because the conducted studies *in vitro* and *in vivo* show significant results in the conservation, cultivation, and protection of agricultural products, it can be expected that nanotechnology can significantly influence the economics of agriculture by promoting farming fields. In this case, nanobiopolymers can become one of the primers for creating environmentally friendly nanochemicals. They can be used to grow crops, control pests, save seeds, *etc.* In addition, indiscriminate synthetic chemicals in agriculture to manage microbial growth damage the environment and threaten our existence. So, in this case, the biopolymer-based nanocomposite can be one of the effective ways to reduce the use of pesticides and environmental pollution. As a result of this review study, it can be summarized that in recent years, systems for the controlled release of agrochemicals have been developed, and various nanocarriers have been proposed, which make it possible to obtain a plant protection drug with desired properties. Among them, the investigation of the photostability of the active substance in the structure of the nanoformulations occupies a special place. Most of the studies were carried out on model substances that are most sensitive to the influence of sunlight. But there is reason to believe that these conclusions can be valid for all other cases. Summing up, we can conclude that the increase in the photostability of active substances in nanoparticles is provided by the protective properties of the nanocarrier (as a result of coating) or by adding special substances that are capable to absorb ultraviolet irradiation to the structure of nanocomposites. Although the development of smart nanocarriers that can be produced from renewable and inexpensive materials is theoretically well-developed, it remains a challenge for agricultural applications.

ACKNOWLEDGMENT

This research was supported by the Moldovan National Research Program [projects № 20.80009.5007.17 and № 23.00208.5007.04].

REFERENCES

Amaral, I. F., Granja, P. L., & Barbosa, M. A. (2005). Chemical modification of chitosan by phosphorylation: An XPS, FT-IR, and SEM study. *Journal of Biomaterials Science. Polymer Edition*, *6*(12), 1575–1593. doi:10.1163/156856205774576736 PMID:16366338

Aranaz, I., Harris, R., & Heras, A. (2010). Chitosan amphiphilic derivatives. Chemistry and applications. *Current Organic Chemistry*, *14*(3), 308–330. doi:10.2174/138527210790231919

Ardila, N., Daigle, F., Heuzey, M. C., & Ajji, A. (2017). Effect of chitosan physical form on its antibacterial activity against pathogenic bacteria. *Journal of Food Science*, *82*(3), 679–686. doi:10.1111/1750-3841.13635 PMID:28140469

Arias-Estevez, M., Lopez-Periago, E., Martinez-Carballo, E., Simal-Gandara, J., Mejuto, J. C., & Garcia-Rio, L. (2008). The mobility and degradation of pesticides in soils and the pollution of groundwater resources. *Agriculture, Ecosystems & Environment*, *123*(4), 247–260. doi:10.1016/j.agee.2007.07.011

Ashrafi, M., Bayat, M., Mortazavi, P., Hashemi, S. J., & Meimandipour, A. (2020). Antimicrobial effect of chitosan–silver–copper nanocomposite on *Candida albicans*. *Journal of Nanostructure in Chemistry*, *10*(1), 87–95. doi:10.100740097-020-00331-3

Asrar, J., & Ding, Y. (2003). *US Patent WO2003005816A1*. Monsanto Technology Lic.

Asrar, J., & Ding, Y. (2010). *US Patent 7771749 B2*. Monsanto Technology Lic.

Asrar, J., & Ding, Y. (2016). US *Patent 9,445,599 B2*. Monsanto Technology Lic.

Bagheri-Khoulenjani, S., Taghizadeh, S. M., & Mirzadeh, H. (2009). An investigation on the short-term biodegradability of chitosan with various molecular weights and degrees of deacetylation. *Carbohydrate Polymers*, *78*(4), 773–778. doi:10.1016/j.carbpol.2009.06.020

Bhattarai, N., Gunn, J., & Zhang, M. (2010). Chitosan-based hydrogels for controlled, localized drug delivery. *Advanced Drug Delivery Reviews*, *62*(1), 83–99. doi:10.1016/j.addr.2009.07.019 PMID:19799949

Blizzard, T. A., Mrozik, H., & Fisher, M. H. (1988). Chemical degradation of avermectin B1. *Tetrahedron Letters*, *29*(26), 3163–3166. doi:10.1016/0040-4039(88)85111-6

Bobu, E., Raluca, N., Lupei, M., Ciolacu, F. L., & Desbri'eres, J. (2011). Synthesis and characterization of N-alkyl chitosan for papermaking applications. *Cellulose Chemistry and Technology*, *45*(9-10), 619–625.

Campos, E. V. R., de Oliveira, J. L., & Fraceto, L. F. (2014). Applications of controlled release systems for fungicides, herbicides, acaricides, nutrients, and plant growth hormones: A review. *Advanced Science, Engineering and Medicine*, *6*(4), 373–387. doi:10.1166/asem.2014.1538

Carr, M. E., Doane, W. M., Wing, R. E., & Bagley, E. B. (1993). *Patent US 5183690.* The United States of America as represented by the Secretary of Agriculture.

Chen, H., Cui, S., Zhao, Y., Wang, B., Zhang, S., & Peng, X. (2012). O-Alkylation of chitosan for gene delivery by using ionic liquid in an in-situ reactor. *Engineering (London)*, *04*(10, 10B), 114–117. doi:10.4236/eng.2012.410B029

Chen, L., Lin, Y., Zhou, H., Hao, L., Chen, H., & Zhou, X. (2021). A stable polyamine-modified zein-based nanoformulation with high foliar affinity and lowered toxicity for sustained avermectin release. *Pest Management Science*, *77*(7), 3300–3312. doi:10.1002/ps.6374 PMID:33763979

Crouch, L. S., Feely, W. F., Arison, B. H., VandenHeuvel, W. J. A., Colwell, L. F., Stearns, R. A., Kline, W. F., & Wislocki, P. G. (1991). Photodegradation of avermectin B1a thin films on glass. *Journal of Agricultural and Food Chemistry*, *39*(7), 1310–1319. doi:10.1021/jf00007a024

Cui, J., Mo, D., Jiang, Y., Gan, C., Li, W., Wu, A., Li, X.-Y., Xiao, J.-A., Hu, Q., Yuan, H.-Y., Lu, R., & Huang, Y. (2019). Fabrication, Characterization, and Insecticidal Activity Evaluation of Emamectin Benzoate-Sodium Lignosulfonate Nanoformulation with pH-responsive. *Industrial & Engineering Chemistry Research*, *58*(43), 19741–19751. doi:10.1021/acs.iecr.9b03171

Dai, X., Chen, L., Zhou, H., & Zhou, X. (2022). Development of stable and efficient zein based nanopesticides through green and simple ways for improving utilization efficiency. *Journal of Asia-Pacific Entomology*, *25*(2), 101906. doi:10.1016/j.aspen.2022.101906

de Oliveira, J. L., Campos, E. V. R., Pereira, A. E. S., Pasquoto, T., Lima, R., Grillo, R., Andrade, D. J., Santos, F. A., & Fraceto, L. F. (2018). Zein nanoparticles as eco-friendly carrier systems for botanical repellents aiming sustainable agriculture. *Journal of Agricultural and Food Chemistry*, *66*(6), 1130–1340. doi:10.1021/acs.jafc.7b05552 PMID:29345934

Dellicolli, H. T., Dilling, P., & Falkehag, S. I. (1980). *US Patent 4,3184,866.* Westvaco.

Ding, K., Shi, L., Zhang, L., Zeng, T., Yin, Y., & Yi, Y. (2016). Synthesis of photoresponsive polymeric propesticide micelles based on PEG for the controlled release of a herbicide. *Polymer Chemistry*, *7*(4), 899–904. doi:10.1039/C5PY01690H

Doane, W. M., Maiti, S., & Wing, R. E. (1990). *US Patent 4,911,952 A.* The United States of America as represented by the Secretary of Agriculture.

Dong, J., Chen, W., Qin, D., Chen, Y., Li, J., Wang, C., Yu, Y., Feng, J., & Du, X. (2021a). Cyclodextrin polymer-valved MoS_2-embedded mesoporous silica nanopesticides toward hierarchical targets via multidimensional stimuli of biological and natural environments. *Journal of Hazardous Materials*, *419*, 126404. doi:10.1016/j.jhazmat.2021.126404 PMID:34153613

Dong, J., Liu, X., Chen, Y., Yang, W., & Du, X. (2021b). User-safe and efficient chitosan-gated porous carbon nanopesticides and nanoherbicides. *Journal of Colloid and Interface Science*, *594*, 20–34. doi:10.1016/j.jcis.2021.03.001 PMID:33744730

Eidelman, N., & Simon, C. G. (2004). Characterization of combinatorial polymer blend composition gradients by FTIR microspectroscopy. *Journal of Research of the National Institute of Standards and Technology*, *09*(2), 219–231. doi:10.6028/jres.109.014 PMID:27366606

Ferna'ndez-Pe'rez, M., Flores-Ce'spedes, F., Daza-Fernández, I., Vidal-Peña, F., & Villafranca-Sánchez, M. (2014). Lignin and lignosulfonate-based formulations to protect pyrethrins against photodegradation and volatilization. *Industrial & Engineering Chemistry Research*, *53*(35), 13557–13564. doi:10.1021/ie500186e

Fernandes, C., Encarnacao, I., Gaspar, A., Garrido, J., Borges, F., & Garrido, E. M. (2014). Influence of hydroxypropyl-β-cyclodextrin on the photostability of fungicide pyrimethanil. *International Journal of Photoenergy*, *2014*, 1–8. doi:10.1155/2014/489873

Florêncio, M. H., Pires, E., Castro, A. L., Nunes, M. R., Borges, C., & Costa, F. M. (2004). Photodegradation of Diquat and Paraquat in aqueous solutions by titanium dioxide: Evolution of degradation reactions and characterization of intermediates. *Chemosphere*, *55*(3), 345–355. doi:10.1016/j.chemosphere.2003.11.013 PMID:14987933

Food and Agriculture Organization of the United Nations (FAO). (2018). *FAOSTAT – Pesticide use (Food and Agriculture Organization of the United Nations: Rome, Italy).* FAO. http://www.fao.org/faostat/en/#data/RP

Gogos, A., Knauer, K., & Bucheli, T. D. (2012). Nanomaterials in plant protection and fertilization: Current state, foreseen applications, and research priorities. *Journal of Agricultural and Food Chemistry*, *60*(39), 9781–9792. doi:10.1021/jf302154y PMID:22963545

Hao, L., Gong, L., Chen, L., Guan, M., Zhou, H., Qiu, S., Wen, H., Chen, H., Zhou, X., & Akbulut, M. (2020). Composite pesticide nanocarriers involving functionalized boron nitride nanoplatelets for pH-responsive release and enhanced UV stability. *Chemical Engineering Journal*, *396*, 125233. doi:10.1016/j.cej.2020.125233

Hao, L., Lin, G., Lian, J., Chen, L., Zhou, H., Chen, H., Hu, H., & Zhou, X. (2019). Carboxymethyl cellulose capsulated zein as pesticide nano-delivery system for improving adhesion and anti-UV properties. *Carbohydrate Polymers*, *231*, 115725. doi:10.1016/j.carbpol.2019.115725 PMID:31888813

He, S., Zhang, W., Li, D., Li, P., Zhu, Y., Ao, M., Ki, J., & Cao, Y. (2013). Preparation and characterization of double-shelled avermectin microcapsules based on copolymer matrix of silica–glutaraldehyde–chitosan. *Journal of Materials Chemistry. B, Materials for Biology and Medicine*, *1*(9), 1270–1278. doi:10.1039/c2tb00234e PMID:32260800

Hofmann, T., Lowry, G. V., Ghoshal, S., Tufenkji, N., Brambilla, D., Dutcher, J. R., Gilbertson, L. M., Giraldo, J. P., Kinsella, J. M., Landry, M. P., & Lowell, W. (2020). Technology readiness and overcoming barriers to sustainably implement nanotechnology-enabled plant agriculture. *Nature Food*, *1*(7), 416–425. doi:10.103843016-020-0110-1

Huang, G., Qian, G., Yan, Y., Xu, D., Xu, Ch., Fu, L., & Lin, B. (2019). A super long-acting and anti-photolysis pesticide release platform through self-assembled natural polymer-based polyelectrolyte. *Reactive & Functional Polymers*, *146*, 104429. doi:10.1016/j.reactfunctpolym.2019.104429

Huang, Y., Li, X., Xiong, Q., Chen, Y., Peng, Z., Chen, J., Li, J., Zhang, Y., & Cui, J. (2022). Preparation and insecticidal activity evaluation of emamectin–lignin sulfonic acid conjugate with antiphotolysis property. *ACS Omega*, *7*(33), 29046–29053. doi:10.1021/acsomega.2c02883 PMID:36033669

Jayakumar, R., Nwe, N., Tokura, S., & Tamura, H. (2007). Sulfated chitin and chitosan as novel biomaterials. *International Journal of Biological Macromolecules*, *40*(3), 175–181. doi:10.1016/j.ijbiomac.2006.06.021 PMID:16893564

Jayakumar, R., Prabaharan, M., Reis, R. L., & Mano, J. F. (2005). Graft copolymerized chitosan - present status and applications. *Carbohydrate Polymers*, *62*(2), 142–158. http://repositorium.sdum.uminho.pt/bitstream/1822/20220/1/file.pdf. doi:10.1016/j.carbpol.2005.07.017

Kah, M., Beulke, S., Tiede, K., & Hofmann, T. (2013). Nanopesticides: State of knowledge, environmental fate, and exposure modeling. *Critical Reviews in Environmental Science and Technology*, *43*(16), 1823–1867. doi:10.1080/10643389.2012.671750

Kah, M., & Hofmann, T. (2014). Nanopesticide research: Current trends and future priorities. *Environment International*, *63*, 224–235. doi:10.1016/j.envint.2013.11.015 PMID:24333990

Kah, M., Kookana, R. S., Gogos, A., & Bucheli, T. D. (2018). A critical evaluation of nanopesticides and nanofertilizers against their conventional analogues. *Nature Nanotechnology*, *13*(8), 677–684. doi:10.103841565-018-0131-1 PMID:29736032

Kah, M., Weniger, A., & Hofmann, T. (2016). Impacts of (nano)-formulations on the fate of an insecticide in soil and consequences for environmental exposure assessment. *Environmental Science & Technology*, *50*(20), 10960–10967. doi:10.1021/acs.est.6b02477 PMID:27648740

Karimi, K., & Taherzadeh, M. J. (2016). A critical review on analysis in pretreatment of lignocelluloses: Degree of polymerization, adsorption/desorption, and accessibility. *Bioresource Technology*, *203*, 348–356. doi:10.1016/j.biortech.2015.12.035 PMID:26778166

Kaur, A., Kumar, P., Kaur, L., Sharma, R., & Kush, P. (2020). Thiolated chitosan nanoparticles for augmented oral bioavailability of gemcitabine: Preparation, optimization, in vitro and in vivo study. *Journal of Drug Delivery Science and Technology*, *61*, 102169. doi:10.1016/j.jddst.2020.102169

Kaur, P., Thakur, R., Bernela, M., Chopra, M., Manuja, A., & Chaudhury, A. (2014). Synthesis, characterization and in vitro evaluation of cytotoxicity and antimicrobial activity of chitosan-metal nanocomposites: Antimicrobial & biocompatible chitosan-metal nanocomposites. *Journal of Chemical Technology and Biotechnology*, *90*(5), 867–873. doi:10.1002/jctb.4383

Kaziem, A. E., Gao, Y., Zhang, Y., Qin, X., Xiao, Y., Zhang, Y., You, H., Li, J., & He, S. (2018). α-Amylase triggered carriers based on cyclodextrin anchored hollow mesoporous silica for enhancing insecticidal activity of avermectin against *Plutella xylostella*. *Journal of Hazardous Materials*, *359*, 213–221. doi:10.1016/j.jhazmat.2018.07.059 PMID:30036751

Khor, E., & Lim, L. Y. (2003). Implantable applications of chitin and chitosan. *Biomaterials*, *24*(13), 2339–2349. doi:10.1016/S0142-9612(03)00026-7 PMID:12699672

Kim, I. Y., Seo, S. J., Moon, H. S., Yoo, M. K., Park, I. Y., Kim, B. C., & Cho, C. S. (2008). Chitosan and its derivatives for tissue engineering applications. *Biotechnology Advances*, *26*(1), 1–21. doi:10.1016/j.biotechadv.2007.07.009 PMID:17884325

Knidri, H. E., Laajeb, A., & Lahsini, A. (2020). Chitin and chitosan: Chemistry, solubility, fiber formation, and their potential applications. *Handb. Chitin Chitosan*, *1*, 35–57. doi:10.1016/B978-0-12-817970-3.00002-X

Kookana, R. S., Boxall, A. B. A., Reeves, P. T., Ashauer, R., Beulke, S., Chaudhry, Q., Cornelis, G., Fernandes, T. F., Gan, J., Kah, M., Lynch, I., Ranville, J., Sinclair, C., Spurgeon, D., Tiede, K., & Van den Brink, P. J. (2014). Nanopesticides: Guiding principles for regulatory evaluation of environmental risks. *Journal of Agricultural and Food Chemistry*, *62*(19), 4227–4240. doi:10.1021/jf500232f PMID:24754346

Kumari, S., Singh, R. P., Chavan, N. N., & Annamalai, P. K. (2016). Chitosan-based bionanocomposites for biomedical application. *Bioinspired Biomim. Nanobiomater*, *7*(4), 219–227. doi:10.1680/jbibn.15.00015

Li, Z.-Z., Xu, S.-A., Wen, L.-X., Liu, F., Liu, A.-Q., Wang, Q., Sun, H.-Y., Yu, W., & Chen, J.-F. (2006). Controlled release of avermectin from porous hollow silica nanoparticles: Influence of shell thickness on loading efficiency, UV-shielding property and release. *Journal of Controlled Release*, *111*(1-2), 81–88. doi:10.1016/j.jconrel.2005.10.020 PMID:16388871

Liang, Y., Gao, Y., Wang, W., Dong, H., Tang, R., Yang, J., Niu, J., Zhou, Z., Jiang, N., & Cao, Y. (2020). Fabrication of smart stimuli-responsive mesoporous organosilica nano-vehicles for targeted pesticide delivery. *Journal of Hazardous Materials*, *389*, 122075. doi:10.1016/j.jhazmat.2020.122075 PMID:31972522

Liang, Y., Wang, S., Jia, H., Yao, Y., Song, J., Dong, H., Cao, Y., Zhu, F., & Huo, Z. (2022). Pectin functionalized metal-organic frameworks as dual-stimuli-responsive carriers to improve the pesticide targeting and reduce environmental risks. *Colloids and Surfaces. B, Biointerfaces*, *219*, 112796. doi:10.1016/j.colsurfb.2022.112796 PMID:36063717

Liu, W. G., Zhu, X. Y., Yin, M. M., & Chen, F. L. (2012). Photodegradation of three avermectin liquid formulations. *Journal of Environmental Sciences (China)*, *31*, 1906–1912.

Lu, C., Yu, J., Wang, C., Wang, J., & Chu, F. (2018). Fabrication of UV-absorbent cellulose-rosin based thermoplastic elastomer via "graft from" ATRP. *Carbohydrate Polymers*, *188*, 128–135. doi:10.1016/j.carbpol.2018.01.062 PMID:29525148

Luo, Y., & Wang, Q. (2014). Recent development of chitosan-based polyelectrolyte complexes with natural polysaccharides for drug delivery. *International Journal of Biological Macromolecules*, *64*, 353–367. doi:10.1016/j.ijbiomac.2013.12.017 PMID:24360899

Macaev, F., Boldescu, V., Geronikaki, A., & Sucman, N. (2013). Recent advances in the use of cyclodextrins in antifungal formulations. *Current Topics in Medicinal Chemistry, 13*(21), 2677–2683. doi:10.2174/1568026611313660194 PMID:24083792

Madhusudhan, R. A., Karthikeyan, R., Vejandla, R. S., Divya, G., & Babu, P. S. (2017). Controlled release matrix drug delivery system – a review. *Int. J. Allied Med. Sci.Clin. Res., 5*(2), 384–398.

Malakof, D., & Stokstad, M. (2013). Infographic – Pesticide planet. *Science, 341*(6147), 730–731. doi:10.1126cience.341.6147.730 PMID:23950524

Morillo, E. (2006). Application of cyclodextrins in agrochemistry. In: cyclodextrins and their complexes. In H. Dodziuk (Ed.), *Chemistry, analytical methods, applications* (pp. 459–467). Wiley-VCH.

Mourya, V. K., Inamdar, N. N., & Tiwari, A. (2010). Carboxymethyl chitosan and its applications. *Advanced Materials Letters, 1*(1), 11–33. doi:10.5185/amlett.2010.3108

Narayanan, M., Loganathan, S., Valapa, R. B., Thomas, S., & Varghese, T. O. (2017). UV protective poly(lactic acid)/rosin films for sustainable packaging. *International Journal of Biological Macromolecules, 99*, 37–45. doi:10.1016/j.ijbiomac.2017.01.152 PMID:28212931

Neri-Badang, M. C., & Chakraborty, S. (2019). Carbohydrate polymers as controlled release devices for pesticides. *Journal of Carbohydrate Chemistry, 38*(2), 1–19. doi:10.1080/07328303.2019.1568449

Neubauer, M. P., Poehlmann, M., & Fery, A. (2014). Microcapsule mechanics: From stability to function. *Advances in Colloid and Interface Science, 207*, 65–80. doi:10.1016/j.cis.2013.11.016 PMID:24345731

Nguyen, H. M., Hwang, I.-C., Park, J.-W., & Park, H.-J. (2012). Photoprotection for deltamethrin using chitosan-coated beeswax solid lipid nanoparticles. *Pest Management Science, 68*(7), 1062–1068. doi:10.1002/ps.3268 PMID:22383397

Nuruzzaman, M., Rahman, M. M., Liu, Y. J., & Naidu, R. (2016). Nanoencapsulation, nano-guard for pesticides: Anew window for safe application. *Journal of Agricultural and Food Chemistry, 64*(7), 1447–1483. doi:10.1021/acs.jafc.5b05214 PMID:26730488

Otero, R., Fernandez, J. M., Ulibarri, M. A., Celis, R., & Bruna, F. (2012). Adsorption of nonionic pesticide S-Metolachlor on layered double hydroxides intercalated with dodecylsulfate and tetradecanedioate. *Applied Clay Science, 65–66*, 72–79. doi:10.1016/j.clay.2012.04.022

Pang, L., Gao, Z., Feng, H., Wang, S., & Wang, Q. (2019). Cellulose based materials for controlled release formulations of agrochemicals: A review of modifications and applications. *Journal of Controlled Release, 316*, 105–115. doi:10.1016/j.jconrel.2019.11.004 PMID:31704109

Park, S. Y., Jun, S. T., & Marsh, K. S. (2001). Physical properties of PVOH/chitosan-blended films cast from different solvents. *Food Hydrocolloids, 15*(4-6), 499–502. doi:10.1016/S0268-005X(01)00055-8

Peng, R., Yang, D., Qiu, X., Qin, Y., & Zhou, M. (2020). Preparation of self-dispersed lignin-based drug-loaded material and its application in avermectin nano-formulation. *International Journal of Biological Macromolecules, 151*, 421–427. doi:10.1016/j.ijbiomac.2020.02.114 PMID:32061696

Perlatti, B., de Souza Bergo, P. L., Silva, M. F. G. F. d., Fernandes, J. B., & Forim, M. R. (2013). Polymeric Nanoparticle-Based Insecticides: A Controlled Release Purpose for Agrochemicals. In Insecticides - Development of Safer and More Effective Technologies. IntechOpen. https://doi.org/ doi:10.5772/53355

Ragno, G., Risoli, A., Ioele, G., & Luca, M. D. (2006). Photo- and Thermal-Stability Studies on Benzimidazole Anthelmintics by HPLC and GC-MS. *Chemical & Pharmaceutical Bulletin*, *54*(6), 802–806. doi:10.1248/cpb.54.802 PMID:16755047

Rai, P. K., Kumar, V., Lee, S., Raza, N., Kim, K.-H., Ok, Y. S., & Tsang, D. C. W. (2018). Nanoparticle-plant interaction: Implications in energy, environment, and agriculture. *Environment International*, *119*, 1–19. doi:10.1016/j.envint.2018.06.012 PMID:29909166

Rice, C., Young, R. J., Zan, R., Bangert, U., Wolverson, D., Georgiou, T., Jalil, R., & Novoselov, K. S. (2013). Raman-scattering measurements and first-principles calculations of strain-induced phonon shifts in monolayer MoS2. *Physical Review B: Condensed Matter and Materials Physics*, *87*(8), 081307. doi:10.1103/PhysRevB.87.081307

Sébille, B. (1987). Cyclodextrin derivatives. In D. Duchêne (Ed.), *Cyclo- dextrins and their industrial uses* (pp. 351–393). Edition de Santé.

Shariatinia, Z. (2019). Pharmaceutical applications of chitosan. *Advances in Colloid and Interface Science*, *263*, 131–194. doi:10.1016/j.cis.2018.11.008 PMID:30530176

Shih, P.-Y., Liao, Y.-T., Tseng, Y.-K., Deng, F.-S., & Lin, C.-H. (2019). A potential antifungal effect of chitosan against *Candida albicans* is mediated via the inhibition of SAGA complex component expression and the subsequent alteration of cell surface integrity. *Frontiers in Microbiology*, *10*, 602. doi:10.3389/fmicb.2019.00602 PMID:30972050

Shrestha, S., Kognou, A. L. M., Zhang, J., & Qin, W. (2021). Different Facets of Lignocellulosic Biomass Including Pectin and its Perspectives. *Waste and Biomass Valorization*, *12*(9), 4805–4823. doi:10.100712649-020-01305-w

Shukla, S. K., Mishra, A. K., Arotiba, O. A., & Mamba, B. B. (2013). Chitosan-based nanomaterials: A state-of-the-art review. *International Journal of Biological Macromolecules*, *59*, 46–58. doi:10.1016/j.ijbiomac.2013.04.043 PMID:23608103

Singh, A., Dhiman, N., Kar, A. K., Singh, D., Purohit, M. P., Ghosh, D., & Patnaik, S. (2020). Advances in controlled release pesticide formulations: Prospects to safer integrated pest management and sustainable agriculture. *Journal of Hazardous Materials*, *385*, 121525. doi:10.1016/j.jhazmat.2019.121525 PMID:31740313

Song, S., Wang, Y., Xie, J., Sun, B., Zhou, N.-L., Shen, H., & Shen, J. (2019). Carboxymethyl chitosan modified carbon nanoparticle for controlled emamectin benzoate delivery: Improved solubility, ph-responsive release, and sustainable pest control. *ACS Applied Materials & Interfaces*, *11*(37), 34258–34267. doi:10.1021/acsami.9b12564 PMID:31461267

Stubenrauch, J., Ekardt, F., Heyl, K., Garske, B., Schott, V. L., & Ober, S. (2021). How to legally overcome the distinction between organic and conventional farming—Governance approaches for sustainable farming on 100% of the land. *Sustainable Production and Consumption, 28*, 716–725. doi:10.1016/j.spc.2021.06.006

Szejtli, J. (2004). Past, present, and future of cyclodextrin research. *Pure and Applied Chemistry, 76*(10), 1825–1845. doi:10.1351/pac200476101825

Szente, L., & Szejtli, J. (1996). Cyclodextrins in pesticides. In J. Szejtli & T. Osa (Eds.), *Comprehensive supramolecular chemistry* (Vol. 3, pp. 503–514). Pergamon.

Tardy, B. L., Mattos, B. D., Otoni, C. G., Beaumont, M., Majoinen, J., Kämäräinen, T., & Rojas, O. J. (2021). Deconstruction and reassembly of renewable polymers and biocolloids into next generation structured materials. *Chemical Reviews, 121*(22), 14088–14188. doi:10.1021/acs.chemrev.0c01333 PMID:34415732

Tleuova, A. B., Wielogorska, E., Talluri, V. S. S. L. P., Štěpánek, F., Elliott, C. T., & Grigoriev, D. O. (2020). Recent advances and remaining barriers to producing novel formulations of fungicides for safe and sustainable agriculture. *Journal of Controlled Release, 326*, 468–481. doi:10.1016/j.jconrel.2020.07.035 PMID:32721524

Torres, E., Marín, V., Aburto, J., Beltr´an, H.I., Shirai, K., Villanueva, S., & Sandoval, G. (2012) Enzymatic modification of chitosan with quercetin and its application as antioxidant edible films. *Appl. Biochem. Microbiol., 48*(12, 2), 175-182. . doi:10.1134/S0003683812020123

Verlee, A., Mincke, S., & Stevens, C. V. (2017). Recent developments in antibacterial and antifungal chitosan and its derivatives. *Carbohydrate Polymers, 164*, 268–283. doi:10.1016/j.carbpol.2017.02.001 PMID:28325326

Walker, G. W., Kookana, R. S., Smith, N. E., Kah, M., Doolette, C. L., Reeves, P. T., Lovell, W., Anderson, D. J., Turney, T. W., & Navarro, D. A. (2018). Ecological risk assessment of nano-enabled pesticides: A perspective on problem formulation. *Journal of Agricultural and Food Chemistry, 66*(26), 6480–6486. doi:10.1021/acs.jafc.7b02373 PMID:28812885

Wang, L., Chen, Y. C., & Rojanasakul, Y. (2014). Protein nanoparticles as drug delivery carriers for cancer therapy. *BioMed Research International, 2014*, 180549. PMID:24772414

Wang, N., Ji, Y., Zhu, Y., Wu, X., Mei, L., Zhang, H., Deng, J., & Wang, S. (2020). Antibacterial effect of chitosan and its derivative on *enterococcus faecalis* associated with endodontic infection. *Experimental and Therapeutic Medicine, 19*, 3805–3813. doi:10.3892/etm.2020.8656 PMID:32346445

Worrall, E. A., Hamid, A., Mody, K. T., Mitter, N., & Pappu, H. R. (2018). Nanotechnology for Plant Disease Management. *Agronomy (Basel), 8*(12), 285. doi:10.3390/agronomy8120285

Wu, L., Pan, H., Huang, W., Wang, M., Hu, Z., & Zhang, F. (2022). Self-assembled degradable iron-doped mesoporous silica nanoparticles for the smart delivery of prochloraz to improve plant protection and reduce environmental impact. *Environmental Technology & Innovation, 28*, 102890. doi:10.1016/j.eti.2022.102890

Xu, C., Cao, L., Bilal, M., Cao, C., Zhao, P., Zhang, H., & Huang, Q. (2021). Multifunctional manganese-based carboxymethyl chitosan hydrogels for pH-triggered pesticide release and enhanced fungicidal activity. *Carbohydrate Polymers*, *262*, 117933. doi:10.1016/j.carbpol.2021.117933 PMID:33838810

Yan, Y., Hou, H., Ren, T., Xu, Y., Wang, Q., & Xu, W. (2013). Utilization of environmental waste cyanobacteria as a pesticide carrier: Studies on controlled release and photostability of avermectin. *Colloids and Surfaces. B, Biointerfaces*, *102*, 341–347. doi:10.1016/j.colsurfb.2012.07.043 PMID:23006573

Younes, I., & Rinaudo, M. (2015). Chitin and chitosan preparation from marine sources. structure, properties and applications. *Marine Drugs*, *13*(3), 1133–1174. doi:10.3390/md13031133 PMID:25738328

Yu, M., Yao, J., Liang, J., Zeng, Z., Cui, B., Zhao, X., Sun, Ch., Wang, Y., Liu, G., & Cui, H. (2017). Development of functionalized abamectin poly(lactic acid) nanoparticles with regulatable adhesion to enhance foliar retention. *RSC Advances*, *7*(19), 11271–11280. doi:10.1039/C6RA27345A

Zhang, L., Peng, X., Zhong, L., Chua, W., Xiang, Z., & Sun, R. (2019). Lignocellulosic biomass derived functional materials: Synthesis and applications in biomedical engineering. *Current Medicinal Chemistry*, *26*(14), 2456–2474. doi:10.2174/0929867324666170918122125 PMID:28925867

Zhang, X., Tang, X., Zhao, C., Yuan, Z., Zhang, D., Zhao, H., Yang, N., Guo, K., He, Y., He, Y., Hu, J., He, L., He, L., & Qian, K. (2022). A pH-responsive MOF for site-specific delivery of fungicide to control citrus disease of *Botrytis cinerea*. *Chemical Engineering Journal*, *431*, 133351. doi:10.1016/j.cej.2021.133351

Zhang, X., Xu, Z., Wu, M., Qian, X., Lin, D., Zhang, H., Tang, J., Zeng, T., Yao, W., Filser, J., Li, L., & Sharma, V. K. (2019). Potential environmental risks of nanopesticides: Application of $Cu(OH)_2$ nanopesticides to soil mitigates the degradation of neonicotinoid thiacloprid. *Environment International*, *129*, 42–50. doi:10.1016/j.envint.2019.05.022 PMID:31108392

Zhao, M., Chen, Z., Hao, L., Chen, H., Zhou, X., & Zhou, H. (2023). CMC based microcapsules for smart delivery of pesticides with reduced risks to the environment. *Carbohydrate Polymers, 300,* 120260, doi:10.1016/j.carbpol.2021.117749 PMID:33673979

Zhao, M., Zhou, H., Hao, L., Chen, H., & Zhou, X. (2021). Natural rosin modified carboxymethyl cellulose delivery system with lowered toxicity for long-term pest control. *Carbohydrate Polymers*, *259*, 117749. doi:10.1016/j.carbpol.2021.117749

Zhao, P., Cao, L., Ma, D., Zhou, Z., Huang, Q., & Pan, C. (2018). Translocation, distribution and degradation of prochloraz-loaded mesoporous silica nanoparticles in cucumber plants. *Nanoscale*, *10*(4), 1798–1806. doi:10.1039/C7NR08107C PMID:29308814

Zhao, R., Torley, P., & Halley, P. J. (2008). Emerging biodegradable materials: Starch- and proteinbased bio-nanocomposites. *Journal of Materials Science*, *43*(9), 3058–3071. doi:10.100710853-007-2434-8

Zhu, J., He, Y., Gao, M., Zhou, W., Hu, Y., Shen, J., & Zhu, Y. Ch. (2011). Photodegradation of emamectin benzoate and its influence on efficacy against the rice stem borer, *Chilo suppressalis*. *Crop Protection (Guildford, Surrey)*, *30*(10), 1356–1362. doi:10.1016/j.cropro.2011.06.006

Chapter 12
Technological and Environmental Factors Impact on the Antioxidation Mechanism of Oil Lipids

Violina Popovici
https://orcid.org/0000-0001-5393-1181
Technical University of Moldova, Moldova

Rodica Sturza
https://orcid.org/0000-0002-2412-5874
Technical University of Moldova, Moldova

Aliona Ghendov-Mosanu
Technical University of Moldova, Moldova

ABSTRACT

This chapter is related to the general characteristics of vegetable oils; the oxidative stability and the mechanism of lipid oxidation; the factors affecting the mechanism of lipid oxidation; the impact of lipid oxidation on the sensory properties of complex food products; and characteristics of the antioxidant complex from local horticultural products. Oil oxidation also destroys essential fatty acids and produces toxic compounds and oxidized polymers. It also affects the taste, nutritional quality, and toxicity of edible oils. Different chemical mechanisms, autoxidation, and photooxidation, are responsible for the oxidation of edible oils during processing and storage, depending on oxygen types. Two oxygen types can react with edible oils: atmospheric triplet oxygen, 3O2, and singlet oxygen, 1O2.

DOI: 10.4018/979-8-3693-0512-6.ch012

INTRODUCTION

Lipid oxidation is one of the major causes of quality deterioration in natural and processed foods. The oxidation process negatively influences the quality parameters of food products such as flavor, color, texture, and nutritional value of foods. During oxidation, different lipid oxidation products are formed that are potentially toxic compounds for human health.

BACKGROUND

Oxidative stability of vegetable oils

Oxidative stability is the resistance of oils to oxidation during processing and storage (Guillén & Cabo, 2002). It is an important parameter to evaluate oil quality and shelf life because low molecular weight compounds are produced during oxidation. Oxidation products make the oil less acceptable or unacceptable for consumers or for industrial use as a food ingredient. Oil oxidation also destroys essential fatty acids and produces toxic compounds and oxidized polymers. Oil oxidation is very important in terms of taste, nutritional quality, and toxicity of edible oils. Different chemical mechanisms, autoxidation, and photooxidation, are responsible for the oxidation of edible oils during processing and storage, depending on the types of oxygen. Two types of oxygen can react with edible oils: atmospheric triplet oxygen, $^3O_2$, and singlet oxygen, $^1O_2$. $^3O_2$ attacks lipid hydrocarbon chains and causes autoxidation, which is a free radical chain reaction. Oxidation of edible oils is influenced by energy input such as light or heat, fatty acid composition, oxygen species, and minor compounds such as metals, pigments, phospholipids, fatty acids, mono- and diacylglycerols, thermally oxidized compounds, and antioxidants. Many efforts have been made to improve the oxidative stability of oils through systematic studies on the effects of these factors (Akoh & Min, 2002).

Mechanism of Lipid Oxidation

Oil oxidation, the chain reaction of free radicals, includes the stages of initiation, propagation, and interruption (termination) shown in Table 1.

Table 1. The general mechanism of lipid oxidation (Frankel, 2012)

Oxidation stage	Mechanism
Initiation	RH → R·+H·
Propagation	R·+$^3O_2$ → ROO· ROO·+RH → ROOH+R·
Termination	ROO·+R· →ROOR R· + R· →RR (R: alkyl lipid)

The autoxidation of oils is conditioned by the appearance of free radicals of fatty acids or acylglycerols. Fatty acids or acylglycerols are in nonradical singlet states, and the reaction of fatty acids with the radical state of atmospheric $^3O_2$ is thermodynamically unfavorable due to electronic spin conservation (Zielinski & Pratt, 2017). The hydrogen atom from fatty acids or acylglycerols is removed and lipid alkyl radicals are produced in the initiation step. Heat, metal catalysts, and UV light can accelerate the formation of free radicals from fatty acids or acylglycerols. The energy required to remove hydrogen from fatty acids or acylglycerols depends on the position of the hydrogen. Hydrogen atoms in allylic positions are easily cleaved. The hydrogen at C11 of linoleic acid is removed at 50 kcal/mol. The energy required to remove hydrogen at C8 and C14 of linoleic acid is 75 kcal/mol and the homolytic dissociation energy between hydrogen and C17 or C18 is about 100 kcal/mol (Akoh & Min, 2002). After splitting the hydrogen from the favorable C11 allylic position, the stable free radical is formed. Resonance hybrids according to figure 1 explain the change in the position of double bonds and the formation of conjugated systems.

Figure 1. Formation of hydroperoxides from linoleic acid during autoxidation
(Collin, 2019)

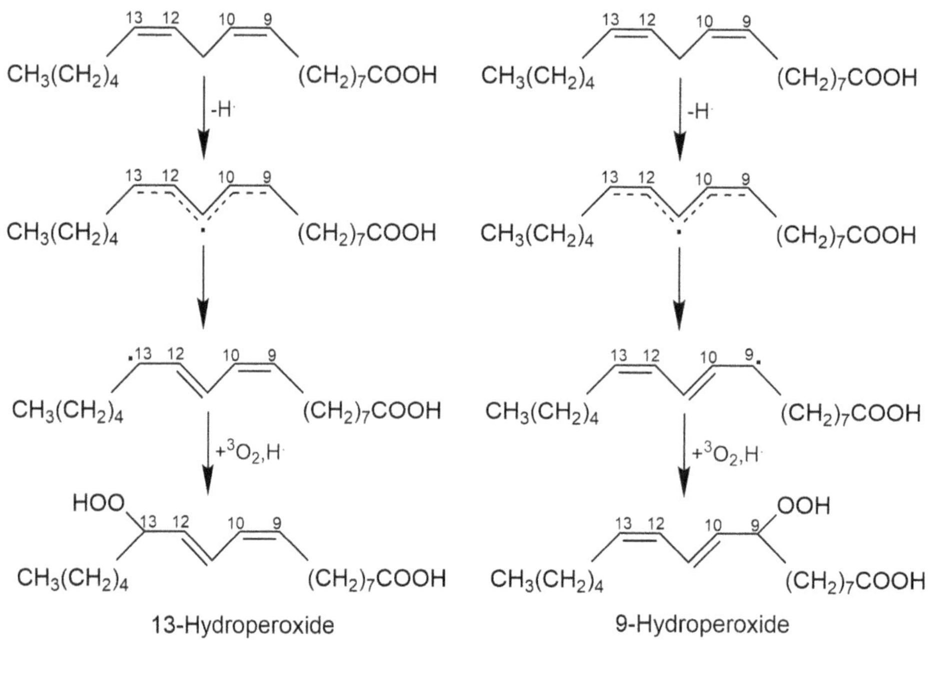

Along with the change in position, there is also a change in the geometrical configuration at the double bond from cis- to trans-. Autooxidation of linoleic and linolenic acids produces only products with conjugated double bonds. The amount of hydroperoxide positional isomers formed in the autoxidation of oleic, linoleic and linolenic acids are presented in Table 2. The lipid alkyl radical formed (with conjugated double bonds) reacts with $^3O_2$ and forms a lipid peroxide radical, another reactive radical. The reaction between lipid alkyl radical and $^3O_2$ occurs very rapidly at normal oxygen pressure and, consequently, the concentration of lipid alkyl radical is much lower than of lipid peroxide radical (Collin, 2019). The lipid peroxidic radical abstracts hydrogen from other lipid molecules and reacts with the hydrogen to form hydroperoxide and another lipid alkyl radical. These radicals catalyze the oxidation reaction, and autoxidation is called a free radical chain reaction. Figure 1 shows the formation of hydroperoxide in the autoxidation of linoleic acid.

Table 2. Hydroperoxides formed during autoxidation (Frankel, 2012)

Fatty acid	Hydroperoxides at	Relative amount, %
Oleic acid	C8	26-28
	C9	22-25
	C10	22-24
	C11	26-28
Linoleic acid	C9	48-53
	C13	48-53
Linolenic acid	C9	28-35
	C12	8-13
	C13	10-13
	C16	28-35

The rates of lipid peroxidic radical and hydroperoxide formation depend only on oxygen availability and temperature (Velasco et al., 2004). When radicals react with each other, non-radical species are produced and the reaction stops. The primary oxidation products, lipid hydroperoxides, are relatively stable at room temperature and in the absence of metals. However, in the presence of metals or at high temperatures, they are easily decomposed into alkoxy radicals and then form aldehydes, ketones, acids, esters, alcohols and short-chain hydrocarbons. The most likely way of decomposition of hydroperoxide is a homolytic cleavage of the bonds between the oxygen atoms, resulting in the production of alkoxy and hydroxy radicals. The activation energy for breaking the oxygen-oxygen bond is 46 kcal/mol lower than that for breaking the oxygen-hydrogen bond (Frankel, 2012).

The alkoxy radical then undergoes homolytic β-scission of the carbon-carbon bond and produces oxo compounds and saturated or unsaturated alkyl radicals as shown in Figure 2. After electronic rearrangement, addition of the hydroxyl radical, or hydrogen transfer, the final lipid oxidation byproducts are mostly low-molecular-weight aldehydes, ketones, alcohols, and short-chain hydrocarbons, as shown in Table 3. The time for the formation of the by-product from the main oxidation product, hydroperoxide, varies within different edible oils. Oxidation by-products are formed immediately after hydroperoxide formation in olive and canola oils. However, in sunflower and safflower oils, oxidation by-products are formed when the concentration of hydroperoxides is appreciable (Guillén & Cabo, 2002).

Most of the hydroperoxide breakdown products are responsible for the rancid flavor in oxidized edible oil. Aliphatic carbonyl compounds have a greater influence on the flavor of oxidized oil due to their low values (Steenson et al., 2002). Pentanal, hexanal, propenal and deca-2,4-dienal are present in large amounts in vegetable oil stored uncovered at 60°C (Gawrysiak-Witulska et al., 2012). The scientists reported that trans, cis-deca-2,4-dienal was the most significant compound in determining the oxidized aroma of the oil, followed by trans-deca-2,4-dienal, trans,cis-hepta-2, 4-dienal, oct-1-en-3-ol, butanal

Table 3. By-products of fatty acid oxidation (Choe & Min, 2006)

Class	Oleic acid	Linoleic acid	Linolenic acid
Aldehydes	Octanal	Pentanal	Propanal
	Nonanal	Hexanal	Butanal
	Dec-2-enal	Oct-2-enal	But-2-enal
		Non-2-enal	Pent-2-enal
		Deca-2,4-dienal	Hex-2-enal
			Nona-3,6-dienal
			Decatrienal
Derivatives of carboxylic acids	Methyl heptanoate	Methyl heptanoate	Methyl heptanoate
	Methyl octanoate	Methyl octanoate	Methyl octanoate
	Methyl 8-oxooctanoate	Methyl 8-oxooctanoate	Methyl nonanoate
	Methyl 9-oxononanoate	Methyl 9-oxononanoate	Methyl 9-oxononanoate
	Methyl 10-oxodecanoate	Methyl 10-oxodecanoate	Methyl 10-oxodecanoate
	Methyl 10-oxodec-8-enoate		
	Methyl 11-oxoundec-9-enoate		
Alcohols	Heptan-1-ol	Pentan-1-ol	
		Oct-1-en-3-ol	
Hydrocarbons	Heptane	Pentane	Ethan
	Octane		pentane

and hexanal (Frankel, 2012). Hexanal, pentanal and deca-2,4-dienal indicators were used as parameters to determine the degree of oil oxidation (Xu et al., 2017). Trans-hex-2-enal and trans-, cis-deca-2,4,7-trienal and oct-1-en-3-one have been reported to give grassy and fishy flavor in oxidized oil (Frankel, 2012). No single aromatic compound is primarily responsible for the oxidized flavor of vegetable oils.

Oil oxidation is accelerated by light, especially in the presence of sensitizers such as chlorophylls. Singlet state sensitizers absorb light very quickly, in picoseconds, and become excited. Excited singlet sensitizers can return to their ground state by light emission, internal conversion, or intersystem crossing. Fluorescence and heat are produced by light emission and internal conversion, respectively. The intersystem crossing results in the excited triplet state of the sensitizers. Excited triplet sensitizers can accept hydrogen or an electron from the substrate and produce radicals (type I). Excited triplet sensitizers react with $^3O_2$ and produce superoxide anion by electron transfer. The superoxide anion produces hydrogen peroxide, one of the reactive oxygen species by spontaneous dismutation, and the reaction of hydrogen peroxide with superoxides results in the formation of singlet oxygen through the Haber-Weiss reaction in the presence of transition metals such as iron or copper (Krumova & Cosa, 2016). When the hydroperoxide is formed, migration of the double bond and formation of the trans fatty acid occur, producing both conjugated and unconjugated hydroperoxides as shown in table 4.

The production of non-conjugated hydroperoxides is not observed in autoxidation. Figure 3 shows the oxidation pathway of linoleic acid with $^1O_2$. Hydroperoxides formed by $^1O_2$ oxidation are decomposed by the same mechanisms as for hydroperoxides formed by $^3O_2$ in autoxidation. The content of dec-2-nal and octane formed during oleic acid oxidation with $^1O_2$ is higher than during autoxidation (Frankel, 2012).

Figure 2. Mechanism of decomposition of hydroperoxides and formation of secondary products (Choe & Min, 2006)

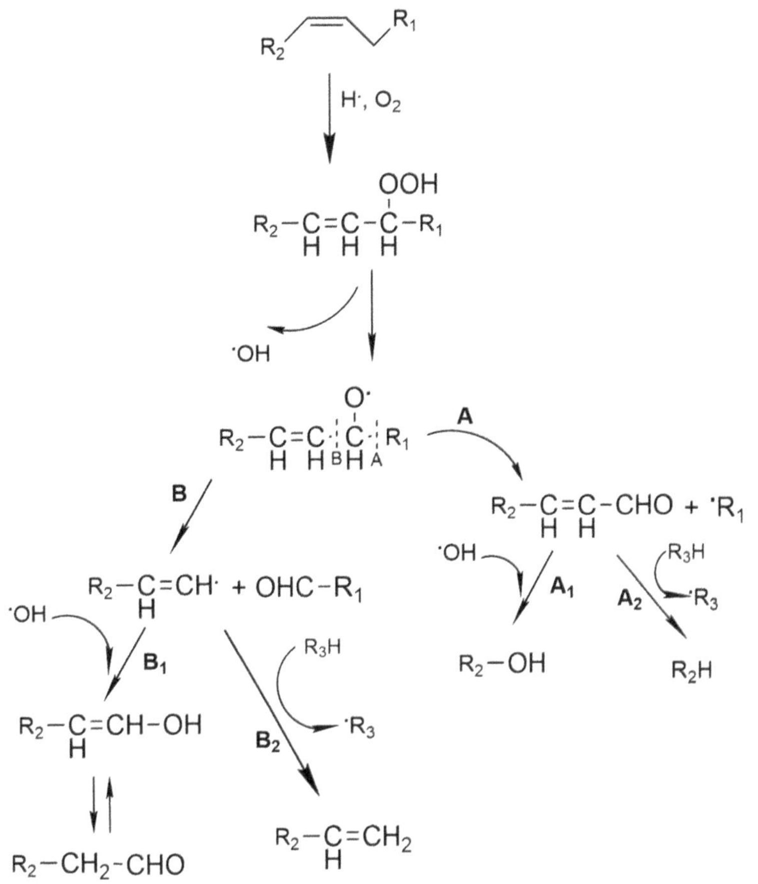

Table 4. Hydroperoxides of fatty acids formed by oxidation of singlet oxygen (Choe & Min, 2006)

Fatty acids	Hydroperoxides at	Relative amount, %	Acid type
Oleic acid	C9	48	
	C10	52	
Linoleic acid	C9	32	Conjugated
	C10	17	Unconjugated
	C12	17	Unconjugated
	C13	34	Conjugated
Linolenic acid	C9	23	Conjugated
	C10	13	Unconjugated
	C12	12	Conjugated
	C13	14	Conjugated
	C15	13	Unconjugated
	C16	25	Conjugated

The content of octanal and 10-oxodecanoate in autoxidized oleate was higher than that of $^1O_2$ oxidized oleate. Hept-2-enal and but-2-enal were visible in $^1O_2$-oxidized linoleic and linolenic acids, while they were negligible in autoxidized linoleic and linolenic acids. Heptenal was formed in edible oil oxidized with $^1O_2$ in the presence of chlorophyll and light. A bean flavor, which is a unique and undesirable flavor in edible oil with a low peroxide value, has been a problem internationally (Wszelaki et al., 2005).

Researchers have strongly indicated that the unpleasant flavor of edible oil can be reduced or eliminated by removing chlorophylls from the oil during processing. Currently, the soybean oil industry effectively removes chlorophylls from soybean oil using bleaching materials during the refining process, and bean flavor is no longer a serious problem in soybean oil (Min et al., 2003).

Figure 3. Mechanism of oxidation of linoleic acid with singlet oxygen
(Velasco et al., 2004)

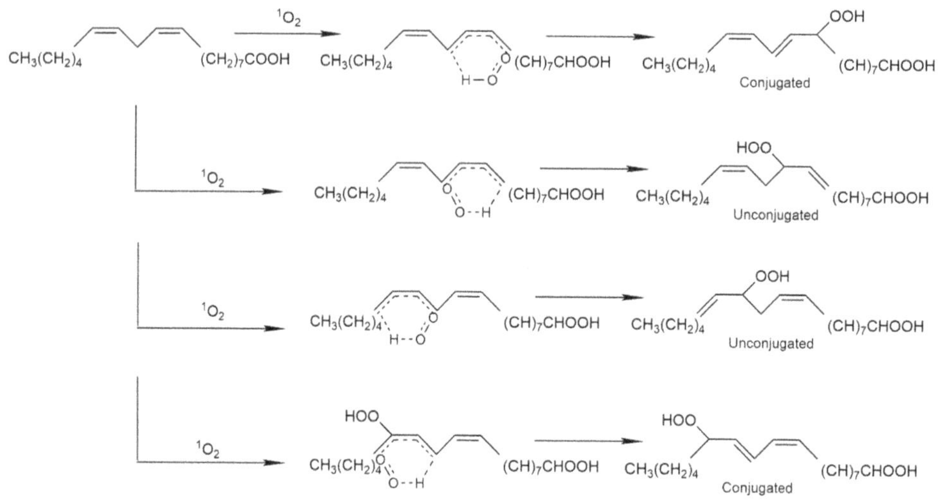

INFLUENCE OF CHEMICAL FACTORS ON VEGETABLE OILS STABILITY

Oil oxidation is influenced by the fatty acid composition of the oil, oil processing, heat or light energy, concentration and type of oxygen and free fatty acids, mono- and diacylglycerols, transition metals, peroxides, thermally oxidized compounds, pigments and antioxidants. These factors interactively affect oil oxidation and it is not easy to differentiate the individual effect of the factors.

Composition of Fatty Acids in Vegetable Oils

Oils that are more unsaturated are oxidized faster than less unsaturated oils. As the degree of unsaturation increases, both the rate of formation and the amount of primary oxidation products accumulated at the end of the induction period increase (Martín-Polvillo et al., 2004). Sunflower, soybean or safflower oil (iodine values > 130) stored in the dark showed a significantly shorter induction period ($P < 0.05$) than coconut or palm oil whose iodine value is lower of 20 (Tan et al., 2002).

The rate of autoxidation depends largely on the rate of formation of fatty acid or alkyl acylglycerol radicals, and the rate of free radical formation depends mainly on the type of fatty acid or the type of acylglycerols. The relative rate of autoxidation of oleic, linoleic, and linolenic acids has been reported from 1:40 to 50:100 based on oxygen uptake (Frankel, 2012). The difference in rate of oxidation by O_2 between fatty acids is smaller than for autoxidation. The type of polyunsaturated fatty acids, unconjugated or conjugated dienes or trienes, has reduced effect on the reaction between lipids and $^1O_2$ (Chen et al., 2011).

Free Fatty Acids, Mono- and Diacylglycerols

Crude oil contains free fatty acids, and oil processing, such as refining, lowers the free fatty acid content. Crude vegetable oil contains about 0.7% free fatty acids; however, refined oil contains 0.02% free fatty acids (Rakotondramavo et al., 2019). Free fatty acids are more susceptible to autoxidation than esterified fatty acids. Free fatty acids act as prooxidants in edible oil (Yun & Surh, 2012). They have hydrophilic and hydrophobic groups in the same molecule and prefer to be concentrated on the surface of edible oils. Hydrophilic carboxy groups of free fatty acids will not dissolve easily in the hydrophobic edible oil and are present on the surface of the edible oil. Researchers have reported that free fatty acids lower the surface tension of edible oil and increase the diffusion rate of oxygen in the oil to accelerate oil oxidation.

Mono- and diacylglycerol, typically present at 0.07% to 0.11% and 1.05% to 1.20% in vegetable oil, respectively, acted as prooxidants to increase oxidation at 55°C in the dark (Kim & Choe, 2012). Mono- and diacylglycerols, which have free hydrophilic hydroxyl groups and a hydrophobic hydrocarbon residue, lower the surface tension of edible oil and increase the diffusion rate of oxygen to the oil to accelerate its oxidation. Mono- and diacylglycerols must be removed from the oil during the oil refining process to improve the oxidative stability of edible oils (Kim & Choe, 2012).

Metals

The oil contains transition metals such as iron or copper. For example, cold-pressed sunflower oil contains copper – 5.2 ppb and iron – 0.26 ppm, respectively. However, refining process reduces their content. Edible oils manufactured without refining, such as extra virgin olive oil and sesame oil, contain relatively high amounts of transition metals as shown in table 5. Metals increase the rate of oil oxidation due to the reduction of the activation energy of the initiation step in autoxidation up to 63-104 kJ/mol (Zielinski & Pratt, 2017). Metals react directly with lipids to produce lipid alkyl radicals. They also produce reactive oxygen species such as $^1O_2$ and hydroxy radical from $^3O_2$ and hydrogen peroxide, respectively (Carvajal et al., 2009).

Lipid alkyl radicals and reactive oxygen species accelerate oil oxidation. The reaction of hydrogen peroxide with Fe^{3+} is significantly slower as compared to Fe^{2+}:

$$Fe^{3+} + RH \rightarrow Fe^{2+} + R^{-} + H^{+} \tag{1}$$

$$Fe^{2+} + {}^{3}O_2 \rightarrow Fe^{3+} + O_2^{-}\bullet \tag{2}$$

Table 5. Iron and copper content in edible oils (Choe & Min, 2006)

Edible oils	Metal content	
	Copper (ppb)	*Iron (ppm)*
Cold pressed sunflower oil	5,2	0,26
Refined olive oil	15	0,08
Refined soybean oil	2,5	0,20
Cold pressed sesame seed oil	16	1,16
Raw soybean oil	13,2	2,80
Virgin olive oil	9,8	0,73

$$O_2^-\bullet + O_2^-\bullet + 2H^+ \rightarrow {}^1O_2 + H_2O_2 \quad (3)$$

$$H_2O_2 + O_2^- \xrightarrow{\text{metal catalysis}} HO\bullet + OH^- + {}^1O_2 \quad (4)$$

$$Fe^{2+} + H_2O_2 \rightarrow Fe^{3+} + OH^- + HO\bullet \quad (5)$$

Also, metals accelerate oil autoxidation by decomposing hydroperoxides (Liu, 2018):

$$ROOH + Fe^{2+}\left(\text{or } Cu^+\right) \rightarrow RO\bullet + Fe^{3+}\left(\text{or } Cu^{2+}\right) + OH^- \quad (6)$$

$$ROOH + Fe^{3+}\left(\text{or } Cu^{2+}\right) \rightarrow ROO\bullet + Fe^{2+}\left(\text{or } Cu^+\right) + H^+ \quad (7)$$

Fe^{2+} is much more active at a rate of $1{,}5\times10^3 M^{-1}s^{-1}$ (Halliwell & Gutteridge, 2015) than Fe^{3+} in breaking down lipid hydroperoxides to catalyze autoxidation (Liu, 2018). Fe^{3+} also causes the decomposition of phenolic compounds such as caffeic acid in olive oil and decreases the oxidative stability of the oil (Choe & Min, 2006). Scientists reported that the pro-oxidant activity of iron was suppressed by lactoferrin in the oxidation of vegetable oil from 50°C to 120°C, possibly due to the binding ability of lactoferrin by iron. Depletion of ambient iron by lactoferrin can suppress iron-catalyzed oil oxidation (Shiota et al., 2006).

Phospholipids

Crude oils contain phospholipids such as phosphatidylethanolamine, phosphatidylcholine, phosphatidylinositol, phosphatidylserine and phosphatidic acid, but most of them are removed by processing such as degumming. Crude vegetable oil contains 500.8 and 213.6 ppm of phosphatidylcholine and phosphatidylethanolamine, respectively. However, the refined, decolorized and deodorized oil contains between 0.86 and 0.12 ppm of phosphatidylcholine and phosphatidylethanolamine, respectively (Chandran et al., 2017). Phospholipids act as antioxidants and prooxidants depending on their concentration and the presence of metals. The mechanism of the antioxidant effects of phospholipids has not yet been elucidated in detail, but the polar group plays an important role, and nitrogen-containing phospholipids such as phosphatidylcholine and phosphatidylethanolamine are effective antioxidants under most conditions (Burri et al., 2012).

Phospholipids slow down oil oxidation by sequestering metals, the maximum antioxidant activity was between 3 and 60 ppm. The oxidation of vegetable oil decreased with the addition of 5-10 ppm phospholipids and a higher amount of phospholipids acted as prooxidants. It was reported that phospholipids acted as antioxidants only in the presence of Fe^{2+} by iron chelation. In the purified, metal-free oil, phospholipids functioned as prooxidants. Phospholipids have hydrophilic and hydrophobic groups in the same molecule. The hydrophilic groups of phospholipids are on the surface of the oil and the hydrophobic groups are in the edible oil. Phospholipids lower the surface tension of edible oil and increase the rate of diffusion of oxygen into the oil to accelerate its oxidation. Oil oxidation in the dark at 60°C was more reduced with phosphatidic acid and phosphatidylethanolamine, followed by phosphatidylcholine, phosphatidylglycerol and phosphatidylinositol (Sousa et al., 2017).

INFLUENCE OF PHYSICAL FACTORS ON THE STABILITY OF VEGETABLE OILS

The method of oil processing affects its oxidative stability. Crude oil was the most stable to oxidation, followed by deodorized, degummed, refined and decolorized oil during 6-day storage at 55°C in the dark (Zebib et al., 2015). It has been suggested that the higher oxidative stability of the crude oil than the refined oil is partly due to the higher concentrations of tocopherols in the crude oil (1670 ppm) than the refined oil (1546 ppm). The induction time in the oxidation of vegetable oil extracted with hexane at 90°C was 10.5±1.9 hours compared to an induction time of 8.1±0.7 hours for pressed vegetable oil (Tynek et al., 2012). Roasting sunflower seeds prior to oil extraction improved the oxidative stability of oils (Lee, 2004; Zebib et al., 2015), partly due to Maillard products released during roasting. Some Maillard reaction products have been reported as antioxidants. Oxidative stability increased with increasing roasting and seed expulsion temperatures.

Temperature and Light

Oil autoxidation and hydroperoxide decomposition increase with the increase of temperature (Gharby et al., 2016; Domínguez et al., 2019). The formation of autoxidation products during the induction period is slow at low temperature (Velasco et al., 2004). The concentration of hydroperoxides increases until the advanced stages of oxidation. The content of polymerized compounds increases significantly at the end of the autoxidation induction period (Lise Halvorsen & Blomhoff, 2011). The rate of decomposition of hydroperoxide in oil stored at 50°C in the dark was higher than the rate of formation of hydroperoxide. The opposite phenomena were observed in the same oil at 0 or 20°C in the dark (Choe & Min, 2006). Temperature has reduced effect on $^1O_2$ oxidation due to the low activation energy from 0 to 6 kcal/mol (Domínguez et al., 2019). Light is much more important than temperature in $^1O_2$ oxidation. Light with shorter wavelengths had more damaging effects on oils than longer wavelengths (Negash et al., 2019). It appears that the effect of light on oil oxidation becomes less as the temperature increases (Velasco et al., 2004).

INFLUENCE OF OXIDANTS AND ANTIOXIDANTS ON THE STABILITY OF VEGETABLE OILS

Oxygen

Oil oxidation can often occur when oil, oxygen and catalysts are in contact. Both the concentration and type of oxygen affect the oxidation of oils. The oxygen concentration in the oil depends on the partial pressure of oxygen in the oil (Carvajal et al., 2009). A greater amount of oxygen is dissolved in the oil when the oxygen partial pressure is high. Oil oxidation increased with the amount of dissolved oxygen (Kim & Choe, 2012). One gram of vegetable oil dissolves 55 µg of oxygen at room temperature (Biglar et al., 2012). The amount of oxygen dissolved in the oil is sufficient to oxidize the oil to a peroxide value of about 10 mEq/kg in the dark (Frankel, 2012). The effect of oxygen concentration on oil oxidation increased at high temperatures, in the presence of light and metals such as iron or copper. A higher oxygen dependence in oil oxidation at high temperatures is due to the low solubility of oxygen in oil at temperature (Gawrysiak-Witulska et al., 2012). Oxygen must be carried into the oil by diffusion when the oil is not agitated, such as during low temperature storage. Convection is another important pathway for oxygen to enter the oil from the surface when the oil is agitated, for example during high temperature processing.

Oxygen and edible oil can react more efficiently when the oil sample has a very large surface area to volume ratio. When the surface-to-volume ratio increases, the relative rate of oxidation is less dependent on oxygen. The surface area of the container can act as a reduction catalyst, and its effect has been shown to be proportional to the area of the container in contact with the oils (Choe & Min, 2006). The interaction between temperature and oxygen concentration affects the formation of volatile compounds in rapeseed oil in the dark; the production of pent-2-enal and pent-1-en-3-one correlated positively with oxygen concentration at 50 °C, but negatively at 35 °C (Carvajal et al., 2009). The rate of reaction between lipids and oxygen is dependent on the type of oxygen; the reaction rate of $^1O_2$ with lipids is much higher than that of $^3O_2$ because $^1O_2$ can react directly with lipids. $^3O_2$ reacts with the radical state of lipids.

The Influence of Bioactive Compounds on the Oxidative Stability of Vegetable Oils

Bioactive compounds are extra-nutritional constituents that usually occur in small amounts in food. They are being studied intensively to assess health effects. Research has shown the protective effects of plant-based diets on cardiovascular disease and cancer (Septembre-Malaterre et al., 2018). Bioactive compounds are able to modulate metabolic processes and demonstrate positive properties such as antioxidant effect, inhibition of receptor activity, inhibition or induction of enzymes, and induction and inhibition of gene expression (Carbonell-Capella et al., 2014). The diversity of chemical structures of bioactive compounds influences bioavailability and biological properties, while anti-nutritional factors can decrease the bioavailability of certain compounds or inhibit digestion enzymes (Septembre-Malaterre et al., 2018).

The Influence of Synthetic Antioxidants on the Lipid Oxidation Process

Butylated hydroxytoluene (BHT), butylated hydroxyanisole (BHA), tert-butylhydroquinone (TBHQ) and α-tocopherol, are the most commonly used antioxidants in the food industry to prevent polyunsaturated fatty acids (PUFA) oxidation (McClements, 2016). The researchers determined that natural and synthetic antioxidants provide variable protection against light-induced oxidation of vegetable oils, where synthetic antioxidants provide better protection throughout treatment and storage (Ullah et al., 2003).

BHA has a slightly higher stability and is therefore more effective especially in protecting the flavor and color of food. Furthermore, BHA is more effective in preserving foods containing animal fats with short-chain fatty acids, such as those found in walnut, coconut, and palm oils used in confectionery. As BHT and BHA are quite volatile with their boiling points around 265°C and 268°C respectively, they are also used in food packaging materials either by direct addition to waxed inner liners or indirectly as emulsions, from which they migrate into the food. BHT and BHA have been shown to have synergistic effects when used in combination, particularly in nuts and nut products (Rangan & Barceloux, 2009).

Tert-butylhydroquinone (TBHQ) is another important synthetic antioxidant most commonly used in the preservation of food products containing frying oils. Chemically, TBHQ is a diphenol and is commercially available as a beige solid, i.e., fat soluble like BHT and BHA due to similar structural characteristics. Having a boiling point range of about 271.3–311.3°C, it is less volatile compared to BHT and BHA. TBHQ is an effective adjunct to the liquid oil hydrogenation process to provide increased oxidative stability and color improvement. Its stabilizing effects on dietary lipids can also be enhanced when used in combination with chelating agents such as citric acid in substances such as vegetable oils or animal fats. But, the use of TBHQ with propyl gallate (PG) is prohibited (Rangan & Barceloux, 2009).

There are only a few synthetic antioxidants such as BHA, BHT, PG and TBHQ currently allowed for use in food as preservatives, as the toxic effects of their degradation products are still under study (Pravst et al., 2010). Even for the synthetic antioxidants mentioned above, the Food and Drug Administration (FDA) requires that their presence be noted on food labels along with an explanation of their intended use. Allowable levels of synthetic antioxidants in food are established based on the fat content of the recipient food product and are usually limited to 0.02% total antioxidants (Poljsak et al., 2021).

The Influence of Some Natural Bioactive Compounds on the Lipid Oxidation Process

Edible oils naturally contain antioxidants such as tocopherols, tocotrienols, carotenoids, phenolic compounds and sterols. Antioxidants are sometimes intentionally added to the oil to improve oxidative stability. Antioxidants are compounds that prolong the induction period of oxidation or slow down the rate of oxidation. Antioxidants scavenge free radicals such as lipid alkyl radicals or lipid peroxy radicals, control transition metals, inhibit singlet oxygen, and inactivate sensitizers. Antioxidants can donate hydrogen atoms to free radicals, converting them into more stable non-radical products (Decker et al., 2017).

The main hydrogen-donating antioxidants are monohydroxy or polyhydroxy phenolic compounds with various substitutions in the aromatic ring. Any compound whose reduction potential is lower than that of a free radical can donate hydrogen to that radical, unless the reaction is kinetically unfavorable (Choe & Min, 2006). Any phenoxy radical produced from the reaction of the antioxidant with the lipid peroxide radical is lower in energy than the lipid peroxide radical. Phenoxy radicals are stabilized by electromeric effects, generating resonance structures as shown in figure 4.

Figure 4. Resonance stabilization of the phenoxy radical
(Choe & Min, 2006)

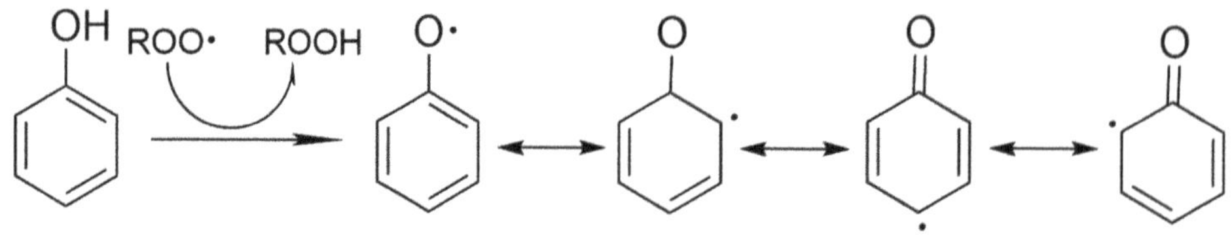

Metal chelators such as phosphoric acid, citric acid, ascorbic acid and EDTA (ethylenediaminetetraacetic acid) reduce oil oxidation in an indirect way. They can convert iron or copper ions into insoluble complexes or prevent the formation of complexes between metals and lipid hydroperoxides (Poljsak et al., 2021). Citric acid improved the sensory quality of soybean oil containing 1 ppm iron during storage at 55°C. Citric acid is often added to oil in the industry to reduce its oxidation during storage before processing. (Choe & Min, 2006) reported that the antioxidant effects of citric acid increased with increasing its content; and more than 150 ppm citric acid was required to overcome the catalytic effect of 1 ppm iron.

When the vegetable oil contains 0.1 ppm or less iron, such as in deodorized refined oil, the common practice of adding about 150 ppm citric acid to the oil to improve oxidative stability is unnecessary (Choe & Min, 2006). Some antioxidants inhibit $^1O_2$ or excited sensitizers. $^1O_2$ is physically and chemically inhibited. In physical inhibition, $^1O_2$ is converted to $^3O_2$ by either energy transfer or charge transfer, and there is no oxidation of antioxidants. In chemical inhibition, antioxidants react with $^1O_2$ and produce oxidized antioxidants (Akoh & Min, 2002).

Chlorophylls

Chlorophylls are common pigments present in edible oil. Virgin olive oil and rapeseed oil contain chlorophyll from 5 or 10 ppm to 35 ppm (Kim et al., 2003). Chlorophylls are generally removed during oil processing, particularly in the bleaching process as shown in table 6.

Table 6. Chlorophyll content in vegetable oil during processing (ppm) (Choe & Min, 2006)

Oil	Chlorophyll α	Phaeophytin α	Phaeophytin β	Pyropheophytin α	Pyropheophytin β
Extract	1.88	3.31	1.34	16.57	3.13
Degummed	0.27	7.16	1.07	9.40	1.84
Refined	0.22	6.27	1.12	9.13	1.79
Bleached	-	0.56	0.32	0.21	0.25

Chlorophylls and chlorophyll derivatives (phaeophytins, pheophorbides) act as sensitizers to produce $^1O_2$ in the presence of light and atmospheric $^3O_2$ and accelerate oil oxidation (Kashyap et al., 2021). Phaeophytins have a higher sensitizing activity than chlorophylls, but lower than that of pheophorbides (Frankel, 2012). Vegetable oil purified by silicic acid column chromatography did not contain chlorophylls and did not produce volatiles under light at 10 °C, but vegetable oil purified with added chlorophyll and refined vegetable oil produced space volatiles under the same experimental conditions (Lee, 2004). Also, appreciable amounts of volatiles were produced in the chlorophyll-containing oil only in the light but not in the dark. The formation of volatiles in vegetable oil under light at 10°C increased with increasing chlorophyll concentration. It was shown that the oxidation of virgin olive oil containing pheophytin was accelerated by fluorescent light illumination. Although chlorophylls are potent prooxidants under light, acting as a sensitizer to produce $^1O_2$, they act as antioxidants in the dark, possibly donating hydrogen to free radicals (Kashyap et al., 2021).

Tocopherols

Tocopherols are the most important antioxidants present in edible oil. Animal fats also contain tocopherols, but in lower concentrations as shown in table 7. Sunflower, soybean, canola and corn vegetable oils contain relatively high amounts of tocopherols. The tocopherol content of edible oils is affected by oil cultivar, processing and storage (Deiana et al., 2002). The tocopherol content of sesame oil varies between 404 and 540 ppm depending on the varieties (Hsu & Parthasarathy, 2017). The tocopherol content of rapeseed oil is 794 and 749 ppm for hexane-extracted oil and pressed oil, respectively. In virgin olive oil, the concentration of α-tocopherol decreased with the time of oil storage in the dark (Deiana et al., 2002).

Table 7. Tocopherol content in fats (Lee, 2004)

Oil	Tocopherol (ppm)				
	α	β	γ	δ	Total
Sunflower	613.0	17.0	18.9	-	648.9
Soy	116.0	34.0	737.0	275.0	1162
Rape	272.1	0.1	423.2	-	695.4
Maize	134.0	18.0	412.0	39.0	603
Fried Sesame	4	-	584	9	597
Rapeseed (technical)	252	-	314	-	566
Saffron	386-520	8.6-12.4	2.4-7.7	-	397-540
Olives	168-226	-	-	-	168-226
Lard	18.0	-	-	-	18.0

Tocopherol was no longer detected in olive oil stored in the dark at room temperature for 12 months (Morelló et al., 2004). Tocopherols can transfer a hydrogen atom to the 6-hydroxy group of its chroman ring in lipid peroxy radicals and scavenge peroxy radicals. Tocopherol (T), with a reduction potential of 500 mV, donates hydrogen to lipid peroxide radicals (ROO·) which have a reduction potential of 1000 mV and produces lipid hydroperoxide (ROOH) and tocopheroxic radicals (T·). Tocopheroxy radicals are more stable than lipid peroxide radicals due to their resonance structures. This ultimately slows down the rate of oil oxidation in the propagation stage of autoxidation. Tocopheroxy radicals can interact with other compounds or with each other, depending on the rates of lipid oxidation. Tocopheroxy radicals react with each other at low rates of lipid oxidation and produce tocopheryl-quinone and tocopherol.

At higher oxidation rates, tocopheroxy radicals can react with lipid peroxy radicals and produce tocopherol-lipid peroxy complexes ([T-OOR]); they can be hydrolyzed into tocopheryl-quinone and lipid hydroperoxide (Mollica et al., 2020):

$$T + ROO^{-} \rightarrow T \bullet + ROOH \quad (8)$$

$$T \bullet + T \bullet \rightarrow T + Tocopheryl\ quinone \quad (9)$$

$$T \bullet + ROO \bullet \rightarrow \left[T - OOR\right] \rightarrow Tocopheryl\ quinone + ROOH \quad (10)$$

The effectiveness of tocopherols as antioxidants depends on the isomers and the concentration of the tocopherols. The antiradical activity of tocopherols is highest in δ-tocopherol followed by γ-, β- and α-tocopherol. α-Tocopherol was more effective in reducing oil oxidation than γ-tocopherol up to 200 ppm and less effective above this concentration (Hira et al., 2019). The optimal concentration of tocopherols as antioxidants is dependent on their oxidative stability; the lower the oxidative stability of tocopherol, the lower the optimal concentration of tocopherol for maximum antioxidant activity. α-Tocopherol, which is the least stable of the tocopherol isomers, showed maximum antioxidant activity at 100 ppm in the oxidation of soybean oil at 55°C in the dark, but the optimal concentrations of γ- and δ-tocopherols as antioxidants were 250 ppm and 500 ppm respectively (Biglar et al., 2012).

When the concentration of the lipid peroxide radical is very low, the tocopheroxy radical abstracts hydrogen from lipids and produces tocopherol radical and lipid alkyl; however, the reaction rate is very slow. The formation of lipid alkyl radicals by tocopherols accelerates lipid oxidation, which is called tocopherol-mediated peroxidation (Jerzykiewicz et al., 2013). The addition of 100 ppm α-tocopherol increased the oxidation of refined olive oil at an early stage of autoxidation; however, α-tocopherol added to refined olive oil or moderately oxidized lard significantly decreased oil oxidation. The limit value for α-tocopherol as a prooxidant in the oxidation of virgin olive oil was 60 to 70 ppm. When less α-tocopherol was present in the oil, the faster the limit value was reached, and the oxidation was enhanced (Deiana et al., 2002). The pro-oxidant activity of α-tocopherol decreased with increasing temperature even at a high concentration (Hira et al., 2019). Ascorbic acid (Asc) can reduce the tocopheroxy radical and prevent tocopherol-mediated peroxidation (Choe & Min, 2006):

$$T + RH \rightarrow T + R \bullet \quad (11)$$

$$R \bullet + O_2 \rightarrow ROO \bullet \quad (12)$$

$$T \bullet + Asc \rightarrow T + Dehydroascorbic\ acid \quad (13)$$

Oxidized tocopherols increase the oxidation of soybean oil, and the prooxidant activity was highest in oxidized α-tocopherol, followed by oxidized γ- and δ-tocopherol (Altemimi et al., 2017). Prevention of tocopherol oxidation and removal of oxidized tocopherols during oil processing are recommended to improve the oxidative stability of the oil.

In addition to free radical-inhibiting activity, tocopherols decrease the oxidation of soybean oil under light only by $^{1}O_2$ inhibition (Kashyap et al., 2021). Tocopherols can form a charge transfer complex ($[T^{+}-{}^{1}O_2]^{1}$) with $^{1}O_2$ by donating electrons to $^{1}O_2$. The singlet state of the tocopherol – $^{1}O_2$ complex undergoes an intersystem crossing into the triplet state ($[T^{+}-{}^{1}O_2]^{3}$) and produces less reactive $^{3}O_2$ and tocopherol (Rastogi et al., 2014). Since this does not involve chemical reactions between tocopherol and $^{1}O_2$, it is called physical inhibition:

$$T + {}^{1}O_2{}^{-} \rightarrow [T + {}^{-1}O_2]^{1-} \rightarrow [T + {}^{-1}O_2]^{3-} \rightarrow T + {}^{3}O_2 \quad (14)$$

Carotenoids

Carotenoids are a group of tetraterpenoids consisting of isoprenoid units. The double bonds in carotenoids are conjugated and usually all trans forms. β-Carotene is one of the most studied carotenoids. Edible oils, especially unrefined oils, contain β-carotene. Virgin olive oil contains 1.0 to 2.7 ppm β-carotene as well as 0.9 to 2.3 ppm lutein (Mordi et al., 2020). β-Carotene can slow oil oxidation by mild filtration, $^{1}O_2$ inhibition, sensitizer inactivation, and free radical scavenging. It was reported that the oxidation of olive oil containing only β-carotene under the action of light at 25°C was reduced by filtering some light energy, mainly between 400 and 500 nm (Rastogi et al., 2014). In the co-presence of chlorophylls, β-carotene decreased the oxidation of vegetable oil stored under light by inhibiting $^{1}O_2$ (Mordi et al., 2020).

Inhibition of $^{1}O_2$ by carotenoids is mainly through energy transfer from $^{1}O_2$ to carotenoids without generating oxidized products. Excited carotenoids return to the ground state by releasing heat:

$$^{1}O_2 + 1Carotenoid^{-} \rightarrow 3O_2 + 3Carotenoid{*}3Carotenoid^{-} \rightarrow Carotenoid + Q \quad (15)$$

Carotenoids that have at least 9 conjugated double bonds act as an effective $^{1}O_2$ inhibitor: *β*-carotene, lycopene and lutein are good $^{1}O_2$ inhibitors. The $^{1}O_2$-inhibitory activity of carotenoids increases with increasing number of conjugated double bonds. Lycopene and α-carotene demonstrate greater $^{1}O_2$ inhibition capacity than β-carotene (Young & Lowe, 2018). Carotenoids inactivate physically excited sensitizers by absorbing energy from the sensitizers. Excited carotenoids return to the ground state by energy transfer to the oil (Lee, 2004) reported that β-carotene with a high one-electron reduction potential of 1060 mV had great difficulty donating hydrogen to the radical alkyl ($E^{o} = 600$ mV) or peroxy radical ($E^{o} = 770$ to 1440 mV) of the polyunsaturated fatty acid. Nuclear magnetic resonance study showed that β-carotene did not donate a hydrogen atom to inhibit these free radicals (Lee, 2004). However, β-carotene can donate hydrogen to the hydroxy radical, which has a high reduction potential (2310 mV), and produces a carotene radical (Car·). A carotene radical is a fairly stable species due to the delocalization of unpaired electrons through its conjugated polyene system. At low oxygen concentration, a carotene radical can react with other radicals such as lipoxy peroxy radicals and form non-radical products (Choe & Min, 2006):

$$Car + HO\cdot \longrightarrow Car\cdot + H_2O \quad (16)$$

$$Car\cdot + ROO\cdot \longrightarrow Car - OOR \quad (17)$$

A lipid peroxy radical can interact with β-carotene and produce carotene peroxy radical (ROO-Car·), especially at more than 150 mmHg of oxygen (Heymann et al., 2015). Carotene peroxy radical reacts with $^3O_2$ and then with lipid molecules and produces lipid alkyl radicals, which propagate the chain reaction of lipid oxidation (Petrou et al., 2018):

$$ROO\cdot + Car \rightarrow ROO - Car\cdot \quad (18)$$

$$ROO - Car\cdot + {}^3O_2 \rightarrow ROO - Car - OO\cdot \quad (19)$$

$$ROO - Car - OO\cdot + RH \rightarrow ROO - Car - OOH + R\cdot \quad (20)$$

Antioxidant activity of β-carotene was not detected in vegetable oil stored in the dark (Lee, 2004). During oil photooxidation in open vessels, β-carotene enhanced oil oxidation, but the co-presence of tocopherols decreased oil oxidation (Jerzykiewicz et al., 2013). The antioxidant activities of carotenoids via hydrogen donation remain controversial. β-Carotene can also donate electrons to free radicals and become a β-carotene cation radical, Car·+ (Choe & Min, 2006). The transfer of hydrogen or electrons from carotenoids to free radicals depends on the reduction potentials of free radicals and the chemical structures of carotenoids, especially the presence of oxygen-containing functional groups (Che et al., 2016). The β-carotene cation radical can interact with alkyl, alkoxy, or peroxy radicals formed in vegetable oil during oxidation (Mordi et al., 2020)

Phenolic Compounds

Phenolic compounds other than tocopherols in edible oils also exert antioxidant activity. Sesame oil, which contains a large amount of unsaturated fatty acids (Iodine Index=109), has good oxidative stability (Tan et al., 2002). The autoxidation rate of sesame oils at 60°C was much lower than that of corn oil, safflower oil and a mixture of soybean and rapeseed oils (Prescha et al., 2014). The remarkable oxidative stability of sesame oil is due to the presence of lignan compounds as well as tocopherols (Bhat et al., 2014). Lignan compounds in sesame oil include sesamin, sesamol, sesamolin, sesaminol, and sesamolinol. Sesamin is the predominant lignan compound in unroasted sesame oil (474 ppm) followed by sesamolin (159 ppm) (J. Y. Kim et al., 2003).

Olive oil, which is very stable to autoxidation (Guillén & Cabo, 2002), contains phenolic compounds and tocopherols. Phenolic compounds in olive oil include tyrosol (4-hydroxyphenylethanol), hydroxytyrosol (3,4-dihydroxyphenylethanol), hydroxybenzoic acids, oleuropein, caffeic acid, vanillic acid, *p*-coumaric acid, and derivatives of tyrosol and hydroxytyrosol (Javidipour et al., 2017), presented in table 8. Phenolic compounds in olive oil acted as antioxidants mainly in the initial stage of autoxidation (Deiana et al., 2002) by scavenging free radicals and chelating metals (Bektaşoğlu et al., 2006).

Among phenolic compounds, *o*-diphenols such as caffeic acid are oxidized to quinones by ferric ions and become ineffective in inhibiting iron-dependent free radical chain reactions in oil (Choe & Min, 2006). Chlorogenic acid and caffeic acid are the main phenolic acids found in sunflower oil. Prooxidant effects of polar phenolic acids such as quercetin and gallic acid have been reported on rapeseed oil in the presence of iron at pH=3.0, due to the increased capacity of iron reduction (Osborne & Akoh, 2003).

Table 8. Phenolic compounds in vegetable oil (Servili & Montedoro, 2002)

Phenolic alcohols	3,4-dihydroxyphenylethanol (3,4-DHPEA, hydroxytyrosol), p-hydroxyphenylethanol (*p*-HPEA, tyrosol), 3,4-dihydroxyphenylethanol glucoside
Phenolic acid and its derivatives	Vanillic acid, syringic acid, p-coumaric acid, o-coumaric acid, gallic acid, protocatechuic acid, p-hydroxybenzoic acid, ferulic acid, cinnamic acid, 4-(acetoxyethyl)-1,2-dihydroxybenzene, benzoic acid
Secoiridoids	Dialdehyde form of elenoic acid linked to 3,4-DHPEA, Dialdehyde form of elenoic acid linked to p-HPEA, Derivatives of p-HPEA
Lignans	1-acetoxypinoresinol, 1-hydroxypinoresinol
Flavonoids	Apigenin, luteolin

OXIDATIVE STABILITY OF VEGETABLE OILS IN COMPLEX ENVIRONMENTS AND THEIR IMPACT ON QUALITY

The quality of the food product is very important that influences the behavior of the consumer, especially when he comes to buy and consume this product. From the point of view of the consumer of food products, the aroma is one of the most important parameter of the quality of the food product. The consumption statistics of countries with a high level of development firmly prove this. The aroma of different food products can undergo major changes during the processing of the raw material, processing and obtaining the finished product, as well as during storage. The components that form the primary aroma can be destroyed, and secondary ones can be formed, usually for the benefit of the final product obtained, as an example, we can mention the maturation processes. However, quite often such changes lead to the loss of important sensory qualities of the food product. This refers in particular to products with high lipid content, which are prone to oxidation. Oxidation of lipids is followed by an unpleasant off-flavor and odor that indicates poor food quality. Unsaturated fatty acids, which are prone to autoxidation, predominate in commercial edible oils which affect their quality and shelf life.

The edible oil is easily perishable due to environmental factors, the presence of metals. In pure form, oils can be protected by the use of antioxidants. More complicated is for the oils in the food matrix, where the intensification of oxidative processes takes place due to enzymes (lipoxygenase, lipase), technological factors etc. The oxidation of fatty acids in food leads to the formation of volatile compounds, most of which have an unpleasant odor and are responsible for altering the flavor in the food industry (Wąsowicz et al., 2004).

As a result of the autoxidation of fatty acids, monohydroperoxides are formed which are initially odorless, which later decompose into volatile compounds. This group includes: aldehydes, ketones, alcohols, acids, hydrocarbons, etc. Due to the low odor resistance of most components, the presence of volatile hydroperoxide degradation products even in a small amount influences the sensory properties of oils and products with high lipid content. Increasing the concentration of carbonyls in the presence of pro-oxidant copper, the original smell of the oil is replaced by an altered, spoiled, unpleasant smell.

According to Lawless, the human ability to smell is considered basic in establishing the quality of food in terms of flavor and odor (Lawless & Heymann, 2010). Volatile compound analysis, often second after sensory analysis, is used to assess lipid oxidation in terms of their utility. Any chemical method used to determine their shelf life must be strictly correlated with sensory analysis. Only sensory analysis can detect aromas and odors that show us or not the presence of oxidative or non-oxidative degradation processes (Wąsowicz et al., 2004).

Emulsions are thermodynamically unstable systems that require stabilizers, such as emulsifiers or texture modifiers, to prevent gravitational separation, flocculation, coalescence, and Ostwald ripening (McClements, 2016). Conventional oil/water emulsions are often prepared by homogenizing oil and a water-soluble emulsifier solution. Homogenizers can be high shear shakers, high pressure homogenizers, colloid mills, ultrasonic homogenizers, and membrane homogenizers. Desired droplet characteristics, such as size of oil-in-water droplets, can be manipulated by changes in homogenizer type, homogenizer operating conditions, and emulsifier type and concentration.

It is important to ensure the oxidative stability of food by using antioxidants. Synthetic antioxidants are used instead of natural ones because they are cheap, widely distributed and accessible (Saad et al., 2007). There are several published studies that indicate a relationship between long-term intake of synthetic antioxidants and some health problems, such as skin allergies, gastrointestinal tract problems, and in some cases, increased cancer risk (Engin et al., 2011). An alternative would be to use natural antioxidants from renewable sources. Natural antioxidants are compounds derived from plants and other living organisms with a strong potential to inhibit oxidative stress by controlling the formation of free radicals, scavenging free radicals, interrupting free radical-mediated chain reactions, and preventing lipid peroxidation (Ramana et al., 2018).

Edible oil often contains multicomponent antioxidants and demonstrates the interactions between them. Metal chelator and free radical scavenger demonstrate better antioxidant activities in oil autoxidation when present together than when used separately.

The enhanced antioxidant capacity of the oil by a combination of metal chelator and free radical scavenging antioxidant is mainly caused by the action of the chelator at the oxidation initiation stage and the free radical inhibition at the propagation stage. Since tocopherols are the most common antioxidants in edible oil, they have been used in research on antioxidant interactions. α-Tocopherol showed synergistic effects with β-carotene to decrease autoxidation and photosensitized oxidation of soybean oil (Frankel, 2012). The synergistic effects of tocopherol and β-carotene have been suggested to be due to protection of β-carotene from degradation by tocopherols (Shibasaki-Kitakawa et al., 2004). α-Tocopherol shows synergism with ascorbic acid and phospholipids. Ascorbic acid donates hydrogen to the tocopheroxy radical produced by the reaction between the peroxy lipoxy radical and tocopherol and then regenerates the tocopherol. The nitrogen moiety of phosphatidylethanolamine and phosphatidylcholine, or phosphatidylinositol reducing sugars, donates hydrogen or electrons to tocopherols and delays the oxidation of tocopherols to tocopheryl-quinone (Bhat et al., 2014).

Synergism between antioxidants is affected by hydroperoxide concentration. Autooxidation of olive oil was lower when α-tocopherol and phenolic compounds were used together than when used separately (Velasco et al., 2004); however, α-tocopherol exerted a negative effect on the antioxidation of olive oil by phenolic compounds such as tyrosol or oleuropein during the period of low hydroperoxide formation, for example, less than 20 meq (Saad et al., 2007).

In recent decades, the number of studies on low-fat edibles has increased, with a considerable focus on diseases such as obesity, cardiovascular disease, and cancer (Chang et al., 2017). Lipids, as a substantial component of food, have been noted as the main source of energy and satiety. Gradually, other benefits of edible products as carriers of lipophilic vitamins and major sources of essential fatty acids have been recognized. However, due to lifestyle changes and a lack of balance between energy intake and expenditure, obesity has increased globally. Considering excessive fat consumption as a decisive factor in obesity, the production of low-fat foods has stimulated many research interests (Ma & Boye, 2013).

CONCLUSION

This study carried out the analysis of the oxidative stability of edible oils and the impact of plant sources with antioxidant potential on slowing down the oxidative processes of food products that allowed the following conclusions to be formulated:

- The oxidation process is one of the most widespread causes of deterioration of food products, especially those with increased lipid content. The process can be influenced by light or heat, fatty acid composition, oxygen species and minor compounds such as metals, pigments, phospholipids etc.
- Synthetic antioxidants (BHT, BHA, TBHQ) are the most commonly used antioxidants in the food industry to prevent the oxidation processes, although the toxic effects of their degradation products are still under study.
- Bioactive compounds from plant-based foods provide antioxidant effects and inhibit oxidative processes. They can influence the bioavailability and biological properties of foods providing protective effects. Edible oils naturally contain antioxidants such as tocopherols, tocotrienols, carotenoids, phenolic compounds, and sterols. Antioxidants are sometimes intentionally added to the oil to improve oxidative stability.
- Foods containing high lipid content are easily perishable due to the intensification of the oxidative processes that take place due to enzymes (lipoxygenase, lipase), technological factors, ingridients etc. The formation of volatile compounds leads to unpleasant odor and decreases the quality of food products.
- It is important to develope the technologies to obtain lipophilic extracts from vegetable powders with antioxidant capacity that represents a good possibility to replace synthetic additives with natural ones in order to obtain safe complex food products with an increased shelf life.

ACKNOWLEDGMENT

This research has been performed under the Moldova National Research Project Nr. 20.80009.5007.27 “Physical-chemical redox processes with electron transfer in the vital, technological and environmental processes”.

REFERENCES

Akoh, C. C., & Min, D. B. (Eds.). (2002). Food lipids: Chemistry, nutrition, and biochemistry (2nd ed., rev.expanded). M. Dekker.

Altemimi, A., Lakhssassi, N., Baharlouei, A., Watson, D., & Lightfoot, D. (2017). Phytochemicals: Extraction, isolation, and identification of bioactive compounds from plant extracts. *Plants*, *6*(4), 42. doi:10.3390/plants6040042 PMID:28937585

Bektaşoğlu, B., Esin Çelik, S., Özyürek, M., Güçlü, K., & Apak, R. (2006). Novel hydroxyl radical scavenging antioxidant activity assay for water-soluble antioxidants using a modified CUPRAC method. *Biochemical and Biophysical Research Communications*, *345*(3), 1194–1200. doi:10.1016/j.bbrc.2006.05.038 PMID:16716257

Bhat, K. V., Kumari, R., Pathak, N., & Rai, A. K. (2014). Value addition in sesame: A perspective on bioactive components for enhancing utility and profitability. *Pharmacognosy Reviews*, *8*(16), 147. doi:10.4103/0973-7847.134249 PMID:25125886

Biglar, M., Khanavi, M., Hajimahmoodi, M., Hassani, S., Moghaddam, G., Sadeghi, N., & Oveisi, M. R. (2012). Tocopherol content and fatty acid profile of different Iranian date seed oils. *Iranian Journal of Pharmaceutical Research: IJPR*, *11*(3), 873–878. PMID:24250514

Burri, L., Hoem, N., Banni, S., & Berge, K. (2012). Marine omega-3 phospholipids: Metabolism and biological activities. *International Journal of Molecular Sciences*, *13*(12), 15401–15419. doi:10.3390/ijms131115401 PMID:23203133

Carbonell-Capella, J. M., Buniowska, M., Barba, F. J., Esteve, M. J., & Frígola, A. (2014). Analytical methods for determining bioavailability and bioaccessibility of bioactive compounds from fruits and vegetables: A review: Bioavailability of bioactive compounds…. *Comprehensive Reviews in Food Science and Food Safety*, *13*(2), 155–171. doi:10.1111/1541-4337.12049 PMID:33412647

Carvajal, A. K., Rustad, T., Mozuraityte, R., & Storrø, I. (2009). Kinetic studies of lipid oxidation induced by hemoglobin measured by consumption of dissolved oxygen in a liposome model system. *Journal of Agricultural and Food Chemistry*, *57*(17), 7826–7833. doi:10.1021/jf9013394 PMID:19691337

Chandran, J., Nayana, N., Roshini, N., & Nisha, P. (2017). Oxidative stability, thermal stability and acceptability of coconut oil flavored with essential oils from black pepper and ginger. *Journal of Food Science and Technology*, *54*(1), 144–152. doi:10.100713197-016-2446-y PMID:28242912

Chang, C., Li, J., Li, X., Wang, C., Zhou, B., Su, Y., & Yang, Y. (2017). Effect of protein microparticle and pectin on properties of light mayonnaise. *Lebensmittel-Wissenschaft + Technologie*, *82*, 8–14. doi:10.1016/j.lwt.2017.04.013

Che, P., Zhao, Z.-Y., Glassman, K., Dolde, D., Hu, T. X., Jones, T. J., Gruis, D. F., Obukosia, S., Wambugu, F., & Albertsen, M. C. (2016). Elevated vitamin E content improves all- *trans* β-carotene accumulation and stability in biofortified sorghum. *Proceedings of the National Academy of Sciences of the United States of America*, *113*(39), 11040–11045. doi:10.1073/pnas.1605689113 PMID:27621466

Chen, B., McClements, D. J., & Decker, E. A. (2011). Minor components in food oils: A critical review of their roles on lipid oxidation chemistry in bulk oils and emulsions. *Critical Reviews in Food Science and Nutrition*, *51*(10), 901–916. doi:10.1080/10408398.2011.606379 PMID:21955091

Choe, E., & Min, D. B. (2006). Mechanisms and factors for edible oil oxidation. *Comprehensive Reviews in Food Science and Food Safety*, *5*(4), 169–186. doi:10.1111/j.1541-4337.2006.00009.x

Collin, F. (2019). Chemical basis of reactive oxygen species reactivity and involvement in neurodegenerative diseases. *International Journal of Molecular Sciences*, *20*(10), 2407. doi:10.3390/ijms20102407 PMID:31096608

Decker, E. A., McClements, D. J., Bourlieu-Lacanal, C., Durand, E., Figueroa-Espinoza, M. C., Lecomte, J., & Villeneuve, P. (2017). Hurdles in predicting antioxidant efficacy in oil-in-water emulsions. *Trends in Food Science & Technology*, *67*, 183–194. doi:10.1016/j.tifs.2017.07.001

Deiana, M., Rosa, A., Cao, C. F., Pirisi, F. M., Bandino, G., & Dessì, M. A. (2002). Novel approach to study oxidative stability of extra virgin olive oils: Importance of α-tocopherol concentration. *Journal of Agricultural and Food Chemistry*, *50*(15), 4342–4346. doi:10.1021/jf020033t PMID:12105968

Domínguez, R., Pateiro, M., Gagaoua, M., Barba, F. J., Zhang, W., & Lorenzo, J. M. (2019). A Comprehensive review on lipid oxidation in meat and meat products. *Antioxidants*, *8*(10), 10. doi:10.3390/antiox8100429 PMID:31557858

Engin, A. B., Bukan, N., Kurukahvecioglu, O., Memis, L., & Engin, A. (2011). Effect of butylated hydroxytoluene (E321) pretreatment versus l-arginine on liver injury after sub-lethal dose of endotoxin administration. *Environmental Toxicology and Pharmacology*, *32*(3), 457–464. doi:10.1016/j.etap.2011.08.014 PMID:22004966

Frankel, E. N. (2012). Hydroperoxide decomposition. In *Lipid Oxidation* (pp. 67–98). Elsevier. doi:10.1533/9780857097927.67

Gawrysiak-Witulska, M., Rudzińska, M., Wawrzyniak, J., & Siger, A. (2012). The effect of temperature and moisture content of stored rapeseed on the phytosterol degradation rate. *Journal of the American Oil Chemists' Society*, *89*(9), 1673–1679. doi:10.100711746-012-2064-4 PMID:22923815

Gharby, S., Harhar, H., Matthäus, B., Bouzoubaa, Z., & Charrouf, Z. (2016). The chemical parameters and oxidative resistance to heat treatment of refined and extra virgin Moroccan Picholine olive oil. *Journal of Taibah University for Science : JTUSCI*, *10*(1), 100–106. doi:10.1016/j.jtusci.2015.05.004

Guillén, M. D., & Cabo, N. (2002). Fourier transform infrared spectra data versus peroxide and anisidine values to determine oxidative stability of edible oils. *Food Chemistry*, *77*(4), 503–510. doi:10.1016/S0308-8146(01)00371-5

Halliwell, B., & Gutteridge, J. M. C. (2015). *Free Radicals in Biology and Medicine*. Oxford University Press. doi:10.1093/acprof:oso/9780198717478.001.0001

Heymann, T., Heinz, P., & Glomb, M. A. (2015). Lycopene inhibits the isomerization of β-carotene during quenching of singlet oxygen and free radicals. *Journal of Agricultural and Food Chemistry*, *63*(12), 3279–3287. doi:10.1021/acs.jafc.5b00377 PMID:25803572

Hira, S., Saleem, U., Anwar, F., Sohail, M. F., Raza, Z., & Ahmad, B. (2019). β-Carotene: A natural compound improves cognitive impairment and oxidative stress in a mouse model of streptozotocin-induced Alzheimer's disease. *Biomolecules*, *9*(9), 441. doi:10.3390/biom9090441 PMID:31480727

Hsu, E., & Parthasarathy, S. (2017). Anti-inflammatory and antioxidant effects of sesame oil on atherosclerosis: A descriptive literature review. *Cureus*. doi:10.7759/cureus.1438 PMID:28924525

Javidipour, I., Erinç, H., Baştürk, A., & Tekin, A. (2017). Oxidative changes in hazelnut, olive, soybean, and sunflower oils during microwave heating. *International Journal of Food Properties*, *20*(7), 1582–1592. doi:10.1080/10942912.2016.1214963

Jerzykiewicz, M., Ćwieląg-Piasecka, I., & Jezierski, A. (2013). Pro- and antioxidative effect of α-tocopherol on edible oils, triglycerides and fatty acids. *Journal of the American Oil Chemists' Society, 90*(6), 803–811. doi:10.100711746-013-2227-y PMID:23710072

Kashyap, A., Ramasamy, E., Ramalingam, V., & Pattabiraman, M. (2021). Supramolecular control of singlet oxygen generation. *Molecules (Basel, Switzerland), 26*(9), 2673. doi:10.3390/molecules26092673 PMID:34063309

Kim, J. Y., Choi, D. S., & Jung, M. Y. (2003). Antiphoto-oxidative activity of sesamol in methylene blue- and chlorophyll-sensitized photo-oxidation of oil. *Journal of Agricultural and Food Chemistry, 51*(11), 3460–3465. doi:10.1021/jf026056p PMID:12744684

Kim, N., & Choe, E. (2012). Effects of monoacylglycerols on the oil oxidation of acidic water/perilla oil emulsion under light in the presence of chlorophyll. *Food Science and Biotechnology, 21*(1), 183–189. doi:10.100710068-012-0023-3

Krumova, K., & Cosa, G. (2016). Overview of reactive oxygen species. In S. Nonell & C. Flors (Eds.), *Comprehensive Series in Photochemical & Photobiological Sciences* (Vol. 1, pp. 1–21). Royal Society of Chemistry. doi:10.1039/9781782622208-00001

Lawless, H. T., & Heymann, H. (2010). *Sensory Evaluation of Food: Principles and Practices*. Springer New York. doi:10.1007/978-1-4419-6488-5

Lee, Y. (2004). Chemical composition and oxidative stability of safflower oil prepared from safflower seed roasted with different temperatures. *Food Chemistry, 84*(1), 1–6. doi:10.1016/S0308-8146(03)00158-4

Lise Halvorsen, B., & Blomhoff, R. (2011). Determination of lipid oxidation products in vegetable oils and marine omega-3 supplements. *Food & Nutrition Research, 55*(1), 5792. doi:10.3402/fnr.v55i0.5792 PMID:21691461

Liu, X. (2018). Progress in the mechanism and kinetics of Fenton reaction. *MOJ Ecology & Environmental Sciences, 3*(1). doi:10.15406/mojes.2018.03.00060

Ma, Z., & Boye, J. I. (2013). Advances in the design and production of reduced-fat and reduced-cholesterol salad dressing and mayonnaise: A Review. *Food and Bioprocess Technology, 6*(3), 3. doi:10.100711947-012-1000-9

Martín-Polvillo, M., Márquez-Ruiz, G., & Dobarganes, M. C. (2004). Oxidative stability of sunflower oils differing in unsaturation degree during long-term storage at room temperature. *Journal of the American Oil Chemists' Society, 81*(6), 577–583. doi:10.100711746-006-0944-1

McClements, D. J. (2016). Food emulsions: Principles, practices, and techniques (Third edition). CRC Press, Taylor & Francis Group.

Min, D. B., Callison, A. L., & Lee, H. O. (2003). Singlet oxygen oxidation for 2-pentylfuran and 2-penteny-furan formation in soybean oil. *Journal of Food Science, 68*(4), 1175–1178. doi:10.1111/j.1365-2621.2003.tb09620.x

Mollica, F., Lucarini, M., Passerini, C., Carati, C., Pavoni, S., Bonoldi, L., & Amorati, R. (2020). Effect of antioxidants on high-temperature stability of renewable bio-oils revealed by an innovative method for the determination of kinetic parameters of oxidative reactions. *Antioxidants*, *9*(5), 399. doi:10.3390/antiox9050399 PMID:32397271

Mordi, R. C., Ademosun, O. T., Ajanaku, C. O., Olanrewaju, I. O., & Walton, J. C. (2020). Free radical mediated oxidative degradation of carotenes and xanthophylls. *Molecules (Basel, Switzerland)*, *25*(5), 1038. doi:10.3390/molecules25051038 PMID:32110916

Morelló, J.-R., Motilva, M.-J., Tovar, M.-J., & Romero, M.-P. (2004). Changes in commercial virgin olive oil (cv Arbequina) during storage, with special emphasis on the phenolic fraction. *Food Chemistry*, *85*(3), 357–364. doi:10.1016/j.foodchem.2003.07.012

Negash, Y. A., Amare, D. E., Bitew, B. D., & Dagne, H. (2019). Assessment of quality of edible vegetable oils accessed in Gondar City, Northwest Ethiopia. *BMC Research Notes*, *12*(1), 793. doi:10.118613104-019-4831-x PMID:31801628

Osborne, H., & Akoh, C. C. (2003). Effects of natural antioxidants on iron-catalyzed lipid oxidation of structured lipid-based emulsions. *Journal of the American Oil Chemists' Society*, *80*(9), 847–852. doi:10.100711746-003-0784-z

Petrou, A., Petrou, P., Ntanos, T., & Liapis, A. (2018). A possible role for singlet oxygen in the degradation of various antioxidants. A meta-analysis and review of literature Data. *Antioxidants*, *7*(3), 35. doi:10.3390/antiox7030035 PMID:29495515

Poljsak, B., Kovač, V., & Milisav, I. (2021). Antioxidants, food processing and health. *Antioxidants*, *10*(3), 433. doi:10.3390/antiox10030433 PMID:33799844

Pravst, I., Žmitek, K., & Žmitek, J. (2010). Coenzyme Q10 contents in foods and fortification strategies. *Critical Reviews in Food Science and Nutrition*, *50*(4), 269–280. doi:10.1080/10408390902773037 PMID:20301015

Prescha, A., Grajzer, M., Dedyk, M., & Grajeta, H. (2014). The antioxidant activity and oxidative stability of cold-pressed oils. *Journal of the American Oil Chemists' Society*, *91*(8), 1291–1301. doi:10.100711746-014-2479-1 PMID:25076788

Rakotondramavo, A., Ribourg, L., Meynier, A., Guyon, C., de Lamballerie, M., & Pottier, L. (2019). Monitoring oxidation during the storage of pressure-treated cooked ham and impact on technological attributes. *Heliyon*, *5*(8), e02285. doi:10.1016/j.heliyon.2019.e02285 PMID:31485513

Ramana, K. V., Reddy, A. B. M., Majeti, N. V. R. K., & Singhal, S. S. (2018). Therapeutic potential of natural antioxidants. *Oxidative Medicine and Cellular Longevity*, *1–3*, 1–3. doi:10.1155/2018/9471051 PMID:30186551

Rangan, C., & Barceloux, D. G. (2009). Food additives and sensitivities. *Disease-a-Month*, *55*(5), 292–311. doi:10.1016/j.disamonth.2009.01.004 PMID:19362177

Rastogi, A., Yadav, D. K., Szymańska, R., Kruk, J., Sedlářová, M., & Pospíšil, P. (2014). Singlet oxygen scavenging activity of tocopherol and plastochromanol in *Arabidopsis thaliana*: Relevance to photooxidative stress: $^{1}O_{2}$ scavenging by tocopherol and plastochromanol. *Plant, Cell & Environment*, *37*(2), 392–401. doi:10.1111/pce.12161 PMID:23848570

Saad, B., Sing, Y., Nawi, M., Hashim, N., Mohamedali, A., Saleh, M., Sulaiman, S., Talib, K., & Ahmad, K. (2007). Determination of synthetic phenolic antioxidants in food items using reversed-phase HPLC. *Food Chemistry*, *105*(1), 389–394. doi:10.1016/j.foodchem.2006.12.025

Septembre-Malaterre, A., Remize, F., & Poucheret, P. (2018). Fruits and vegetables, as a source of nutritional compounds and phytochemicals: Changes in bioactive compounds during lactic fermentation. *Food Research International*, *104*, 86–99. doi:10.1016/j.foodres.2017.09.031 PMID:29433787

Servili, M., & Montedoro, G. F. (2002). Contribution of phenolic compounds to virgin olive oil quality. *European Journal of Lipid Science and Technology*, *104*(9-10), 602–613. doi:10.1002/1438-9312(200210)104:9/10<602::AID-EJLT602>3.0.CO;2-X

Shibasaki-Kitakawa, N., Kato, H., & Yonemoto, T. (2004). Oxidation kinetics of β—Carotene in oleic acid solvent with addition of an antioxidant, α-tocopherol. *Journal of the American Oil Chemists' Society*, *81*(4), 389–394. doi:10.100711746-004-0911-x

Shiota, M., Uchida, T., Oda, T., & Kawakami, H. (2006). Utilization of lactoferrin as an iron-stabilizer for soybean and fish oil. *Journal of Food Science*, *71*(3), C120–C123. doi:10.1111/j.1365-2621.2006.tb15606.x

Sousa, B. C., Pitt, A. R., & Spickett, C. M. (2017). Chemistry and analysis of HNE and other prominent carbonyl-containing lipid oxidation compounds. *Free Radical Biology & Medicine*, *111*, 294–308. doi:10.1016/j.freeradbiomed.2017.02.003 PMID:28192230

Steenson, D. F., Lee, J. H., & Min, D. B. (2002). Solid phase microextraction of volatile soybean oil and corn oil compounds. *Journal of Food Science*, *67*(1), 71–76. doi:10.1111/j.1365-2621.2002.tb11361.x

Tan, C. P., Che Man, Y. B., Selamat, J., & Yusoff, M. S. A. (2002). Comparative studies of oxidative stability of edible oils by differential scanning calorimetry and oxidative stability index methods. *Food Chemistry*, *76*(3), 385–389. doi:10.1016/S0308-8146(01)00272-2

Tynek, M., Pawłowicz, R., Gromadzka, J., Tylingo, R., Wardencki, W., & Karlovits, G. (2012). Virgin rapeseed oils obtained from different rape varieties by cold pressed method—Their characteristics, properties, and differences. *European Journal of Lipid Science and Technology*, *114*(3), 357–366. doi:10.1002/ejlt.201100296

Ullah, J. M. H. T. A. M. A., &. M. Z. (2003). Effect of light, natural and synthetic antioxidants on stability of edible oil and fats. *Asian Journal of Plant Sciences*, *2*(17), 1192–1194. doi:10.3923/ajps.2003.1192.1194

Velasco, J., Andersen, M. L., & Skibsted, L. H. (2004). Evaluation of oxidative stability of vegetable oils by monitoring the tendency to radical formation. A comparison of electron spin resonance spectroscopy with the Rancimat method and differential scanning calorimetry. *Food Chemistry*, *85*(4), 623–632. doi:10.1016/j.foodchem.2003.07.020

Wąsowicz, E., Gramza, A., Hęś, M., Jeleń, H. H., Korczak, J., Małecka, M., Mildner-Szkudlarz, S., Rudzińska, M., Samotyja, U., & Zawirska-Wojtasiak, R. (2004). Oxidation of lipids in food. *Polish Journal of Food and Nutrition Sciences*, *54*(1s), 87–100.

Wszelaki, A. L., Delwiche, J. F., Walker, S. D., Liggett, R. E., Miller, S. A., & Kleinhenz, M. D. (2005). Consumer liking and descriptive analysis of six varieties of organically grown edamame-type soybean. *Food Quality and Preference*, *16*(8), 651–658. doi:10.1016/j.foodqual.2005.02.001

Xu, L., Yu, X., Li, M., Chen, J., & Wang, X. (2017). Monitoring oxidative stability and changes in key volatile compounds in edible oils during ambient storage through HS-SPME/GC–MS. *International Journal of Food Properties, 20*(sup3), S2926–S2938. doi:10.1080/10942912.2017.1382510

Young, A., & Lowe, G. (2018). Carotenoids—Antioxidant Properties. *Antioxidants*, *7*(2), 28. doi:10.3390/antiox7020028 PMID:29439455

Yun, J.-M., & Surh, J.-H. (2012). Fatty acid composition as a predictor for the oxidation stability of Korean vegetable oils with or without induced oxidative stress. *Preventive Nutrition and Food Science*, *17*(2), 158–165. doi:10.3746/pnf.2012.17.2.158 PMID:24471078

Zebib, H., Bultosa, G., & Abera, S. (2015). Physico-chemical Properties of Sesame (<i>Sesamum indicum</i> L.) Varieties Grown in Northern Area, Ethiopia. *Agricultural Sciences*, *06*(02), 238–246. doi:10.4236/as.2015.62024

Zielinski, Z. A. M., & Pratt, D. A. (2017). Lipid Peroxidation: Kinetics, Mechanisms, and Products. *The Journal of Organic Chemistry*, *82*(6), 2817–2825. doi:10.1021/acs.joc.7b00152 PMID:28248497

Chapter 13
Combined Ferrite Treatment of Multi-Component Wastewaters

Gheorghe Duca
Institute of Chemistry, Moldova State University, Moldova

Olga Covaliova
Institute of Chemistry, Moldova State University, Moldova

Victor Covaliov
Moldova State University, Moldova

Lidia Romanciuc
https://orcid.org/0000-0002-4555-1194
Institute of Chemistry, Moldova State University, Moldova

Ashok Vaseashta
https://orcid.org/0000-0002-5649-006
International Clean Water Institute, USA

ABSTRACT

Ferrite treatment of wastewater is a promising approach, making it possible to resolve environmental and economic problems, due to the high degree of water purification and final formation of chemically stable sludge with crystal oxide structure and ferromagnetic properties. This chapter discusses the specifics of ferrite treatment of wastewater containing heavy metal ions and its advantages with regard to other physical-chemical technologies. The role of ζ-potential of microparticles Fe3O4 in the formation of spinel structures is discussed. The mechanism of ferritization reactions is connected with the catalytic nature of the running processes and the probability of the intermediate active radical's formation. To make the ferrite treatment more flexible and effective, it can be combined with photocatalytical treatment, which allows acceleration of the oxidation processes during the ferrite sludge formation, and at the same time promotes the destruction of organic contaminants due to the formation of free radicals in treated water.

DOI: 10.4018/979-8-3693-0512-6.ch013

INTRODUCTION

The application of ferritization methods for wastewater treatment is promising since it allows for solving a number of important technological, environmental, and socio-economic problems. These methods allow removing the heavy metal ions from industrial wastewater, which are in the form of water-soluble complexes within all pH ranges, for example, ammonia complexes (Covaliov & Caraghiaur, 1985; Demirel et al., 1999; Hua et al., 2012; Kefeni et al., 2017; Masunga et al., 2019; Sharma et al., 2022; Singh et al., 2020; Qiao et al., 2019). Another problem solved by ferrite technology is the possibility of easy removal of sludge from treated wastewater both in gravitational and magnetic fields, due to the crystalline structure of sludge particles and their ferromagnetic properties. It is known (Yakovlev et al., 1999) that in real treatment facilities, the expenditures for treated water clarifying, water treating with coagulants to settle the sediments in settling tanks, and their subsequent dehydration, amount up to 60% of capital and operating costs. The third important problem to be solved in ferrite water treatment technology is the obtaining of metal oxide structures that are chemically resistant against leaching, as compared to metal hydroxides (Covaliova & Covaliov, 2004; Pynaev, 2006; Duca & Vaseashta, 2021, Duca et al, 2022). On the one hand, due to this specific property of forming sludge, a 1.5-2.0 times increase in the wastewater treatment degree from heavy metal ions can be reached, and on the other hand, ferrite treatment solves an important environmental problem of preventing the leaching of sludge components during the storage. In addition, the ferritization technology opens new ways to utilize the sludge with the oxide-spinel structure, whereas the hydroxide compounds of heavy metals formed as a result of wastewater treatment by traditional technologies, despite the great efforts of researchers to identify the ways of their proper management, still do not have real ways of safe storage and utilization. Due to these advantages, ferritization treatment is more suitable for heavy metal extraction from wastewater, as compared to other traditional technologies such as ion exchange, ultrafiltration, adsorbtion, etc.

BACKGROUND

The research teams of Bybik and Makarov (Bybik et al., 1979; Makarov et al., 2015), studying the kinetics of ferrite formation from the colloidal particles of iron (II) and (III) hydroxides during the chemical production of magnetic fluids, have observed some incubation period of ferritized particles crystallization, which is usually typical for the induction of autocatalytic reactions. Semenov (Semenov et al., 2005), while developing the technology for stabilizing long-term stored heavy metal hydroxide sludge, in order to reduce the metal ions "washout" during storage, have proposed to transform the hydroxide sludge into a more chemically stable oxide (ferrite) form. To do this, hydroxide sludge was subjected to preliminary dissolution in sulfuric acid with the subsequent introduction of alkaline chemicals, iron (II) sulfate, and previously obtained magnetite-containing sludge, and heating the mixture to 70–90°C, which ensured ferritization of the entire mass of the sludge. The positive effect of ferritized sludge additions on the process was explained only from the standpoint of improving of its technological properties, without explaining the mechanism of ferritization.

Analyzing the state-of-the-art of ferritization technology, Ryzo (Ryzo et al., 2006) considers it to be the most universal and progressive among the other water treatment methods. It includes many aspects and a series of simultaneously occurring electrochemical and reduction-oxidation processes, as well as phase-dispersed transformations, and other processes. However, there is still a lack of research and theoretical justification of ferrite method which prevents its further development. Specifically, most of the ferritization processes require high-temperature conditions and are, therefore, energy-intensive, as treatment processes are running when the wastewater or condensed sludge is heated to 65-70°C and above.

De Satour (1973) put forward an assumption about the radical nature of the chemical interaction mechanism of the reaction products. It was assumed that iron (II) and (III) hydroxides, during the initial period of their formation, are forming the intermediate radical ions $\bullet FeOH^{+}$ and $\bullet FeOH^{2+}$, which, as a result of their interaction when the suspension is heated, are forming the magnetite:

$$\bullet FeOH^{+} + 2\bullet FeOH^{2+} + O^{2-} \rightarrow Fe_3O_4 + 3H^{+} \quad (1)$$

However, without taking into account the values of the electrokinetic potential of particles formed in this process, it was difficult to explain the interaction mechanism of two positively charged radical ions with each other. The probability of radical ions formation can be sufficiently justified on the basis of the Fenton mechanism of redox processes involving iron compounds (Duca, 2012). Recently, the behavior of various aqueous iron(III) complexes has become the subject of numerous studies of the photocatalytic degradation of persistent organic compounds. On the other hand, it has also been shown that the formation of hydroxyl radicals depends not only on the presence of iron(III) particles and iron-containing complexes in water solution but also on the wavelength of UV irradiation and pH.

In the presence of oxygen, ferrous iron (Fe^{2+}) is oxidized to the ferric state (Fe^{3+}), a process that is highly dependent on pH. An increase in Fe(II) oxidation rate with the pH change is attributed to the higher reactivity of oxygen with respect to $Fe(OH)^{+}$ and $Fe(OH)_2$ forms than with respect to the Fe^{2+} form. This process results in the formation of a superoxide radical ($\bullet O^{2-}$), the more reactive reduced form of oxygen. This radical, in turn, can interact with ferrous iron, leading to the formation of H_2O_2, which, due to the Fenton reaction, makes it possible to obtain $\bullet OH$- radicals, which are more active in oxidation reactions.

Under these conditions, the reaction is running following the photocatalytic mechanism, associated with the hydroxyl radical formation, formed as a result of photodissociation of $Fe(OH)^{2+}$ due to the presence of the hydroxyl group (OH^{-}) in the composition of iron water complexes:

$$Fe(III) - OH \underset{[O_2]}{\overset{h\nu}{\rightleftharpoons}} Fe(II) + {}^{\bullet}OH \quad (2)$$

The hydroxyl radicals ($\bullet OH$) have the increased value of free negative energy (263 kJ/mol), therefore, thermodynamically they exhibit high reactive (oxidizing) properties with respect to the organic substances, oxidizing their molecules, following the mechanism of hydrogen atom abstraction, with the formation of water molecule, according to the general reaction (Covaliova, 2010):

$$RH + {}^{\bullet}OH \rightarrow {}^{\bullet}R + H_2O \quad (3)$$

The $\bullet OH$ radical can react with Fe(II) to form Fe(III). The resulting ferric ion (Fe^{3+}) can further participate in various reactions, first of all undergoing the hydrolysis being precipitated in the form of iron(III) hydroxide $Fe(OH)_3$. The process of active radical formation can be significantly accelerated when hydrogen peroxide is introduced into the treated water solution. At the same time, active radicals' consumption in the redox processes can be lower than the stoichiometric one.

In (Casbeer, 2012; Lou et al., 2009; Surech et al., 2021; Wang et al., 2022), the possibilities to combine the ferritization and photocatalytic processes are described; however, the explanation of the processes mechanism running under the room temperatures is rather ambiguous. Despite the experimental data available and the practical importance of this process, its theoretical aspects have not been sufficiently explained and justified.

Therefore, recognizing the complexity of this task, we have made an attempt to explain the mechanism of catalytic processes occurring during the ferrite treatment of industrial wastewater, and on this basis to propose technological and instrumental solutions to reduce the energy intensity of low-temperature ferrite processes.

MATERIALS AND METHODS

For the research the hydroxide suspensions of sediments were taken, formed during the electrocoagulation treatment of wastewater from the galvanic production, containing amorphous products of metal hydroxides with the following characteristics:

Table 1. The composition of hydroxide suspensions resulted from electrocoagulation processes

Parameters	Values
Contents in solid phase, %	3.1 - 3.25
pH	7.1 - 7.5
Ions contents in the initial wastewater, mg/L:	
Fe^{3+}/Fe^{2+}	45.3 - 51.5
Ni^{2+}	16.2 - 17.5
Zn^{2+}	6.2 - 6.7
Cr(VI)	8.8 - 9.5

The electrokinetic properties of sediment particles were measured by classical electroosmotic method (Pysarenko et al., 1964), taking into account the electroosmosis rate, measured values of the liquid viscosity, the distance between the electrodes, and the applied potential difference. The value of ζ-potential was calculated using the Helmholtz-Smoluchowski equation, in Volts (Diehl & Levin, 2006):

$$\zeta = 4\pi\eta\upsilon/H\varepsilon, \tag{4}$$

where η - is the viscosity coefficient; υ is the electroosmosis velocity gradient; H - is the potential gradient of the external field, expressing the potential drop (U) per 1 cm of tube length, i.e., H=U/l; ε is the dielectric constant. In this case, the measured value of the viscosity of the studied dispersion medium was (1.03-1.035) $\bullet 10^{-3}$ N•sec/m^2 (Newton-second per square meter).

Magnetometric studies of dispersed ferrite sludge were carried out using the laboratory setup at the St. Petersburg Technological University at the Department of Colloidal Chemistry (Bybik et al., 1980), and using the setup specially developed at the "Technology" Institute (Chisinau, Republic of Moldova) for measuring the magnetic parameters using a measuring cell described by us in (Covaliov & Covaliova, 2004).

RESULTS AND DISCUSSION

Heavy metal hydroxide compounds in suspensions were formed as a result of electrochemical and redox processes, with the approximate molar of ratio Fe(II) and Fe(III) hydroxides was 1: 2. During the anodic process, iron electrodes were dissolved with the release of Fe^{2+} ions into the solution, whereas electrolysis of water molecules occurred at the cathode with the release of H_2 gas and formation of free OH- ions, which in turn contributed to alkalization and formation of metal hydroxides in the treated medium. Formation of $Fe(OH)_3$ under these conditions is ensured by redox hydrolysis processes in the presence of Cr(VI) ions and at the expense of atmospheric oxygen.

In the initial state, electrocoagulation sludge does not have magnetic properties, which practically are not changed with time (Fig. 1, curve 1). Formation of highly dispersed crystalline spinel particles with magnetic properties (magnetite particles) can occur by ferritization of the initial amorphous sludge in two ways: either by thermal heating of the sludge to 70°C, or by dosing of a preliminarily obtained magnetite suspension into the electrocoagulation reactor under the normal room temperature (20-25°C). In both cases, the formation of ferromagnetic spinel structures proceeds according to the following:

$$2Fe(OH)_3 + Fe(OH)_2 = Fe_3O_4 + 4H_2O, \tag{5}$$

$$nMe^{2+} + (3\text{-}n)Fe^{3+} + 6OH^- \rightarrow Me_nFe_{3-n}(OH)_6, \tag{6}$$

$$Me_nFe_{3-n}(OH)_6 + 1/2O_2 \rightarrow Me_nFe_{3-n}O_4 + 3H_2O \tag{7}$$

As evidenced by the data obtained (Fig.1), the formation of ferrite particles, recorded by magnetic measurements of the suspension, occurs with a certain delay in time, determined by the induction period of particle nucleation for 20–40 seconds, after which the ferritization process accelerates and practically ends within 5-10 minutes. At the same time, the induction period of the process with the dosing of magnetite-containing sludge, as well as the nature of ferritization (curve 3), proceeds noticeably faster than under conditions of heating of the hydroxide suspension (curve 2). This suggests that the induction period is associated with the autocatalytic nature of the processes occurring in this case, since after its completion, an intensive increase in the suspension ferritization rate begins, which practically ends in 3-4 minutes.

To explain the mechanism of interaction between colloidal hydroxide structures and to relate the electrokinetic properties of the dispersed system to the processes of sludge ferritization, the dependence of the electrokinetic potential of magnetic particles on the pH was studied. As can be seen from Fig.2, the ξ-potential in the acidic pH area is positive, and as its values increase, it goes into the negative region. In this case, the isoelectric point is within pH $\approx$ 6.5. This is explained by the nature of the dispersed phase and dispersion medium and is associated with the adsorption of adsorbate molecules and the selective adsorption of electrolyte ions.

Figure 1. Change in the magnetization of the suspension depending on the time of ferritization of amorphous sludge resulted from the electrocoagulation wastewater treatment: 1 - initial sludge; 2 - when the initial sludge is heated up to 70°C; 3 - when magnetite-containing sludge is added to the hydroxide suspension

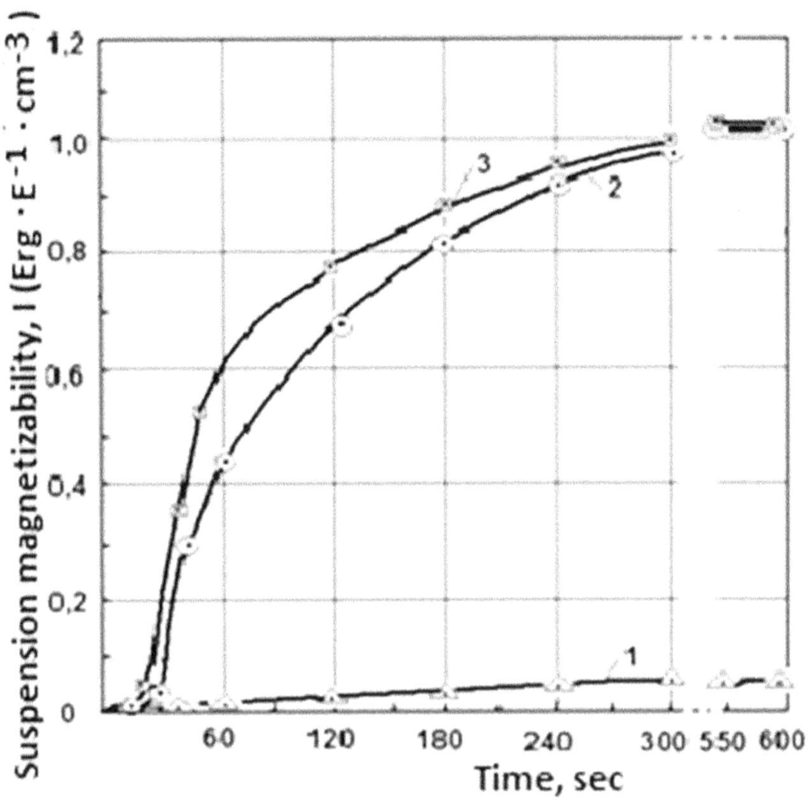

Catalytic activity of magnetite particles can be explained on the basis of the structure of its crystal lattice, which can be expressed by the formula $Fe^{III}Fe^{II}[Fe^{III}O_4]$, in which the Fe^{2+} and Fe^{3+} ions located outside the $[FeO_4]$ structural groups easily exchange with charges and have a mobile electron. This, in particular, explains the good electrical conductivity of magnetite. Positive charge of magnetite particles in the acidic pH region is due to the adsorption of H^+ ions on their surface, and the negative charge of hydroxide sludge at a higher pH value is probably due to both hydroxide ions (OH^-), which are involved in the completion of the crystal lattice of magnetite, and other spinel structures and mobile free electrons. This makes it possible to explain the mechanism of the heterogeneous interaction of amorphous colloidal particles of metal hydroxides, which, apparently, are characterized by a positive charge due to the nature of the charge of iron hydroxide molecules.

During the ferrite wastewater treatment at room temperature, introduced particles of magnetite Fe_3O_4 play the role of active catalytic centers for the reactions of the interaction of radical ions of iron(II) and iron(III) hydroxides with the formation of sludge of a spinel crystal structure with stable ferromagnetic properties. This process is autocatalytic, in which the resulting magnetite particles are catalysts for new

Figure 2. Dependence of the electrokinetic potential ξ of Fe_3O_4 particles in suspension on the pH value

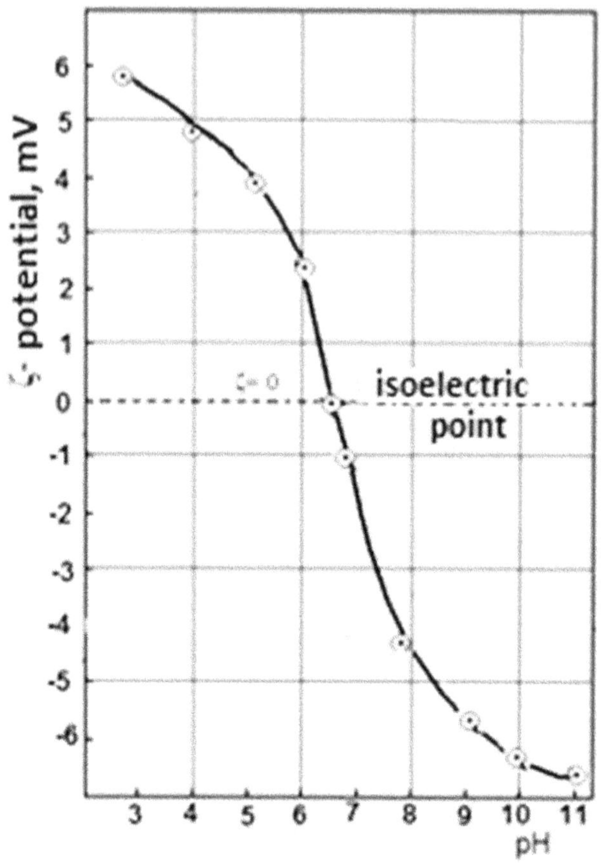

interfacial interactions in the bulk of treated water. The process of sludge ferritization proceeds within all temperature ranges, starting from room temperature and tending to accelerate with temperature increase.

At the same time, this process has an induction period typical for all autocatalytic processes, making 0.5-1.0 minutes, which depends on a number of factors - temperature, current density, pH value, content, and the ratio of ferrous and ferric ions in the suspension, which should be, according to the stoichiometry of the reaction, within 1:2. Under these conditions, the magnetization of the resulting suspension has an optimum within the low alkaline pH range of 6.5-8.5 and is equal to 1800-2500 Oersted.

The resulting ferritized sludge containing complex oxide compounds of heavy metals is more chemically stable in water as compared to hydroxide compounds, therefore, more efficient water purification from heavy metal ions is provided.

Comparative experimental results have shown that the conditions proposed for the treatment of wastewater from heavy metal ions, even without the presence of chromium ions in them, provide almost 99.9-100% degree of water purification, while the specific consumption of anodically dissolved iron is reduced by 2-3 times, and, accordingly, this triggers the reduced power consumption. The resulting sludge has a crystalline structure and ferromagnetic properties, so the treated water can be easily clarified.

In addition to colloidal structures, Fe^{3+} ions form a number of water-soluble hydrated complexes with water molecules, which determine their positive charge, including the $[Fe(H_2O)_6]^{3+}$ monomer, which includes 6 water molecules, as well as other complexes determined (Sommer & Margerum, 1970) by spectrophotometric studies:

- Monomers: $[Fe(H_2O)_5(OH)]^{2+}$, or $Fe(OH)^{2+}$; $[Fe(H_2O)_4(OH)_2]^{2+}$, or $Fe(OH)_2^{+}$;
- Dimers: $[(Fe)_2\,(H_2O)_8(OH)_2]^{4+}$, or $Fe(OH)_2^{+}$;
- Trimers $[(Fe)_3(OH)_3(H_2O)_{12}]^{6+}$, or $[(Fe)_3(OH)_3]^{6+}$; $[(Fe)_3(OH)_3(H_2O)_{14}]^{5+}$. $[(Fe)_3(OH)_3(H_2O)_{12}]^{6}$;
- Other polymer hydroxo-complexes with general formula $[Fe_n(H_2O)_x(OH)_m]^{(3n-m)+}$.

Thus, all this variety of hydro-complexes of iron (III) bears a positive charge. A similar positive charge is carried by iron (II) hydroxide molecules and its hydro-complexes. For several reasons, in particular, in the presence of oxidizing agents (O_2, H_2O_2) or UV light exposure, iron compounds in water can form the appropriate radical ions with the same positive charge.

According to the Derjaguin-Landau theory of interaction in colloidal systems (Derjaguin & Landau, 1941), two simultaneously charged colloidal particles, approaching to each other, are subject to the simultaneous action of two opposite forces: molecular forces of attraction and electrostatic forces of repulsion. The energy of intermolecular attraction appears at a distance approximately equal to the radius of the particles themselves.

At large distances, determined by the size of colloidal structures (micelles), the energy of intermolecular attraction sharply decreases, and the forces of electrostatic repulsion of particles from each other prevail. In this case, the so-called "disjoining pressure", and a thin layer of liquid appear between the particles, trying to push them away from each other. Therefore, the approaching particles of iron hydroxides (II) and (III), as well as their positively charged radicals, have repulsive forces that prevent their phase-dispersed interaction and the formation of ferritized particles of the spinel structure.

Under these conditions, at least two options for reducing the electrostatic repulsive forces are possible: 1) increasing the temperature, loosening the hydration shell of micelles of iron hydroxides, or 2) creating micro-heterogeneous systems by introducing microparticles, in particular, negatively charged particles of magnetite, which have the opposite surface ξ-potential, contributing to the recharging of sols and leading to a change in the charge of colloidal structures.

Low-temperature ferritization proceeds according to the second of these options, according to which the introduced magnetite particles (Fe_3O_4), due to the negative value of the electrokinetic potential in the low-alkaline pH range, exhibit a catalytic effect in the ferritization process (Covaliov & Covaliova, 2000). This process has an induction period typical for autocatalytic reactions, within 15-20 seconds, and further develops quite quickly, according to the chain mechanism, under the normal temperature conditions, within the entire bulk of solution:

$$\cdot FeOH^{+} + 2\cdot FeOH^{2+} + O^{2-} \rightarrow Fe_3O_4 + 3H^{+} \quad (8)$$

In this way, the conditions are provided for the purification of multicomponent wastewater according to the scheme of low-temperature technological ferritization.

Figure 3 shows a general scheme for the complex treatment of multicomponent wastewater based on an electrocoagulator with plate soluble iron electrodes (Covaliov & Covaliova, 2002). In the wastewater flow and under the application of direct current, anodic dissolution of iron occurs and redox processes proceed according to reactions of the general form:

$$Fe^0 + 2e \rightarrow Fe^{2+} \text{ (on the anode)}, \tag{9}$$

and then

$$Cr_2O_7^{-2} + 6Fe^{+2} + 14H^+ \rightarrow 6Fe^{+3} + 2Cr^{+3} + 7H_2O \tag{10}$$

During the water electrolysis on the cathode following the reaction:

$$H_2O \rightarrow ½\, H_2\uparrow + OH^- \tag{11}$$

alkalization of the treated water occurs and the formation of a suspension of heavy metal hydroxides proceeds.

Due to the dosing of a suspension of magnetite (Fe_3O_4) into the electrocoagulation process, an autocatalytic ferritization reaction begins to develop, according to the previously described general reaction:

$$2Fe(OH)_3 + Fe(OH)_2 = Fe_3O_4 + 4H_2O \tag{12}$$

In the presence of a number of non-ferrous metal ions, the formation of more complex crystalline structures such as $Me_nFe_{3-n}O_4$ spinels occurs, where Me is the ions of iron (Fe^{2+}), copper (Cu^{2+}), zinc (Zn^{2+}), nickel (Ni^{2+}), chromium (Cr^{3+}) and others. At the moment of their formation, such oxides have elevated sorption properties, due to which water can be simultaneously purified from organic compounds. Meanwhile, water mineralization and hardness index are decreased.

The process of wastewater treatment according to the presented scheme (Fig.3) takes place in an electrocoagulator 1 of a flow type, where multicomponent wastewater is supplied by a pump 4. As a result of electrode, redox processes and phase-dispersed transformations, the resulting ferritized suspension is discharged into a thin-layer settling tank 5, where it is clarified in a gravitational field. Clarified water is discharged to a magnetic filter for post-treatment and reuse.

The condensed concentrate of the ferritized sludge accumulating in a thin-layer sump is divided into two parts using a divider, the smaller part of which is fed to the dosing pump for its secondary introduction into the above-electrode space, and the larger part is diverted for dehydration and recycling is periodically diverted to the magnetic separator 7 for dehydration and recycling.

In the upper zone of this separator, a hollow rotating drum made of a diamagnetic material is installed, enveloping a block of magnetic circuits with magnetization coils placed in it at 1/2-2/3 of the drum circumference. Due to the emerging electromagnetic field, ferromagnetic particles are attracted to its outer surface and are transferred to the upper zone of this drum. Then, having passed the upper zone, where the magnetic field is removed due to the absence of a magnetic circuit in it, the dehydrated sludge is removed from the drum using a scraper device. The pre-ferritized magnetite-containing sludge is dosed into the mixer 9 with the help of a dosing pump 8 for mixing with the initial wastewater, ensuring the initiation of the autocatalytic ferritization process due to the interaction of freshly formed hydroxyl compounds of metals in the entire volume of the electrocoagulator.

Figure 3. Schematic diagram of electrocoagulation-ferritization treatment of multicomponent wastewater containing chromium ions: 1 – electrocoagulator; 2 – block of soluble iron electrodes; 3 - rectifier; 4 - pump; 5 – thin-layer concentrator; 6 – magnetic filter; 7 – magnetic sediment separator; 8 - dosing pump; 9 – mixer
(Covaliov & Covaliova, 2002)

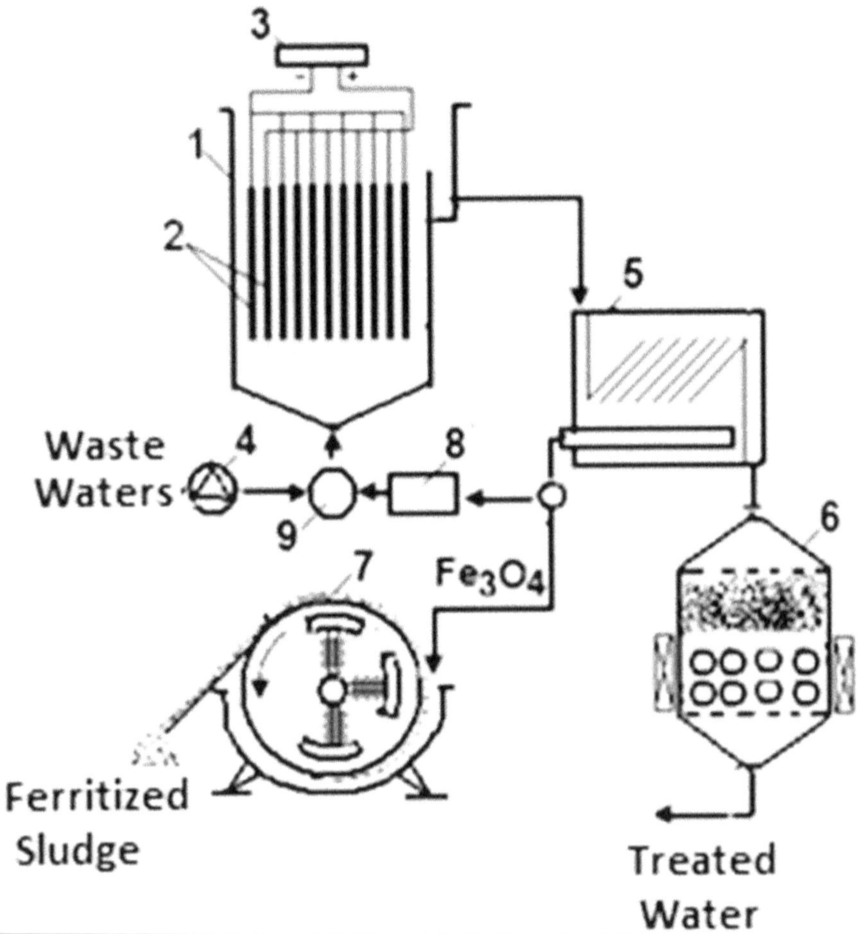

The high chemical resistance of the oxides formed in the composition of ferritized sediments compared to heavy metal hydroxides contributes to an improvement in the degree of wastewater treatment from them by more than an order of magnitude. After post-treatment on a magnetic filter, purified water can be reused for technological needs.

In the absence of Cr(VI) ions in wastewater as an oxidizing agent of electrochemically dissolved ferrous ions, there is a problem associated with the need for a sufficiently rapid oxidation of them to the trivalent state and the implementation of appropriate ferritization processes for water purification.

To ensure the efficiency of treatment of such wastewater containing heavy metals and organic pollutants, we have proposed a process (Covaliova, 2008), which provides for the possibility of oxidizing Fe^{2+} ions to Fe^{3+} by introducing into electrochemically treated water, along with dispersed magnetite, also hydrogen peroxide in the following amounts:

Hydrogen peroxide (H_2O_2, 33%), ml/L 0.5-1.0
Dispersed magnetite (Fe_3O_4), g/L 0.5-2.0.

In the presence of a number of other heavy metal ions, more complex spinel-type crystal structures are formed according to the general reactions:

$$xMe^{2+} + (3\text{-}x)Fe^{2+} + 6OH^- \rightarrow Me_xFe_{3-x}(OH)_6 + ½\,O_2 \rightarrow Me_xFe_{3-x}O_4 + H_2O \quad (13)$$

The introduction of hydrogen peroxide into the treated medium, which provides the required degree of oxidation of Fe^{2+} ions to Fe^{3+}, under the UV irradiation, promotes the running of photocatalytic redox processes according to the Fenton mechanism with the formation of a number of active radicals (Tsuneda, 2020).

Possessing high values of free negative energy, active radicals thermodynamically exhibit high reactive oxidizing properties with respect to organic substances, oxidizing complex organic molecules to simpler compounds by the mechanism of abstraction of the hydrogen atom with the formation of water molecule according to the general reaction:

$$RH + \bullet OH \rightarrow {}'R + H_2O \quad (14)$$

or by another type reactions. Out of these, the radical •OH has the highest activity, contributing to the photocatalytic destruction of organic compounds according to the scheme:

$$RH + \bullet OH \rightarrow \bullet R + H_2O \quad (15)$$

up to their complete mineralization.

However, the hydroxy-free radical •OH- is short-lived, while the superoxide $\bullet O_2^-$, which has both oxidizing and reducing properties, is longer-living compared to other active radicals, therefore it has an increased reactivity and the ability to decompose the organic macromolecules of polyphenol type, leading to the breaking of their bonds and destruction.

The formation of active radicals also proceeds due to the radiolysis of water molecules under the action of powerful ultraviolet radiation, carried out at wavelengths in the range of 180-253 nm, according to the reaction:

$$H_2O + \gamma\text{-rays} \rightarrow {}^{\bullet}OH + e_{aq}. \quad (16)$$

In this case, the hydrated electron (e_{aq}) interacts with hydrogen peroxide, also forming active radicals:

$$e_{aq} + H_2O_2 \rightarrow OH^- + {}^{\bullet}OH \quad (17)$$

This process associated with the formation of active radicals can also involve mobile electrons in the crystal lattice of magnetite (Fe_3O_4), in which the Fe^{2+} and Fe^{3+} ions located outside the [FeO_4] structural groups easily exchange with charges and have a mobile electron.

Thus, in this complex homogeneous-heterogeneous process, active radicals can be formed:

$$H_2O_2 + Fe(OH)_2 \rightarrow {}^{\bullet}FeOH^+ + OH^- + H_2O \quad (18)$$

$$2{}^{\bullet}FeOH^+ + \tfrac{1}{2}\,O_2 \rightarrow 2{}^{\bullet}Fe(OH)^{2+} + O^{2-} \quad (19)$$

which ultimately determine the development of ferrite formation by autocatalytic mechanism:

$$ {}^{\bullet}FeOH^+ + 2{}^{\bullet}FeOH^{2+} + O^{2-} \xrightarrow{cat.Fe_3O_4} Fe_3O_4 + 3H^+ \quad (20)$$

Tests of this technology were carried out on a model wastewater solution with a volume of 10 L, containing 50 mg/l of zinc, nickel and copper ions, respectively, which was treated under the flow-through conditions at a linear flow rate of 1 cm/s in an electrolyzer with soluble iron electrodes made of low-carbon steel. Before each test, the electrodes were anodically treated in a 3% NaCl solution to eliminate the passive film from their surface passivation. The electrolysis was carried out at a current density of 1 A/dm^2. UV lamps with a wavelength of 180–400 nm were used as a source of UV radiation.

The test results have shown that, the conditions proposed for the purification of wastewater from heavy metal ions provide almost 99.8-100% degree of water purification, while the specific consumption of anodically dissolved iron is reduced by 2-3 times as compared to the standard indicators for this process, and, consequently, reduced energy consumption. The resulting sludge had a crystalline structure and ferromagnetic properties, so the treated water is quickly clarified. Under these conditions, the magnetization of the resulting sludge had an optimum in the low alkaline pH range of 6.5-8.5, and made 1800-2500 Oersted.

CONCLUSION

The analysis of the state of developments in the field of ferritization methods for the purification of multicomponent wastewater from heavy metal ions made it possible to assess this direction as the most promising due to undeniable technical, economic, and environmental advantages compared to other known technologies.

The results of experimental studies have made it possible to propose a new interpretation of the mechanism of ferritization processes based on heterogeneous catalysis, the participation of active radicals, and the role of photocatalysis in increasing the efficiency of these technologies. From these positions, the participation of additives of magnetite and other spinel structures in wastewater treatment processes provides the development of low-temperature ferritization processes and the production of sludge with magnetic properties, which simplify and reduce the cost of subsequent technological operations of water clarification, thickening and dehydration of sludge. On this basis, new improvements in ferrite technologies for water purification are proposed.

ACKNOWLEDGMENT

This research has been performed under the Moldova National Research Project Nr. 20.80009.5007.27 "Physical-chemical redox processes with electron transfer in the vital, technological and environmental processes".

REFERENCES

Bybik, E. E., Buzunov, O. V., Gribanov, N. M., & Lavrov, I. S. (1979). Study of the kinetics of colliodal magnetite particles (Rus.). *Journal of Applied Chemistry*, *7*, 1631–1632.

Bybik, E. E., Buzunov, O. V., & Skobochkin, V. E. (1980). Magnetometric method of fine ferromagnetics study. *Industrial Laboratory*, *46*(7), 618–619.

Casbeer, E., Sharma, V. K., & Li, X. (2012). Synthesis and photocatalytic activity of ferrites under visible light: A review. *Separation and Purification Technology*, *87*, 1–14. doi:10.1016/j.seppur.2011.11.034

Covaliov, V., & Covaliova, O. (2000). Autocatalysis in the processes of wastewaters treatment. *Intellectus*, *4*, 55–60.

Covaliov, V., & Covaliova, O. (2002). *MD Patent Nr. 2427*. Reactor for magnetic treatment of liquids.

Covaliov, V., & Covaliova, O. (2004). *Theoretical and practical aspects of electrochemical water treatment*. CEP USM.

Covaliov, V.V., Caraghiaur, V.G. (1985). Intensification of electrocoagulation wastewater treatment/ Electromagnetite treatment. *Chisinau, VINITI. 85*(18).

Covaliova, O. (2008). *MD Patent nr. 3912*. Method of electrochemical purification of waste waters from heavy metal ions.

Covaliova, O. (2010). Role of iron compounds in photocatalytic destruction of 2-aminobenzothiazole in water environment. *Studia Universitatis Moldaviae. Nature and Science*, *1*(31), 199–206.

Covaliova, O., & Covaliov, V. (2004). *Fundamentals of intensification of electrochemical processes of wastewaters treatment and sludge ferritization in plating industry. Ecologie-Acoperiri Metalice*. ICTCM.

De Satour, C. (1973). Magnetic separation in water pollution control. *IEEE Transactions on Magnetics*, *3*(3), 314–316. doi:10.1109/TMAG.1973.1067685

Demirel, D., Yenigün, O., & Bekbölet, M. (1999). Removal of Cu, Ni and Zn from wastewater by the ferrite process. *Environmental Technology*, *20*(9), 963–970. doi:10.1080/09593332008616891

Derjaguin, B., & Landau, L. D. (1941). Theory of the stability of strongly charged lyophobic sols and of the adhesion of strongly charged particles in solutions of electrolytes. *Acta Phys. Chim.*, *14*, 633–662.

Diehl, A., & Levin, Y. (2006). Smoluchowski equation and the colloidal charge reversal. *The Journal of Chemical Physics*, *125*(5), 054902. doi:10.1063/1.2222372 PMID:16942253

Duca, Gh. (2012). Heterogeneous catalysis with metal complexes: fundamentals and applications. Springer. doi:10.1007/978-3-642-24629-6

Duca, G., & Vaseashta, A. (Eds.). (2020). *Handbook of Research on Emerging Developments and Environmental Impacts of Ecological Chemistry*. IGI Global. doi:10.4018/978-1-7998-1241-8

Hua, M., Zhang, S. J., Pan, B. C., Zhang, W. M., Lv, L., & Zhang, Q. X. (2012). Heavy metal removal from water/wastewater by nanosized metal oxides: A review. *Journal of Hazardous Materials, 211–212*, 317–331. doi:10.1016/j.jhazmat.2011.10.016 PMID:22018872

Kefeni, R. R., Mamba, B. B., & Msagati, T. A. M. (2017). Application of spinel ferrite nanoparticles in water and wastewater treatment: A review. *Separation and Purification Technology, 188*, 399–422. doi:10.1016/j.seppur.2017.07.015

Lou, J.-C., Huang, Yu.-J., & Han, J.-Yu. (2009). Treatment of printed circuit board industrial wastewater by ferrite process combined with Fenton method. *Journal of Hazardous Materials, 170*(2-3), 620–626. doi:10.1016/j.jhazmat.2009.05.020 PMID:19493613

Makarov, V.M., Kalaeva, S.Z., Zakharova, I.N., Nevzorov, I.A., Maltseva, M.S., Shipilin, A.M., Krzhizhanovskaya, M.G. (2015). Magnetic and X-ray studies of nanodispersed magnetite synthesized from chrome containing galvanic sludge. *Journal of Nano- and Electronic Physics, 7*(4), 04010.

Masunga, N., Kelebogile, O., Kefeni, K. K., & Mamba, B. B. (2019). Recent advances in copper ferrite nanoparticles and nanocomposites synthesis, magnetic properties and application in water treatment [Review]. *Journal of Environmental Chemical Engineering, 7*(3), 103179. doi:10.1016/j.jece.2019.103179

Pynaev, A. V. (2006). Study of heavy metal ions leaching from the ferritized plating sludge under the field conditions (Rus.). *Proceedings of Universities from North-Caucasus Region. Technical Sciences., 4*, 46–48.

Pysarenko, A. E., Pospelova, K. A., & Yakovlev, A. G. (1964). The course of colloidal chemistry. M: Vysshaia Shkola (High School).

Qiao, R., Tian, W., Bai, J., Wang, L., Zhao, J., Du, Zh., & Gong, X. (2019). Application of magnetic adsorbents based on iron oxide nanoparticles for oil spill remediation: A review. *Journal of the Taiwan Institute of Chemical Engineers, 97*, 227–236. doi:10.1016/j.jtice.2019.01.029

Ryzo, E. G., Liubman, I. A., & Dmitriev, S. N. (2006). Universal technology of heavy metal ions sedimentation from wastewaters – ferritization technology. *Water and Ecology: Problems and Solutions, 3*, 32–37.

Semenov, V. V., Varlamova, S. I., & Klimov, E. S. (2005). Utilization of plating sludge by ferritization method. *University News, 48*(2), 111–112.

Sharma, V., Singh, H., Guleria, S., Bhardwaj, N., Puri, S., Arya, S. K., & Khatri, M. (2022). Application of superparamagnetic iron oxide nanoparticles (SPIONs) for heavy metal adsorption: A 10-year meta-analysis. *Environmental Nanotechnology, Monitoring & Management, 18*, 100716. doi:10.1016/j.enmm.2022.100716

Singh, H., Bhardwaj, N., Arya, S. K., & Khatri, M. (2020). Environmental impacts of oil spills and their remediation by magnetic nanomaterials. *Environmental Nanotechnology, Monitoring & Management, 14*, 100305. doi:10.1016/j.enmm.2020.100305

Sommer, B. A., & Margerum, D. W. (1970). Kinetic study of the hidroxoiron(III) dimmer. *Inorganic Chemistry, 9*(11), 2517–2521. doi:10.1021/ic50093a028

Suresh, R., Saravanan, R., Ponnusamy, S. K., Vo, D.-V., & Cornejo-Ponce, L. (2021). Recent advancements of spinel ferrite based binary nanocomposite photocatalysts in wastewater treatment. *Chemosphere, 274*, 129734. doi:10.1016/j.chemosphere.2021.129734 PMID:33548641

Tsuneda, T. (2020). Fenton reaction mechanism generating no OH radicals in Nafion membrane decomposition. *Scientific Reports, 10*(1), 18144. doi:10.103841598-020-74646-0 PMID:33097759

Vaseashta, A., Duca, G., Covaliova, O., & Romanciuc, L. (2022). Water Safety, Security, and Sustainability: Emerging Trends and Future Pathways. In A. Vaseashta, G. Duca, & S. Travin (Eds.), *Handbook of Research on Water Sciences and Society* (pp. 1–36). IGI Global. doi:10.4018/978-1-7998-7356-3.ch001

Wang, L., Luo, Z., Hong, Y., Chelme-Ayala, P., Meng, L., Wu, Z., & El-Din, M. G. (2022). The treatment of electroplating wastewater using an integrated approach of interior micro electrolysis and Fenton combined with recycle ferrite. *Chemosphere, 286*(1), 131543. doi:10.1016/j.chemosphere.2021.131543 PMID:34303045

Yakovlev, S.V., Volkov, L.S., Voronov, Yu.V. (1999). *Treatment and utilization of sludge from industrial waste water.* (Rus.). M.: Chemistry.

Chapter 14
Redox Technologies in Wastewater Treatment

Viorica Parvulescu
Institute of Physical Chemistry of Romanian Academy, Romania

ABSTRACT

Various physico-chemical and biological methods have been developed for the removal of organic and inorganic pollutants from wastewater. The application of these techniques, their combination, and the development of hybrid technologies to remove the persistent pollutants from water are presented in this chapter. The advanced oxidation processes (AOPs) such as photocatalysis, modified Fenton reactions, electro- and photoelectrochemical oxidation are evidenced as the main applied technologies. The combination and effects of direct and indirect oxidation methods and of the alternative greener technologies with traditional methods are presented as a function of pollutants' nature, volume, and concentration. For each method, results were obtained for various pollutants such as pesticides, hydrocarbons, dyes, surfactants, aromatic compounds, and pharmaceutical residues. The main objective of this chapter is the application of redox technologies in wastewater treatment by correlating redox processes with working conditions, pollutants, and oxygen reactive species.

INTRODUCTION

The acceleration of industrialization and urbanization generated a large amount of wastewater. Wastewater from various human activities such as industrial, agricultural, and domestic contains various organic compounds, heavy metals, and other chemicals that degrade the environment and represent human and animal health hazards. Therefore, water pollution was defined as the decreasing of water quality due to the release without treatment of physical, chemical, or biological contaminants directly from industries or from other sources. The pollutants were classified as biodegradable, with short-term effects, and non-biodegradable such as heavy metals, plastics, and chemicals that cause long time toxic effects. The combination of pollutants occurs in many cases, producing a synergetic effect that makes the water treatment process more difficult. Over the years, various physicochemical including adsorption, membrane-based processes, electrocoagulation, electrodialysis, reverse osmosis ion exchange, and chemi-

DOI: 10.4018/979-8-3693-0512-6.ch014

cal oxidation have been developed in the removal of organic and inorganic pollutants from wastewater. The combination of these techniques with the biological methods resulted in hybrid technologies, which allowed the development of advanced methods for wastewater treatment. The conventional techniques of wastewater remediation like chemical precipitation, coagulation and flocculation, flotation, filtration, adsorption, ion exchange, and chemical and biological treatment are very predominant. Among these, the long list of chemical technologies based solely on redox chemistry stands out. This chapter presents the applications of redox technologies in wastewater treatment by correlating the redox processes with the working conditions and the nature of the pollutants. Redox processes and generated technologies as well as the synergistic aspect of their action in hybrid processes are discussed in accordance with the most advanced technologies. Since the progress in wastewater decontamination is also the result of development of the new materials, the role of nanomaterials on the future performances of advanced oxidation processes on synergistic effects of combined advanced technologies will also be evidenced.

REDOX PROCESSES FOR WASTEWATER TREATMENT

Chemical methods for wastewater treatment are redox processes in many cases. There are applications where both oxidation and reduction occur due to the involvement of multiple species. The oxidizer tends to accept electrons and has a high electric potential. The reducer donates electrons and has a low potential. The oxidation processes can be direct or indirect. Each of these can be simple or combined. Direct oxidation takes place with oxidizing agents and indirect oxidation by generating active radicals that react with organic compounds. The resulting organic radicals propagate reactions in which water and oxygen play an important role. Most organic pollutants decompose into carbon dioxide and water, resulting in water with a high content of nitrites, nitrates, and phosphates that can be removed in subsequent stages (Shabir et al., 2022). A large number of recent publications has generated multiple review articles that highlight the use of combined or hybrid technologies in wastewater treatment (Qiao & Xiong, 2021; Lin et al., 2022; Ewuzie et al., 2022; Aniipsitakis & Dionysiou, 2006; Duca et al, 2021). They present the advantages and disadvantages of these techniques (Saravanan et al., 2022) but are focused either on the nature of the pollutants (Shabir et al., 2022; Weng et al., 2023) or on the type of technique used. In most cases, conventional techniques are only mentioned. Thus, most direct oxidation processes are now advanced oxidation processes (AOPs) because they use combinations of oxidants or are associated with oxidic materials as a catalyst.

Oxidative processes used in the treatment of wastewater are either biological aerobic or chemical, in the presence of oxidizing agents or combined. The most used oxidizing agents in conventional redox technologies are oxygen gas, ozone (Martínez et al., 2011), hypochlorite and chlorite (Cheng et al., 2022), potassium permanganate (Wang et al., 2023), hydrogen peroxide (Ksibi, 2006), peroxy sulphates (Nidheesh et al., 2022; Maifadi et al., 2022). In the case of these reactions, the thermodynamic suggests that the efficiency is determined by the redox potential of the oxidant. The traditional and conventional methods used to treat wastewater have proven, to some extent, effective and are still used to remove most of the pollutants from wastewater. Different contaminants, resistant to degradation, such as pesticides, aromatic and aliphatic hydrocarbons, phthalates, phenols, cyanides, textile dyes, surfactants, antibiotics, inorganic chemicals, heavy metals, aromatic compounds present in the wastewater are not completely removed using conventional treatment methods. In recent years, various techniques have been developed for the removal from wastewater of these pollutants and advanced oxidation processes (AOPs) have received increased interest as promising, powerful, and ecological methods (Saravanan et al., 2022; Li et al., 2022; Shabir et al., 2022). The high performances of AOPs were attributed to the generation of various reactive oxygen species (ROS) such as •OH, HO_2•, •O_2^-, SO_4•$^-$ and $1O_2$. ROS are highly active for the removal of a wide range of organic pollutants (Weng et al., 2023) while direct or conventional redox reactions are effective in transforming inorganic contaminants. Different types of advanced oxidation methods were applied. Each method requires certain parameters such as pH, temperature, time, and type of reactor for better performance. As well the combination of UV or visible light treatments with these methods has a high efficiency (Dhaka et al., 2018). Therefore, there are two main categories of AOPs: photochemical and non-photochemical. AOPs have, compared with conventional technologies, a stronger oxidation capacity, faster reaction speed, less secondary pollution, milder reaction conditions, and wider applications in the field of water treatment (Saravanan et al., 2022; Pavanello et al., 2022; Kokkinos et al., 2021; Turchi & Ollis, 1990).

In recent years there have developed new and alternative greener technologies, based on redox processes along with the synthesis of active multifunctional materials (semiconductor, composite and hybrid materials) (Qiao & Xiong, 2021; Hong et al., 2022; Das et al., 2021; Liu et al., 2021; Vaseashta et al, 2022; Sibhatu et al., 2022; Hoang et al., 2022). Various composite or hybrid materials have also been obtained, with active metallic or bio compounds immobilized on inorganic or polymeric supports, to develop heterogeneous methods for the degradation of pollutants from water. Depending on the type of pollutants, the extent of pollution and the volume to be treated advanced oxidation methods are used separately or combined with others (hybrid technologies) leading to a maximization of contaminants removal (Liu et al., 2022). Photocatalysis, photo- and electro-Fenton, bio-electrochemical, electro-, and photoelectrochemical oxidation are AOPs much studied. The latter is of interest for wastewater treatment and electricity generation, methods (Mohyudin et al., 2022; Fernandez-Ibanez et al., 2021; Gul et al., 2021). Therefore, new hybrid technologies were developed in which photocatalytic, photo- and electro-Fenton, sonophotocatalytic, electrophotocatalytic, photoelectrocatalytic, biophotoelectrocatalytic, photocatalytic hybrid Z processes have allowed advanced water purification and electricity generation by non-polluting methods (Goncharuk et al., 2010).

DIRECT OXIDATION TECHNIQUES

Oxidation reactions are better known compared to reduction (Liu et al., 2022) because it represents a favorable direction in water treatment known as mineralization of total or dissolved organic carbon by transformation into CO_2. Many of the chemical wastewater treatment technologies are based on the activity of oxidizing agents or reactive oxidation species generated in situ. Conventional redox technologies include ozonation, chlorination, and permanganate addition. Besides these technologies UV radiation plus ozone (UV/O_3) or hydrogen peroxide (UV/H_2O_2), ozone in combination with hydrogen peroxide (O_3/H_2O_2), Fenton's reagent (H_2O_2/Fe^{2+}) and photo-Fenton (UV/H_2O_2/Fe^{2+}) are considered to be effective for the oxidation of effluents from various industries.

Ozonation

Ozonation is a chemical procedure used to treat wastewater. This consists of the infusion of ozone into the aqueous medium. Ozone (O_3) possesses a high oxidation-reduction potential (E_0=2.07 V) (Liu et al.,

2022; Covinich et al., 2014) and is typically used as a disinfecting and oxidizing agent. It has been used worldwide for over a century in drinking water treatment plants (Pariente et al., 2022). Ozone oxidizes preferentially electron-rich molecules containing carbon-carbon double bonds and aromatic alcohols, but not those that have single bonds (Ghazal et al., 2022). Ozone is used as an oxidant to eliminate many toxic elements from wastewater as pesticides, dyes, detergents, phenols, aromatic and chlorinated hydrocarbons. The major disadvantage of using O_3 is the formation of toxic byproducts. Its activity is also affected by the presence of salts, the pH value, and the increase in temperature (Martínez et al., 2011; Tufail et al., 2020). O_3 is poorly soluble in water and in contact with water, it becomes highly unstable and generates active radicals through a series of reactions determined by pH. Therefore, depending on the behavior of the ozone in the aqueous environment, the selectivity of ozonation may be high (direct reaction) or low (indirect reaction). The direct pathway normally occurs under acidic conditions ($pH < 4$) and the indirect pathway above pH 10. In acidic conditions, the selective reaction of ozonation involves only the ozone with unsaturated organic compounds. Ozonation's best performance occurs at alkaline pH. The higher pH (> 9) favors the oxidation of phenolic compounds. Phenols promote O_3 decomposition and increase the effect of •OH radicals.

Ozone can oxidize the compounds present in water (organic and inorganic compounds) both directly and indirectly. The efficiency of direct ozone treatment was investigated for the recovery and reuse of water in plant irrigation. The obtained results showed the removal of heavy metals (Cd, Cu, Fe, Zn) and a significant decrease in bromide, nitrite, and nitrate concentration from water. Thus, the nitrite–nitrate contents of purified water treated with ozone complied with the standards for irrigation water (0–10 mg/L). Ozonation is more advantageous in terms of results compared to halogenated compounds or UV treatment. This method is applied with moderation due to the high costs. That is why the studies were oriented toward the use of combined processes to improve the efficiency of wastewater treatment. Ozone alone does not cause complete oxidation of some refractory organic compounds and has a low reaction rate. This is why ozone is combined with H_2O_2, UV light, catalysts, and ultrasound to enhance the generation of hydroxyl radicals and increase the efficiency of the treatment process. Ozone absorbs strongly in the UV range of 200-300 nm and much more than hydrogen peroxide at 254 nm. Ozone can be coupled with H_2O_2 and the process is named peroxonation or with catalytic materials and the process is named catalytic ozonation (Liu et al., 2022). When O_3 is coupled with transition metal ions, it is homogeneous catalytic ozonation (Malik et al., 2018) and when is coupled with metal oxides, it is heterogeneous catalytic ozonation (Quan et al., 2017). Some of the photochemical treatments with O_3 reported in the literature are those implemented with ultraviolet light such as UV/ozone (UV/O_3) and UV/hydrogen peroxide/ozone ($UV/H_2O_2/O_3$) which are efficient technologies in the complete mineralization of some pollutants and reducing of toxicity (Amaro-Soriano et al., 2021). This can be accomplished in a variety of ways (Pandey et al., 2021). As well, the reaction of active hydroxyl radical, generated in UV/O_3 and $UV/H_2O_2/O_3$ technologies, is shown below (Liu et al., 2022). UV activation of ozone leads to the formation of highly reactive species, such as $O(^1D)$, which form hydrogen peroxide and hydroxyl radicals. As well, the initial and formed H_2O_2 form hydroxyl radicals under UV activation. Ozone together with hydrogen peroxide generates •OH, as a result of the successive reactions with the formation of ions and active radicals:

UV/O_3 UV/H_2O_2 O_3/H_2O_2

$O_3 + h\nu \rightarrow O_2 + O(^1D)$ $H_2O_2 + h\nu \rightarrow 2{\bullet}OH$ $H_2O_2 + 2O_3 \rightarrow 2{\bullet}OH + 3O_2$

$O(^1D) + H_2O \rightarrow H_2O_2$

$H_2O_2 + hv \rightarrow 2{\bullet}OH$

Oxidation occurs as a result of multiple reactions with hydroxyl radicals, an oxidizing agent with a higher oxidation potential (E_0=2.72 V) (Liu et al., 2022), forming the superoxide anion radical and the hydroproxyl radical that produces the oxidation of organic pollutants. The efficiency of this step depends on the amount of ozone, the time present in the reaction, and the basic nature of the effluent. Ozone activation-based AOPs are systems such as O_3/UV, O_3/H_2O_2, O_3/Fe(II), O_3/metal oxide catalyst, O_3/activated carbon, O_3/ultrasound, O_3/Fenton (Jiang et al., 2021; Rekhate & Srivastava, 2020; Rossi et al., 2021). They usually generate highly reactive oxygen species (ROS), such as •OH, $O_2{\bullet}^-$, and $1O_2$ which have been proven as an efficient and environmentally friendly method for the removal of refractory pollutants. A wide range of industrial effluents contains recalcitrant organic pollutants such as pharmaceuticals, pesticides, phenols, dyes, etc. Another type of reaction is catalytic ozonation (Covinich et al., 2014; Chokshi & Ruparelia, 2021). This method is more efficient for removing total organic carbon. Several metal ions such as manganese, ferrous, ferric, and metal oxides such as aluminum and titanium oxides are the preferences for the catalyst. Redox properties of the catalysts can provide excellent active sites for O_3 decomposition and ROS generation. Independently, ozone has shown selectivity for the nucleophilic sites of the aromatic rings. By contrast with catalytic ozonation, especially for supported bimetallic oxides, it was highlighted accelerating the decomposition of O_3 into •OH radicals which improves the elimination of pollutants. Compared with independent ozonation the ozonation in the presence of catalysts significantly increased. In an O_3/H_2O_2 system, drug mineralization could reach nearly 100%. This phenomenon was attributed to the combined use of O_3 and H_2O_2, which leads to the generation of •OH radicals with higher activity, enhancing the mineralization effect of the reaction. The synthesis and use of different catalysts have also proven to be a way to successfully use ozone in wastewater decontamination. It inspected the catalytic mechanism and confirmed that the number of free radicals •OH and superoxide increase with the surface of the used catalyst which improves the degradation effect (Liu et al., 2021).

Hydrogen Peroxide Treatment

The simplest and best-known oxidant is oxygen, which is very important in nature. However, its molecular form is inert under normal conditions. The formation processes of the active species consist of several stages involving the formation and transformation of hydrogen peroxide (Duca, 2021). The redox potential of H_2O_2 is comparable with most of the oxidizing agents (ozone, persulfate, and permanganate anions). This explains their ability to oxidize different organic compounds as carbonates, bicarbonates, and chlorides in addition to dyes from textile effluents (Asghar et al., 2015). These inorganic anions reduce the degradation efficiency of the dyes by scavenging the oxidizing agents. One of the most scavenging phenomena is the formation of H_2O_2 in the presence of excessive •OH. Two phenomena occur in the presence of $SO_4{\bullet}^-$ radicals. Sulfate ions are scavenging holes and •OH with a productive role in the formation of active sulfate radicals. In addition, it is well known that the use of ozone or molecular oxygen as an oxidant can generate solubility problems while treatment with H_2O_2 is greatly attractive and is more feasible among AOPs. Hydrogen peroxide is known to be a strong oxidizing agent. It is used for the oxidation of many types of substances in both organic and inorganic waters. Also, in wastewater, hydrogen peroxide can decompose into water and oxygen. Although the cost of removing organic pollutants by chemical oxidation with hydrogen peroxide is usually higher than by physical or biological means, there are still specific situations that justify its use such as the presence in moderate or high amounts of compounds that inhibit biological treatment such as pesticides, plasticizers, resins, coolants, and dyestuffs. For the oxidation of some inorganic compounds such as sulfur (sulphides or thiosulphate), only hydrogen peroxide can be used effectively. Another direct oxidation reaction with H_2O_2 is hydrogen sulfide removal (El Brahmi & Abderafi, 2021). Sulfides are oxidized to elemental sulfur due to the dissolved oxygen from peroxide decomposition.

The moderate activation of hydrogen peroxide can be achieved with alkalis (generation of the per hydroxyl ion, OOH-), in the presence of transition metals (W, V, Mo), with which they form reactive peroxometals, or with mineral acids (e.g., sulfuric) with which it forms peroxy acidic derivatives (*peroxydisulfate,* persulphate) (Martínez et al., 2011; Chen et al., 2020). For more recalcitrant organics, such as chlorinated solvents, highly reactive free radicals are required and for this there were developed advanced oxidation processes (AOPs). In AOPs hydrogen peroxide is activated by additional components, such as iron salts (Fenton's reagent), ozone, or ultraviolet light. The AOPs technologies are the most commonly applied approaches for wastewater treatment. It is successfully used for the degradation of recalcitrant pollutants in various refinery effluents and wastewater, especially from chemical or pharmaceutical facilities. The high efficiency of this process is based on the oxidation power of highly reactive hydroxyl radicals, which are formed by the activation of hydrogen peroxide. Such radicals can be obtained using Fenton's reagent. Therefore, hydrogen peroxide has a direct participation in numerous redox processes due to its dual nature as an oxidant or reducer. It can be decomposed into oxygen and water. This reaction passes through a series of intermediate steps that may differ with conditions (temperature, pH, catalysts, UV irradiation, water chemical, and biological composition). In conditions of moderate temperature and without the presence of catalysts the reaction of H_2O_2 self-decomposition is very slow. Because almost all types of catalysts can accelerate the transformation of H_2O_2 into active species decomposition, hydrogen peroxide was involved in various technologies for wastewater treatment by direct or indirect oxidation (AOPs catalytic, photocatalytic or biochemical) processes (Li et al., 2021; Jalali et al., 2021). AOPs technologies were developed in the last forty years. These are oxidation processes carried out by reactive oxygen species (ROS) such as hydroxyl (•OH), superoxide radical anion ($O_2•^-$), singlet oxygen ($1O_2$) generated through a series of complex, physical, chemical, and biological processes (Aniipsitakis & Dionysiou, 2006). Degradation of organic compounds with •OH takes place by two key reactions such as addition and hydrogen abstraction (reactions 1).

$$R-H+OH \rightarrow R+H2O; ROH+OH \rightarrow R1COOH; R+OH \rightarrow ROH \quad (1)$$

The resulting organic radicals are formed into organic molecules such as ketones, aldehydes, and alcohols. These compounds are then oxidized to small molecule organic acids, carbon dioxide, and water (reactions 2).

$$H2O \rightarrow ROH + H+; R + O2 \rightarrow ROO; 2ROO \rightarrow 2RO + O2; RO + R \rightarrow R2COR \rightarrow R2COOH \rightarrow CO2 + H2O$$

R•+ (2)

Compared to traditional oxidation techniques, AOPs show some advantages such as higher reaction speed, stronger oxidation capacity, and lower secondary pollution, they occur in milder reaction conditions and can be applied in wider areas of water treatment.

Generation of •OH is commonly accelerated by combining oxidizing agents. Among these treatments, UV radiation plus hydrogen peroxide (UV/H_2O_2), Fenton's reagent (H_2O_2/Fe^{2+}), photo-Fenton (UV/ H_2O_2/Fe^{2+}), and ozone in different combinations (O_3/UV; O_3/H_2O_2) are considered to be effective for the oxidation of effluents from various industries (Mierzwa et al., 2018). Generation of •OH is commonly accelerated by combining some oxidizing agents such as ozone (O_3), hydrogen peroxide (H_2O_2), UV radiation, and ferrous and ferric salts (Fe^{2+} and Fe^{3+}) (Jiang et al., 2021), and also by radiation sources such as ultrasound, visible, solar and thermal energy. In more advanced processes, the radicals can be generated by gamma radiation, microwave, or electron beams (Tufail et al., 2020). The high reactivity of these radicals and their low oxidation selectivity are useful attributes that made hydrogen peroxide based AOPs a promising technology for the treatment of effluents containing refractory organic compounds (Asghar et al., 2015; Samal et al., 2022).

AOPs exploit the high energy of hydroxyl radicals, which attack most organic molecules: aromatic rings, polyphenols, halogenated compounds, resin acids, unsaturated fatty acids, and resulting compounds from the decomposition process of organic compounds. Therefore, AOPs can produce total mineralization, transforming recalcitrant compounds into inorganic substances (CO_2 and H_2O), or partial mineralization, transforming them into more biodegradable substances. They can destroy organic contaminants, such as halogenated hydrocarbons, aromatic compounds, and pesticides, as well as inorganic contaminants such as cyanides, sulfides, and nitrites. Thus, various H_2O_2 or UV/H_2O_2 processes were applied for the degradation of pharmaceutical wastewater contaminants such as acetaminophen. Several studies have also been made on the use of Fenton and photo-Fenton reactions for the degradation of penicillin. AOPs with ex-situ or in-situ production of H_2O_2 were used for the removal of antibiotics from wastewater effluents. Hydrogen peroxide can be obtained *in situ* by electrochemical, photochemical, bio-electrochemical, and photo-electrochemical methods. Ex-situ H_2O_2-based AOPs are Fenton or Fenton combined with ultrasound, photocatalytic, and ozonization processes (Wang et al., 2022).

Hydrogen peroxide was also used in AOPs associated with persulfate and the activation of hydroxyl ($HO\text{–}OH\text{–} \rightarrow 2OH$) and sulfate ($\text{–}SO3\ \text{–}O\ \text{–}O\ \text{–}SO3\text{–} \rightarrow 2O\ \text{–}SO3\text{–}$) radicals was achieved by microwave, ultrasound or transition metals (Chen et al., 2020). •OH is non-selective in the oxidation of organic substances, unlike the sulfate radical with a lower redox potential, under certain conditions, is more selective for the degradation of unsaturated bonds and aromatic substances. •OH and SO_4•$^-$ were used together to degrade organics, but SO_4•$^-$ played a predominant role. The benzene structure and functional groups (diazo and nitro groups) were considerably degraded by the two active species. The microwave/persulfate process had a higher efficiency than the O_3/H_2O_2 process, but the O_3/H_2O_2 process was more efficient. Another activation method used was carried out in visible light and with photocatalytic nanomaterials ((ZnO@D-Y_2O_3 and Sm-ZnO@D-Y_2O_3) thus achieving the accelerated activation of H_2O_2 and persulfate to degrade ionic dyes from wastewater (Darwish et al., 2022). A synergistic contribution of three reactive oxidizing species (SO_4•$^-$, h^+, and •OH) was evidenced. Therefore, the Sm-ZnO@D-Y_2O_3/PS/Vis system proved to be an outstanding candidate for the complete destruction of organic pollutants in the presence of H_2O_2 and persulfate. In photocatalytic reactions, hydrogen peroxide can intercept the free electrons from the valence band and thereby prevent recombination effects. H_2O_2 reacts with the free electrons to form hydroxide ions and hydroxyl radicals ($H_2O_2 + e^- \rightarrow OH^- +$ •OH). The combination between the hydrogen peroxide effect and photocatalysis evidenced a synergetic degradation of organic pollutants.

Other Oxidizing Agent-Based Technologies

Quinones, an important group of humic acids, can be reversibly oxidized and reduced, so it has the potential to act as an external redox mediator. The addition of this redox mediator, which is even a pollutant in some wastewater, can catalyze the degradation of polluting organic materials and generate electricity (Li et al., 2022). Therefore, quinones, from medicinal wastewaters, were used as redox mediators for the synergistic removal of azo dye in microbial fuel cells. It was evidenced by both influences on electricity generation and pollutant removal. Potassium permanganate (KMnO4), is a green and non-toxic oxidant, largely employed in wastewater and sludge treatments (Wang et al., 2023) with strong oxidizability including in pollutants removal, decrement of hazardous gas footprint, improvement of sludge dewatering and resource recovery. In the last time were developed $KMnO_4$-based advanced oxidation processes with higher performances in wastewater treatment. It is worth noting that KMnO4 is suitable for a wide range pH condition. The oxidative reactivity of $KMnO_4$ in acidic-neutral pH (E_0=1.7 V) is higher than that in strong acid (E_0=1.51 V) or alkaline (E_0=0.59 V) conditions. Mn(VII) was transferred into the solid phase (Figure 1) after sludge oxidation with $KMnO_4$ relatively stable and nontoxic, posing negligible environmental risks to final sludge disposal. MnO_2 generated in situ could catalyze KMnO4 reaction, but it can also contribute to adsorption, coagulation, and oxidation (Sun et al., 2021).

In wastewater treatment, direct KMnO4 oxidation is primarily applied for industrial applications. During KMnO4-based AOPs, reactive manganese species such as soluble Mn(III), Mn(V), and Mn(VI), are essential intermediates extremely reactive toward a range of pollutants. Also, the activation of MnO_4 with bisulfite to oxidize large amounts of resistant pollutants (as nitrobenzene, 4-nitrophenol, benzene, and caffeine) attracted great attention. In the bisulfite/$KMnO_4$ system, bisulfite and Mn (VII) were considered as a mutual catalytic process with generation of various reactive Mn species (Mn (III), Mn(V), Mn(VI)) and free radicals (*SO3–*, *SO4–*, *SO5–*) (Guan et al.,2022).

Figure 1. Effect of pH conditions on manganate reactive species transformation

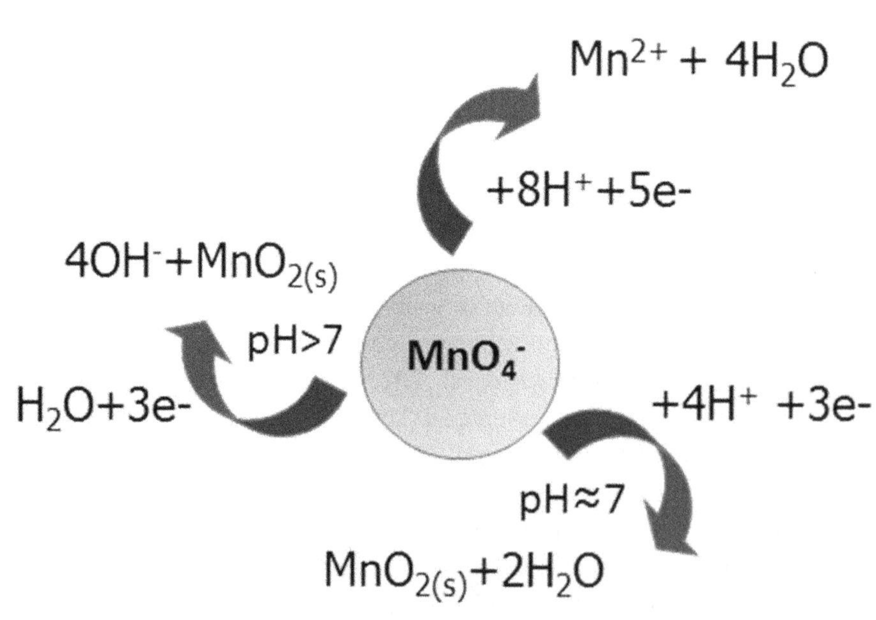

When $KMnO_4$ was combined with ultrasonic irradiation it was reduced to colloidal MnO_2. This MnO_2 was also a catalyst in the oxidative processes of permanganate (Sun et al., 2021). Likewise, $KMnO_4$ was activated by UV photolysis at 254 nm. In the first step of UV/KMnO4 process was obtained Mn (V) that was converted to Mn(III) and finally to MnO_2. The cooperation between oxidative (•OH) and adsorptive (MnO_2) species was evidenced (Wei et al., 2021). Other methods used in the activation of $KMnO_4$ were activation by carbon-based catalysts and by humic acids. Humic acids (HA) promote the degradation of organic pollutants with $KMnO_4$ by different mechanisms (Xu et al., 2018): HA serves as a ligand and binds the oxidative intermediate Mn(III), and the resulting complex causes organic degradation at a high rate; HA promotes organic pollutants degradation through increasing their electron density; as reductant HA accelerate $KMnO_4$ reduction and formation in-situ of MnO_2 with catalytic activity on $KMnO_4$ oxidation.

Other oxidizing species are hypochlorites (including HClO, NaClO, and $Ca(ClO)_2$). Hypochlorites are strong oxidizers, frequently used due to their accessibility and low price. They are able to generate various active species (ClO^-, $Cl^{-\cdot}$ and Cl_2^-) with strong oxidizing abilities (Cheng et al., 2022). Hypochlorite technology is effective for removing pollutants from both wastewater and sludge. Moreover, the effectiveness of hypochlorite-based treatment on pollutant removal is comparable to or even better than those of conventional strategies, such as the H_2O_2.

The interest in hypochlorite technology in wastewater treatment results from its high efficiency in the removal of typical pollutants (organic phenols, antibiotics, and heavy metals). The performance of hypochlorite-based treatment was correlated with critical operational conditions, such as pH, temperature, and coexisting ions (Cl^-, Br^-, SO_4^{2-}, NO_3^-, HCO_3^-) in wastewater. The generation of free radicals was stimulated by treatment with UV light, by electro-oxidation. Under UV irradiation were generated both reactive oxygen radicals (•OH and O_2•$^-$) and reactive chlorine species (Cl•, Cl_2–•, ClO•), which could increase free radical attacks on the targeted pollutants.

INDIRECT OXIDATION TECHNIQUES FOR WASTEWATER TREATMENT

Fenton Wastewater Treatments

Among AOPs, Fenton processes are the most widely used wastewater treatment methods due to their high oxidation efficiency, low cost, and simplicity. The conventional Fenton reaction uses iron (II) salts as the homogeneous catalyst to activate under acidic conditions to produce highly reactive hydroxyl radicals which can oxidize organic compounds in solution (Gan et al., 2023). Fenton system consists of Fe^{2+}/Fe^{3+} combined with hydrogen peroxide under acidic conditions which form a high concentration of •OH ($Fe^{2+} + H_2O_2 \rightarrow Fe^{3+}$ + •OH + OH–) (Wang et al., 2022; Phan Quang et al., 2022). •OH is scavenger on H_2O_2 in excess (•OH + $H_2O_2 \rightarrow H_2O$ + •HO_2). The hydroperoxide radicals react with Fe(II) and result hydroperoxide ions (• $HO_2 + Fe^{2+} \rightarrow HO_2^- + Fe^{3+}$). The use of the Fenton process combined with ozonation increases the number of hydroxyl radicals favoring the efficiency of the treatment. $O_3/Fe^{2+}/H_2O_2$ process can be carried out at room temperature with high effectiveness of the detoxification of heavy metals in wastewater (Rekhate & Srivastava, 2021). The combination of this process with light effect and other conventional techniques is one of the combined solutions for advanced wastewater treatment (Figure 2). Fenton processes can be utilized after primary or secondary treatments. In most situations are used traditional processes such as coagulation, flocculation, sedimentation, and filtration are both before and after chemical oxidation processes. Iron species such as FeOOH could be excited by irradiation to produce photogenerated electrons for Fe^{3+} reduction and H_2O_2 decomposition, thus promoting the production of •OH (Martínez et al., 2011). In mixed oxides, hydroxyl radicals are formed not only in the presence of Fe^{2+}ions, but can also be formed in the presence of other metal cations. This indirect oxidation reaction has the disadvantage of generating large amounts of iron sludge at the optimal pH range which limits its application in wastewater treatment. To overcome this disadvantage, in the last years Fenton reactions were modified by combining it with a physical process, by using an alternative catalyst or source of H_2O_2 (peroxides). There are many modified Fenton reactions like Heterogeneous Fenton, Photo-Fenton, Electro-Fenton, Bioelectro-Fenton, Sono-Fenton, Galvano-Fenton, Ionizing Radiation-Fenton, fluidized bed-Fenton (Ziembowicz & Kida, 2022).

Figure 2. Hybrid wastewater treatment based on Fenton processes

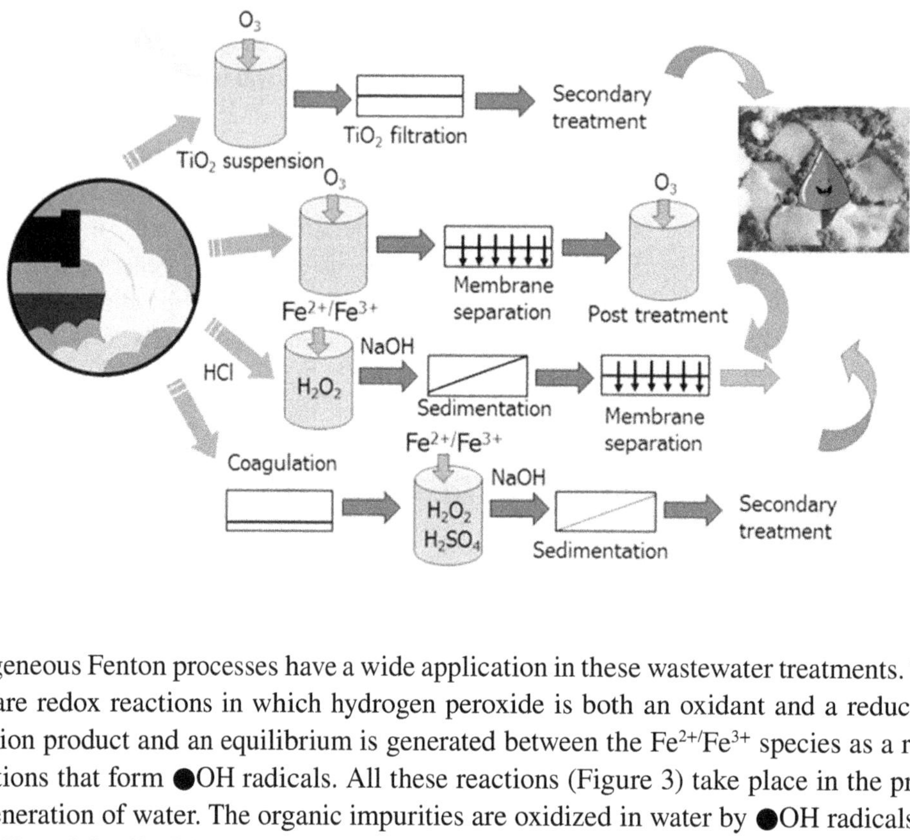

Heterogeneous Fenton processes have a wide application in these wastewater treatments. The Fenton processes are redox reactions in which hydrogen peroxide is both an oxidant and a reducer, reactant and a reaction product and an equilibrium is generated between the Fe^{2+}/Fe^{3+} species as a result of the redox reactions that form ●OH radicals. All these reactions (Figure 3) take place in the presence and with the generation of water. The organic impurities are oxidized in water by ●OH radicals to RCHO, RCH_3COOH, and finally CO_2 and H_2O. In the presence of light, both Fe^{3+} and H_2O_2 generate hydroxyl radicals, thus increasing the content of active species. In the heterogeneous Fenton catalysts, Fe^{2+} was replaced by many kinds of iron-based solid materials. Iron compounds are used individually (zerovalent iron, goethite-α-FeOOH, akageneite-β-FeOOH, lepidocrocite γ-FeOOH, ferroxyhite δ-FeOOH, hematite α-Fe_2O_3, magnetite-Fe_3O_4) (Martínez et al., 2011; Chu et al., 2012; Wang et al., 2022), as bimetallic oxide (Fe-Cu, Fe-Co, Fe-Ce) (Chu et al., 2012; Phan Quang et al., 2022) or supported (FeOOH/Bi_2WO_6, FeO/MoS_2/ Fe_2O_3/graphite, $CoFe_2O_4$/ carbon, Fe0/Fe_3O_4, Fe0/CeO_2, Fe_2O_3/TiO_2) (Ewuzie et al., 2022; Rekhate & Srivastava, 2021; Wang et al., 2022; Chu et al., 2012). Thus, Fe^{2+} and Co^{2+} react with H_2O_2 to form •OH, while Fe^{3+} and Co^{3+} react with H_2O_2 to form •OH according to the Fenton mechanism (Chu et al.,

2012). Cu-based heterogeneous catalysts are recently attracting attention due to their catalytic activities in the Fenton-like reactions compared to Fe-based catalysts. Although Cu^{2+}/Cu^{+} ions have a similar reactivity with hydrogen peroxide (H_2O_2) as Fe^{3+}/Fe^{2+} ions, the catalytic activities of copper were higher. The conductive carbon species as g-C_3N_4 from the interface of FeOOH promoted the photogenerated electron transfer, which resulted in rapid conversion of Fe(III)/Fe(II), achieving a high percentage of organic compounds degradation and mineralization (Meijide et al., 2022; Oturan & Aaron, 2014). In electrochemical technology, OH• radicals were produced at the anode surface. OH• radicals, on the other hand, are produced in anodic oxidation *via* electro-Fenton or sonoelectrolysis (Al-Tohamy et al., 2022).

Figure 3. Redox reactions in the heterogeneous Fenton reactions from water

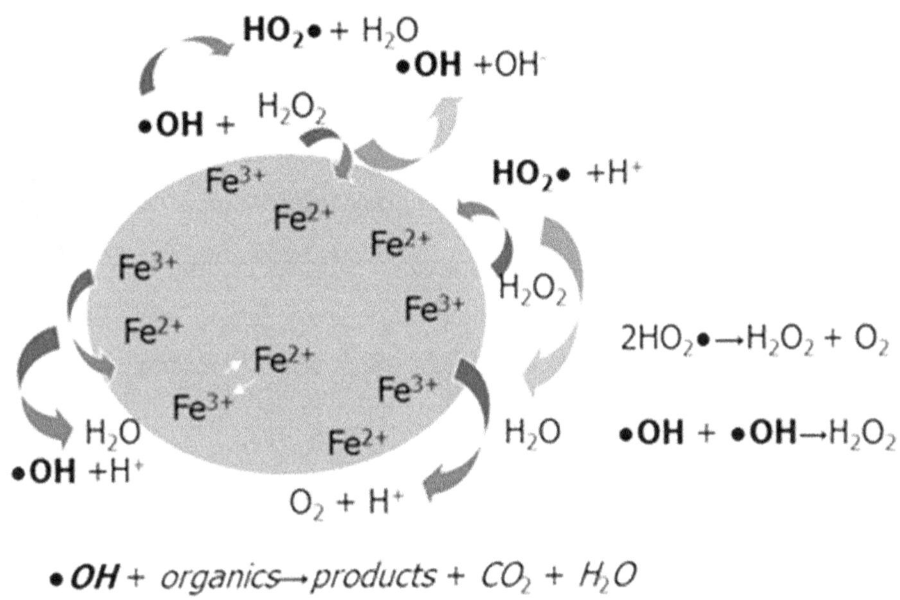

Other modification of heterogeneous Fenton catalysts was pre-magnetization that increased the Fe0 corrosion rate to generate Fe^{2+} owing to the "magnetic memory", chelating agents for Fe-ions (as EDTA), thus preventing their precipitation at neutral pH conditions, promoting Fenton reaction performances (Chu et al., 2012). In heterogeneous electro-Fenton systems were used mixed oxides such as ZnO–CeO_2, $NiMn_2O_4$, $CoFe_2O_4$ as co-catalytic electrodes (Phan Quang et al., 2022). The synergistic effects of oxidation, coagulation, and adsorption also exist in the photo-Fenton process. The common feature of these methods is that they allow the generation of highly reactive hydroxyl (•OH) radicals.

Photocatalytic Wastewater Treatments

Photocatalysis is another advanced oxidation process. The photocatalytic reaction relies on highly reactive oxidation species (ROS) such as superoxide radicals (•O_2-), hydroxyl radicals (•OH), sulfate radicals (SO_4•$^-$), and generated electron holes (h^+) (Hong et al., 2022). The generation of active ROS is result of photocatalyst activation in 5 steps: 1) Absorption of photons which determines the electrons transfer from the valence band (VB) to the conduction band (CB) with formation of an electron-hole; 2) The oxygen from water molecule is attracted to the H^+ on the photocatalyst surface, the water molecule donates an electron to the photocatalyst neutralizing the surface and results a •OH radical; 3) Superoxide $O_2{\cdot}^-$ radical is formed from molecular oxygen dissolved in water with the electron from CB; 4) $O_2{\bullet}^-$ radicals react with other water molecules and produce more •OH radicals, as well as H_2O_2; 5) H_2O_2 reacts with CB electrons producing more •OH radicals (*Ghazal et al., 2022)*. This means that the pollutants present in the wastewater are oxidized especially by •OH radicals. The mechanism of ROS generation and oxidation in photocatalytic processes can be modified depending on the photocatalyst's composition. Figure 4 presents the mechanism of ROS generation and photocatalytic degradation of organic compounds on $Ag/Ag_2O/TiO_2$-activated carbon composite (Negoescu et al., 2022). Under UV light the electrons from the valence band (VB) of TiO_2 are excited and transferred to the conduction band (CB). The activated carbon acts as an electron sink and Ag nanoparticles (AgNPs) can favor the capture of electrons due to inherent interband electronic states in metallic Ag, both determining and decreasing the recombination of photogenerated charge carriers.

Figure 4. Mechanism of the photocatalytic degradation of organic pollutants on $Ag/Ag_2O/TiO_2$-activated carbon
Reprinted from Negoescu et al., 2022 with permission from Elsevier

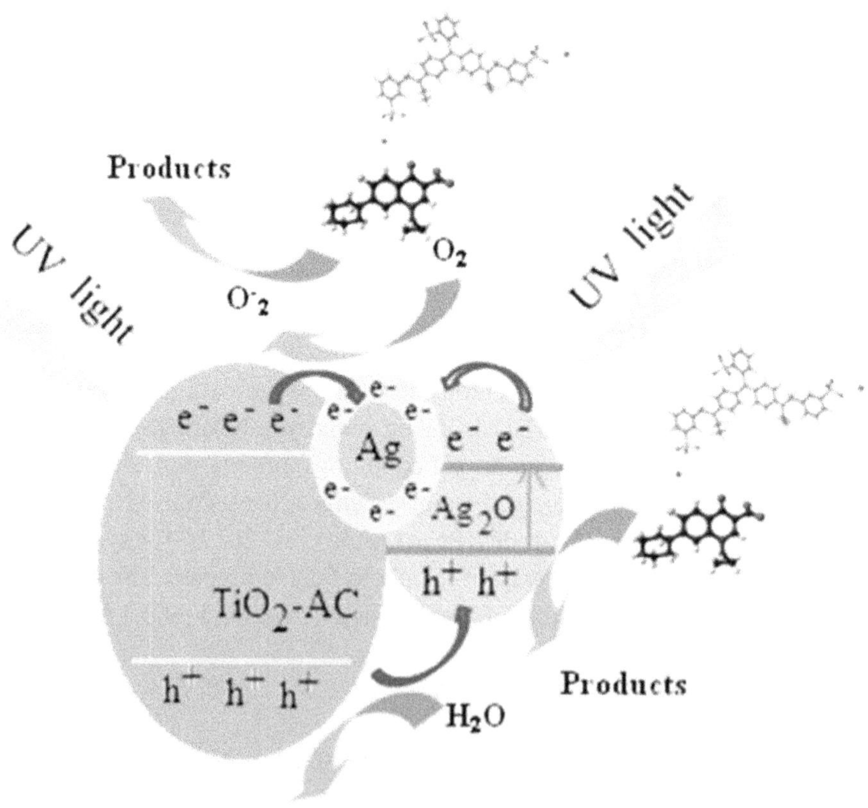

The photo-generated electrons are trapped in less negative Ag^0 which then reacted with surface-adsorbed O_2 generating superoxide radicals (O_2•−). The photo-generated holes on VB of TiO_2 could partly combine with H_2O or –OH to generate •OH, while others are transferred to Ag_2O forming similar active radicals (•OH) for oxidative degradation of organic pollutants. For other photocatalytic materials such as Au-modified TiO_2 supported on silica supports (Petcu et al., 2022) the mechanism can vary under UV and visible light (Figure 5). When the photocatalytic experiment was carried out under UV irradiation (Figure 5A), Au NPs act as electron trap, improving the separation of e^-/h^+ pairs from titania. From unpaired electrons resulted from superoxide species (O_2•$^-$) through the reduction of O_2 molecules adsorbed on the photocatalyst surface, while unpaired holes ensure the oxidation process of hydroxyl ions from aqueous medium to •OH radicals, reactive oxygen species. In visible light (Figure 5B), Au NPs species are activated through the surface plasmon resonance effect.

Figure 5. Mechanisms proposed for the photocatalytic amoxicillin degradation on Au-Ti/silica supports under UV (A) and visible light irradiation (B)
Reprinted from Petcu et al., 2022 with permission from MDPI.

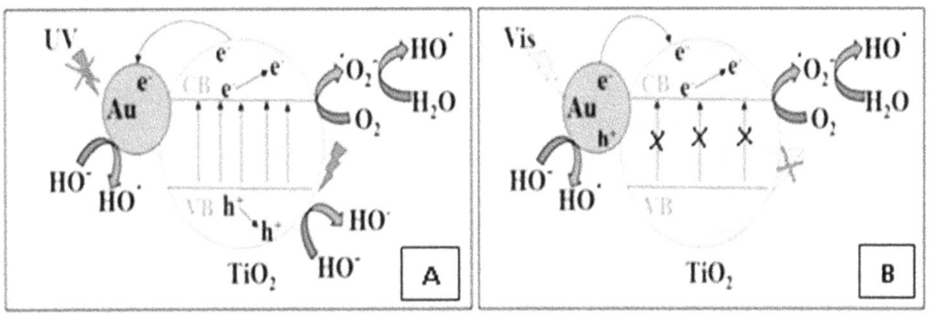

The photogenerated electrons from gold nanoparticles are injected into the conduction band of TiO_2 and used to obtain highly reactive oxygen species as O_2•$^-$ and •OH after the reduction of adsorbed oxygen molecules. The h^+ species generated in Au NPs nanoparticles after electrons transfer can oxidize hydroxyl ions to •OH radicals. The generation of the sulfate radical (SO_4•$^-$) in the photocatalytic reactions is the result of the peroxymonosulfate activation. Photocatalytic degradation has been utilized to treat contaminants ranging from dyes, chemical precursors, and pharmaceuticals to diverse organic and inorganic waste (Vaseashta et al, 2021). The performance of photocatalytic processes was influenced by some basic parameters, like the type of pollutant and concentration, photocatalyst concentration, pH, light intensity, irradiation time, reaction temperature, inorganic ions, and dissolved oxygen (Das et al., 2021). The advantages of photocatalytic technologies are an environmentally friendly process with less consumption of energy and materials, non-toxic by-products, small reaction volume, control of the phases contact by flow volume, and high efficiency of the pollutant's degradation. Photocatalytic oxidation using semiconductors as photocatalysts has become an attractive candidate as a green technology for environmental applications, such as wastewater treatment. The enhancement of semiconductors activity and selectivity, the application of nanomaterials, development of new composite and hybrid materials have received significant attention for environmental and energy applications. Among the semiconductors, TiO_2, titania-based materials, and ZnO have many attractive properties (nontoxicity, less costly, advanced optical properties, high stability) for photodegradation process. TiO_2, as a heterogeneous photocatalyst, has been used for the complete mineralization of organic substances like chloroform, trichloroethylene, dichloromethane, and carbon tetrachloride for water purification (Das et al., 2021). The formation of nanocomposite with TiO_2 (Negoescu et al., 2022), and the modifications of TiO_2 through functionalization have been carried out to overcome the limitations and improve photocatalytic performance under visible light irradiation (Petcu et al., 2022). As a result of the investigation, many other solid semiconductors have been used in photocatalytic heterogeneous wastewater treatment such as cerium oxide, zinc oxide, zirconium oxide, iron oxide, and tungsten oxide. Doping heteroatoms in the semiconductor is a commonly used method to add additional energy levels in the middle of the bandgap, leading to an increase in the absorption range of light. Oxygen vacancies can also enhance the photocatalytic behavior of semiconductors by inducing defect energy levels, and by serving as adsorption sites for organic molecules (Hong et al., 2022). The effect of different TiO_2 supports (graphite, ceramic, silica, zirconia, zeolite, alumina, polymers) was evidenced on photodecolorization of various dyes.

The photocatalytic processes aided by ultrasound (sonophotocatalysis) increased activity for ZnO nanoparticles and carbon in dyes degradation as a result of the electron–holes separation at the hetero-interface and the formation of more reactive radicals as well as the increase of the active surface area. Perovskites are applied as photocatalysts to the treatment of wastewater with organic pollutants, such as methylparaben, and pharmaceutical wastewater including doxorubicin, ciprofloxacin, and salicylic acid (Lin et al., 2022).

An efficient alternative to other types of photocatalysts based on semiconductors is organic photocatalysts. They absorb visible light, are usually cheap, and do not release metals in the natural environment. Even if in general, they do not produce directly hydroxyl radicals (•OH), electron transfer reactions from their excited states could be successfully applied in the field of wastewater treatment. The oxidizing and reducing species are generated by the activation of organic catalysts under light irradiation (Pavanello et al. 2022). When, •OH is the main radical of the advanced oxidation process, $SO_2{\bullet}^-$ is the most active radical in advanced reduction processes (reduction potential of -0.66 V). Other reducing radicals are sulfite radical anion ($SO_3{\bullet}^-$), sulfur dioxide radical anion ($SO_2{\bullet}^-$), the hydrated electron and H•. The radicals were obtained by various activation processes such as ultraviolet light, electron beam or ionizing radiation. Advanced reduction processes have been applied to treat halogenated organic compounds or organic compounds combined with bromated, perchlorate, nitrate, drugs or pesticides. The use of organic catalysts in visible light was an innovative alternative to reduction of contaminants from wastewater. Moreover, the photocatalytic processes were intensified by adding an oxidizing agent such as H_2O_2. Photocatalytic processes using TiO_2 as well as modified TiO_2 are, together with Fenton methods, the most powerful technologies used for wastewater treatment.

Electrochemical Wastewater Treatments

Electrochemical technology, especially three-dimensional electrochemical technology, has been focused on wastewater treatment with very good results. This technology is efficient, clean, and sustainable for removing a wide range of organic, inorganic, or ionic species from wastewater. They only need electricity to produce the reactive species in situ. The electrochemical reactor was primarily applied in the pretreatment of refractory wastewater in order to save operation costs. The Electrochemical Advances Oxidation Process includes the oxidation of pollutants in an electrolytic cell by direct electron transfer between the molecule of contaminant and anode or by indirect oxidation with heterogeneous radicals formed from water discharge at the anode (Samal et al., 2022). Compared with the fixed bed, the fluidized bed electrochemical reactor has better solid-liquid mixing, higher mass transfer, high active electrode area, and less chemical consumption. This technique was successfully used in the efficient degradation of organic compounds from wastewater (Liu et al., 2022). The new fluidized three-dimensional electrochemical treatment process was obtained by integration with activated carbon adsorption. In this process, activated carbon was polarized in a fluidized bed electrochemical reactor to enhance the direct electro-oxidation and •OH radicals oxidation. A synergistic effect of effective adsorption and electrochemical oxidation was obtained to increase the efficiency of degradation.

An integrated biological-electrocatalytic process was developed by Wu et al. in the treatment of coking wastewater (Wu et al., 2021). Coking wastewater contains various kinds of pollutants such as phenolic compounds, cyanide, poly nuclear aromatic hydrocarbons, most of which are toxic, mutative, and carcinogenic. The results of the integrated system recommended its application in the treatment of highly toxic wastewater. Significant removal efficiencies were obtained. Thus, compounds with carbon and nitrogen were mainly removed by direct oxidation at electrodes or indirect oxidation of •OH (Deng et al., 2019). Another technique reverse electrodialysis was successfully used at the pilot scale to evaluate the effect of redox processes and other operating parameters in the simultaneous generation of electricity and the treatment of wastewater contaminated with resistant pollutants (D'Angelo et al. 2017). The electrochemical techniques used in environmental remediation are electrocatalysis and photoelectrocatalysis. The electrocatalyst plays a key role during the redox pair reaction process because the rate, efficiency, and selectivity of the reactions are determined by an anode-like catalyst (Deng et al., 2020). Photo-catalytic fuel cell (PFC) is a promising and sustainable technology that allows the simultaneous degradation of organic pollutants from wastewater and the recovery of their chemical energy. Degradation of the pollutants occurs at the photoanode through a photocatalytic process, while the photogenerated electrons pass through the cathode by an external circuit. The materials for the anode are, in many cases, metal oxide semiconductors. Thus, CeO_2 heterojunction with TiO_2 has been extensively investigated. For Ce-modified TiO_2 photocatalysts anode material, the production of more electrons and holes and better conductivity was explained by Ti^{3+} formation which creates a more abundant oxygen vacancy and produces more electrons and holes (Mureseaunu et al., 2021). In phenol degradation was compared the standard redox potential of O_2/O_2^- oxidizing species with the position of the CB of the tested photocatalysts that must be at more negative potential than the targeted redox couple potential (Figure 6). The addition of cerium-led to the change of the CB position to the more negative which favored the generation of active species.

Among the new and attractive electrochemical technologies developed recently due to the advantages in treating wastewater are based on bioelectrochemical processes. Among them, microbial fuel cell (MFC) technology was used for pollution remediation and power production from waste.

The main characteristics of the MFC technology are the presence of two different redox environments in the cathode and anode. The electrodes are used as inexhaustible electron acceptors or donors having enormous opportunities to utilize microbial communities as biocatalysts for the complete removal of contaminants from wastewater (Kaushik et al., 2020). In general, a bioelectrochemical system consists of two chambers. One is called an anodic chamber, containing a positive electrode (anode) and biocatalysts such as microorganisms and enzymes. The other is called a cathodic chamber containing a negative electrode (cathode), and it accepts oxygen during the process. These two chambers are separated by a proton exchange membrane. The substrate is oxidized in the anodic chamber, contributing to the flow of electrons to the cathode that produces electricity directly or reduces electron acceptor chemicals reducing the toxicity of water (Kumar Gupta et al., 2022).

Figure 6. Diagrams of energy levels in TiO_2-CeO_2 photocatalysts with various Ce loading (TGP-0%, TGP2.5C-2.5%, TGP5C5%)
Reprinted from Mureseanu et al., 2021 with permission from Elsevier.

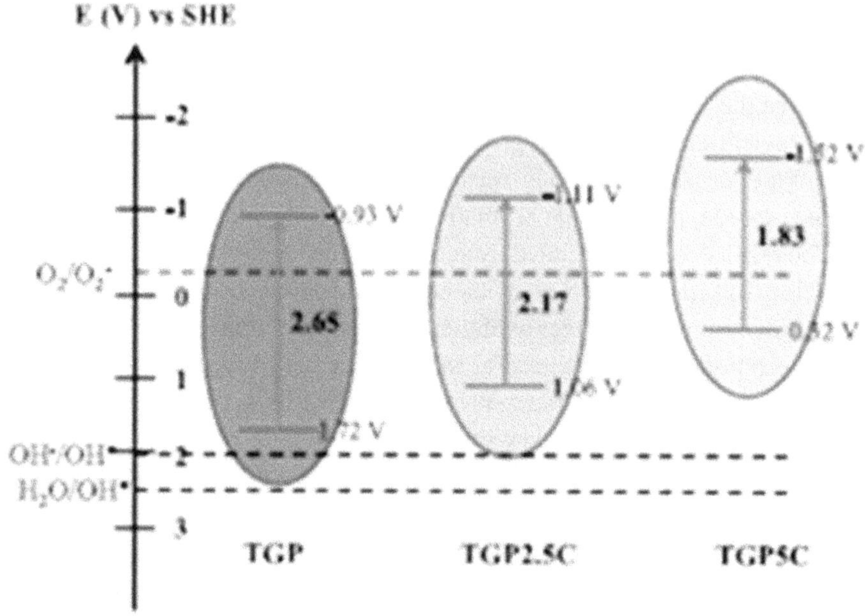

MFC performance and electricity generation are influenced by various factors such as electrode and membrane material, temperature, pH, substrate loading rate, biofilm, and external resistance. Different materials were used for MFC electrodes. Among the materials used for anodes can be mentioned carbon or carbon-based materials (oxides of Ti, Fe, Sn, Mn on carbon, active polymers as polyaniline). For the cathode, the materials were based on graphite and stainless steel. Although oxygen is preferred as an electron acceptor due to its higher redox potential, its use is limited due to the reduced contact surface with the electrode and the high energy consumption required to feed it.

The diversity of the wastewater composition and the interest in pollutants' total degradation has determined the increasing use of advanced oxidation processes and their association with various other techniques. Therefore, the degradation and mineralization of some persistent pollutants can be improved by combining two or more processes due to synergistic effects and using the advantages of the individual treatment method (*Ghazal et al., 2022*). Although AOPs have been successfully applied for the removal of refractory pollutants, through their oxidation by hydroxyl radicals, the main disadvantages of their wide application are the frequent production of more stable and toxic by-products and the high operational costs (Al-Tohamy et al., 2022). The limitations of individual processes can be overcome by combining physical and/or chemical methods with biological treatment, where AOPs can be used either as pre-treatment or as post-treatment to biological techniques (Pariente et al., 2022). In this way, various pilot stations have been created that use both combined and hybrid wastewater treatment technologies. Thus, the number and diversity of the studied and applied technologies on a pilot scale or for large quantities is continuously increasing.

CONCLUSION

Redox technologies have a decisive role in the treatment of wastewater. This chapter presents most of the used redox technologies and highlights the reactive oxidative species and their action in both direct and indirect oxidation processes. The association of several oxidation techniques leads to combined oxidation technologies or hybrid technologies that favor advanced wastewater purification and can be developed in the future. The high variety of techniques, which in most cases combine water decontamination methods, made classification difficult. In this chapter, the redox techniques are presented as direct and indirect oxidation processes. In the case of direct oxidation techniques, the techniques that use the most active oxidizing agents are discussed. For indirect oxidation, techniques were presented methods that generate the active oxidizing species, their nature and their action. In the case of each presented technique were evidenced changes of efficiency by combining them with other processes in correlation with the nature of the contaminants. The chapter ends with a short presentation of combined oxidation methods and hybrid techniques used in wastewater treatment.

REFERENCES

Akhter, M., Habib, G., & Qamar, S. U. (2018). Application of Electrodialysis in Wastewater Treatment and Impact of Fouling on Process Performance. *Journal of Membrane Science & Technology*, *8*(2), 1000182. doi:10.4172/2155-9589.1000182

Al-Tohamy, R., Ali, S. S., Li, F., Okasha, K. M., Mahmoud, Y. A.-G., Elsamahy, T., Jiao, H., Fu, Y., & Sun, J. (2022). A critical review on the treatment of dye-containing wastewater: Ecotoxicological and health concerns of textile dyes and possible remediation approaches for environmental safety. *Ecotoxicology and Environmental Safety*, *231*, 113160. doi:10.1016/j.ecoenv.2021.113160 PMID:35026583

Amaro-Soriano, A., Hernández-Aldana, F., & Rivera, A. (2021). Photochemical treatments (UV/H_2O_2, UV/O_3 and UV/H_2O_2/O_3) and inverse osmosis in wastewater: Systematic review. *World Journal of Advanced Research and Reviews*, *10*(2), 229–240. doi:10.30574/wjarr.2021.10.2.0231

Aniipsitakis, G. P., & Dionysiou, D. D. (2006) Chemically Induced Redox Reactions in Water Treatment: A Summary of Advanced and Direct Technologies. *World Environmental and Water Resource Congress 2006*, (pp. 1-10). ASCE. 10.1061/40856(200)31

Asghar, A., Abdul Raman, A. A., & Wan Daud, W. M. A. (2015). Advanced oxidation processes for in-situ production of hydrogen peroxide/hydroxyl radical for textile wastewater treatment: A review. *Journal of Cleaner Production*, *87*, 826–838. doi:10.1016/j.jclepro.2014.09.010

Chaturvedi, P., Giri, B. S., Shukla, P., & Gupta, P. (2021). Recent advancement in remediation of synthetic organic antibiotics from environmental matrices: Challenges and perspective. *Bioresource Technology*, *319*, 124161. doi:10.1016/j.biortech.2020.124161 PMID:33007697

Chen, W., Luo, Y., Ran, G., & Li, Q. (2020). Microwave-induced persulfate-hydrogen peroxide binary oxidant process for the treatment of dinitrodiazophenol industrial wastewater. *Chemical Engineering Journal*, *382*, 122803. doi:10.1016/j.cej.2019.122803

Cheng, X., Wang, S., Huang, W., Wang, F., Fang, S., Ge, R., Zhang, Q., Zhang, L., Du, W., Fang, F., Feng, Q., Cao, J., & Luo, J. (2022). Current status of hypochlorite technology on the wastewater treatment and sludge disposal: Performance, principals and prospects. *The Science of the Total Environment*, *803*, 150085. doi:10.1016/j.scitotenv.2021.150085 PMID:34525771

Chokshi, N. P., & Ruparelia, J. P. (2021). Synthesis of nano Ag-La-Co composite metal oxide for degradation of RB 5 dye using catalytic ozonation process. *Ozone Science and Engineering*, *44*(2), 182–195. doi:10.1080/01919512.2021.1901070

Chu, L., Wang, J., Dong, J., Liu, H., & Sun, X. (2012). Treatment of coking wastewater by an advanced Fenton oxidation process using iron powder and hydrogen peroxide. *Chemosphere*, *86*(4), 409–414. doi:10.1016/j.chemosphere.2011.09.007 PMID:22014660

Covinich, L. G., Bengoechea, D. I., Fenoglio, R. J., & Area, M. C. (2014). Advanced Oxidation Processes for Wastewater Treatment in the Pulp and Paper Industry: A Review. *American Journal of Environmental Engineering*, *4*, 56–70. doi:10.5923/j.ajee.20140403.03

D'Angelo, A., Tedesco, M., Cipollina, A., Galia, A., Micale, G., & Scialdone, O. (2017). Reverse electrodialysis performed at pilot plant scale: Evaluation of redox processes and simultaneous generation of electric energy and treatment of wastewater. *Water Research*, *125*, 123–131. doi:10.1016/j.watres.2017.08.008 PMID:28843152

Darwish, A. S., Attia, S. K., & Osman, D. I. (2022). Accelerated activation of H_2O_2 and persulfate by Sm-doped ZnO@highly- defective layered yttria nanocomposite under visible-light irradiation for dyeing wastewater treatment: Comprehensive dominance of oxygen vacancies in photocatalytic advanced oxidation processes. *Journal of Alloys and Compounds*, *925*, 166742. doi:10.1016/j.jallcom.2022.166742

Das, A., Adak, M. K., Mahata, N., & Biswas, B. (2021). Wastewater treatment with the advent of TiO_2 endowed photocatalysts and their reaction kinetics with scavenger effect. *Journal of Molecular Liquids*, *338*, 116479. doi:10.1016/j.molliq.2021.116479

Deng, Y., Chen, N., Feng, C., Chen, F., Wang, H., Kuang, P., Feng, Z., Liu, T., Gao, Y., & Hu, W. (2019). Treatment of organic wastewater containing nitrogen and chlorine by combinatorial electrochemical system: Taking biologically treated landfill leachate treatment as an example. *Chemical Engineering Journal*, *364*, 349–360. doi:10.1016/j.cej.2019.01.176

Dhaka, S., Kumar, R., Lee, S., Kurade, M. B., & Jeon, B.-H. (2018). Degradation of ethyl paraben in aqueous medium using advanced oxidation processes: Efficiency evaluation of UVC supported oxidants. *Journal of Cleaner Production*, *180*, 505–513. doi:10.1016/j.jclepro.2018.01.197

Duca, G. (2021). Hydrogen peroxide in ecological and environmental chemistry, *Chemistry Journal of Moldova. General. Industrial and Ecological Chemistry*, *16*, 28–45.

Duca, G., & Vaseashta, A. (Eds.). (2020). *Handbook of Research on Emerging Developments and Environmental Impacts of Ecological Chemistry*. IGI Global. doi:10.4018/978-1-7998-1241-8

El Brahmi, A., & Abderafi, S. (2021). Hydrogen sulfide removal from wastewater using hydrogen peroxide insitu treatment: Case study of Moroccan urban sewers. *Materials Today: Proceedings*, *45*, 7424–7427. doi:10.1016/j.matpr.2021.01.641

Ewuzie, U., Saliu, O. D., Dulta, K., Ogunniyi, S., Bajeh, A. O., Iwuozori, K. O., & Ighalo, J. O. (2022). A review on treatment technologies for printing and dyeing wastewater (PDW). *Journal of Water Process Engineering*, *50*, 103273. doi:10.1016/j.jwpe.2022.103273

Fernandez-Ibanez, P., McMichael, S., Cabanillas, A. R., Alkharabsheh, S., Moranchel, A. T., & Abyrne, J. (2021). New trends on photoelectrocatalysis (PEC): Nanomaterials, wastewater treatment and hydrogen generation. *Current Opinion in Chemical Engineering*, *34*, 100725. doi:10.1016/j.coche.2021.100725

Gan, P., Sun, Y., Li, Y., Liu, W., Ye, J., Tong, M., & Liang, J. (2023). The degradation of municipal solid waste incineration leachate by UV/ persulfate and UV/H_2O_2 processes: The different selectivity of SO_4•$^-$ and •OH. *Chemosphere*, *311*, 137009. doi:10.1016/j.chemosphere.2022.137009 PMID:36326516

Ghazal, H., Koumaki, E., Hoslett, J., Malamis, S., Katsou, E., Barcelo, D., & Jouhara, H. (2022). Insights into current physical, chemical and hybrid technologies used for the treatment of wastewater contaminated with pharmaceuticals. *Journal of Cleaner Production*, *361*, 132079. doi:10.1016/j.jclepro.2022.132079

Goncharuk, V. V., Bagrii, V. A., Melnik, L. A., Chebotareva, R. D., & Bashtan, S. Y. (2010). The Use of Redox Potential in Water Treatment Processes. *Journal of Water Chemistry and Technology*, *32*(1), 1–9. doi:10.3103/S1063455X10010017

Guan, C., Guo, Q., Wang, Z., Wei, X., Han, B., Luo, X., Pan, H., & Jiang, J. (2022). Bisulfite activated permanganate for oxidative water decontamination. *Water Research*, *216*, 118331. doi:10.1016/j.watres.2022.118331 PMID:35358879

Gul, H., Raza, W., Lee, J., Azam, M., Ashraf, M., & Kim, K.-H. (2021). Progress in microbial fuel cell technology for wastewater treatment and energy harvesting. *Chemosphere*, *281*, 130828. doi:10.1016/j.chemosphere.2021.130828 PMID:34023759

Gupta, K., Rachna, Singh, B., Mungray, A. K., Bharti, R., Nema, A. K., Pant, K. K., & Mulla, S. I. (2022). Bioelectrochemical technologies for removal of xenobiotics from wastewater. *Sustainable Energy Technologies and Assessments*, *49*, 101652. doi:10.1016/j.seta.2021.101652

Hoang, H., & Quang, P. ThiDinh, N., Ngoc, T., Thi, T., Tran, L., Bao, N., Yuvakkumar, R., & Nguyen, V.-H. (2022) Fe2+, Fe3+, Co2+ as highly efficient cocatalysts in the homogeneous electro-Fenton process for enhanced treatment of real pharmaceutical wastewater. *Journal of Water Process Engineering*, *46*, 102635.

Hong, J., Cho, K.-H., Presser, V., & Su, X. (2022). Recent advances in wastewater treatment using semiconductor photocatalysts. *Current Opinion in Green and Sustainable Chemistry*, *36*, 100644. doi:10.1016/j.cogsc.2022.100644

Jalali, S., Ardjmand, M., Ramavandi, B., & Nosratinia, F. (2021). Removal of amoxicillin from wastewater in the presence of H_2O_2 using modified zeolite Y- MgO catalyst: An optimization study. *Chemosphere*, *274*, 129844. doi:10.1016/j.chemosphere.2021.129844 PMID:33582537

Kaushik, A., & Aradhana Singh, M. (2020). Metal removal and recovery using bioelectrochemical technology: The major determinants and opportunities for synchronic wastewater treatment and energy production. *Journal of Environmental Management*, *270*, 110826. doi:10.1016/j.jenvman.2020.110826 PMID:32721300

Kokkinos, P., Venieri, D., & Mantzavinos, D. (2021). Advanced Oxidation Processes for Water and Wastewater Viral Disinfection. A Systematic Review. *Food and Environmental Virology*, *13*(3), 283–302. doi:10.100712560-021-09481-1 PMID:34125359

Ksibi, M. (2006). Chemical oxidation with hydrogen peroxide for domestic wastewater treatment. *Chemical Engineering Journal*, *119*(2-3), 161–165. doi:10.1016/j.cej.2006.03.022

Li, R., Speed, D., Siriwardena, D., Fernando, S., Thagard, S. M., & Holsen, T. M. (2021). Comparison of hydrogen peroxide-based advanced oxidation processes for the treatment of azole-containing industrial wastewater. *Chemical Engineering Journal*, *425*, 131785. doi:10.1016/j.cej.2021.131785

Li, T., Yang, X.-L., Song, H.-L., Xu, H., & Chen, Q.-L. (2022). Quinones contained in wastewater as redox mediators for the synergistic removal of azo dye in microbial fuel cells. *Journal of Environmental Management*, *301*, 113924. doi:10.1016/j.jenvman.2021.113924 PMID:34731947

Lin, N., Gong, Y., Wang, R., Yin Wang, Y., & Zhang, X. (2022). Critical review of perovskite-based materials in advanced *oxidation system for wastewater treatment: Design, applications and mechanisms. Journal of Hazardous Materials*, *424*, 127637. doi:10.1016/j.jhazmat.2021.127637 PMID:34753649

Lin, R., Li, Y., Yong, T., Cao, W., Wu, J., & Shen, Y. (2022). Synergistic effects of oxidation, coagulation and adsorption in the integrated Fenton-based process for wastewater treatment: A review. *Journal of Environmental Management*, *306*, 114460. doi:10.1016/j.jenvman.2022.114460 PMID:35026715

Liu, H., Gao, Y., Wang, J., Pan, J., Gao, B., & Yue, Q. (2022). Catalytic ozonation performance and mechanism ofMn-CeOx@γ-Al2O3/O3 in the treatment of sulfate-containing hypersaline antibiotic wastewater. *The Science of the Total Environment*, *807*, 150867. doi:10.1016/j.scitotenv.2021.150867 PMID:34627913

Liu, L., Chen, Z., Zhang, J., Shan, D., Wu, Y., Bai, L., & Wang, B. (2021). Treatment of industrial dye wastewater and pharmaceutical residue wastewater by advanced oxidation processes and its combination with nanocatalysts: A review. *Journal of Water Process Engineering*, *42*, 102122. doi:10.1016/j.jwpe.2021.102122

Liu, X., Chen, Z., Du, W., Liu, P., Zhang, L., & Shi, F. (2022). Treatment of wastewater containing methyl orange dye by fluidized three-dimensional electrochemical oxidation process integrated with chemical oxidation and adsorption. *Journal of Environmental Management*, *311*, 114775. doi:10.1016/j.jenvman.2022.114775 PMID:35245840

Maifadi, S., Mhlanga, S. D., Nxumalo, E. N., Motsa, M. M., & Kuvarega, A. T. (2022). Treatment of salon wastewater by peroxydisulfate based advanced oxidation process (PDS-AOP) under solar light: Synergy through integrated technologies. *Journal of Water Process Engineering*, *49*, 103062. doi:10.1016/j.jwpe.2022.103062

Malik, S. N., Ghosh, P. C., Vaidya, A. N., & Mudliar, S. N. (2018). Catalytic ozone pretreatment of complex textile effluent using Fe2+and zero-valent iron nanoparticles. *Journal of Hazardous Materials*, *357*, 363–375. doi:10.1016/j.jhazmat.2018.05.070 PMID:29909170

Martínez, S. B., Pérez-Parra, J., & Suay, R. (2011). Use of Ozone in Wastewater Treatment to Produce Water Suitable for Irrigation. *Water Resources Management, 25*(9), 2109–2124. doi:10.100711269-011-9798-x

Meijide, J., Lama, G., Pazos, M., Sanroman, M. A., & Dunlop, P. S. M. (2022). Ultraviolet-based heterogeneous advanced oxidation processes as technologies to remove pharmaceuticals from wastewater: An overview. *Journal of Environmental Chemical Engineering, 10*(3), 107630. doi:10.1016/j.jece.2022.107630

Mierzwa, J.C., Rodrigues, R., & Teixeira, A.C. (2018) UV-hydrogen peroxide processes. *Advanced oxidation processes for wastewater treatment, 3*, 13–48.

Mohyudin, S., Farooq, R., Jubeen, F., Rasheed, T., Fatima, M., & Sher, F. (2022). Microbial fuel cells a state-of-the-art technology for wastewater treatment and bioelectricity generation. *Environmental Research, 204*, 112387. doi:10.1016/j.envres.2021.112387 PMID:34785206

Mureseanu, M., Chivu, V., Osiac, M., Ciobanu, M., Bucur, C., Parvulescu, V., & Cioatera, N. (2021). New photoactive mesoporous Ce-modified TiO_2 for simultaneous wastewater treatment and electric power generation. *Catalysis Today, 366*, 164–176. doi:10.1016/j.cattod.2020.09.035

Negoescu, D., Atkinson, I., Gherendi, M., Culita, D. C., Baran, A., Petrescu, S., Trica, B., Pelinescu, D., Ionescu, R., Bratan, V., & Parvulescu, V. (2023). Brij 58–activated carbon assisted synthesis of $Ag/Ag_2O/TiO_2$-AC photocatalysts for efficient organic pollutants degradation. *Journal of Alloys and Compounds, 931*, 167528. doi:10.1016/j.jallcom.2022.167528

Nidheesh, P. V., Divyapriya, G., Titchou, F. E., & Hamdani, M. (2022). Treatment of textile wastewater by sulfate radical based advanced oxidation processes. *Separation and Purification Technology, 293*, 121115. doi:10.1016/j.seppur.2022.121115

Oturan, M. A., & Aaron, J.-J. (2014). Advanced oxidation processes in water/wastewater treatment: Principles and applications. A review. *Critical Reviews in Environmental Science and Technology, 44*(23), 2577–2641. doi:10.1080/10643389.2013.829765

Pandey, A. K., Kumar, R. R., Kalidasan, B., Laghari, I. A., Samykano, M., Kothari, R., Abusorrah, A. M., Sharma, K., & Tyagi, V. V. (2021). Utilization of solar energy for wastewater treatment: Challenges and progressive research trends. *Journal of Environmental Management, 297*, 113300. doi:10.1016/j.jenvman.2021.113300 PMID:34293672

Pariente, M. I., Segura, Y., Álvarez-Torrellas, S., Casas, J. A., de Pedro, Z. M., Diaz, E., García, J., López-Muñoz, M. J., Marugán, J., Mohedano, A. F., Molina, R., Munoz, M., Pablos, C., Perdigón-Meló, J. A., Petre, A. L., Rodríguez, J. J., Tobajas, M., & Martínez, F. (2022). Critical review of technologies for the on-site treatment of hospital wastewater: From conventional to combined advanced processes. *Journal of Environmental Management, 320*, 115769. doi:10.1016/j.jenvman.2022.115769 PMID:35944316

Pavanello, A., Miranda, M. A., & Marin, M. L. (2022). Organic photoredox catalysts for wastewater remediation: Beyond the established advanced oxidation processes. *Chemical Engineering Journal Advances, 11*, 100296. doi:10.1016/j.ceja.2022.100296

Petcu, G., Anghel, E. M., Buixaderas, E., Atkinson, I., Somacescu, S., Baran, A., Culita, D. C., Trica, B., Bradu, C., Ciobanu, M., & Parvulescu, V. (2022). Au/Ti Synergistically Modified Supports Based on SiO_2 with Different Pore Geometries and Architectures. *Catalysts*, *12*(10), 1129. doi:10.3390/catal12101129

Phan Quang, H. H., Dinh, N. T., Tran Thi, T. N., Ngoc Bao, L. T., Yuvakkumar, R., & Nguyen, V.-H. (2022). Fe^{2+}, Fe^{3+}, Co^{2+}as highly efficient cocatalysts in the homogeneous electro-Fenton process for enhanced treatment of real pharmaceutical wastewater. *Journal of Water Process Engineering*, *46*, 102635. doi:10.1016/j.jwpe.2022.102635

Qiao, J., & Xiong, Y. (2021). Electrochemical oxidation technology: A review of its application in high-efficiency treatment of wastewater containing persistent organic pollutants. *Journal of Water Process Engineering*, *44*, 102308. doi:10.1016/j.jwpe.2021.102308

Quan, X., Luo, D., Wu, J., Li, R., Cheng, W., & Ge. (2017). Ozonation of acid red 18 wastewater using O3/Ca (OH) 2 system in a micro bubble gas-liquid reactor. *Journal of Environmental Chemical Engineering*, *5*(1), 283–291. doi:10.1016/j.jece.2016.12.007

Rekhate, C. V., & Srivastava, J. K. (2020). Recent advances in ozone-based advanced oxidation processes for treatment of wastewater- A review. *Chemical Engineering Journal Advances*, *3*, 100031. doi:10.1016/j.ceja.2020.100031

Rekhate, C. V., & Srivastava, J. K. (2021). Effectiveness of O3/Fe2+/H2O2 process for detoxification of heavy metals in municipal wastewater by using RSM. *Chemical Engineering and Processing*, *165*, 108442. doi:10.1016/j.cep.2021.108442

Samal, K., Mahapatra, S., & Ali, M. H. (2022). Pharmaceutical wastewater as Emerging Contaminants (EC): Treatment technologies, impact on environment and human health. *Energy Nexus*, *6*, 100076. doi:10.1016/j.nexus.2022.100076

Saravanan, A., Deivayanai, V. C., Senthil Kumar, P., Rangasamy, G., Hemavathy, R. V., Harshana, T., Gayathri, N., & Alagumalai, K. (2022). A detailed review on advanced oxidation process in treatment of wastewater: Mechanism, challenges and future outlook. *Chemosphere*, *308*, 136524. doi:10.1016/j.chemosphere.2022.136524 PMID:36165838

Shabir, M., Yasin, M., Hussain, M., Shafiq, I., Akhter, P., Nizami, A.-S., Jeon, B.-H., & Park, Y.-K. (2022). A review on recent advances in the treatment of dye-polluted wastewater. *Journal of Industrial and Engineering Chemistry*, *112*, 1–19. doi:10.1016/j.jiec.2022.05.013

Sibhatu, A. K., Weldegebrieal, G. K., Sagadevan, S., Tran, N. N., & Hessel, V. (2022). Photocatalytic activity of CuO nanoparticles for organic and inorganic pollutants removal in wastewater remediation. *Chemosphere*, *300*, 134623. doi:10.1016/j.chemosphere.2022.134623 PMID:35439489

Simonic, M. (2021). Reverse Osmosis Treatment of Wastewater for Reuse as Process Water—A Case Study. *Membranes*, *11*(12), 976. doi:10.3390/membranes11120976 PMID:34940477

Sun, B., Zhang, Y., Gong, Z., Zhang, J., & Zhang, J. (2021). Reducing substances-enhanced degradation of pollutants by permanganate: The role of in situ formed colloidal MnO_2. *Chemosphere*, *276*, 130203. doi:10.1016/j.chemosphere.2021.130203 PMID:33725625

Tufail, A., Price, W. E., Mohseni, M., Pramanik, B. K., & Hai, F. I. (2020). A critical review of advanced oxidation processes for emerging trace organic contaminant degradation: Mechanisms, factors, degradation products, and effluent toxicity. *Journal of Water Process Engineering*, *40*, 101778. doi:10.1016/j.jwpe.2020.101778

Turchi, C. S., & Ollis, D. F. (1990). Photocatalytic Degradation of Organic Water Contaminants: Mechanisms Involving Hydroxyl Radical Attack. *Journal of Catalysis*, *122*(1), 178–192. doi:10.1016/0021-9517(90)90269-P

Vaseashta, A., Duca, G., & Travin, S. (Eds.). (2022). *Handbook of Research on Water Sciences and Society* (Vols. 1–2). IGI Global. doi:10.4018/978-1-7998-7356-3

Vaseashta, A., Gevorgyan, G., Kavaz, D., Ivanov, O., Jawaid, M., & Vasović, D. (2021). Exposome, Biomonitoring, Assessment and Data Analytics to Quantify Universal Water Quality. In A. Vaseashta & C. Maftei (Eds.), *Water Safety, Security and Sustainability. Advanced Sciences and Technologies for Security Applications*. Springer. doi:10.1007/978-3-030-76008-3_4

Wang, X., Jing, J., Zhou, M., & Dewil, R. (2022). (in press). Recent advances in H_2O_2-based advanced oxidation processes for removal of antibiotics from wastewater. *Chinese Chemical Letters*. doi:10.1016/j.cclet.2022.06.044

Wang, Y., Zheng, K., Guo, H., Tian, L., He, Y., Wang, X., Zhu, T., Sun, P., & Liu, Y. (2023). Potassium permanganate-based advanced oxidation processes for wastewater decontamination and sludge treatment: A review. *Chemical Engineering Journal*, *452*, 139529. doi:10.1016/j.cej.2022.139529

Wei, W., Guo, K., Kang, X., Zhang, J., Li, C., & Fang, J. (2021). Fang, J., Complete Removal of Organoarsenic by the UV/Permanganate Process via HO• Oxidation and in Situ- Formed Manganese Dioxide Adsorption. *ACS ES&T Engineering*, *1*(4), 794–803. doi:10.1021/acsestengg.1c00004

Weng, H., Yang, Y., Zhang, C., Cheng, M., Wang, W., Song, B., Luo, H., Qin, D., Huang, C., Qin, F., & Li, K. (2023). Insight into FeOOH-mediated advanced oxidation processes for the treatment of organic polluted wastewater. *Chemical Engineering Journal*, *453*, 139812. doi:10.1016/j.cej.2022.139812

Wu, Z.-Y., Zhu, W.-P., Liu, Y., Zhou, L.-L., Liu, P.-X. J., & Xu, J. (2021). An integrated biological-electrocatalytic process for highly efficient treatment of coking wastewater. *Bioresource Technology*, *339*, 125584. doi:10.1016/j.biortech.2021.125584 PMID:34303099

Xu, K., Dong, H., Li, M., & Qiang, Z. (2018). Quinone group enhances the degradation of levofloxacin by aqueous permanganate: Kinetics and mechanism. *Water Research*, *143*, 109–116. doi:10.1016/j.watres.2018.06.026 PMID:29940356

Ziembowicz, S., & Kida, M. (2022). Limitations and future directions of application of the Fenton-like process in micropollutants degradation in water and wastewater treatment: A critical review. *Chemosphere*, *296*, 134041. doi:10.1016/j.chemosphere.2022.134041 PMID:35189198

Chapter 15
Redox Processes in Grapes Processing

Rodica Sturza
https://orcid.org/0000-0002-2412-5874
Technical University of Moldova, Moldova

Iurie Scutaru
Technical University of Moldova, Moldova

Gheorghe Duca
Moldova State University, Moldova

ABSTRACT

The redox processes are an unavoidable and very complex part of winemaking. The chapter examines the main redox processes that take place during grape subsequent technological processes, maturation, and aging of wine. Some of them lead to the ennobling of wines and their stabilization, while others cause the alteration, and degradation of wines, and the appearance of numerous defects of wines. The consecutive, parallel, and spontaneous development of multiple redox processes, and their dependence on pedoclimatic, agro-technical, and technological conditions influence the structure, appearance, aroma/bouquet, taste, and stability of wines. The knowledge of the production mechanisms, the influencing factors, of the particularities allows the realization of precision vinification, characterized by the reduction of the additive and subtractive technological interventions, of the quantities of adjuvants, and the production of ecological wines. Monitoring redox processes during alcoholic fermentation through oxygen management is essential for the quality of wines.

INTRODUCTION

The processing of the grapes themselves involves redox processes, especially in the case of damaged and contaminated grapes, when the contact of oxygen in the air with the grape juice is facilitated (Briedis et al., 2003). The substances, which form redox couples in grapes, must and wine, are numerous - polyphe-

DOI: 10.4018/979-8-3693-0512-6.ch015

nols/quinones, GSH / GSSG, endiol / diketones, acetoin/diacetyl, anthocyanins / reduced anthocyanins, dioximaleic acid / dioxitartaric acid, etc., and their influence on the product depends on their share in the general transformations during the life of the wine and become responsible for their self-life (Zironi et al., 2010). In oxidative-reducing chemical transformations, an essential role belongs to the pairs of transition metals Fe^{3+}/Fe^{2+}, and Cu^{2+}/Cu^{+}, naturally present in wines, but also in many enzymes (until their inactivation by ethanol and elimination by technological processes). The presence of phenolic substances in grapes, must, and wine largely determines the physicochemical properties and quality of the finished product (Tarko et al., 2020). In seeds, their oxidative transformations are catalyzed by oxidase enzymes, present in healthy grapes. Monophenols are converted to *ortho*-diphenols by oxidation in the presence of tyrosinase (monophenoloxidase), while di- and triphenols are oxidized to quinones in the presence of polyphenol oxidase (PPO). This ferment is quite specific to *ortho*-diphenols, which are oxidized to *ortho*-quinones, compounds highly reactive (Sun et al., 2011). They form brown polymers in the must, which affect the color of white wines. Grapes affected by *Botrytis cinerea* contain a much more active yeast produced by gray rot, which oxidizes a much wider spectrum of substances in must and wine. All of these oxidative enzymes contain active sites of Cu^{2+} ions (de Simone et al., 2020).

Oxidations are favored by high temperatures, high pH, and oxygen content (Scutaru et al., 2022). At the stage of processing white grapes, the inactivation of these enzymes is crucial, which is achieved by the addition of grapes or must the antioxidants, SO_2, and ascorbic acid. Monitoring redox processes during alcoholic fermentation by oxygen management is essential for the resulting quality of dry wines. Oxidations in the wine production process led to a decrease in varietal aromas (Kanavouras et al., 2019). At the same time, the reducing conditions are responsible for possible undesirable aromas of disulfides and hydrogen sulfide (Bekker et al., 2016). In particular, ensuring the reduced environment in bottled white wines affects the quality of white wines, in which, after several months, atypical, reduced aromas can be attested (Sturza et al., 2021). Ensuring strongly reducing bottling conditions by applying inert gases and sealing, contributes, in the long run, to achieving reactions with the participation of varietal thiols.

The quality of wines can be disastrously affected by the phenomenon of redox photoactivation of wine bottled in transparent vessels, exposed for a certain period (a few hours being sufficient) to natural or artificial light (Celotti et al., 2022). As a result, the redox state is modified, which causes the modeling of the organoleptic properties of wines and causes the olfactory defect called "light taste" (sunlight flavor, light-struck gout). It is accompanied by the smell of boiled cabbage, onions, damp wool, and rubber. At the same time, the redox balance of wines is disturbed by the simultaneous formation of peroxides, glyoxylates and acrolein. In the production of red wines, the redox processes are no less important, although the impact is less significant (Márquez et al., 2019).

During alcoholic fermentation, the presence of oxygen contributes to the activity of yeast cells by stimulating the production of sterols (Saerens et al., 2010, Piva et al., 2022). At the same time, at this stage, oxygen influences the color of the must, contributing to its stability, including the formation of tannin-anthocyanin compounds. In the post-fermentation phases, redox processes with the participation of oxygen can have both a positive impact (maturation) and a negative impact (uncontrolled oxidations) (Saerens et al., 2010). Redox phenomena are extremely important for the specific winemaking processes of production of fortified wines - Heres, Marsala, Madeira, Portwein, etc (Abreu et al., 2021).

The protective role of SO_2 against the oxidation of musts and wines is well known (Danilewicz, 2016). It has a combined action on dissolved oxygen and allows avoiding wine oxidation. Sulfur dioxide forms naturally in wines, in low concentrations of 10–30 mg/L. The added concentration in commercial wines varies between 50–200 mg/L. Sulfur dioxide plays a critical role in preventing microbial spoilage and is important as an antioxidant. By reacting with hydrogen peroxide generated from oxygen through a series of redox steps, as well as with quinones. Sulfur dioxide contributes to the regeneration of polyphenols (Waterhouse & Laurie, 2006). Although SO_2 influences the loss of thiol compounds after bottling, acting as a sacrificial nucleophile, an equivalent alternative for it has not yet been established (Nikolantonaki et al., 2014). In summary, SO_2 preserves the wines from too intense oxidation of the phenolic compounds and certain elements of the aroma, favorable to the development of the sensory qualities of the wines during the technological processes and aging of the wine (Gutiérrez-Escobar et al., 2021).

The quality of the wine reaching the consumer is decisively determined by the redox processes, carried out in the fermentation processes and the maturation of the wine (Kukec et al., 2002). Thus, the objective of this chapter is to elucidate the main redox processes that take place during the winemaking process and their role in ensuring wine quality. The chapter includes the presentation of participants in redox processes in must and wine: oxidizable substances in musts and wines; the role of transition metal ions in redox processes in wines; causes and aggravating factors of must and wine oxidation; oxidative enzymes; oxidation mechanisms and impact factors.

OXIDIZABLE SUBSTANCES IN MUSTS AND WINES

The main consumers of oxygen in wine are phenolic compounds - 64.6%, then ethyl alcohol - 23.3%, and SO_2 - 12.1% (At the same time, oxygen consumption does not depend on the total concentration of polyphenols in the wine, but on their chemical structure. The phenolic substances found in white wines are part of the C6-C3 group, called cinnamates. Oxygen oxidizes the phenolic substances in the wine to ketones, which are also removed by gravity sedimentation during the technological process. The formed ketones have an increased reactivity and very easily form polymeric combinations, which represent the brown color (Gutiérrez-Escobar et al., 2021). As a result, this subgroup of phenolic compounds causes the oxidation of white wines.

Hydroxycinnamic acids are one of the most representative classes of phenolic acids found in both grapes and wine. *P*-coumaric, caffeic, ferulic, and sinapic acids are some of the most commonly referred compounds. These cinnamic acids are also associated with the pinking of wine, being precursors of volatile phenolic compounds. Cinnamic acids can present isomeric forms -*cis*- or -*trans* - the latter being the most abundant in nature. These isomers are convertible either enzymatically or by the action of light. Natural hydroxycinnamic acids usually occur as esters or glucose molecules. However, these types of derivatives are not present in grapes and derived products (except Riesling wine) being tartaric esters or diesters (hydroxycinnamoyltartaric esters) (Visioli et al., 2020). Hydroxycinnamic acids are considered precursors of some volatile phenols with a negative impact on the aromatic complex of wines. The total concentration of volatile phenols is within the limits of 10-150 mg/l. The cinnamic substances of wines represent several beneficial qualities. In addition to strong antioxidant properties, some of them also exhibit antiseptic action. These compounds are known to be partially responsible for the astringent properties of both grapes and wine. Although white wines have a fairly low concentration of phenolic compounds compared to red wines, white wines in turn show a fairly high degree of hydroxycinnamoyltartaric acids (e.g., caftaric acid). Hydroxycinnamic acids and their tartaric esters are the main class of phenolic compounds in white wines. During the wine fermentation process, free hydroxycinnamic acids result from the partial hydrolysis of these types of esters, which are then transformed into ethyl esters, namely ethyl coumarate and ethyl caffeate.

Dihydroxybenzenes (benzenediols) are organic chemical compounds in which two hydroxyl groups are placed on a benzene ring. There are three isomers: 1,2-dihydroxybenzene (the ortho isomer) is known as catechol, 1,3-dihydroxybenzene (the meta isomer) is known as resorcinol, and 1,4-dihydroxybenzene (the para isomer) is usually known as the name hydroquinone. Similar to other phenols, the hydroxyl groups on the aromatic ring of a benzenediol are weakly acidic. Each benzenediol can lose an H^+ from one of its hydroxyls to form a type of phenolate ion. The Dakin oxidation is an organic redox reaction in which an *ortho-* or *para-*hydroxylated phenylaldehyde or ketone reacts with hydrogen peroxide in the base to form a benzenediol and a carboxylate. In general, the carbonyl group is oxidized and the hydrogen peroxide is reduced.

*Ortho-*diphenols are oxidized to quinone and hydrogen peroxide or lead to a dimer that can in turn react with ortho-diphenol to form polymers that are brown pigments. Hydrogen peroxide reacts with ethanol which turns into ethanal or with tartaric acid which turns into glyoxylic acid. These reactions are very sensitive to the presence of ferric ions, ferrous ions, and cupric ions, which act as catalysts (Nikolantonaki et al., 2014). *O-*quinones can cause the oxidation of other substrates with lower redox potential, such as other phenolic substances, ascorbic acid (AA), and SO_2, and during this process, quinones are reduced to phenols. AA is present in grapes, in amounts up to 50 mg/L, being completely degraded in the process of alcoholic fermentation. Added together with SO_2 in white grapes must, AA provides protection from initial oxidations by reducing the o-quinones. The oxidation of ethanol by the Fenton reaction can be prevented by the interaction of hydrogen peroxide and quinones resulting from the oxidation of catechol with sulfur dioxide (Figure 1).

Figure 1. Catechol oxidation and the preventive action of sulfur dioxide
(Oliveira, 2011)

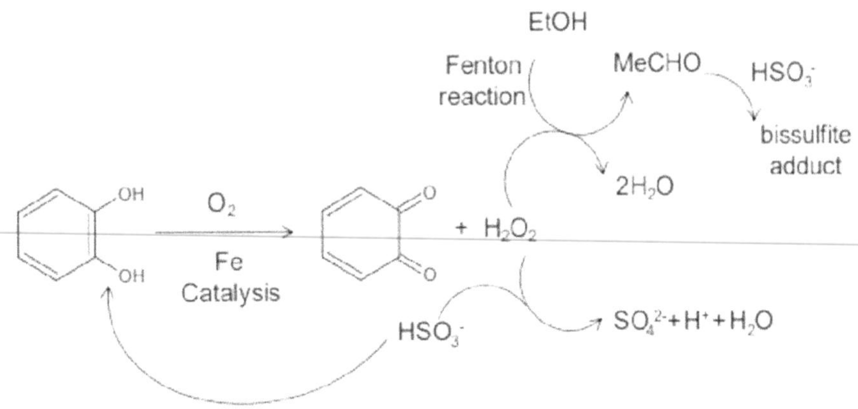

Quinones can be also quenched by nucleophiles, preventing oxidative damage because of the loss of desirable flavor molecules (Ma, 2018). The quinones can also polymerize and condense with many other chemical compounds and lead to the formation of brown pigments. Quinones are strong oxidants, indirectly responsible for the yellowing and degradation of flavors, due to possible interactions, which can lead to the formation of numerous adducts (Figure 2).

Figure 2. Possible interactions of o-quinones with wine compounds (Ito, 2020)

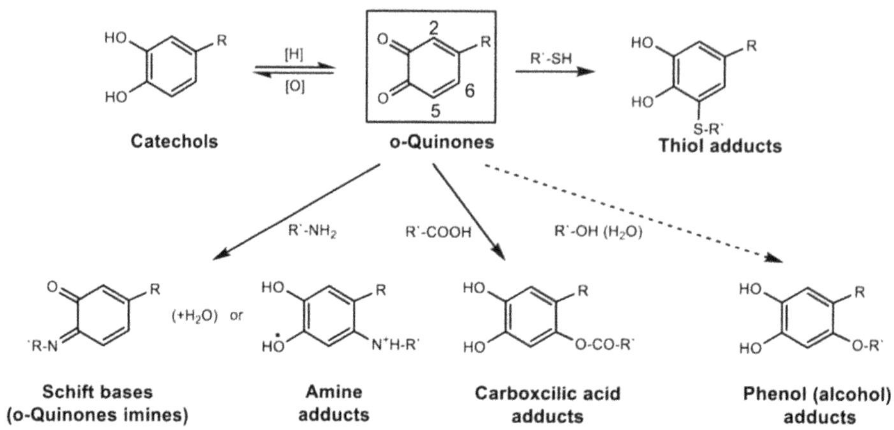

O-quinone, formed from *trans*-caftaric acid, reacts with flavonoids and condenses with other polyphenolic substances to form polymers. These polymers are brown and participate in the oxidative color change of white wines. They determine the organoleptic alteration of wines. *O*-quinones interact with volatile sulfur compounds. This can influence both positively and negatively the aromatic profile of wines: it diminishes varietal aromas, but also reduces unwanted reducing aromas. At the same time, their interaction with amino acids results in the aromatic shades of oxidation through the formation of aldehydes. Reactions with wine tannins cause changes in structure and taste sensations of astringency and bitterness.

Monophenols can also be converted to brown products, but first, they are converted to *o*-diphenols. Below is the general scheme of enzymatic oxidation of phenolic compounds (Figure 3), where the final result is the browning of wines (Tarko et al., 2020).

Figure 3. General scheme of enzymatic oxidation of phenolic compounds

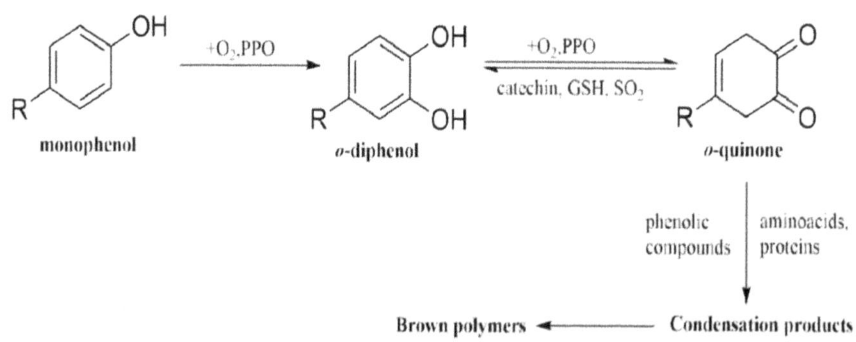

Catechins are found in the seeds and skin of grapes, their extraction in the must depends on the temperature, the period during which the must is in contact with the solid part, and the activity of enzymes and SO_2. The oxidation of catechins depends on their distribution in must and wine. When wines mature, the concentration of catechins decreases or even disappears, because they oxidize during the winemaking process forming brown polymers. Catechins, having two hydroxyl groups present in ring B in the neighboring positions 3' and 4', are oxidized forming brown polymers, according to the chemical reactions presented previously. These reactions are catalyzed in the must until fermentation, by oxidases, and after fermentation when the enzymes have been inactivated, catalysis is ensured by Fe^{2+} and Cu^{2+} transition metal ions (Hernanz et al., 2007).

Low molecular weight catechins and hydroxycinnamic acids form the substrate of polyphenoloxidase. From their oxidation, quinones are formed from which the reaction chain leads to pinking and qualitative degradation of the future wine, therefore the elimination of quinones would increase the life of the wine. The colored compounds that oxidize into macromolecular compounds, sediment, and the wines become clear with the changed color. Along with the degradation of the color, the organoleptic characteristics of the wines are also negatively affected: the taste takes on nuances of cooking, becomes bland, and sometimes bitter. In the aroma, there are nuances of breath, the specific aroma of dried fruits (Sun et al., 2011).

A significant role in the quality of wine, their physico-chemical and organoleptic qualities, in the keeping capacity and the evolution of wines belongs to tannins. In white wines, the concentrations found vary between 0.1-0.5 g/l, and in red wines, their content is 2-5 g/l. Tannin plays an antioxidant role, therefore red wines are more stable against oxidation (Márquez et al., 2019, Piva et al., 2022).

Figure 4. Procyanidin oxidation

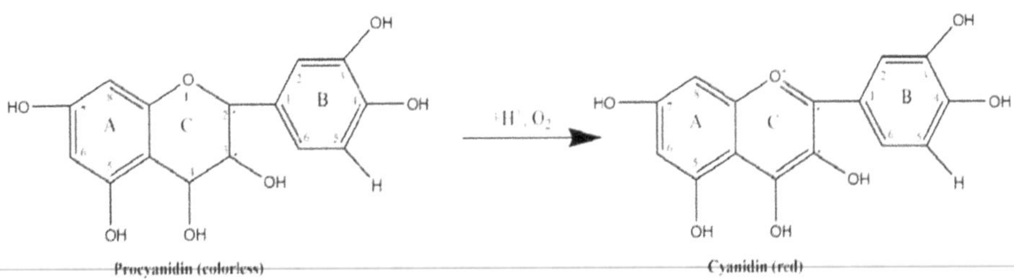

Colorless procyanidins in the presence of the source of active oxygen (singlet oxygen, hydrogen peroxide) in an acidic environment form cyanidin, are responsible for the pinking effect (Figure 4). The transformation of procyanidins into cyanidins depends on the pH of the wine and the concentration of dissolved oxygen (Enaru et al., 2021). Such transformations are imminent in wines produced under reducing or cryomaceration conditions. This presents a real danger for white wines, because, unlike rosé and red wines, their trichromatic changes are essential - the synthesis of already 0.3 mg/l of anthocyanins causes the appearance of a pink tint, perceptible by consumers (Andrea-Silva, 2014).

The easily oxidizable substances in white wines are the polyphenolic substances from the cinnamate group, including caffeic acid, and *p*-coumaric acid, where in white wines they are found in concentrations that do not exceed 100 mg/l but can rarely exceed this value. Cinnamates are chemicals with strong antioxidant properties, but with their oxidation, they lead to negative changes in the organoleptic characteristics (color, taste, aroma) of the wine. The higher the concentration of cinnamic substances (200 mg/l), the greater the risk of oxidation is. Consequently, enzymes belonging to the oxidoreductase class can cause the oxidation of these compounds (especially hydroxybenzoic and hydroxycinnamic acid esters such as coumaric acid and caftaric acid) (Sova & Saso, 2020, Tarko et al., 2020).

Hydrogen peroxide, in the presence of Fe^{2+}, initiates the Fenton reaction (Singleton 1987) and gives rise to very high energy radicals, the hydroxyl radicals $OH^{\bullet}$. Hydrogen peroxide combined with iron ions generates a hydroxyl radical $OH^{\bullet}$ that is capable of oxidizing almost every organic molecule and many inorganic compounds, for example, ethyl alcohol, tartaric acid, glycerol, sugars, and organic acids. Oxidation of ethanol leads to the formation of acetaldehyde, while tartaric acid oxidizes to dihydroxy-fumaric acid, which forms yellow xanthan ions by reaction with (+)-catechin (Petrozziello et al., 2018).

In conclusion, the constituents of wines are capable of reacting with significant amounts of oxygen, polyphenols being among them the most easily oxidized wine constituents. Non-enzymatic oxidation of wine predominates in fermented wine and begins by oxidizing polyphenols containing a catechol or galloyl group. These phenolic reactions result in byproducts called quinones. In non-enzymatic oxidation, oxygen does not directly react with phenolic compounds. Oxidation can be induced by transition metal ions. The initial electron transfer leads to the formation of superoxide radical anion in protonated form (HOO•). The transfer to the second electron will produce hydrogen peroxide (H_2O_2). The next reduction step creates an even more reactive oxidant, the hydroxyl radical (HO•), which can abstract a hydrogen atom from organic compounds (Danilewicz, 2012).

THE ROLE OF TRANSITION METAL IONS IN REDOX PROCESSES IN WINES

In the post-fermentative stages, transition metal ions, with variable degrees of oxidation, act as catalysts. Both according to the concentrations normally found in wines and according to the catalytic capacity, the main transition metals are Fe and Cu, the ionic species of which, under wine conditions, are Fe^{2+}/Fe^{3+} and Cu^{+}/Cu^{2+}, respectively. Metal concentrations, sufficient for the catalytic effect, are lower than the permissible limit concentrations - up to 1 mg/L. The efficiency of the catalysts depends on the oxidation-reduction state of the wine, and the nature and concentration of the metal compounds: with some components of the wines, the metal ions form inactive chelates (Piva et al., 2022).

Iron reduces oxygen to the hydroperoxyl radical or protonated superoxide (Figure 5). The reaction rate of each phenolic compound depends on its ability to form a stable product radical. Compounds containing a 1,2,3-trihydroxyl group (pyrogallol) and a 1,2-aromatic dihydroxy ring (catechol) are the easiest to oxidize because the resulting phenoxyl semiquinone radical can be stabilized by a second oxygen atom (Waterhouse, 2006).

Most of the phenolics in wine are highly reactive toward the hydroperoxyl radical. When hydrogen peroxide reacts with ferrous iron to yield hydroxyl radical, this highly unstable radical reacts almost immediately. Thus, it does not react selectively with antioxidants such as phenolics, but instead reacts with all substances present in the solution, in proportion to their concentration and the redox potential.

Copper is an essential element for the activity of tyrosinase and laccase. To reduce the activity of polyphenoloxidase, it must be removed as quickly as possible to significantly reduce its effects and produce a wine that is more stable to oxidation, fresher in aroma and color, a lasting longer (Márquez et al., 2019). Oxidases are composed of manganese compounds, capable of activating oxygen: manganese

Figure 5. Reductive oxidation pathway of o-phenolics and primary oxidation products (Waterhouse, 2006)

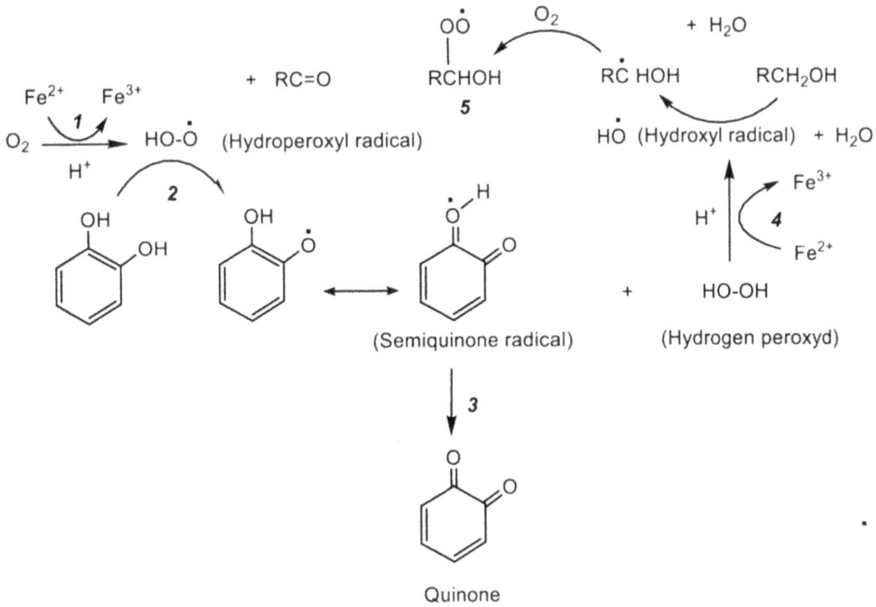

oxide, e.g., it readily autoxidizes to manganese dioxide, which can release atomic oxygen to another substance, reducing to manganese oxide. The iron can be alternatively transformed from a divalent metal (reduction and fixation of molecular oxygen) to a polyvalent metal (oxidation and withdrawal of atomic oxygen). Catalysts can be of inorganic origin (copper or iron ions) or organic (polyphenoloxidases, e.g., laccases are enzymes that use copper); without these catalysts the oxidation of polyphenols is low. All these reactions contribute to the formation of derivatives that participate in the reduction of fruity or floral notes and that make heavier notes appear (honey, sweetness, etc.). Antioxidant properties in a broad sense can be associated with any activity that reduces, in particular, polyphenols and transition metals such as copper and iron. Since copper is a catalyst for chemical oxidation, its use in the vineyard must be as limited as possible but adapted to the protection of the grapes against the development of the pathogenic fungus (Echave et al., 2021).

The copper found naturally in wine has mostly been reported to be formed by yeast during fermentation. However, residual copper concentrations are believed to increase after fermentation depending on the yeast strain used. Improvements in machinery in recent decades have largely eliminated heavy contamination of grapes and wine with iron, except for some accounts of mechanical or automatic harvesting equipment. Depending on the vineyard area, natural iron concentrations found in wine can easily exceed 5 mg/l. There are several other metals in wine (Al^{3+}, Mn^{2+}, and Zn^{2+}) that have been shown to contribute to wine oxidation. Manganese ion, in combination with residual concentrations of copper and iron, was found to further facilitate the oxidative process. Early removal of copper and iron from wine presents an advantage in terms of limiting the wine's capacity for oxidation during vinification and maturation (Pérez-Navarro et al., 2020).

The copper ion reacts with the oxygen directly when the oxygen is dissolved in the wine in the first stage of oxidation. While oxygen does not readily oxidize phenolic compounds including catechin or catechol, it can react directly with the copper ion, initiating a major oxidation process involving the iron ion and di-phenol. Concentrations of up to 0.05 mg/l of copper proved to have a good impact on the development of oxidative processes and the acceleration of the oxidation reaction. Also, increasing copper concentration has been shown to directly influence the rate of oxidation of polyphenolic compounds. The iron ion is the second catalyst in the chemical oxidation of di-phenols in wine. After Cu^{+} is oxidized to Cu^{2+}, Cu^{2+} reacts with Fe^{2+} to form Fe^{3+}. This form of iron can go on to oxidize di-phenols, forming quinones. In addition, this reaction can continue to form acetaldehyde, which contributes to the formation of brown pigments and the loss of flavor and freshness. In the presence of sulfite, these reactions recycle, leading to a build-up of bound sulfite.

The impact of oxygen exposure during winemaking on metal ion concentrations in wine was determined during the winemaking process in white wines. Oxygen exposure affected 13 metal ions (Al^{3+}, Ca^{2+}, Co^{2+}, Cr^{3+}, Cu^{2+}, Fe^{2+}, K^{+}, Mg^{2+}, Mn^{2+}, Na^{+}, Ni^{2+}, Sn^{2+} and Zn^{2+}) at different stages of winemaking (Bekker et al., 2019). However, only Cr^{3+}, Cu^{2+}, and Fe^{2+} concentrations were affected by early oxygen exposure during pressing, with lower Cr^{3+} and Cu^{2+} concentrations in wines that were aerobically pressed, and lower Fe^{2+} concentrations in wines that were inertly pressed. The contact of metal ions such as Al^{3+}, Cu^{2+}, Fe^{2+}, Ni^{2+}, and Zn^{2+} with wine yeast was affected by oxygen treatment. These metal ions have the ability to catalyze redox reactions related to wine oxidation, as well as turbidity and disorder formation. The metal ion that was most affected by oxygen exposure during pressing is the Cu^{2+} ion, whose high concentrations have negative implications for wine taste and aroma. Oxygen management during winemaking has a significant impact on metal ion concentrations in yeast and wine, which can decrease the risk of defects (Estela-Escalante et al., 2012).

Although the connection between oxygen reduction and polyphenol oxidation by iron was considered, the hypothesis that oxygen consumption rates would depend on iron recycling by phenolics was disproved: oxygen consumption was found to occur more rapidly at higher pH and with the inclusion of copper (Nguyen T. H., 2021). The pseudo-first-order rate constants determined for oxygen consumption were lower than those for iron(III) reduction, suggesting that iron(II) oxidation is the rate-determining reaction for wine oxidation.

The metal ion composition of wines can be used to determine wine authenticity, making it essential to understand how winemaking practices can influence residual metal ion profiles in the finished wine. Cu^{2+}, Fe^{2+}, and Mn^{2+} concentrations changed significantly during the fermentation and storage of red wines. Similarly, exogenous factors such as packaging material and storage conditions influence the metal ion concentration of wines, i.e., through the use of screw caps (Sn^{2+}, V^{3+}), stainless steel storage vessels (Cr^{3+}), as well as the storage temperature (Cu^{2+}, Pb^{2+}, Sn^{2-}, V^{3+}) (Hopfer et al., 2012). Since the concentrations of many metal ions are susceptible to changes during winemaking, it is important not to include these metal ions in studies aimed at establishing the geographical origin of a wine. The binding capacity and affinity, the oxidation state of metal ions, as well as the stability of metal ion complexes can all be subject to change because of varying oxygen exposure regimes, which will ultimately affect the final residual concentration of metal ions in wines (Bekker et al., 2019). The increased concentrations of Cu^{2+} and K^{+} at the end of pressing compared to those measured in the must before pressing may occur because of increased extraction from natural metal ions obtained from grape berries or artificial sources such as residual pesticides, fungicides and/or fertilizers introduced into the grape skin during crushing.

In conclusion, iron (II) oxidation is the reaction that determines the rate of oxidation of the wine. A determining factor is the amount of oxygen that can be consumed by the wine before it reaches a certain chemical or sensory threshold. A measure of oxidizing capacity should consider both phenolic and nucleophilic reactivity, presumably based on iron(III) reduction.

PREFERMENTATIVE OXIDATION OF MUST: CAUSES AND AGGRAVATING FACTORS

Prefermentative operations must be carried out as quickly as possible to limit exposure to oxygen, except in cases where hyperoxygenation of the must is used as a technique to remove oxidizable compounds before vinification (Zironi et al., 2010). Although low temperatures slow down oxidative processes, their effect must be taken into account when increasing oxygen solubility. Too low a temperature causes an increase in oxygen solubility, while too high a temperature favors oxidative reactions and the development of microorganisms. Oxygen triggers a series of active catalysts in the presence of air, which are part of the group of polyphenoloxidases, namely laccase and tyrosinase (Killeen et al., 2018). Oxidation begins its activity from the moment the grapes are harvested, especially when the grapes are harvested mechanically, most of the berries are already partially crushed at this stage, starting the oxidative processes, the multiplication of aerobic microorganisms and bacteria. During the transportation of the grapes, the grapes release the must along with the chemicals beneficial to the wine, and by the time they are received at the processing facility, the grapes are damaged. To prevent oxidation and the development of aerobic microorganisms, antioxidant, and antiseptic protection is carried out by administering K metabisulphite in a solid state, in powder form over the grapes in the bunker.

In the case of manual harvesting, the grapes are in a better condition, but it is not excluded that some berries are crushed, and oxidative processes already take place in the grapes. During the transport of the grapes, the risk of oxidation is greater, because due to the weight of the grapes, the skin of the berries bursts, and the must is released, together with the chemical substances that play an important role in the composition of the wines, and from this moment oxygen has free access to the core of the berry. Then processing grapes, the de-stemming/crushing operation presents a high risk of oxidation of the must. For antioxidant but also antiseptic protection of the must, sulfuric anhydride solution in the form of potassium metabisulphite ($K_2S_2O_5$) is administered in doses of 50-75 mg/l, but also ascorbic acid in a dose of 50 mg/l because the antioxidant action of ascorbic acid is much faster than that of metabisulfite (Korenika et al., 2020).

After draining and pressing, the obtained must is circulated in the tank, at the same time there is a very large amount of oxygen in the vessel, so the risk of oxidation of the must is very high. In order to avoid the oxygen source, carbonic ice is administered again, and the must is also sulfite in doses of 25-30 mg/L SO_2 (Esparza et al., 2020). Because the tanks are not hermetically sealed, oxygen will enter the vessel in small quantities, and to stop the oxidative processes, the wort must be at a low but constant temperature, considering that the solubility of oxygen increases when temperatures are low. During decantation, the must is circulated to another tank, in the presence of the inert atmosphere, to consume the oxygen dissolved in the wine and to stop the oxidative processes. To start alcoholic fermentation, small amounts of oxygen are needed to activate the yeasts, which are introduced through micro-oxygenation. After activating the yeasts, it is desirable to avoid contact with oxygen in the must, in order not to oxidize different compounds in the wine, which consequently change the organoleptic profile of the young wine.

During the prefermentative stages, the polyphenoloxidases act as catalysts for the oxidative processes, but during alcoholic fermentation, these catalysts disappear. At the post-fermentative stages, transition metal ions (Fe, Cu, Mn) will be catalysts that will continue the oxidative processes in young wines.

The role of oxygen during winemaking is extremely complex. This is because this element is used by both yeasts and bacteria and is a substrate for numerous chemical transformations of must and wine during maturation. The presence of oxygen during the fermentation phase changes the aromatic profile of the resulting wine. The extent of these changes depends mainly on the grape variety, the composition of the must, and the amount of oxygen introduced. The main precursors of volatile substances with aroma perception are esters of unsaturated fatty acids (Zironi et al., 2010, Tarko et al., 2020). Oxidation of the must can lead to higher concentrations of chemical compounds that favor wine aroma, such as C6 compounds, higher alcohols, fatty acids, acetates, and their ethyl esters. These compounds are responsible for freshness and the formation of fruity notes.

Ascorbic acid (AA) is present in grapes in amounts up to 50 mg/L, being completely degraded in the alcoholic fermentation process. Added together with SO_2 in white grapes (must), AA provides protection against initial oxidations by reducing *o*-quinones formed according to the scheme of action of SO_2 (figure 6).

Figure 6. Protection against initial oxidations by reduction of o-quinones by AA

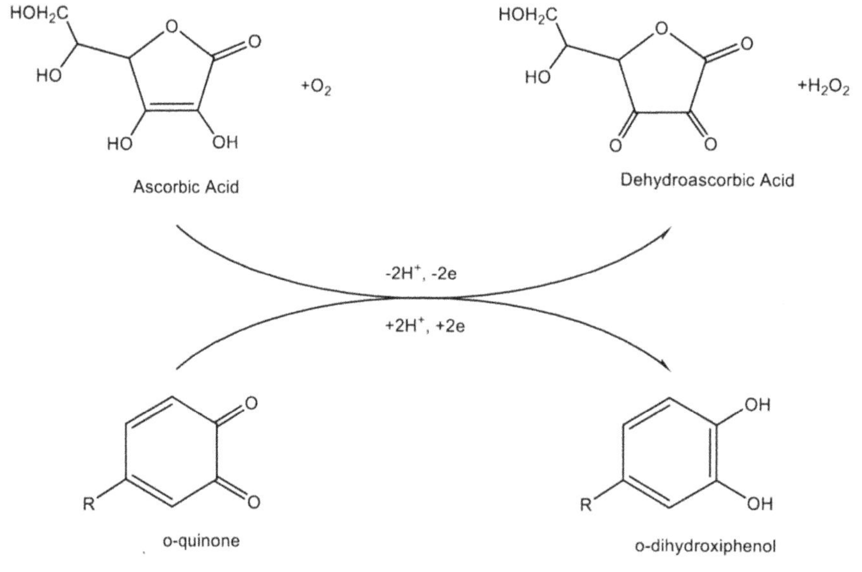

The transformation of ascorbic acid into dehydroascorbic acid is catalyzed by Fe^{2+} and Cu^{2+} ions. AA also has a reducing effect on Fe^{3+} ions, keeping them in their reduced form and preventing the oxidation of polyphenols.

Simultaneously with the oxidation of AA in the presence of oxygen, the formation of hydrogen peroxide takes place, which is a much more active oxidant than the oxygen itself (in the usual triplet form).

So, AA would play the role of a prooxidant, if its administration were not accompanied by additions of SO_2, which are "sacrificed" in the reaction with H_2O_2 and thus confirm the reducing role of AA:

$$SO_2+2H_2O_2 \rightarrow H_2SO_4 + 2H_2O \quad (1)$$

In the absence of sulfur dioxide, the redox processes with the participation of ascorbic acid, catalyzed by metal ions, proceed with the formation of ethoxy radicals, which amplify the oxidation processes (Barril, 2016). The presence of SO_2 causes the formation of the reversible adduct with dehydroascorbic acid, with weak bonds, which degrades into several products such as furoic acid and 3-hydroxy-2-pyrone, but also a number of unidentified minor products (Figure 7).

Figure 7. Reaction of AA with oxygen in the presence of sulfur dioxide
(Barril, 2016)

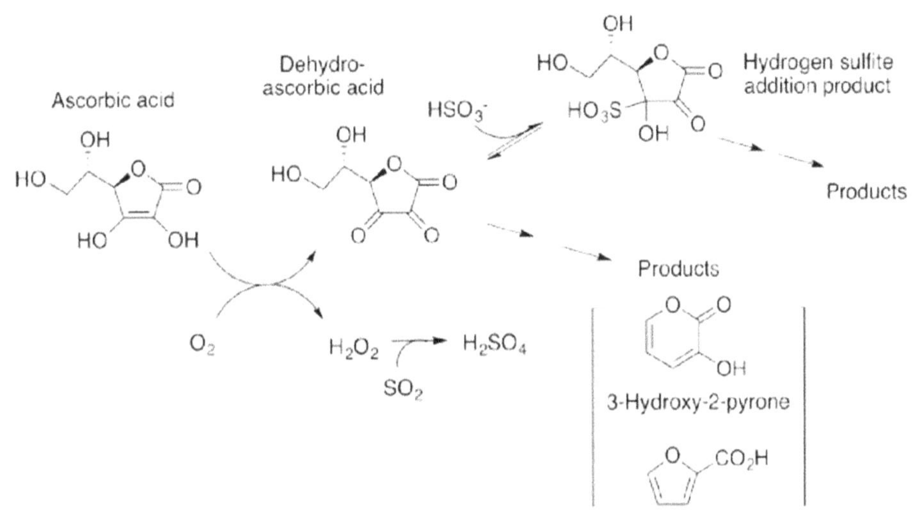

AA reduces the negative olfactory impact of disulfides (dithiols), transforming them into thiols with a higher perception threshold, less irritating and more volatile, so more easily removed from the wine by aeration.

Processing grapes under reducing conditions results in higher concentrations of esters, which support the fruity expression specific to the variety. Processing grape must without sulfur dioxide induces enzymatic oxidation and precipitation of phenolic compounds as insoluble brown pigments. But wines obtained from oxidized musts show more resistance to oxidative quality degradation during aging. Hyperoxygenation makes use of deliberate oxidation before fermentation to improve the shelf life of wines. The sensory results depend in particular on the removal of the involved flavonoids.

OXIDATION OF WINES: MEASURES TO PREVENT/ REDUCE THE EFFECTS OF OXIDATION

Unwanted oxidation processes, which can greatly change the quality of the wine, must be avoided. The oxygen concentration in the wine is measured at different stages (production, storage, bottling). The quality of the wine is influenced especially by its oxygen content. Oxidation can occur in all types of wines and can affect all sensory characteristics (Ribéreau-Gayon et al., 2006):

- Changing the color;
- Development of aerobic microorganisms (yeasts, bacteria);
- Loss of freshness and flavor;
- Fast aging, due to the oxygen present in the bottled wine.

Oxygen in the presence of redox couple Fe^{3+}/Fe^{2+} forms free radicals and quinones, triggering chain reactions that affect polyphenols, alcohols, and aromatic compounds. The formation of acetaldehyde from ethyl alcohol is the quantitatively most important Strecker reaction that contributes to the "oxidized" note of wine (Figure 8).

Figure 8. Scheme of the mechanism of non-enzymatic oxidation of polyphenols, alcohols and aromatic compounds in wine
(Zironi, 2010)

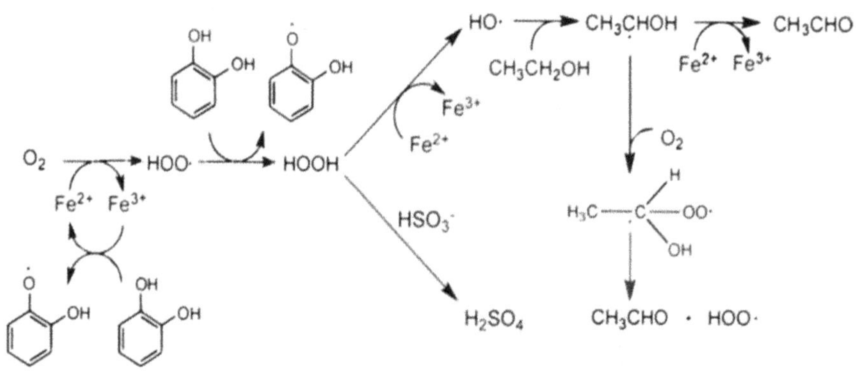

Phenylacetaldehyde and methional, which have a much lower perception threshold than acetaldehyde, are formed by the same chemical mechanism. Oxygen also contributes to the formation of polymeric substances with the subsequent formation of sediments in wine.

Wine oxidation is a very complex phenomenon of chemical reactions caused by the contact of wine with oxygen. The oxidation controlled by the winemaker is a chemical process, which ensures the fixing of the oxygen necessary for various technological processes, which would ensure the improvement of the quality of the wines or the improvement of some processes such as fermentation and maturation. Also, the directed oxidation process is found in the production of Heres skin wines, and the production of Madeira and Marsala special wines (Francesca et al., 2014). It should be noted that, in some cases, oxygen access to the must in a controlled manner can have positive effects. In the case of red wine production, oxygen during the maceration stage allows an increased extraction of phenolic compounds from the berry skin. Exposure to oxygen, although generally considered detrimental to white wine, also has positive effects (Englezos et al., 2018, Tarko et al., 2020).

Contact with oxygen in large doses gives the wine a natural alteration, and in small doses in a short time, oxygen is beneficial to the wine, enriching it with different chemical compounds and favoring the organoleptic characteristics during maturation. In the case of high doses of oxygen, the wine will then lose its freshness and color along with the taste and aroma will change (Petrozziello et al., 2018). In enology, it is commonly accepted that advanced oxidation is rather unfavorable, while the slow and continuous dissolution of oxygen is meant to play a positive role in the evolution of wines (Ailer et al., 2022). Oxygen has an essential role in the evolution of wine, from the beginning of the winemaking process to the moment of consumption. The presence of oxygen directly influences the way the wine evolves, both from a chemical point of view and from the point of view of its sensory attributes (the evolution of its organoleptic characteristics): aroma, taste, structure, and color. Therefore, it is essential to properly check oxygen during the winemaking process, knowing that it is difficult to measure the rate (Müller & Rauhut, 2018). The concentration of dissolved oxygen is a function of the matrix, in which it is dissolved, and the temperature (Saa et al., 2012). The reaction of polyphenols with oxygen is mediated by iron and copper (Figure 9).

Figure 9. Catalytic action of iron and copper ions in the oxidation of catechols to quinones
(Danilewicz et al., 2008)

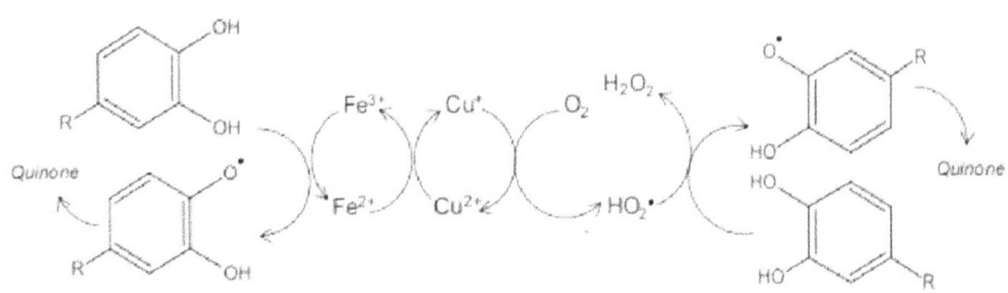

Quinones formed from the oxidation of polyphenols are unstable and may undergo further reactions: can combine spontaneously with nucleophilic compounds, including some phenols, thiols and amines. The oxidation of wine changes its organoleptic qualities, which can be desirable, in the case of maturation, inheritance, or maderization, and undesirable when they already become wine defects, in the case of premature oxidation, oxidative breakdown, or the pinking effect of white wines.

Oxygen is supplied through contact with air, especially during the winemaking operations (crushing, pressing, pumping, decanting, centrifugation, gluing, cooling, etc.). Oxygen is also useful for several elements in winemaking: to multiply yeasts at the initial stage of fermentation, to avoid reduction defects, and to stabilize polyphenols (Mekoue Nguela et al., 2019). Maderization is an oxidation phenomenon that transforms the flavors of wine. The specific aromas are first of green apples (characteristic of acetaldehyde), then green walnuts and dried walnuts. Some wines benefit from taking on oxidized aromas. Among them are natural sweet wines. The term maderization refers to a sweet wine. Indeed, the sugar in the wine balances the oxidation flavors. The initial oxidation then becomes a factor in the complexity of the wine (Bueno et al., 2018).

During alcoholic fermentation, thanks to the nutrition of the yeasts, various reactions of reducing sulfur to hydrogen sulfide take place, since the yeasts are not rich in vitamins and nitrogen (Spiropoulos & Bisson, 2000). At the same time, oxidative processes can occur, which as a result lead to the formation of the browning effect of musts or wines. With the browning of the must/wine, part of the chemical substances sediments, they can be excluded with sorbents, but the exclusion of these substances leads to the impoverishment of the wine in aromatic and taste substances, because the added sorbents, usually less specific, can remove substances that contribute to the aroma, color, taste. Oxidation of wines results in loss of clarity and fruitiness (wines lacking freshness).

Red wine needs a certain amount of oxygen during maturation in order to develop. At the end, more oxygen must be introduced to rejuvenate the red wine, if it has an unpleasant aroma (hydrogen sulfide and its compounds) and if the tannins are hard and harsh. This negative effect appears immediately after alcoholic fermentation. During the bottling stage, no more oxygen needs to be added, but during the storage of the wine in the bottles, it may need oxygen in small quantities for the development of the aromatic complex, therefore the bottles are topped with gas dosing plugs, which have a semi-permeable membrane which allows limited, controlled access of oxygen in the wine (Echave et al., 2021).

Under certain circumstances, oxygen can also have positive effects on the winemaking process:

- at the beginning of alcoholic fermentation, oxygen is used by the yeasts before reacting chemically with the compounds of the wine;
- moderate oxygenation of the wine can prevent the appearance of off-odors, caused by the formation in a strongly reducing environment of sulphide compounds that can mask the secondary aromas of the wine;
- In red wines, acetaldehyde favors the stabilization of anthocyanins and the polymerization of tannins, making the wine softer.

Consequently, total exclusion of oxygen is not a solution for winemaking. Since the oxidation reactions in wine are irreversible, the degraded compounds cannot return to their original state even if the wine is stored in the absence of oxygen, the only way to preserve the quality of a wine is to prevent oxidation.

Unfavorable Oxidation

Oxygen, through its influence on the numerous oxidation-reduction equilibria in wine, is responsible for a good part of the changes in the chemical composition of wine. Other defects may also occur, such as acetic, ferric, polyphenolic, and oxidazic scraping. But in this case, oxygen is a substrate for microorganisms or a promoter of chemical reactions (du Toit et al., 2017).

The wines, which should reach close to their optimal aging, show signs of oxidation. This late phenomenon appears in the bottle, however, at a stage where the producer can no longer act. This manifestation in wine can be attributed to a reduction in antioxidant protection (low level of SO_2, low amount of yeast during vinification, defective corks) and more attacks (agitation during aging in barrels, numerous pumpings, and aerations) (Echave et al., 2021, Tedesco et al., 2022).

Measures to Prevent Oxidation

To prevent the oxidation defect, it is necessary to act from the beginning of the technological process, with the reception of the raw material grapes at the processing unit (Flieger et al., 2021). Prevention of oxidation can be done with the administration of antioxidant agents such as sulfur dioxide or ascorbic acid and dry ice. Carbonic ice is a solid form of CO_2, in the form of granules, at a temperature of -80°C, which dissolves in must or wines, being used both for antioxidant protection, and also to prevent the development of microorganisms and spontaneous fermentation. The administration of this protector takes place as needed in the grape supply hopper for the destemming crusher, where it depends on the physiological state of the grapes, when pumping the must for maceration or pressing, also the administration takes place during draining and pressing.

At the same time, it is important that antioxidant protection also refers to the stage of care and maturation of raw material wines, especially when the wines are matured in oak barrels (Bartkovský et al., 2020). Maturation in oak barrels leads to the release of CO_2 in small quantities and the decrease of the volume of the wine, this way oxygen has access to the barrel, and if a very important technological process is not carried out at the maturation stage, the oxidation of the volume will take place of barrel wine. To prevent this moment, it is necessary to fill the voids, i.e. fill the barrels with wine to block the access of oxygen and avoid oxidation of the wine. The winemaker must pay attention to the storage conditions of the wine, in this sense, in addition to the temperature and ventilation system, the humidity plays a very important role in preventing the oxidation of the wine in the bottles. If the humidity will be higher than 80%, there is a risk of the appearance and development of pathogenic fungi on the cork of the bottle, where the wine is enriched with shades of mold. Likewise, if the humidity will be less than 60%, then the glass stoppers will lose their quality over time due to dry air, the cork will become more sensitive and the bottles will not be hermetically closed, from this moment oxygen will have access to the bottle, as result, over time will cause wine oxidation or pinking and browning effects. To prevent oxidation in the bottle during the wine storage stage, the humidity must be between the optimal values of 65-75%, so that the cork in the bottle is always hermetically closed, to limit the access of oxygen in the gas tank.

From ancient amphorae to 21st-century bottles, producers and consumers have always tried to protect wine from oxygen. In the cellar, the winemaker takes care to keep the barrels always full, especially the barrels that require regular filling. The principle of inerting makes it possible to replace oxygen with another gas, such as carbon dioxide, nitrogen, or argon (CO_2, N_2, Ar), and thus oxidation is avoided. It can be done during pressing. During transfers, it is possible to inert the line before pumping, and pump the wine into the line with the same gases when the operation is finished. To prevent the access of oxygen in tanks that are not full of wine and have a certain void, they can also be filled with neutral gas or dry ice (Carpena et al., 2020).

Measures to Reduce the Effects of Oxidation

At the beginning of alcoholic fermentation, yeasts need oxygen for their multiplication and activation. Oxygen can be added by micro-oxygenation. Micro-oxygenation is a relatively recent technological process (1980s) for a regular and slow, controlled supply of oxygen in wine. In the case of micro-oxygenation, we do not try to remove oxygen from the wine but introduce it in a reasoned way and in small doses over time (of the order of a few milligrams/l/month). This process allows very fine control of the organoleptic qualities of the wine. Micro-oxygenation is known to remove herbaceous and vegetal characteristics, bringing out fruity and varietal characteristics. Micro-oxygenation stabilizes the color and can reduce astringency and bitterness over time. By controlling the oxygen content, this process is also a good ally for the effective reduction of reduced tastes, while respecting the fruitiness of white or red wines (Lesica & Košmerl, 2009).

Microoxygenation can also be carried out after alcoholic and malolactic fermentation, but also during the maturation period of the wines. After alcoholic fermentation, it allows the creation of chemical bonds, e.g., between anthocyanins and tannins, and the reduction of vegetal characteristics. This aspect is very important for structuring and stabilizing red wines, enriching them with a creamy taste and round consistency. The classic values are between 10 mg/l/month and 40 mg/l/month for 3 to 15 days. Micro-oxygenation during the maturation period allows the structuring of the wines, and the polymerization of the phenolic compounds with a stabilization of the color intensity, it also contributes to the aromatic potential of the wines (Cejudo-Bastante et al., 2011).

The dose of added oxygen is a difficult aspect of microoxygenation and involves avoiding the accumulation of dissolved oxygen. High amounts of oxygen in wine can cause (Cano-López et al., 2008):

- oxidation of aldehydes, which can reduce the organoleptic characteristics;
- the accumulation of oxygen in the free space of the tank and the multiplication of harmful aerobic microorganisms.
- oxidative degradation of numerous substances, which affect the color, taste, and stability of the wine.

Below are presented the amounts of oxygen, with which the wines are enriched following some technological operations carried out in the winery (Cano-López et al., 2008):

- careful transvasation in the presence of inert gas – 0.1-0.2 mg/L;
- less careful transvasation in the presence of inert gas – max. 5 mg/L;
- centrifugation – 1 mg/L;
- closed tributary – 2.2 mg/L;
- open flow - 7.4 mg/L;
- bottling – max. 1 mg/L.

During intensive wine aeration in 30 sec., at a temperature of 20°C, it dissolves max of 8.5 mg/L molecular oxygen. This amount increases with decreasing temperature. The molecular oxygen dissolved in the wine is consumed in reactions with various chemical compounds in the wine (Estela-Escalante et al., 2012). The speed of the process depends on the temperature. Thus, after saturation at a temperature of 20°C, oxygen is consumed for 5 days and at a temperature of 13°C – for 25 days.

Depending on the amount of oxygen administered, the process can take place in micro- or macro-oxygenation mode. Conventionally, it has been established that microoxygenation presents a dose of 10 mg/L of oxygen, for a period of 30 days, starting from the 10-30th day after the alcoholic fermentation process. The added oxygen is gradually absorbed and decreases the reduction reactions and the formation of some unwanted chemical compounds. The macro-oxygenation process presents an oxygen dose between 3-5 mg/L/day, for a period of 1-2 days, during and at the end of alcoholic fermentation. Oxygen enables the yeast to finish the alcohol fermentation process, preventing the appearance of unpleasant aromas and tastes. Likewise, it favors the formation of new phenolic substances, thus maintaining the natural color of the wine (Morales et al., 2015), (Tarko et al., 2020).

Hyperoxygenation is the only preferential technological technique in which oxygen is supplied in very high doses (more than 20 mg/L). The principle of hyperoxygenation is to quickly oxidize the phenolic compounds in the must, before alcoholic fermentation, which can later (after vinification) convey herbaceous and bitter notes (Barbateskovic et al., 2021). After this technological technique, the wines turn out to be fresher and more stable in terms of color. A disadvantage of this vinification technique would be that with the oxidation of the polyphenolic compounds, other beneficial chemical substances in the wine are also oxidized, which changes the qualities of the wine from an organoleptic point of view. Hyperoxygenation may be involved in creating the desired sensory characteristics, associated with the removal of flavonoids (such as catechin, epicatechin, and procyanidins B2 and B3) whereas, in white wine, they could also be responsible for the sensation of excessive acidity and astringency (Rihak et al., 2022).

Hyperoxygenation can cause both favorable and unfavorable flavor changes depending on the grape variety. For some types of wine - Chardonnay, Parrelada, and Muscat from Spain, some improvement in the aroma was observed through the formation of higher concentrations of six-carbon compounds (hexan-1-ol, 2-hexenal), higher alcohols, aromatics (2-phenylethanol), fatty acids, acetate and their ethyl esters (hexyl, isoamyl, 2-phenylethyl acetate and butyrate, ethyl decanoate and hexanoate), as well as volatile terpenes that favorably modify the aroma of these wines. These chemical compounds are responsible for the freshness of wines, associated with the formation of fruity notes (Morales et al., 2015). In the case of Chardonnay, Mauzac, and Chenin wines from France, no changes in aroma were observed, while wines produced from several European varieties resulted in lower aroma intensity due to the formation of higher acetates and aldehydes and lower concentrations of higher alcohols (Tarko et al., 2020). On the other hand, other studies have indicated that wines produced using hyperoxygenation may lose their characteristic fruity aroma (Pérez-Navarro et al., 2020).

Hyperoxygenation is one of the methods of preventing the formation of an unfavorable brown color in white wine (Hernanz et al., 2007). It involves the exposure of the must to oxygen, resulting in the rapid conversion of phenolic compounds, especially proanthocyanidins, contained in the white must into brown polymerized compounds, as a result of the activity of polyphenoloxidase. These compounds can then be easily removed during the clarification or fermentation step. The concentration of 9 mg O_2/L is sufficient for the flavonoids to be precipitated at a concentration lower than 100 mg/L. Since the oxidation of phenolic compounds in the finished wine causes a brown color change, which is unacceptable to the consumer, their removal by intentionally initiating the enzymatic oxidation reaction at the pre-fermentation stage removes the substrates involved in non-enzymatic oxidation reactions. White wines produced by this method are resistant to pinking, even after bottling (Schneider, 1998, Rihak et al., 2022).

OXIDATIVE ENZYMES

In the musts, the oxidation process begins with the action of the polyphenol oxidase enzyme (PPO). These are enzymes naturally present in grapes (tyrosinase) or produced by pathogenic fungi on grapes (laccase produced by *Botrytis cinerea*) which in the presence of oxygen are able to oxidize the phenolic compounds of grapes in a few minutes and form quinones. Quinones produced by enzymatic oxidation can oxidize other polyphenols.

An important actor of the redox processes occurring in musts and young wines is glutathione (GSH) - an endogenous sulfur tripeptide of grapes. Possessing the –SH group, glutathione can be present in both reduced form (properly, GSH) and oxidized form (GSSG). The latter is obtained through an oxidative dimerization through the -SH group with the formation of the -S-S- bridge (Figure 10).

Figure 10. Oxidative dimerization of glutathione

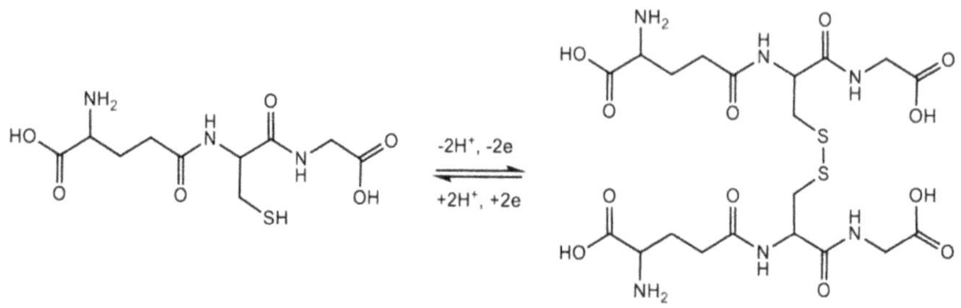

The reducing role of glutathione is particularly relevant in the production of white wines, with a higher content of easily oxidizable polyphenols - cinnamic acids and their tartaric esters, trans-caffeoyltartaric (caftaric) and trans-coumaroyltartaric (cutaric) acids. PPO catalyzes the conversion of these compounds with the incorporation of glutathione into a specific reaction product (Figure 11).

When caftaric acid is oxidized to the corresponding quinone, glutathione will rapidly react with the quinone to form a colorless product called grape reaction product (GRP) (Kritzinger et al., 2012). GRPs are colorless, stable and resistant to the action of PPOs, thus protecting cinnamic acids from oxidation and, implicitly, from the formation of brown oxidized products. Therefore, must glutathione play an essential role in the oxidation mechanism and reduction of wine browning and protection of various wine aroma compounds.

PPOs, which occur naturally in grapevine grapes, have various activities: o-phenol hydroxylase causes the conversion of monophenol to monocatechol, and catecholase participates in the oxidation of catechol to brown melanin, while laccase catalyzes the oxidation of p-hydroquinones to p-benzoquinones (Tarko et al., 2020). The two most common types of oxidative enzymes are peroxidases, which use hydrogen peroxide, and oxidases, which use molecular oxygen. Oxidative enzymes are responsible for the oxidation of the must, in the presence of the oxygen source (Danilewicz, 2016).

PPOs are responsible for the pinking of the must in case of unwanted oxidation during winemaking. The grape's own tyrosinase is largely removed during clarification. Oxidases are very sensitive to heat and are inhibited when the must is heat treated. The laccase species, which occurs when the weather is rainy and the grapes are affected by mold, remains in larger quantities and is also resistant to sulfur dioxide. Oxidases catalyze the oxidation of C–N and C–O bonds at the expense of molecular oxygen, which is reduced to hydrogen peroxide. The three main classes of substrates for oxidase enzymes are amino acids, amines, and alcohols. Amino acids and amines are oxidized to the corresponding imine, while alcohols are oxidized to aldehydes or ketones. Polyphenoloxidases (PPOs) contain Cu^{2+} and catalyze the oxidation of several phenols to o-quinones. In turn, o-quinones are highly reactive molecules that can undergo non-enzymatic side reactions to form complex brown polymers known as melanins and cross-linked polymers with protein functional groups. The conversion of phenolic substrates to o-quinones by PPO occurs via two oxidation steps. The first step is the hydroxylation of the *ortho*-position adjacent to an existing hydroxyl group, "monophenoloxidase" or "monophenolase" activity, also called hydroxylase or cresolase activity. The second step is the oxidation of *o*-dihydroxybenzenes to *o*-benzoquinones, "diphenoloxidase" or "diphenolase activity", also called catecholase activity or oxidase.

Figure 11. Mechanism of oxidation and production of grape reaction product in grape must and wine (Managing wine quality, 2010)

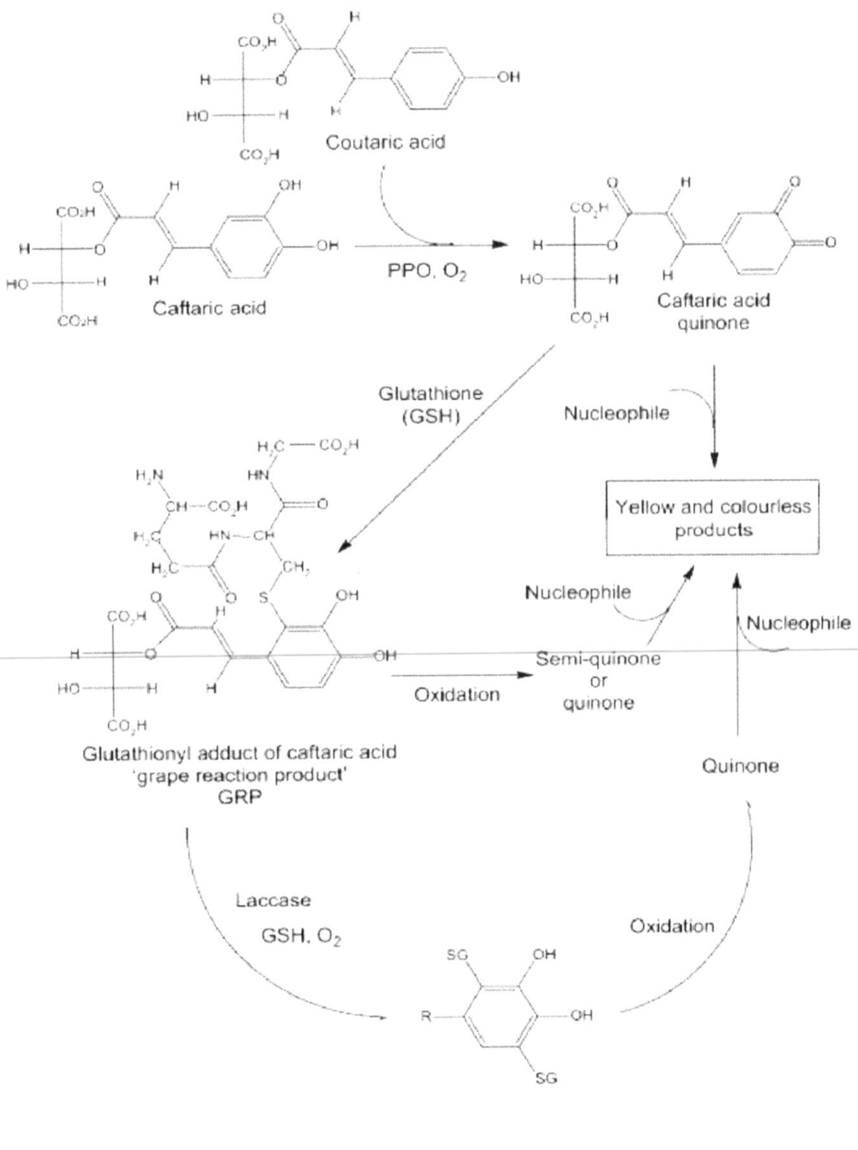

Based on their specific substrate and mechanism of action, PPOs are classified into three different types: tyrosinases, catecholoxidases, and laccases. Tyrosinases have both cresolase and catecholase activities. Catecholoxidases, also known as *o*-diphenoloxidases, catalyze the oxidation of *o*-diphenols to *o*-quinones. Finally, laccases are capable of oxidizing a wide spectrum of aromatic compounds through a radical-catalyzed reaction mechanism. The optimum pH of PPO in grapes depends on the grape variety, ranging from 3.5 to 7.3 and 4.5 to 7.0, respectively. Temperature is another important factor that significantly affects the catalytic activity of PPO, as it influences the solubility of oxygen and can lead to the denaturation of the enzyme. The optimal temperature range of PPO in grapes ranges from 25–45°C (Taranto et al., 2017).

The grape reaction product (2-S glutathione caftaric acid) is an oxidation compound produced by the action of PPO on caftaric acid and found in wine. This compound production is responsible for the lower level of rosiness in certain white wines. Tyrosine is a key compound of the pigmentation reaction through the action of polyphenoloxidase. The discrimination of this pigmentation reaction is responsible for various types of diseases and breakdowns of wins (Ünal & Şener, 2016). In the wine industry, inhibition of tyrosinase is appropriate because tyrosinase catalyzes the oxidation of phenolic compounds in must to quinones, which gives an undesirable taste and color. Must browning can result from enzymatic and non-enzymatic oxidation, both of which result in the formation of *o*-quinones. In general, monomeric procyanidins and catechins are associated with a higher degree of browning than other phenols. Enzymatic browning depends on tyrosinases and laccases, which are produced by grape berry biosynthesis and molds, respectively, and is associated with the content of hydroxycinnamates such as caffeoyltartaric acid (caftaric acid) and *p*-coumaroyltartaric acid (cutaric acid). In addition, peroxidases (PODs) can enhance the degradation of phenols when interacting with PPO. Non-enzymatic oxidation is due to the conversion of polyphenols, such as (+)-catechin or (-)-epicatechin, gallic acid, and caffeic acid, which are the most easily oxidizable wine constituents (Tarko et al., 2020, Gutiérrez-Escobar et al., 2021).

Laccase is found in plants, fungi, and bacteria, but is produced in grapes by *Botrytis cinerea*. The enzyme is considered environmentally friendly because it requires molecular oxygen as a co-substrate for catalysis and gives water as the only by-product. Other laccases produced by fungi can facilitate the biosynthesis of melanin pigments. They are some of the few oxidoreductases commercialized as industrial catalysts (Brugnari et al., 2021). The optimal temperature of fungal laccases can vary in a range of 40-70°C. Wines that are produced from mold-infected grapes risk enzymatic oxidative processes, irreversibly affecting their organoleptic properties. One of the most well-known applications of laccase in the food industry is to stabilize the wine by controlling phenolic compounds. Maintaining flavor in wines is fundamental to quality control. In addition, oxidative reactions that can occur due to the complex mixture of chemicals in wines can even intensify the color, especially in red wines. Chemical methods to remove phenolic compounds, such as the addition of polyvinylpolypyrrolidone (PVPP) and sulfur dioxide, have been used to stabilize wines and prevent loss of flavor and color quality. Since laccase has good stability at low pH, it is possible to reverse its sulfite inhibition, so enzyme treatment of wines is widely used. At the same time, it is also possible to treat white wine with laccase, representing an alternative to ensure the quality of wines that remain stored for a long time, avoiding their damage and reducing their production costs.

PPO from grapes and must show an activity reduction of about 50% after 20 minutes at 65°C and a complete inactivation after 15 minutes at 75°C, but in this case, the wine will receive different organoleptic changes, as e.g., the loss of varietal aromas and the appearance of cooking nuances. β-Cyclodextrin binds non-covalently to PPO better than other polysaccharides and has relevant inhibition capacity under optimal conditions. Chemical inhibitors that act differently, such as reducing compounds (ascorbic acid and sulfites), chelators (ethylenediaminetetraacetate), acid (citric and malic), and complexing compounds (cyclodextrins), inhibit melanin formation by preventing the accumulation of *o*-quinones or by forming stable of colorless products (Scutaru et al., 2018).

Several polyphenols, including flavonoids or stilbenoids, free radical scavengers, and copper chelators inhibit tyrosinase. Tyrosinase inhibitors are as follows: azelaic acid, hydroquinone, arbutin, kojic acid, ascorbic acid, ellagic acid, tranexamic acid, coumarin, vanillic acid, vanillin, and vanillic alcohol (Wang et al., 2022). There are two types of PPO inhibitors, those competitive with the oxygen between the copper reaction of the enzyme and those competitive with phenolic substances. Tropolone is an inhibitor of grape polyphenol oxidase. Another inhibitor of this enzyme is metabisulfite. *O*-quinones can further transform depending on their redox potential and electron affinity. They can react with glutathione (GSH), which occurs naturally in grapes, resulting in the formation of 2-S-glutathionyl caftaric acid (GRP). Non-enzymatic oxidation reactions resulting in the formation of *o*-quinones can also be observed in must, but due to the slow speed of the process, enzymatic oxidation plays an incomparably more important role. The effect on non-enzymatic oxidation reactions is based on the reaction with hydrogen peroxide, but not directly on the reaction with oxygen (at typical wine pH, this reaction occurs very slowly and is not significant) (Kuijpers et al., 2013).

The transformations of flavonoids in non-enzymatic oxidation reactions can lead to the loss of the characteristic aroma. As a result of the oxidation reactions that take place in wine, there are changes in aroma, color, and taste that can have both positive and negative meanings. It is important that the amount of oxygen that comes into contact with the wine is adequate and that the transformations take place over a relatively long period of time. Non-enzymatic oxidation is believed to contribute to the decline in white wine quality, due to pinking and negative flavor changes, but in the case of red wines, it allows an intense and stable color and improves taste properties. Red wines receive favorable transformations when oxygenated with more than 60 mg/l, while their quality decreases after the introduction of 150 mg/l O_2. Non-enzymatic oxidation of polyphenols is a process that occurs due to the presence of compounds with a catecholic ring, such as (+)-catechin, (-)-epicatechin, (+)-catechin gallate, gallic acid, and its esters, caffeic acid and anthocyanins (Gutiérrez-Escobar et al., 2021).

In chemical oxidation, oxygen does not directly react with phenolic compounds. Followed by the addition of a single transition electron of metal ions to oxygen, which leads to the formation of the superoxide anion O^{2-}, which at wine pH is present as a hydroperoxyl radical $HOO^{\bullet}$. This radical causes the oxidation of phenolic compounds to quinones and can be reduced to hydrogen peroxide.

It is generally accepted that the wine oxidation mechanism starts with the oxidation of certain polyphenols, *o*-diphenols, in the presence of Fe^{3+} ions. In a mechanism with many similarities, in musts, enzymes whose active site is a Cu^{2+} ion catalyze this first step. At this point in the mechanism everything becomes complicated because the two products of this oxidation reaction of polyphenols are unstable, they are the ones that will lead to the appearance of signs of oxidation by creating chain reactions. Oxidation mechanisms have been widely studied by scientists and the generally accepted hypothesis is that polyphenols are the molecules that are oxidized first (Perron & Brumaghim, 2009).

COMPOUNDS FROM WHITE WINES SENSITIVE TO OXIDATION

The general characteristic of oxidized white wine is the yellow color, oxidized wine aroma and increasing bitterness. The most common compound associated to oxidized wine aroma is the acetaldehyde that has green apple, nuts aroma due to the chemical oxidation of ethanol. The easily oxidizable substances in white wines are the polyphenolic substances from the cinnamate group, including caffeic acid, and *p*-coumaric acid, where in white wines they are found in concentrations that do not exceed 100 mg/l but can rarely exceed this value. Cinnamates are chemicals with strong antioxidant properties, but with their oxidation, they lead to negative changes in the organoleptic characteristics (color, taste, aroma) of the wine. The higher the concentration of cinnamic substances (200 mg/l), the greater the risk of oxidation is. Consequently, enzymes belonging to the oxidoreductase class can cause the oxidation of these compounds, especially hydroxybenzoic and hydroxycinnamic acid esters such as coumaric acid and caftaric acid (Figure 12) (Sova & Saso, 2020, Tarko et al., 2020).

The production of acetaldehyde by catalase is limited by the availability of hydrogen peroxide, a potentially harmful byproduct of ethanol metabolism by cytochrome P450. Hydrogen peroxide can also come from several other sources, including the enzyme monoamine oxidase, ascorbic acid (vitamin C), and other cytochrome P450 oxidations (Petrozziello et al., 2018). As a result of the oxidation of white wines, their qualities change abundantly, which concerns the visual qualities (wine appearance, color, clarity), gustatory qualities (taste), and olfactory qualities (aroma). The visual and gustatory qualities change during the oxidation process, by changing the color e.g., pinking or browning of the wine, by changing the taste, e.g., flat taste, pickling taste. The olfactory qualities are influenced by the oxidation-reduction process by changing the aroma towards odor defects, e.g., smell of vinegar, boiled eggs, or sauerkraut.

An oxidative risk with essential changes in the organoleptic properties of white wine is presented by its pinking, which can be the result of the rapid conversion of accumulated flavones into red flavylium salts, formed by the hydrolysis of leucoanthocyanins (Figure 12).

Pinking is the terminology used for the red color that can appear in white wines made exclusively from white grape varieties. The contact of the wine with oxygen also has a significant impact on its color. Oxygen in the case of white wine causes unfavorable pinking, which is caused by three oxidation mechanisms (Tarko et al., 2020):

1. It causes the oxidation of phenolic compounds to the corresponding quinones, which are then polymerized, resulting in yellow-brown pigments. This reaction involves iron and copper ions that catalyze oxidation reactions. The main compounds associated with the pinkness of white wine are not only (+)-catechin, (-)-epicatechin, and dimers of proanthocyanidins B1-B4, but also coumarin and stearic, ferulic, and caffeic acids.
2. It oxidizes tartaric acid to glyoxylic acid by forming bonds between phenolic compounds and causing their condensation, and copper and iron ions are catalysts. The colorless mixtures of the two (+)-catechin molecules linked by the carboxymethyl bond, as a result of dehydration, are transformed into yellow-brown xanthylium salts.
3. It oxidizes phenolic compounds, resulting in the generation of hydrogen peroxide, which in turn oxidizes ethanol to acetaldehyde. Acetaldehyde forms bonds between (+)-catechin molecules (giving ethyl-linked dimers) and also between dimers resulting from the oxidation of tartaric acid. This leads to the formation of brown pigments, the intensity of which is proportional to the degree of polymerization.

Figure 12. Origin of pinking in wines
(Andrea-Silva, 2014)

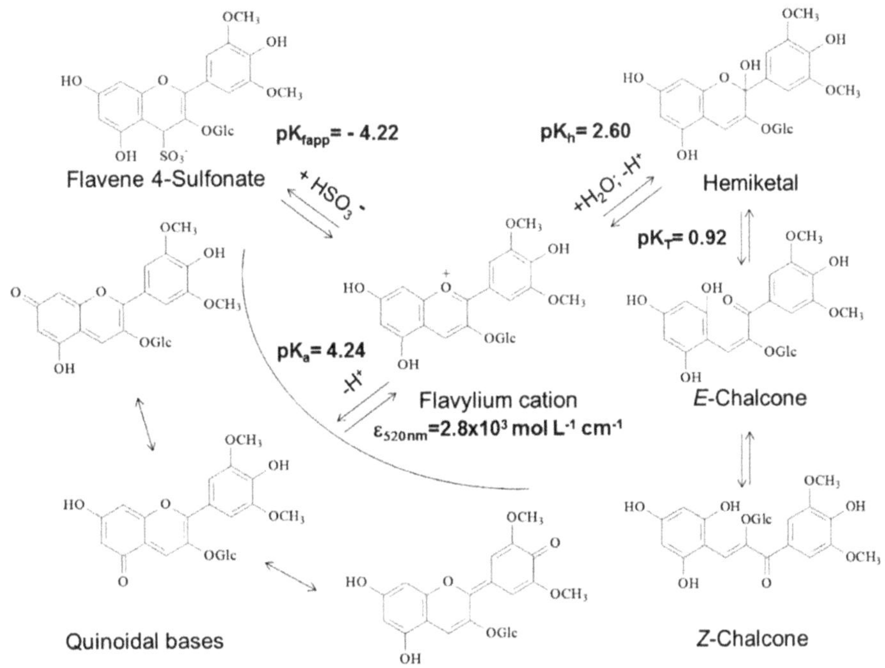

Pink compounds resistant to pH changes and sulfur dioxide decolorization are formed when the level of free sulfur dioxide decreases, leading to an increase in the relative amount of the anthocyanin red flavylium form and its polymerization. However, this staining can be caused by other compounds and polymeric materials. The pink pigments could also be derived from 2-S-glutathione-caftaric acid, formed as a result of the reaction of glutathione with o-quinones (produced by the enzymatic oxidation of caffeoylacetic acid and *p*-coumaric acid). The subsequent aging of such wine in bottles determines its final pinking (Kanavouras et al., 2019).

Due to condensation reactions that are mediated by oxygen, the resulting polymerized colorants fall as sediment and cause unacceptable changes in the appearance of the wine. Oxidized white wine also shows notes of oxidation, perceived olfactory and gustatory, accompanied by the typical bitterness and lacking fruity or floral notes, a color evolution with the appearance of yellow or pinkish shades, an aromatic evolution described as a loss of varietal notes (wood of boxwood) and an appearance of so-called oxidative notes (honey, wax, chocolate, stone, dried fruit, dried grass, nuts, etc.) but sometimes also a change in taste balance with increased astringency or loss of fat (Abreu et al., 2021).

The tendency to pinking is not the same for all white wines. This depends on the variety, the redox potential of the wine, the level of polyphenols, oxidative enzymes, the winemaking system, etc. Leucoanthocyanins play an important role in the appearance of the pink color in white wines. Recent research claims that malvidin-3-O-glucoside is the most abundant monomeric anthocyanin found in pinked wines and could be the cause of pinking.

When whole bunches are pressed, this phenomenon is more likely to occur because the stems have high concentrations of phenolic compounds. Excessive ripening of grapes, low-temperature treatment of grapes and must, reductive vinification, and high pH are risk factors for the further development of the "pinking" phenomenon.

To eliminate the risk of pinking, it is recommended to age the white wine on yeast under reductive conditions, which reduces the sensitivity to pinking. The use of high doses of PVPP (100 g/Hl) reduces the sensitivity of white wines to pinking by 74%. If PVPP is combined with bentonite (50 g/hl) or ascorbic acid (15 mg/l), this reduction can reach 90-98% (Lamuela Raventos, 2001). The use of ascorbic acid at bottling, in a dose of 45 mg/l, completely prevents the risk of the "pinking" phenomenon (Nel, 2020).

CONCLUSION

The processes of oxidation and reduction occur continuously during grape processing and wine production. The presence of oxygen during alcoholic fermentation helps yeast cells' function. But large amounts of oxygen exposure cause the wine to naturally develop; yet brief exposure to oxygen in moderate doses during the maturation process helps the wine retain its organoleptic qualities. While the changes of flavonoids in non-enzymatic oxidation reactions might result in the loss of the characteristic aroma, in red wines they produce an intense and stable color and enhance the flavor. However, redox phenomena are crucial to the specific winemaking procedures used to produce fortified wines.

Oxygen-mediated condensation reactions result in polymerized colorants like brown (from the oxidation of catechins, orthodiphenols, and hydroxycinnamic acids) and red (from the oxidation of procyanidins), which create undesirable changes in the wine's color. Polyphenoloxidases are the most significant enzymes in charge of the enzymatic oxidation of phenolic compounds in must. The oxidation of phenolics in the presence of Fe^{3+} ions and the presence of Cu^{2+} ions in the active site of the enzymes in musts are the first steps in the wine oxidation process.

Acting at the outset of the technological process is essential to preventing the oxidation issue. A lot goes into preventing oxidation, including the harvesting method, light exposure, temperature, ventilation system, and humidity (65-75%). As with micro-oxygenation and sulfur dioxide addition, inerting enables the prevention of oxidation. It is possible to utilize hyperoxygenation to keep white wine from developing an undesirable shade of brown. In the case of faults caused by metal ions, oxygen management is also effective.

The complexity of the mechanisms involved in the oxidation of wine, despite the impressive number of research carried out in this direction, is not yet fully elucidated. It is necessary to identify all mediator reactions and characterize them, including by applying chemometrics. Global warming, which leads to an increase in the sugar content of grapes, brings certain changes in traditional winemaking schemes, which require the reevaluation of redox processes, metabolites and antioxidants used to protect wines against oxidation.

REFERENCES

Abreu, T., Perestrelo, R., Bordiga, M., Locatelli, M., Coïsson, J. D., & Câmara, J. S. (2021). The flavor chemistry of fortified wines - a comprehensive approach. *Foods*, *10*(6), 1239. doi:10.3390/foods10061239 PMID:34072391

Ailer, Š., Jakabová, S., Benešová, L., & Ivanova-Petropulos, V. (2022). Wine Faults: State of Knowledge in Reductive Aromas, Oxidation and Atypical Aging, Prevention, and Correction Methods. *Molecules (Basel, Switzerland), 27*(11), 3535. doi:10.3390/molecules27113535 PMID:35684472

Andrea-Silva, J., Cosme, F., Ribeiro, L. F., Moreira, A. S. P., Malheiro, A. C., Coimbra, M., Domingues, R. M., & Nunes, F. M. (2014). Origin of the Pinking Phenomenon of White Wines. *Journal of Agricultural and Food Chemistry, 62*(24), 5651–5659. doi:10.1021/jf500825h PMID:24857316

Barbateskovic, M., Schjørring, O. L., Krauss, S. R., Jakobsen, J. C., Meyhoff, C. S., Dahl, R. M., Rasmussen, B. S., Perner, A., & Wetterslev, J. (2021). Higher versus lower fraction of inspired oxygen or targets of arterial oxygenation for adults admitted to the intensive care unit. *The Cochrane Library, 33*(4). Advance online publication. doi:10.1002/14651858.CD012631.pub2 PMID:34251144

Barril, C., Rutledge, D. N., Scollary, G. R., & Clarc, A. C. (2016). Ascorbic Acid and White wine production: a review of beneficial versus detrimental impacts, 169-181. Australian Society of Viticulture and Oenology Inc.

Bartkovský, M., Semjon, B., Marcinčák, S., Turek, P., & Baričičová, V. (2020). Effect of wine maturing on the colour and chemical properties of chardonnay wine. *Czech Journal of Food Sciences, 38*(4), 223–228. doi:10.17221/139/2019-CJFS

Bekker, M. Z., Day, M. P., & Smith, P. A. (2019). Changes in metal ion concentrations in a Chardonnay wine related to oxygen exposure during vinification. *Molecules (Basel, Switzerland), 24*(8), 1523. doi:10.3390/molecules24081523 PMID:30999713

Bekker, M. Z., Smith, M. E., Smith, P. A., & Wilkes, E. N. (2016). Formation of hydrogen sulfide in wine: Interactions between copper and sulfur dioxide. *Molecules (Basel, Switzerland), 21*(9), 1214. doi:10.3390/molecules21091214 PMID:27626394

Berovič, M., & Nemanič, J. (2019). The role of redox potential measurement in oak barrel wine maturation. *Chemical and Biochemical Engineering Quarterly, 33*(1), 153–160. Advance online publication. doi:10.15255/CABEQ.2018.1469

Briedis, V., Povilaitytė, V., Kazlauskas, S., & Venskutonis, P. R. (2003). *Polifenolių ir antocianinų kiekis vynuogėse, vynuogių sultyse ir raudonuose vynuose bei jų antioksidacinio aktyvumo įvertinimas.*

Brugnari, T., Braga, D. M., dos Santos, C. S. A., Torres, B. H. C., Modkovski, T. A., Haminiuk, C. W. I., & Maciel, G. M. (2021). Laccases as green and versatile biocatalysts: From lab to enzyme market—an overview. *Bioresources and Bioprocessing, 8*(1), 131. doi:10.118640643-021-00484-1

Bueno, M., Marrufo-Curtido, A., Carrascón, V., Fernández-Zurbano, P., Escudero, A., & Ferreira, V. (2018). Formation and Accumulation of Acetaldehyde and Strecker Aldehydes during Red Wine Oxidation. *Frontiers in Chemistry, 6*, 20. doi:10.3389/fchem.2018.00020 PMID:29492401

Cano-López, M., Pardo-Mínguez, F., Schmauch, G., Saucier, C., Teissedre, P. L., López-Roca, J. M., & Gómez-Plaza, E. (2008). Effect of micro-oxygenation on color and anthocyanin-related compounds of wines with different phenolic contents. *Journal of Agricultural and Food Chemistry, 56*(14), 5932–5941. doi:10.1021/jf8006147 PMID:18558704

Carpena, M., Pereira, A. G., Prieto, M. A., & Simal-Gandara, J. (2020). Wine aging technology: Fundamental role of wood barrels. *Foods*, *9*(9), 1160. doi:10.3390/foods9091160 PMID:32842468

Cejudo-Bastante, M. J., Pérez-Coello, M. S., & Hermosín-Gutiérrez, I. (2011). Effect of wine micro-oxygenation treatment and storage period on colour-related phenolics, volatile composition and sensory characteristics. *Lebensmittel-Wissenschaft + Technologie*, *44*(4), 866–874. doi:10.1016/j.lwt.2010.10.015

Celotti, E., Lazaridis, G., Figelj, J., Scutaru, Y., & Natolino, A. (2022). Comparison of a Rapid Light-Induced and Forced Test to Study the Oxidative Stability of White Wines. *Molecules (Basel, Switzerland)*, *27*(1), 326. doi:10.3390/molecules27010326 PMID:35011556

Danilewicz, J. C. (2012). Review of oxidative processes in wine and value of reduction potentials in enology. *American Journal of Enology and Viticulture*, *63*(1), 1–10. doi:10.5344/ajev.2011.11046

Danilewicz, J. C. (2016). Reaction of oxygen and sulfite in wine. *American Journal of Enology and Viticulture*, *67*(1), 13–17. doi:10.5344/ajev.2015.15069

de Simone, N., Pace, B., Grieco, F., Chimienti, M., Tyibilika, V., Santoro, V., Capozzi, V., Colelli, G., Spano, G., & Russo, P. (2020). Botrytis cinerea and table grapes: A review of the main physical, chemical, and bio-based control treatments in post-harvest. *Foods*, *9*(9), 1138. doi:10.3390/foods9091138 PMID:32824971

du Toit, W. J., Marais, J., Pretorius, I. S., & du Toit, M. (2017). Oxygen in Must and Wine: A review. *South African Journal of Enology and Viticulture*, *27*(1). doi:10.21548/27-1-1610

Echave, J., Barral, M., Fraga-Corral, M., Prieto, M. A., & Simal-Gandara, J. (2021). Bottle aging and storage of wines: A review. *Molecules (Basel, Switzerland)*, *26*(3), 713. doi:10.3390/molecules26030713 PMID:33573099

Enaru, B., Drețcanu, G., Pop, T. D., Stănilă, A., & Diaconeasa, Z. (2021). Anthocyanins: Factors affecting their stability and degradation. *Antioxidants*, *10*(12), 1967. doi:10.3390/antiox10121967 PMID:34943070

Englezos, V., Cravero, F., Torchio, F., Rantsiou, K., Ortiz-Julien, A., Lambri, M., Gerbi, V., Rolle, L., & Cocolin, L. (2018). Oxygen availability and strain combination modulate yeast growth dynamics in mixed culture fermentations of grape must with Starmerella bacillaris and Saccharomyces cerevisiae. *Food Microbiology*, *69*, 179–188. doi:10.1016/j.fm.2017.08.007 PMID:28941899

Esparza, I., Martínez-Inda, B., Cimminelli, M. J., Jimeno-Mendoza, M. C., Moler, J. A., Jiménez-Moreno, N., & Ancín-Azpilicueta, C. (2020). Reducing SO_2 doses in red wines by using grape stem extracts as antioxidants. *Biomolecules*, *10*(10), 1369. doi:10.3390/biom10101369 PMID:32992977

Estela-Escalante, W., Rychtera, M., Melzoch, K., Hatta-Sakoda, B. (2012). Effect of aeration on the fermentative activity of saccharomyces cerevisiae cultured in apple juice. *Revista Mexicana de Ingeniera Quimica, 1*(2).

Flieger, J., Flieger, W., Baj, J., & Maciejewski, R. (2021). Antioxidants: Classification, natural sources, activity/capacity measurements, and usefulness for the synthesis of nanoparticles. *Materials (Basel)*, *14*(15), 4135. doi:10.3390/ma14154135 PMID:34361329

Francesca, N., Sannino, C., Settanni, L., Corona, O., Barone, E., & Moschetti, G. (2014). Microbiological and chemical monitoring of Marsala base wine obtained by spontaneous fermentation during large-scale production. *Annals of Microbiology*, *64*(4), 1643–1657. doi:10.100713213-014-0808-0

Gutiérrez-Escobar, R., Aliaño-González, M. J., & Cantos-Villar, E. (2021). Wine polyphenol content and its influence on wine quality and properties: A review. *Molecules (Basel, Switzerland)*, *26*(3), 718. doi:10.3390/molecules26030718 PMID:33573150

Hernanz, D., Recamales, Á. F., González-Miret, M. L., Gómez-Míguez, M. J., Vicario, I. M., & Heredia, F. J. (2007). Phenolic composition of white wines with a prefermentative maceration at experimental and industrial scale. *Journal of Food Engineering*, *80*(1), 327–335. doi:10.1016/j.jfoodeng.2006.06.006

Hopfer, H., Ebeler, S. E., & Heymann, H. (2012). The combined effects of storage temperature and packaging type on the sensory and chemical properties of chardonnay. *Journal of Agricultural and Food Chemistry*, *60*(43), 10743–10754. doi:10.1021/jf302910f PMID:23035911

Ito, S., Sugumaran, M., & Wakamatsu, K. (2020). Chemical Reactivities of *ortho*-Quinones Produced in Living Organisms: Fate of Quinonoid Products Formed by Tyrosinase and Phenoloxidase Action on Phenols and Catechols. *International Journal of Molecular Sciences*, *21*(17), 6080. doi:10.3390/ijms21176080 PMID:32846902

Kanavouras, A., Karanika, E., Frank, A. C., Kotseridis, Y., & Kallithraka, S. (2019). Preliminary study of flavor compounds as oxidation markers in bottled white wines of Greek origin. *OENO One*, *53*(3). doi:10.20870/oeno-one.2019.53.3.2439

Killeen, D. J., Boulton, R., & Knoesen, A. (2018). Advanced monitoring and control of redox potential in wine fermentation. *American Journal of Enology and Viticulture*, *69*(4), 394–399. Advance online publication. doi:10.5344/ajev.2018.17063

Korenika, A. M. J., Biloš, J., Kozina, B., Tomaz, I., Preiner, D., & Jeromel, A. (2020). Effect of different reducing agents on aromatic compounds, antioxidant and chromatic properties of sauvignon blanc wine. *Foods*, *9*(8), 996. Advance online publication. doi:10.3390/foods9080996 PMID:32722317

Kritzinger, E., Bauer, F., & du Toit, W. (2012). Role of Glutathione in Winemaking: A Review. *Journal of Agricultural and Food Chemistry*, *61*(2), 269–277. doi:10.1021/jf303665z PMID:23240621

Kuijpers, T. F. M., Gruppen, H., Sforza, S., van Berkel, W. J. H., & Vincken, J. P. (2013). The antibrowning agent sulfite inactivates Agaricus bisporus tyrosinase through covalent modification of the copper-B site. *The FEBS Journal*, *280*(23), 6184–6195. doi:10.1111/febs.12539 PMID:24112416

Kukec, A., Berovic, M., Celan, S., & Wondra, M. (2002). The role of on-line redox potential measurement in Sauvignon blanc fermentation. *Food Technology and Biotechnology*, *40*(1).

Lamuela Raventos, R. M., Huix-Blanquera, M., & Waterhouse, A. (2001). Treatments for Pinking Alteration in White Wines. *American Journal of Enology and Viticulture*, *52*(2), 156–158. doi:10.5344/ajev.2001.52.2.156

Lesica, M., & Košmerl, T. (2009). Microoxygenation of red wines. *Acta Agriculturae Slovenica*, *93*(3).

Liu, C.-G., Qin, J.-C., & Lin, Y.-H. (2017). *Fermentation and Redox Potential*. Fermentation Processes. doi:10.5772/64640

Ma, L., & Waterhouse, A. L. (2018). Flavanols react preferentially with quinones through an electron transfer reaction, stimulating rather than preventing wine browning. *Analytica Chimica Acta*, *18*(1039), 162–171. doi:10.1016/j.aca.2018.07.013 PMID:30322547

Márquez, K., Contreras, D., Salgado, P., & Mardones, C. (2019). Production of hydroxyl radicals and their relationship with phenolic compounds in white wines. *Food Chemistry*, *271*, 80–86. doi:10.1016/j.foodchem.2018.07.165 PMID:30236744

Mekoue Nguela, J., Vernhet, A., Julien-Ortiz, A., Sieczkowski, N., & Mouret, J. R. (2019). Effect of grape must polyphenols on yeast metabolism during alcoholic fermentation. *Food Research International*, *121*, 161–175. doi:10.1016/j.foodres.2019.03.038 PMID:31108737

Morales, P., Rojas, V., Quirós, M., & Gonzalez, R. (2015). The impact of oxygen on the final alcohol content of wine fermented by a mixed starter culture. *Applied Microbiology and Biotechnology*, *99*(9), 3993–4003. doi:10.100700253-014-6321-3 PMID:25582558

Müller, N., & Rauhut, D. (2018). Recent developments on the origin and nature of reductive sulfurous off-odours in wine. *Fermentation (Basel, Switzerland)*, *4*(3), 62. doi:10.3390/fermentation4030062

Nel, A. P., du Toit, W. J., & van Jaarsveld, F. P. (2020). Pinking in White Wines -A Review. *South African Journal of Enology and Viticulture*, *41*(2), 151. 10.21548/41-2-3952

Nguyen, T. H., & Waterhouse, A. L. (2021). Redox Cycling of Iron: Effects of Chemical Composition on 2 Reaction Rates with Phenols and Oxygen in Model Wine. [AJEV]. *American Journal of Enology and Viticulture*, *72*(3), 209–216. doi:10.5344/ajev.2021.20024

Nikolantonaki, M., Magiatis, P., & Waterhouse, A. L. (2014). Measuring protection of aromatic wine thiols from oxidation by competitive reactions vs wine preservatives with ortho-quinones. *Food Chemistry*, *163*, 61–67. doi:10.1016/j.foodchem.2014.04.079 PMID:24912696

Oliveira, C. M., Ferreira, A. C. S., De Freitas, V., & Silva, A. M. S. (2011). Oxidation mechanisms occurring in wines. *Food Research International*, *44*(5), 1115–1126. Advance online publication. doi:10.1016/j.foodres.2011.03.050

Pérez-Navarro, J., Izquierdo-Cañas, P. M., Mena-Morales, A., Chacón-Vozmediano, J. L., Martínez-Gascueña, J., García-Romero, E., Hermosín-Gutiérrez, I., & Gómez-Alonso, S. (2020). Comprehensive chemical and sensory assessment of wines made from white grapes of vitis vinifera cultivars albillo dorado and montonera del casar: A comparative study with airén. *Foods*, *9*(9), 1282. doi:10.3390/foods9091282 PMID:32932602

Perron, N. R., & Brumaghim, J. L. (2009). A review of the antioxidant mechanisms of polyphenol compounds related to iron binding. *Cell Biochemistry and Biophysics*, *53*(2), 75–100. doi:10.100712013-009-9043-x PMID:19184542

Petrozziello, M., Torchio, F., Piano, F., Giacosa, S., Ugliano, M., Bosso, A., & Rolle, L. (2018). Impact of increasing levels of oxygen consumption on the evolution of color, phenolic, and volatile compounds of Nebbiolo wines. *Frontiers in Chemistry*, *6*(APR), 137. doi:10.3389/fchem.2018.00137 PMID:29755971

Piva, G. G., Casalta, E., Legras, J. L., Tesnière, C., Sablayrolles, J. M., Ferreira, D., Ortiz-Julien, A., Galeote, V., & Mouret, J. R. (2022). Characterization and Role of Sterols in Saccharomyces cerevisiae during White Wine Alcoholic Fermentation. *Fermentation (Basel, Switzerland)*, *8*(2), 90. doi:10.3390/fermentation8020090

Elsevier. (2010). *Managing Wine Quality: Viticulture and Wine Quality*. Woodhead Publishing Series in Food Science, Technology and Nutrition.

Ribéreau-Gayon, P., Glories, Y., Maujean, A., & Dubourdieu, D. (2006). The chemistry of wine stabilization and treatments. In: Handbook of Enology, (2).

Rihak, Z., Prusova, B., Kumsta, M., & Baron, M. (2022). Effect of must hyperoxygenation on sensory expression and chemical composition of the resulting wines. *Molecules (Basel, Switzerland)*, *27*(1), 235. doi:10.3390/molecules27010235 PMID:35011467

Saa, P. A., Moenne, M. I., Pérez-Correa, J. R., & Agosin, E. (2012). Modeling oxygen dissolution and biological uptake during pulse oxygen additions in oenological fermentations. *Bioprocess and Biosystems Engineering*, *35*(7), 1167–1178. doi:10.100700449-012-0703-7 PMID:22349928

Saerens, S. M. G., Delvaux, F. R., Verstrepen, K. J., & Thevelein, J. M. (2010). Production and biological function of volatile esters in Saccharomyces cerevisiae. *Microbial Biotechnology*, *3*(2), 165–177. doi:10.1111/j.1751-7915.2009.00106.x PMID:21255318

Schneider, V. (1998). Must hyperoxidation: A review. *American Journal of Enology and Viticulture*, *49*(1), 65–73. doi:10.5344/ajev.1998.49.1.65

Scutaru, I., Balanuta, A., & Zgardan, D. (2018). The determination of oxidation behavior of white wines produced from local and European grape varieties using spectrophotometric method. *Journal of Engineering Science*, *25*(4), 82–93.

Scutaru, I., Sturza, R., & Gherdelescu, L. (2022). Oxidative-reducing processes in winemaking. *Modern Technologies, in the Food Industry – 2022: Proc. of the Intern. Conf.* (pp. 84–85). UTM.

Sova, M., & Saso, L. (2020). Natural sources, pharmacokinetics, biological activities and health benefits of hydroxycinnamic acids and their metabolites. *Nutrients*, *12*(8), 2190. doi:10.3390/nu12082190 PMID:32717940

Spiropoulos, A., & Bisson, L. F. (2000). MET17 and hydrogen sulfide formation in Saccharomyces cerevisiae. *Applied and Environmental Microbiology*, *66*(10), 4421–4426. doi:10.1128/AEM.66.10.4421-4426.2000 PMID:11010893

Sturza, R., Scutaru, I., Duca, G. (2021). Quality Management of Wines and Redox Processes. *Advances in Intelligent Systems and Computing*. AISC. . doi:10.1007/978-3-030-49889-4_45

Sun, B., Neves, A. C., Fernandes, T. A., Fernandes, A. L., Mateus, N., de Freitas, V., Leandro, C., & Spranger, M. I. (2011). Evolution of phenolic composition of red wine during vinification and storage and its contribution to wine sensory properties and antioxidant activity. *Journal of Agricultural and Food Chemistry*, *59*(12), 6550–6557. doi:10.1021/jf201383e PMID:21561162

Taranto, F., Pasqualone, A., Mangini, G., Tripodi, P., Miazzi, M. M., Pavan, S., & Montemurro, C. (2017). Polyphenol oxidases in crops: Biochemical, physiological and genetic aspects. *International Journal of Molecular Sciences*, *18*(2), 377. doi:10.3390/ijms18020377 PMID:28208645

Tarko, T., Duda-Chodak, A., Sroka, P., & Siuta, M. (2020). The Impact of Oxygen at Various Stages of Vinification on the Chemical Composition and the Antioxidant and Sensory Properties of White and Red Wines. *International Journal of Food Sciences*, *2020*, 1–11. doi:10.1155/2020/7902974 PMID:32309422

Tedesco, F., Siesto, G., Pietrafesa, R., Romano, P., Salvia, R., Scieuzo, C., Falabella, P., & Capece, A. (2022). Chemical Methods for Microbiological Control of Winemaking: An Overview of Current and Future Applications. *Beverages*, *8*(3), 58. doi:10.3390/beverages8030058

Ünal, M. Ü., & Şener, A. (2016). Correlation between browning degree and composition of important Turkish white wine grape varieties. *Turkish Journal of Agriculture and Forestry*, *40*(1), 62–67. doi:10.3906/tar-1412-54

Vignault, A., Pascual, O., Jourdes, M., Moine, V., Fermaud, M., Roudet, J., Canals, J. M., Teissedre, P. L., & Zamora, F. (2019). Impact of enological tannins on laccase activity. *OENO One*, *53*(1). doi:10.20870/oeno-one.2019.53.1.2361

Visioli, F., Panaite, S. A., & Tomé-Carneiro, J. (2020). Wine's phenolic compounds and health: A pythagorean view. *Molecules (Basel, Switzerland)*, *25*(18), 4105. doi:10.3390/molecules25184105 PMID:32911765

Wang, W., Gao, Y., Wang, W., Zhang, J., Yin, J., Le, T., Xue, J., Engelhardt, U. H., & Jiang, H. (2022). Kojic Acid Showed Consistent Inhibitory Activity on Tyrosinase from Mushroom and in Cultured B16F10 Cells Compared with Arbutins. *Antioxidants*, *11*(3), 502. doi:10.3390/antiox11030502 PMID:35326152

Waterhouse, A. L., & Laurie, V. F. (2006). Oxidation of wine phenolics: A critical evaluation and hypotheses. *American Journal of Enology and Viticulture*, *57*(3), 306–313. doi:10.5344/ajev.2006.57.3.306

Zironi, R., Comuzzo, P., Tat, L., Scobioala, S. (2010). Oxygen and wine. *Internet Journal of Viticulture and Enology, 3*(2).

Chapter 16
Degradation of Rhodamine B Using CdS Quantum Dots Under Solar Light

Sivakumar Krishnamoorthy
https://orcid.org/0000-0002-1930-3250
Sri Chandrasekharendra Saraswathi Viswa Mahavidyalaya (Deemed), India

Poongodi Thanigaivel
Sri Chandrasekharendra Saraswathi Viswa Mahavidyalaya (Deemed), India

ABSTRACT

Plate-like cadmium sulfide (CdS) quantum dots were prepared by using β-cyclodextrin as reductants. Photoluminescence and UV-Visible diffused reflectance spectroscopic techniques are used to investigate the optical property of CdS. The blue-shifted emission of CdS at 469nm than the bulk particle is observed. The calculated band gap of the CdS quantum dot is 2.62 eV showing a difference of 0.2 eV energy greater than that of bulk, indicating the quantum size effect. The synthesized quantum dot has a size in the range of 3.5nm with a zinc blend crystallite structure. The plate-like nanoclusters of CdS are inferred from SEM images. The photocatalytic efficiency of prepared CdS was tested for the degradation of rhodamine B dye using natural solar light. The effect of the weight of the catalyst, concentration of dye, pH of the solution, and reusability of the catalyst were investigated. The existence of isosbestic point observation of a high intense absorbance peak at 550nm at 300min in different matrix circumstances confirms the formation of leuco Rh B.

INTRODUCTION

Background of the problem

Rhodamine B (Rh B) is a dye molecule used for various industrial and medical needs such as dyeing, printing, textiles, pharmaceuticals, and chemicals, etc., Although, Rh B causes adverse effects on the

DOI: 10.4018/979-8-3693-0512-6.ch016

environment the market size of Rh B is predicted to reach $232.5 million in 2027. Hence, developing facile and affordable methods for the degradation of Rh B is important.

Solution to the problem

Using nanoparticles for the decolorization and degradation of dye molecules is one of the widely used techniques. Attainment of quantum confinement in the range of 1–10 nm, imparts the quantum dots a unique optical property (Abdol, 2016, Catherine, 2002, Chen and Zhu, 2005). Quantum confinement is imparted into the bulk semiconductor material by converting it into nanoparticles. When the size of semiconductor material decreases its radius becomes less than the Bohr exciton radius (the distance between the electron and electron-hole). Consequently, quantum dots possess a broader range of absorption range and a narrow emission which directly impacts the band gap energy (Eg(QD)), and hence the size (Dilag, 2009, Li, 2009).

Cadmium sulfide (CdS) quantum dots (QDs) have a wider band gap than their bulk form and hence it absorbs the visible region (Abdol, 2016, Onwudiwe and Ajibade, 2011, Tyagi, 2014). CdS has a variety of potential applications in optical and electronic-related fields that originated uniquely from its band gap (Chauhan, 2013, Rao, 2011). One such application is its photocatalytic property which is highly useful in the complete degradation of organic pollutants like dye molecules rapidly. CdS is one of the widely used photocatalysts due to its stability and low cost (Desai, 2017).

Absorption of light by the photocatalyst (CdS) initiates the charge separation. Consequently, separation of charge leads to the formation of the pairs of electrons and holes and hence the free radicals. These free radicals degrade the pollutants effectively and are hence useful for wastewater treatment containing organic dyes (Bhadwal, 2014). In the photocatalytic degradation process, the energy/wavelength of light absorbed and emitted is a dependent factor of the band gap of the semiconductor (CdS). Semiconductors with a band gap range of 1.8 to 3.1 eV absorb the visible region (300-700 nm) (Chauhan, 2019) and semiconductors with a wider band gap >3.0 eV absorb in the UV region.

Simple sugars and cyclodextrin molecules are used simultaneously as a capping and reducing agent to impart quantum confinement in semiconductors due to their similar functional group (glucose) and solubility in water (Jiexiang, 2009, Gopalan, 2013). β-cyclodextrin [β-CD] is one of the widely used cyclodextrin due to its low cost and the suitable cavity size; 6-6.5 Å (Panigrahi, 2005, Pande, 2017).

In this chapter, we report the complete decolorization of rhodamine B (Rh B) by using the photocatalytic efficiency of CdS under natural solar light.

MATERIAL AND METHODS

Materials

Cadmium sulfide quantum dots (CdS) were prepared by mixing cadmium chloride ($CdCl_2$) and sodium sulfide (Na_2S) in the presence of β-cyclodextrin (β-CD) reductant. Analytical grade solvents and reagents were used for synthesis and all analyses. Double distilled water was used for all experiments.

Methods

0.1 mole of $CdCl_2$ + 0.1 mole of Na_2S were dissolved in 50 ml of double distilled water and stirred continuously for dissolution. Later 10ml of Na_2S was added dropwise into 10ml $CdCl_2$ solution, stirred at 40°C to yield a yellow color precipitate. Stirring continued for 22hrs with the addition of 0.5g of β-CD. High pure CdS quantum dots were obtained by centrifugation process.

Characterization of CdS QDs

UV-Diffuse Reflectance Spectra (DRS) and photoluminescence spectra for CdS were obtained using Jasco UV–Vis spectrophotometer, Model V-670, and Perkin Elmer spectrophotometer, model 204 A. XRD) pattern of CdS was recorded on XPERT analytical diffractometer with Cu Ka as radiation source ($\lambda = 1.5418Å$). SEM images of CdS were obtained using JEOL JSM 5610 LV taken at an accelerated voltage of 20.0kV. FTIR measurement was done by a BRUKER, Germany spectrometer.

Photocatalytic Activities of CdS QDs

Photocatalytic efficiency of the prepared CdS QDs was assessed by using it for the degradation of rhodamine B (Rh B) under natural solar light irradiation. In a typical process, 50 mg of photocatalyst (CdS QDs) was dispersed into the 80ml of Rh B (10ppm) dye solution under constant stirring. Before irradiation, the solution was stirred in the absence of sunlight for 30 minutes to attain an adsorption/desorption equilibrium. After that, the suspension was irradiated under natural sunlight with frequent stirring. All the measurements were carried out on sunny days in between 10am-3pm. The concentration of dye in the solution was determined by UV-Visible spectrophotometer (ANALYTIC JENA Germany) by measuring the absorption intensity at 554 nm for Rh B every 30 minutes. The percentage of dye degraded was estimated using the formula given below,

$$\text{Dye degradation } \% = \frac{(A_0 - A)}{A_0} * 100$$

Where, A - absorbance value at 554 nm for RhB), A_0 is the value before irradiation and A is the value after the irradiation process.

Focus of the Chapter

Optical property

Bulk CdS quantum dot has a direct band gap with an absorption edge 512nm and with band gap energy 2.42 eV (Smotkin, 1988, Singh and Chauhan, 2009). The absorption spectrum of CdS QDs (Fig. 1a) shows a well-defined absorption peak at 469nm. The Bohr radius of the exciton 5.8nm for CdS, Hence, the quantum confinement is feasible at 5-6nm or lesser size crystals (Mishra, 2011). The blue-shifted emission of CdS at 469nm than the bulk particle is observed. The band gap energy of synthesized CdS deduced from *Tauc's* relation,

$\alpha h\nu = A\ (h\nu - E_g)^{1/2}$

Where, h - Planck's constant, ν - frequency of the incident radiation, A - absorption constant, hν - photon energy, E_g - band gap energy. The direct band gap of CdS was determined by extrapolating the linear portion of the plots of $(\alpha h\nu)^2$ against (hν) to the energy axis (Figure 1b). The calculated band gap of 2.62eV showed a difference of 0.2eV energy greater than that of bulk, indicating quantum size effect.

Figure 1. (a) DRS spectra of CdS QDs (b) Tauc's plot obtained for CdS QDs.

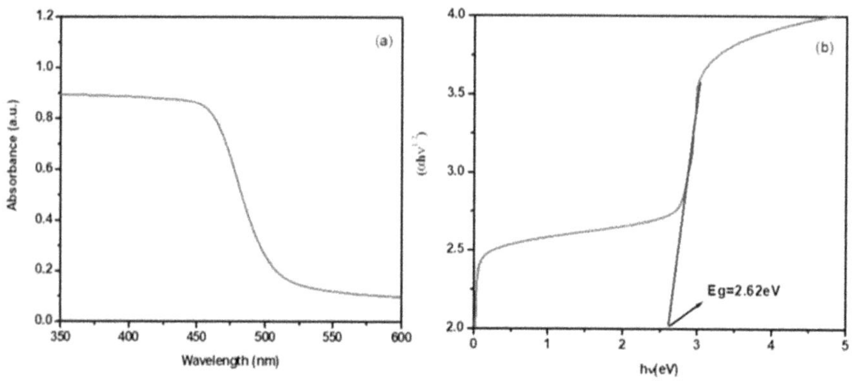

Photo luminescence (PL) is an efficient method in identifying the defects in nanostructures. The CdS quantum dots were investigated by exciting at 469nm excitation wavelength. Emission spectra of CdS QDs showed a broad band with maxima at 678nm (Figure 2). The appearance of broad red emission peak at 678 nm is attributed to the electron hole traps (surface defect emission) (Onwudiwe and Ajibade, 2011, Chauhan, 2013, Samuel, 2018)

Particle size determination

Figure 3 shows the strong diffraction peaks observed in the XRD patterns of CdS QDs at 2θ values; 26.88°, 44.15° and 52.01° closely match with (111), (220) and (311) crystalline planes of the face centered cubic (zincblende) structure of CdS (JCPDS 10-0454) (Sharma, 2016, Balaz, 2003, Tang, 2010) with a crystal lattice parameter 5.8 Å. Debye-Scherer's formula (Mohammed, 2018, Duchaniya, 2014, Alom, 2015) gives the crystallite size,

$$D = K\lambda / \beta \cos\theta$$

Where D - average crystallite size, K - dimensionless shape factor (0.94), λ - X-ray wavelength, β is the Full Width Half of Maximum (FWHM) and θ is the diffraction angle. The average crystallite size is found as 3.5 nm by resolving three peaks, three more peaks are observed for β-cyclodextrin. Lattice parameter is calculated using the formula given below, (Yucel, 2016, Senthamilselvi, 2012).Where, *d* - atomic lattice spacing, *h, k* and *l* are miller indices, *a* is the lattice parameter of the crystal (Table 1).

Figure 2. Photoluminescence spectra of CdS QDs under excitation with UV light 460nm

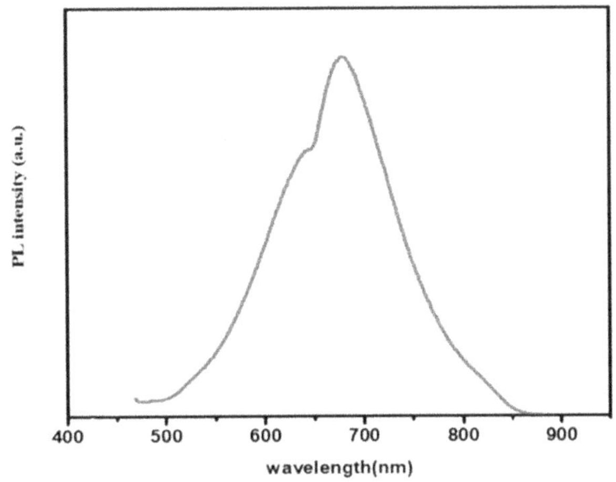

Figure 3. XRD pattern of CdS QDs.

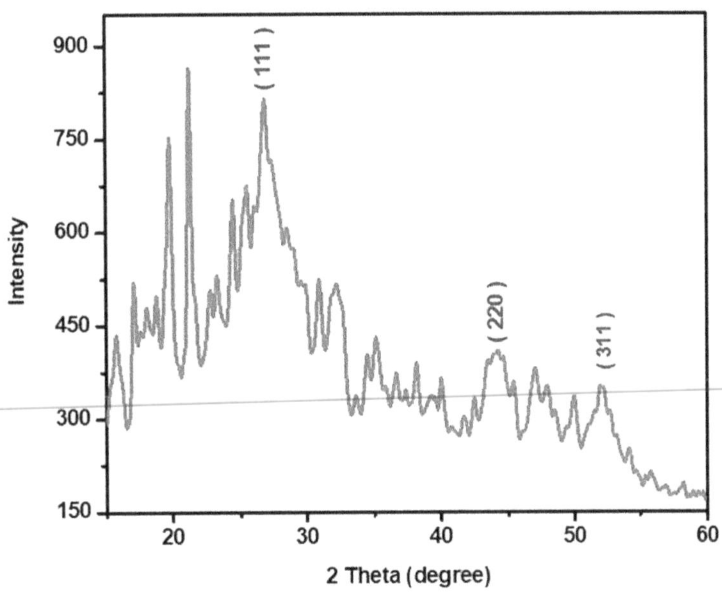

Table 1. Data evaluated from X–ray diffraction pattern for cadmium sulfide nanoparticle

Position (2 Θ)	Inter spacing d (Å)	Indices hkl	FWHM (deg)	a=b=c (Å)	Lattice strain (%)	Crystallite size (nm)
26.884	3.31	1 1 1	4.403	5.74	0.0804	1.94
44.148	2.05	2 2 0	2.220	5.79	0.0239	4.03
52.012	1.76	3 1 1	2.073	5.84	0.0185	4.46

Average = 3.5 nm

Morphology and Elemental Analysis

Figure 4; images clearly show the morphology of the CdS particles. The plate-like nanoclusters of CdS are inferred from different magnifications of SEM images. The grains are aggregated to form clusters[11].

Molecular Confirmation

Figure 5; FTIR spectra of CdS. The most prominent peaks are observed at 3278cm^{-1}, 1635cm^{-1}, 1153cm^{-1}, 1022cm^{-1} and 572cm^{-1} for β-cyclodextrin with CdS quantum dots. Comparison of pure β-CD (Rabadiya, 2013) and β-CD with CdS is shown in Table 2. The small and weak peaks between the range 400-470 cm^{-1} were assigned for Cd-S stretching frequency (Qutub, 2013).

Figure 4. SEM images CdS QDs

Figure 5. FTIR spectra of CdS QDs

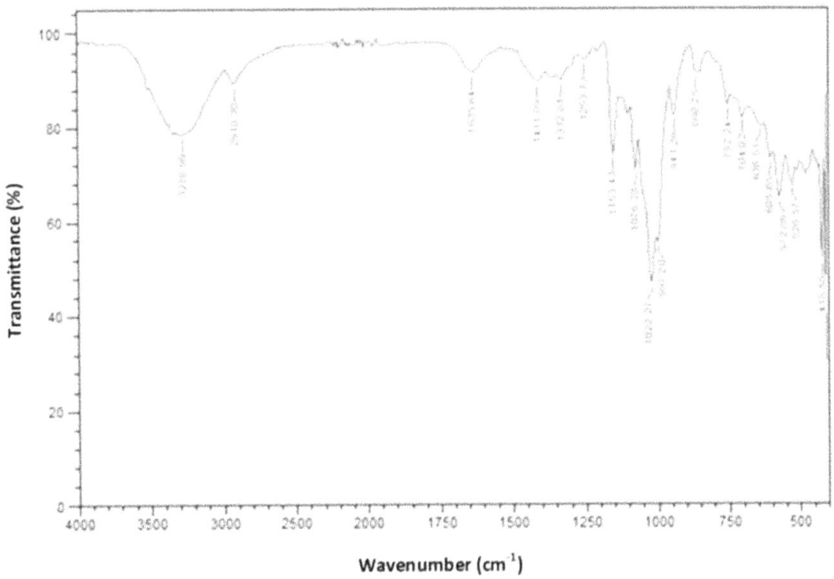

Table 2. Comparative table for FTIR analysis

S.No	Functional group	Pure Beta cyclodextrin (Earlier studies [23]) cm-1	β-CD with CdS QDs (this work) cm^{-1}
1	O-H stretching (broad)	3227	3278
2	O-C (2-bands)	1642	1635
3	O-H bending (in-plane)	1332	1332
4	C-C-C bending	1152	1153
5	O-H (H-bonded), usually broad C-O	1077, 1022	1076, 1022
6	C-H bending & ring puckering	860,753	860,752
7	Cd-S stretching	---	418

Photocatalytic Activity

The effectiveness of CdS QDs as a photocatalyst was evaluated using rhodamine B as a model pollutant under solar light illumination. The photocatalytic efficiency was evaluated by using different experimental conditions such as (i): the effect of photocatalyst amount, (ii): effect of concentration of RhB, (iii): effect of pH of the solution, and (iv) the reusability of CdS QDs. In order to test the photocatalytic activity of CdS on RhB, (i): RhB without CdS and (ii): RhB with 50mg of CdS was placed in sunlight; the absorbance spectra of samples withdrawn in equal interval of time from both test solutions were recorded. In the absence of CdS QDs the absorbance value of RhB solution with respect to time increases (Figure 6a). The reason behind the increase in the absorption intensity of RhB is the evaporation of water during illumination leaving behind a more concentrated solution. In the presence of CdS QDs the absorbance value decreases with respect to time (Figure 6b).

This preliminary observation shows that the RhB dye cannot be degraded just by solar light irradiation without a catalyst. A decrease in the absorbance in the presence of CdS QDs infers the photosensitization role of CdS on RhB degradation. The CdS QDs band gap is 2.62eV which belongs to the range between (1.8 - 3.1eV) which can absorb visible region light from solar radiation. Especially the yellow CdS QDs particularly absorb blue and violet light in the visible region. Color change in RhB dye with CdS QDs under solar light is shown in Figure 6c. Hence, the preliminary results suggest that the synthesized CdS QDs can be used for the effective degradation of RhB under optimized experimental conditions.

Figure 6. Absorption spectra of 10ppm RhB (a) without CdS QDs, and (b) with 50mg CdS QDs under solar light irradiation, (c) images of decolorization of RhB for different exposure time

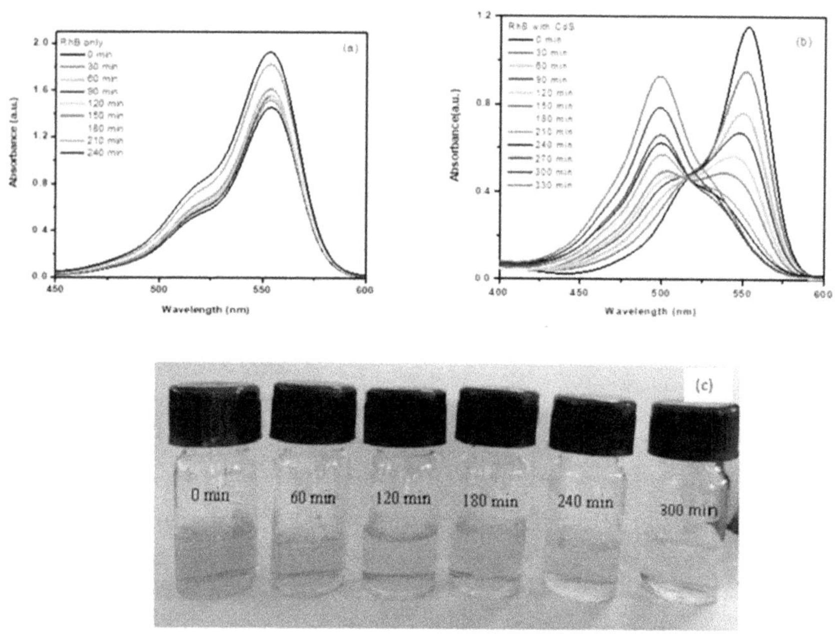

Effect of Catalyst Amount

In order to optimize the weight of the catalyst, different amounts of CdS ranging from 25-150mg were used and absorbance spectra were recorded Figure 7a-f by keeping the dye concentration constant at 10ppm under solar light irradiation. From the analysis, it is inferred that CdS QDs of 50mg degrade Rh B (Figure 7g) to a maximum percentage (52%).

Effect of Dye Concentration

In order to optimize the concentration of Rh B dye, varying concentrations of dye ranging from 4ppm – 12ppm were used and absorbance spectra were recorded Figure 8a-e by keeping the weight of CdS QDs constant as 50mg under solar light irradiation. From the analysis, it is inferred that Rh B concentration of 4ppm degrades (Figure 8f) effectively (52%).

Figure 7. Absorption spectra of 10ppm RhB with different amounts of CdS (a) 25mg (b) 50mg (c) 75mg (d) 100mg (e)125mg (f)150mg, under solar light. (g) Degradation profile of 10ppm RhB with different amounts of catalyst

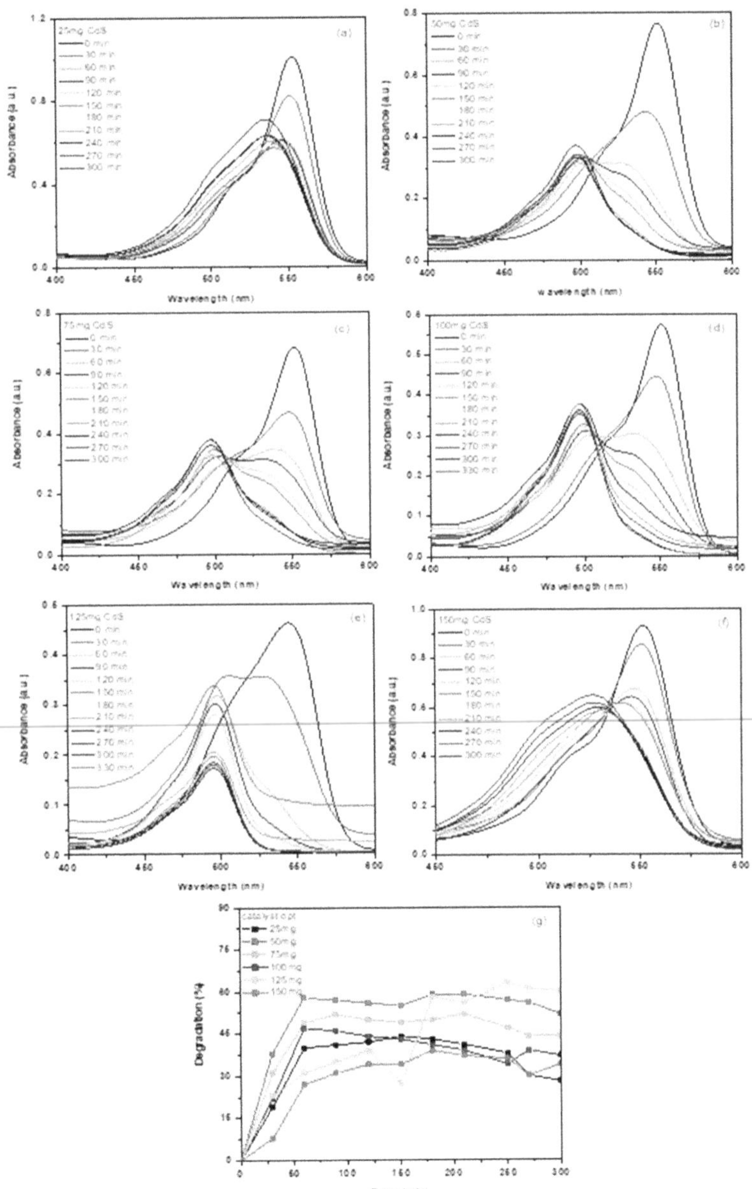

Figure 8. Absorption spectra of different concentrations of RhB with 50mg CdS, (a) 4ppm (b) 6ppm (c) 8ppm (d) 10ppm (e) 12ppm rhodamine B dye under solar light. (f) Degradation profile of RhB for a range (4ppm – 12ppm) of concentration with 50mg of catalyst.

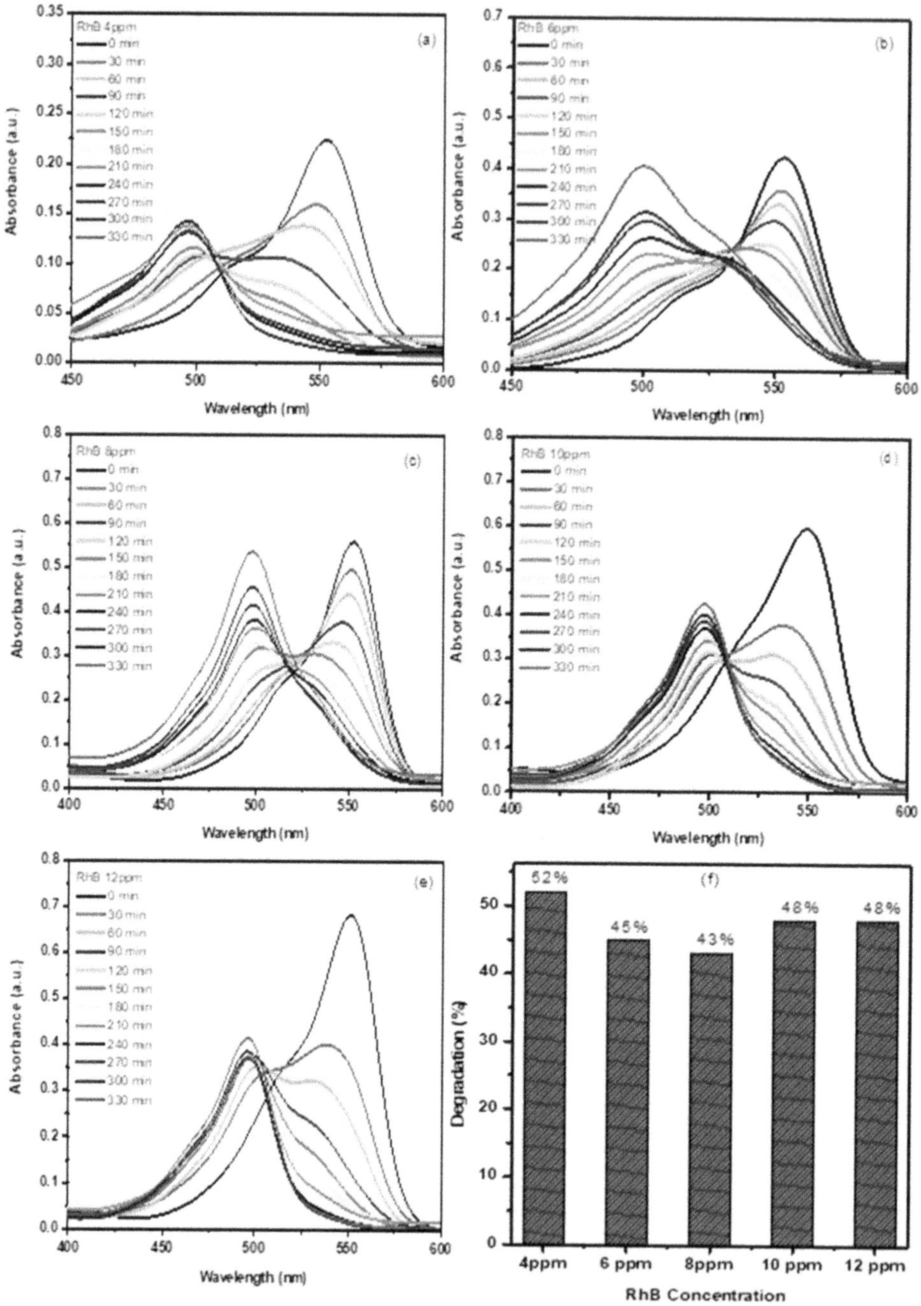

Effect of pH

In order to optimize the effect pH, the absorbance spectra were recorded for the varied pH conditions from pH 3 to 11 using H_2SO_4 and NaOH, by keeping 50mg as CdS weight and concentration of Rh B as 4ppm; Figure 9a-e. From the comparative analysis of results (Figure 9f), it is inferred that degradation is efficient in acidic and neutral mediums and reached maximum degradation at pH 7 (neutral medium). But the degradation efficiency of CdS in the basic medium is poor.

Reusability of the Catalyst

The reusability and stability of photocatalyst material (CdS QDs) is an important parameter for its potential applications in industries. To determine the reusability of CdS used for degradation (50 mg/4ppm RhB sample in neutral solution) was washed thrice with distilled water and then heated at 110°c for 4hrs. It is observed that the catalyst exhibits significant photocatalytic ability even after three runs Figure 10a-d without significant loss of degradation efficiency (Figure 10e).

Isosbestic Point

Overlaid absorption spectra of Rh B in neutral solution unveiling the existence of isosbestic point at ~510nm (Fig. 7b, c, d, 8a, d, e, 9c, 10b, 10c), ~515nm (9a, b, 10a), and ~530nm (7f,8b) in all matrix condition. The presence of isosbestic point in different matrix circumstances confirms the formation of *leuco Rh B* which can be further confirmed by the observation of increased λ_{max} at 550nm. Gradual red shift of λ_{max} from 500nm (*for Rh B*) to 550nm (*leuco Rh B*) and highly intense absorbance is observed in all the spectra recorded. In all the trials, complete decolorization occurs at 300mins (Scheme 1).

Figure 9. Fading of RhB

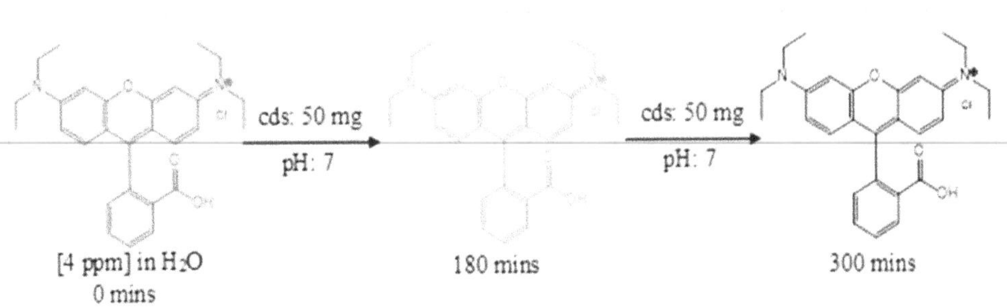

RECOMMENDATIONS

Cadmium sulfide (CdS) quantum dots synthesized by the precipitation method are characterized using spectroscopic and microscopic techniques. The UV- absorption spectrum band edge emission at 469nm and band gap of 2.62eV reveals the increase of 0.2eV band gap of CdS quantum dots than the bulk CdS (512nm with 2.42 eV) and thus confirms the quantum confinement effect. The quantum confinement effect is further confirmed with the photoluminescence spectrum. XRD pattern indicates a zinc blend (fcc) structure with a particle size 3.5nm. SEM analysis suggests a nano plate-like cluster morphology. FTIR confirms the presence of Cd-S species in structure. Optimized experimental conditions for the maximum degradation of RhB using CdS under solar light obtained is (i) weight of CdS catalyst:50mg, (ii) concentration of RhB: 4ppm, (iii) pH of the medium: 7, (iv) duration for complete decolorization: 5hrs, (v) reusability: ~50%. The formation of *leuco Rh B* is confirmed from the observation of a highly intense absorbance peak at 550nm at 300min.

Figure 10. Absorption spectra of 4ppm RhB with 50mg of CdS in different pH, (a) pH 3 (b) pH 5 (c) pH 7 (d) pH 9 (e) pH 11, under solar light (f) Degradation profile of 4ppm RhB with 50mg of CdS in different pH

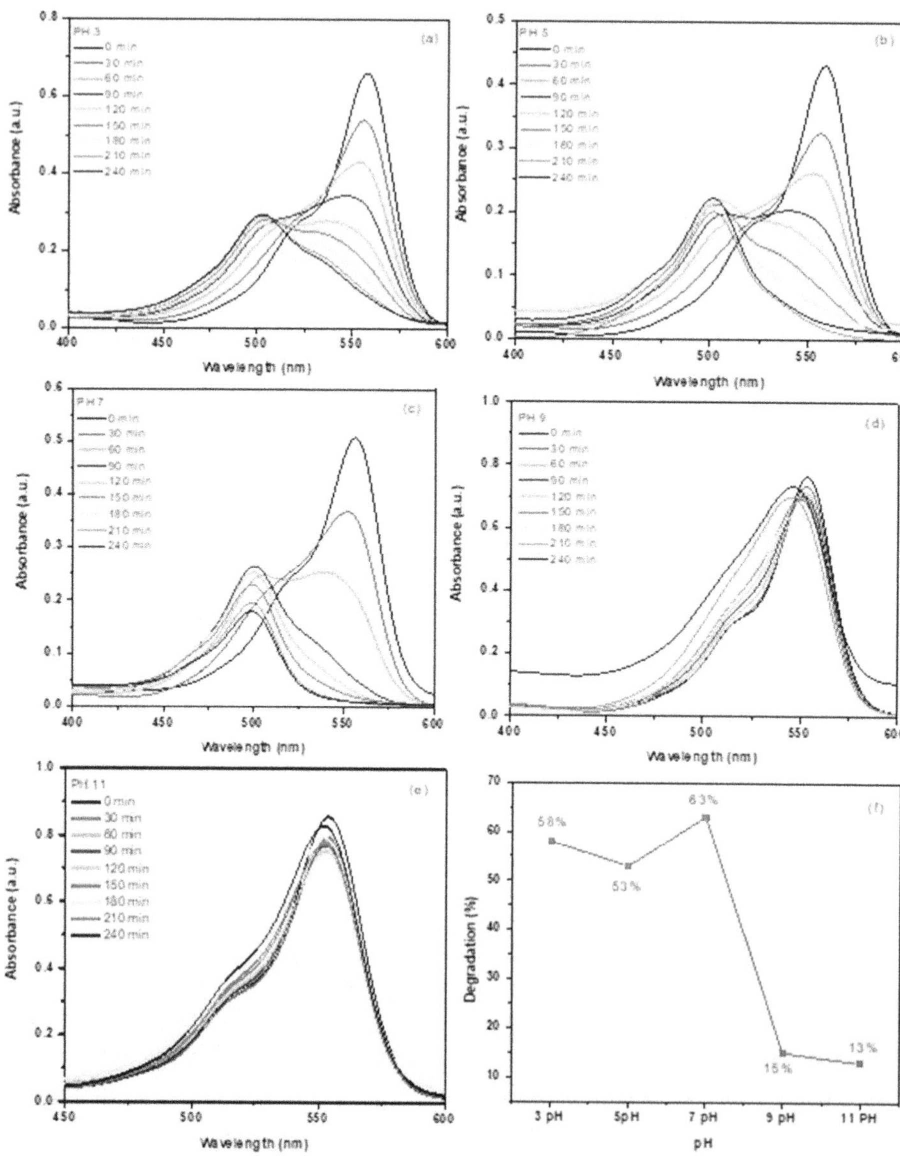

Figure 11. Absorption spectra of 4ppm RhB with 50mg of CdS for different run, (a) Run number 1, (b) Run number 2, (c) Run number 3, (d) Run number 4 under solar light. (e) Reusability profile of 50mg CdS photocatalyst for the phtodegradation of 4ppm RhB dye under solar light

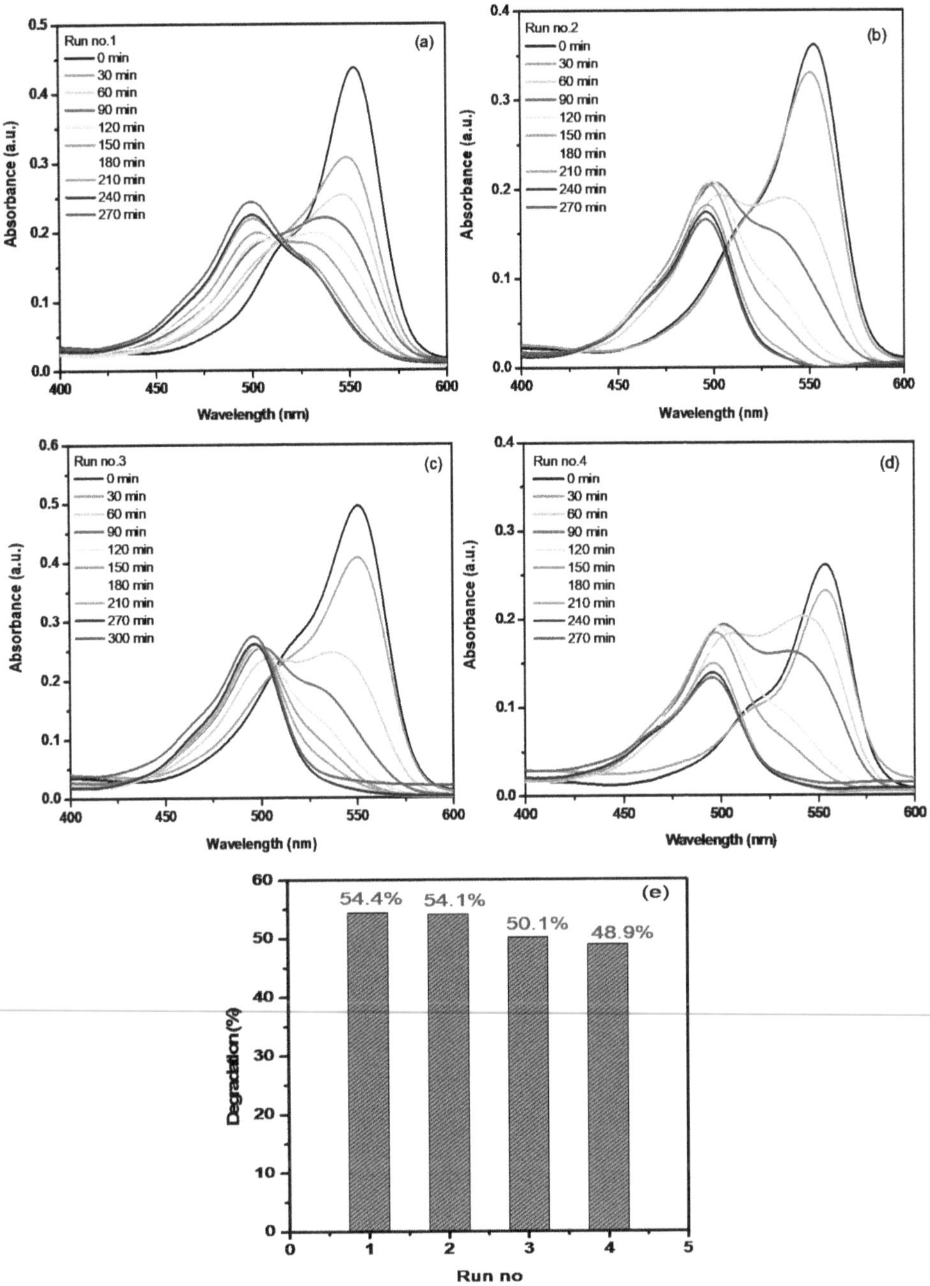

FUTURE RESEARCH DIRECTIONS

In order to enhance the efficiency of reported CdS into a rapidly degrading photocatalyst the reaction conditions of the proposed system would be optimized, investigated, and published elsewhere.

REFERENCES

Abdol, R. S. M., & Mohammad, R. S. M. (2016). In-situ synthesis of highly stable CdS quantum dots and their application for photocatalytic degradation of dyes. *Spectrochimica Acta. Part A: Molecular and Biomolecular Spectroscopy*, *152*, 156–164. doi:10.1016/j.saa.2015.07.052 PMID:26208270

Alom, M. M., Momin, M. A., Udi, M. J., & Islam, M. (2015). Structural and optical evaluation of CdS nanoparticles synthesized by chemical route. *J. Bangladesh Electron.*, *15*, 39–45.

Balaz, P., Boldizarova, E., Godocikova, E., & Briancin, J. (2003). Mechanochemical route for sulphide nanoparticles preparation. *Materials Letters*, *57*(9-10), 1585–1589. doi:10.1016/S0167-577X(02)01037-6

Bhadwal, A. S., Tripathi, R. M., Gupta, R. K., Kumar, N., Singh, R. P., & Shrivastav, A. (2014). Biogenic synthesis and photocatalytic activity of CdS nanoparticles. *RSC Advances*, *4*(19), 9484–9490. doi:10.1039/c3ra46221h

Catherine, J. M. (2002). Optical Sensing with Quantum Dots. *Analytical Chemistry*, *74*(19), 520–526. doi:10.1021/ac022124v PMID:12380801

Chauhan, R., Kumar, A., & Chaudhary, R. P. (2013). Visible-light photocatalytic degradation of methylene blue with Fe doped CdS nanoparticles. *Applied Surface Science*, *270*, 655–660. doi:10.1016/j.apsusc.2013.01.110

Chauhan, S., Darvishzadeh, R., Lu, Y., Stroppiana, M., Pepe, B.M., & Nelson, A. (2019) *Wheat lodging assessment using multispectral UAV Data.*

Chen, J. L., & Zhu, C. Q. (2005). Functionalized cadmium sulfide quantum dots as fluorescence probe for silver ion determination. *Analytica Chimica Acta*, *546*(2), 147–153. doi:10.1016/j.aca.2005.05.006

Desai, K. R., Pathan, A. A., & Bhasin, C. P. (2017). Synthesis, characterization of cadmium sulphide nanoparticles and its application as photocatalytic degradation of Congored. *Int. J. Nanomater. Chem.*, *3*(2), 39–43. doi:10.18576/ijnc/030204

Dilag, J., Kobus, H., & Ellis, A. V. (2009). Cadmium sulfide quantum dot/chitosan nanocomposites for latent fingermark detection. *Forensic Science International*, *187*(1-3), 97–102. doi:10.1016/j.forsciint.2009.03.006 PMID:19356872

Duchaniya, R. K. (2014). Optical studies of chemically synthesized CdS nanoparticles. *International Journal of Minerals Metallurgy and Materials*, *2*, 54–56.

Gopalan, P. R., Annaselvi, A. G., & Subramania, P. (2013). Synthesis and characterization of β-Cyclodextrin capped silver nanoparticles. *Int. J. Nanomater. Biostruct.*, *3*, 26–30.

Jiexiang, Y. (2009). *Synthesis and characterization of cyclodextrin capped Au and Ag nanoparticles*. NUS.

Li, K. G., Chen, J. T., Bai, S. S., Wena, X., Song, S. Y., Yua, Q., & Wanga, Y. Q. (2009). Intracellular oxidative stress and cadmium ions release induce cytotoxicity of unmodified cadmium sulfide quantum dots. *Toxicology In Vitro*, *23*(6), 1007–1013. doi:10.1016/j.tiv.2009.06.020 PMID:19540911

Mishra, S. K., Srivastava, R. K., Prakash, S. G., Yadav, R. S., & Panday, A. C. (2011). Structural, photoconductivity and photoluminescence characterization of cadmium sulfide quantum dots prepared by a co-precipitation method. *Electronic Materials Letters*, *7*(1), 31–38. doi:10.100713391-011-0305-6

Mohammed, H. K., Mamoori, A., Mahdi, D. K., & Alshrefi, S. M. (2018). Synthesis and spectroscopic study of CdS nanoparticle using hydrothermal method. *AIP Conference Proceedings*, 1968.

Onwudiwe, D. C., & Ajibade, P. A. (2011). ZnS, CdS and HgS nanoparticles via alkyl-phenyl dithiocarbomate complexed as single source precursors. *International Journal of Molecular Sciences*, *12*(9), 5538–5551. doi:10.3390/ijms12095538 PMID:22016607

Pande, S., Ghosh, S. K., Praharaj, S., Panigrahi, S., Basu, S., Jana, S., Pal, A., Tsukuda, T., & Pal, T. (2017). Synthesis of Normal and inverted gold-silver core-shell architectures in *â*-cyclodextrin and their applications in SERS. *The Journal of Physical Chemistry C*, *111*(29), 10806–10813. doi:10.1021/jp0702393

Panigrahi, S., Kundu, S., Ghosh, S. K., Nath, S., & Pal, T. (2005). Sugar assisted evolution of mono- and bimetallic nanoparticles. *Colloids and Surfaces. A, Physicochemical and Engineering Aspects*, *264*(1-3), 133–138. doi:10.1016/j.colsurfa.2005.04.017

Qutub, N., & Sabir, S. (2013). Characterization of cadmium sulphide nanoparticles synthesized by chemical precipitation method. *Advanced Science, Engineering and Medicine*, *5*(8), 1–6. doi:10.1166/asem.2013.1350

Rabadiya, B., Thakkar, V., & Rabadiya, P. (2013). Drug excipients interaction and solubility enhancement study of simvastatin. *IJPRBS*, *2*, 168–185.

Rao, B. S., Rajeshkumar, B., Rajagopalreddy, V., & Subbarao, T. (2011). Preparation and characterization of CdS nanoparticles by chemical co-precipitation technique. *Chalcogenide Letters*, *8*, 177–185.

Samuel, B., Mathew, S., Anand, V. R., Correya, A. A., Nampoori, V. P. N., & Mujeeb, A. (2018). Surface defect assisted broad spectra emission from CdSe quantum dots for white LED application. *Materials Research Express*, *5*(2), 025009. doi:10.1088/2053-1591/aaaa83

Senthamilselvi, S., Saravanakumar, K., Jabena Begum, N., Anandhi, R., Ravichandran, A. T., Sakthivel, B., & Ravichandran, K. (2012). Photovoltaic properties of nanocrystalline CdS films deposited by SILAR and CBD techniques—A comparative study. *Journal of Materials Science Materials in Electronics*, *23*(1), 302–308. doi:10.100710854-011-0409-7

Sharma, S., Mothudi, B. M., & Dhlamini, M. S. (2016). Observed coexistence of memristive, memcapacitive and meminductive characteristic in polyvinyl alchol/cadmium sulphide nanocomposites. *Journal of Materials Science Materials in Electronics*, *27*(5), 4551–4558. doi:10.100710854-016-4330-y

Singh, V., & Chauhan, P. (2009). Structural and optical characterization of CdS nanoparticles prepared by chemical precipitation method. *Journal of Physics and Chemistry of Solids*, *70*(7), 1074–1079. doi:10.1016/j.jpcs.2009.05.024

Smotkin, E. S., Lee, C., Brad, A. J., Campion, A., Fox, M. A., Mallouk, T. E., Webber, S. E., & White, J. M. (1988). Size quantization effects in cadmium sulfide layers formed by a langmiur-blodgett technique. *Chemical Physics Letters*, *152*(2-3), 265–268. doi:10.1016/0009-2614(88)87365-2

Tang, T., Teng, T., Hou, Y., Wang, Y., Tan, F., Qu, S., & Wang, Z. (2010). Optical properties and electrical bistability of CdS nanoparticles synthesized in dodecanethiol. *Applied Physics Letters*, *96*(16), 163112–163113. doi:10.1063/1.3402770

Tang, W. Z., & Huang, C. P. (1995). Inhibitory effect of thioacetamide on CdS dissolution during photocatalytic oxidation of 2,4-dichlorophenol. *Chemosphere*, *30*(7), 1385–1399. doi:10.1016/0045-6535(95)00029-8

Tyagi, C., Sharma, A., & Kurchania, R. (2014). Synthesis of CdS Quantum dots using wet chemical co-precipitation method. *Journal of Non-Oxide Glasses*, *6*, 23–26.

Yang, H., Huang, C., Li, X., Shi, R., & Zhang, K. (2005). Luminescent and photocatalytic properties of cadmium sulfide nanoparticles synthesized via microwave irradiation. *Materials Chemistry and Physics*, *90*(1), 155–158. doi:10.1016/j.matchemphys.2004.10.028

Yucel, E., Yucel, Y. Durak, M. (2016). *Process optimization for window material CdS thin films grown by a successive ionic layer adsorption and reaction method using response surface methodology*. Research Gate, 530-537.

Chapter 17
H_2O_2 as Efficient Oxidizing Agent for Wool Processing and the Preparation of Wool-Based Materials

Bogdan-Constantin Condurache
https://orcid.org/0000-0003-3997-3419
Petru Poni Institute of Macromolecular Chemistry, Romania

Petrisor Samoila
https://orcid.org/0000-0003-1782-720X
Petru Poni Institute of Macromolecular Chemistry, Romania

Valeria Harabagiu
https://orcid.org/0000-0002-2512-6609
Petru Poni Institute of Macromolecular Chemistry, Romania

ABSTRACT

Hydrogen peroxide is a compound that plays an important role in the biochemical processes of living organisms and found in a large spectrum of applications from chemistry and soft material processing to cosmetics and medicine. It is present in natural waters and contributes to their purification through reactive radical species. Starting from these considerations, an attempt was made to explain the morphological changes and the properties of the fibrous material achieved by means of the reactions with the reactive radical species that appear following the disproportionation process of the hydrogen peroxide. The influence of different factors, such as temperature, pH, the presence or absence of catalysts, etc. on the complex mechanisms by which morphological changes occur in the wool fibrous materials is discussed by corroborating old and new results collected by different investigation techniques. The reports on hydrogen peroxide involvement in wool processing are summarized and new materials based either on wool waste or on keratin subproducts are described.

DOI: 10.4018/979-8-3693-0512-6.ch017

INTRODUCTION

Wool consists of natural protein fibers produced by sheep having physical and chemical properties that allow it to be quite a versatile material. Based on a hierarchical internal organization of different layers, wool fibers are characterized by both lipophilic and hydrophilic characteristics. These abundant materials are used especially in the textile industry, but other applications, such as sorbents for metal cations or for petrol spills from contaminated waters (Condurache et al., 2022a; 2022b; 2022c) as well as raw materials for the preparation of nano-keratin, following the fiber deconstruction, are also envisaged (Moore et al., 2016). World wool production in 2021 was approximately 2 million tons of wool, which represents approximately 3% of the global textile market, while synthetic materials account for 61% (Rajabinejad et al., 2019; IWTO, 2022). At the EU level, an annual production of approximately 270,000 tons of wool is recorded, and 10% is wool unsuitable for the textile industry (Sulyman et al., 2017). Thus, decreasing interest in the textile industry associated with waste fibers resulting from wool weaving/used clothes and with coarse-grade fibers inappropriate for use in textiles generate important quantities of wool waste. Thus, for both economic and environment-related reasons, innovative modifications and new uses of these wastes should be found.

Oxidative processes have been intensively studied from the perspective of the negative effects they exert on the physiological and/or mechanical properties of different keratinous materials (hair, wool, feathers, or skins). The oxidation of wool takes place in natural conditions under the combined action of oxygen, light, and humidity, when new hydroxyl, carbonyl, and oxygenated acid groups result, leading to an increase in the oxygen content of the materials (Sinkiewicz et al., 2016; Verlackt et al., 2017; Dyer, 2018). Oxidative processes consist of breaking or modification reactions of side substituents of amino acids (Stadtman & Levine, 2006), in the fragmentation of protein chains (Davies, 2003), or reactions at the peptide chain ends terminated in amino or carboxylic groups (Gill et al., 2000). Hydrogen peroxide is an environmentally friendly, colorless, and non-corrosive oxidizing agent, this is why it was chosen as a precursor for the oxidation of wool fibers (St. Laurent et al., 2007).

WOOL FIBER MORPHOLOGY/CHEMICAL STRUCTURE

Wool Fiber Morphology

According to specialized literature, wool fiber can have thicknesses between 10-50 μm and lengths varying between 3 – 25 cm (Rajabinejad et al., 2019; Erdogan et al., 2020). From a morphological point of view, the wool fiber is composed of several layers in the following succession from external to the central part: a hydrophobic layer composed of 18-methyleicosanoic fatty acid (Jones & Rivett, 1997), the cuticle, the cortex, the cell membrane complex and the medullary channel (Jones & Rivett, 1997; Höcker, 2002; Popescu & Wortmann, 2010; Moore et al. 2016).

The cuticle monolayer of a thickness of about 0.5 mm consists of flat, overlapping scales that surround the cortex, having a protective role against physical and chemical agents. Cuticle cells are composed of three sub-lamellar layers: epicuticle, exocuticle with layer A and endocuticle (Bradbury, 1976; Swift & Smith, 2001; Rajabinejad et al., 2019; Erdogan et al., 2020). The unique properties of the wool fiber are given by the cuticle cells (overlapping scales) which have a high cysteine content (high sulfur content) and show hydrophobic properties.

The cortex is the central layer of the fiber, surrounded by the cells of the cuticle, which gives it a series of physical, chemical and mechanical properties. The cortical cells are spindle-shaped and oriented longitudinally, parallel to the fiber axis (Rajabinejad et al., 2018), representing 90% of the fiber mass and having a three-component structure (orthocortex, mesocortex and paracortex) (Höcker, 2002; Rajabinejad et al., 2019). The cortex shows hydrophilic properties (due to the much lower sulfur content than the cuticle) and overall, the wool fiber has a hydrophobic core and hydrophilic shell structure (Popescu & Wortmann, 2010).

The cellular membrane complex is also called intercellular matter (of non-keratinous origin) and the role of a ligand between the cortical and cuticular cells (Bradbury, 1976).

The medullary channel or medulla is the inner layer of the capillary fiber, it is present in low-quality wool (diameters greater than 30 μm), having an amorphous-porous structure (Ghiocel et al., 2005; Erdogan et al., 2020).

Wool fiber can absorb water up to 30 – 45% compared to its own weight (Popescu & Wortmann, 2010). The water content of keratin fibers plays an important role in determining their physical and mechanical properties (Robbins, 2012).

Chemical Composition of Wool Fibers

Keratin is the structural component of all keratinous materials (nails, claws, feathers, hair, horns, wool), representing up to 90% of the mass of wool fibers (Cardamone et al., 2009;). Keratin is a fibrous protein (Wojciechowska et al., 2000), with an average chemical composition of: 50% C; 7% H; 22 – 25% O; 16 – 17% N; 3 – 4% S; 1% mineral substances. The chemical composition of the wool fiber is influenced by several factors, such as: the quality of the feed, the breed of the animals, the climatic conditions, and the way they are cared (Höcker, 2002; Ghiocel et al., 2005; Popescu & Wortmann, 2010; Korjenic et al., 2015).

Wool fibers have fully ionized carboxyl and amino groups at pH less than 7, the wool fiber keratin has a zwitterion (dipolar ion) ($^{-}$OOC-)CHR(-NH_3^{+}) structure corresponding to amino acids, being a biopolymeric ampholyte with acidic and basic active centers (Nikiforova et al., 2019). With its zwitterionic structure, keratin can retain cations on the carboxyl groups in the anionic form and anions on the amine groups in the cationic form (Nikiforova et al., 2019).

More than 20 α-amino acids, bifunctional substances containing an amino group (-NH_2) and a carboxylic group (-COOH) are found in wool fibers (Erdogan et al., 2020; Höcker, 2002; Popescu & Wortmann, 2010).

In the chemical structure of the wool, the amino acids of adjacent polypeptide chains can be linked by different types of chemical bonds. The amino acids that make up the wool fibers have different reactive groups that form intermolecular bonds, which ensure the molecular cohesion of keratin and can participate in chemical reactions (Wojciechowska et al., 2000).

WOOL OXIDATION WITH H_2O_2 IN A BASIC MEDIUM

Generation of Reactive Oxygen Species (ROS)

The decomposition of H_2O_2 takes place through a chain process generating hydroxyl radicals (•OH) and superoxide radical-anion ($\bullet O_2^-$). The radicals formed are very reactive species, capable of reacting with other molecules in order to stabilize them. Hydrogen peroxide undergoes disproportionation, creating two different species of radical oxygen and water. The disproportionation of hydrogen peroxide depends on the concentration of the two ions present in the solution, which is the limiting factor for the reaction rate (Haber & Weiss, 1932).

$$2H_2O_2 \rightarrow HO\cdot + HOO\cdot + H_2O \quad (1)$$

The hydroxyl radical can participate in reactions with hydrogen peroxide giving rise to other hydroperoxyl radicals, which will pass into superoxide radical anions at pH > 4.75, as given by the reactions (2, 3) (Haber & Weiss, 1932; Duca, 2022):

$$2HO\cdot + H_2O_2 \rightarrow HOO\cdot + 2H^+ \quad (2)$$

$$HOO\cdot \xrightleftharpoons{pK_a=4.75} \cdot O_2^- + H^+ \quad (3)$$

The appearance of the two radicals (hydroperoxyl (HOO•) and superoxide ($\bullet O_2^-$)) and their predominance in the solution is influenced by the pH value (Behar et al., 1970). Duca et al. (Duca & Travin, 2020; Duca, 2022) established an equilibrium constant value for the superoxide radical of $pK_a = 11.75$, which proves that at a pH value lower than 11.75 the main species in the system will be the superoxide radical anion, along with the hydroperoxyl radical (Millington, 2013), and at pH values higher than 11.75 the superoxide anion will change to peroxide (O_2^{2-}), according to the reaction (4), being fast protonated by water (Rao & Hayon, 1975):

$$HOO\cdot \xrightarrow{-H^+} \cdot O_2^- \xrightleftharpoons{pK_a=11.75} O_2^{2-} \quad (4)$$

A too high pH value can lead to the degradation of H_2O_2 without the formation of radical species, as the radicals are quickly reacting with water molecules, with the formation of oxygen, hydroxyl anion and reformation of hydrogen peroxide, according to the reaction (5) (Poole, 2004):

$$HOO\cdot + \cdot O_2^- + H_2O \rightarrow H_2O_2 + O_2 + HO^- \quad (5)$$

Free radicals are very reactive and will react with each other or with other non-radical species to reach a thermodynamically stable state (Gupta et al., 2016). The Gibbs free energy of the decomposition of hydrogen peroxide (formation of the hydroperoxyl radical) in an aqueous solution was calculated by considering the redox potentials available in the literature (Bard, 1985; Duca & Travin, 2020; Duca, 2022, Duca et al, 2021, Vaseashta et al, 2022). The obtained value of 138.94 KJ/mol shows that the reaction is not spontaneous and requires external energy to initiate this process. Thus, to achieve the decomposition of hydrogen peroxide, alkaline conditions and temperatures exceeding 50 °C are required; moreover, the presence of easily oxidizable substances (for example NH_4OH) can favor this process (St. Laurent, 2007). For this reason, the reaction system was heated to approximately 95 °C at pH = 11. The temperature effect favors the decomposition of hydrogen peroxide (Giguere, 1975), by a factor of 2.3 times for every 10 °C variation (Khadke et al., 2009) according to Van't Hoff's law and the Arrhenius equation; as well as a higher pH value (Xinhua et al., 2014; Pędziwiatr et al., 2018). The purely thermal decomposition of hydrogen peroxide is a complex reaction composed of several stages (Rarata & Surmacz, 2009).

Calculating the Gibbs free energy (according to the redox potentials available in the literature) of the formation of the superoxide radical from the hydroperoxyl radical, it was proved that the reaction is thermodynamically possible. Moreover, at a pH close to the value of 11.75, a Gibbs free energy values of decomposition of H_2O_2 into ($\bullet O_2^-$) of −66.57 KJ/mol (taking into account the redox potentials) and of −66.46 KJ/mol (taking into account the dissociation constant) were calculated (Duca & Travin, 2020; Duca, 2022). The obtained values are in accordance with the specialized literature (Evans & Uri, 1949) and prove that this reaction is also favored at a high pH. Thus, both species of radical ions ($\bullet O_2^-$, the dominant form) and ($\bullet$OOH) will be present in the system.

The basic pH, temperature, the presence of easily oxidizable substances and cysteine (Goldstone & Voelker, 2000; Rose & Waite, 2005; Gupta et al., 2016), catalyze the decomposition process of hydrogen peroxide, leading to the appearance in the system of new oxygen radical species, according to the equations (6-8) (Haber & Willstätter, 1931; Koppenol, 2001):

$$2H_2O_2 \rightarrow HO\cdot + HOO\cdot + H_2O \tag{6}$$

$$H_2O_2 + HO\cdot \rightarrow H_2O + H^+ + \cdot O_2^- \tag{7}$$

$$H_2O_2 + H^+ + \cdot O_2^- \rightarrow H_2O + HO\cdot + O_2 \tag{8}$$

Kehrer (2000) stated that the latter reaction would not be thermodynamically possible and needs a catalyst (Kehrer, 2000, 2001). However, this equation is thermodynamically possible, having a Gibbs free energy of −46.3 KJ/mol (Koppenol, 1989, 2001).

Influence of Temperature, pH, Catalysts, and Hydrogen Peroxide Concentration on Wool Oxidation

Wool oxidation by hydrogen peroxide was a subject of interest at the beginning of the last century. Smith & Harris (1936a) established the decrease of the cystine and nitrogen content of the wool samples with increasing concentration of hydrogen peroxide, but a sharp decrease is recorded at concentration values between 4-6%. Alkaline treatment of oxidized samples at concentrations higher than 4% H_2O_2 leads to a decrease in the mechanical properties of wool fibers. The increase in the solubility of the treated wool fibers in the alkaline medium, along with the increase in H_2O_2 concentrations, provides data on the degree of oxidation. Increasing the temperature and duration of the treatment with hydrogen peroxide led to a decrease in the cystine content and increased solubility of the samples in the alkaline medium.

The influence of pH on oxidized samples has no appreciable effect at values lower than 7. At pH values between 7 and 10, an increase in fiber solubility in an alkaline medium is observed, along with a decrease in cystine content, and at a pH higher than 10, a decrease in both cystine content and fiber solubility is observed (Smith & Harris, 1936a). So, the degree to which wool fibers are oxidized by hydrogen peroxide solutions is influenced by concentration, temperature, pH and duration of treatment. As mentioned before, the pH value manifestly influences the acid-base dissociation balance, leading to the appearance of hydroperoxyl (•OOH) and superoxide (•O_2^-).

The Haber-Weiss reaction is catalyzed by the presence of iron ions, and in the absence of transition metals the process would be kinetically slow (Haber & Weiss, 1932; Gupta et al., 2016). However, under certain conditions, oxidation-reduction processes can be used without catalyst (transitional metals) for the superficial oxidation of wool fibers, because increased temperature (Pędziwiatr et al., 2018) and basic pH (Xinhua et al., 2014) can stimulate the decomposition process of hydrogen peroxide (Goldstone & Voelker, 2000; Rose & Waite, 2005; Gupta et al., 2016), while some enzimes (piroxiredoxines) also split H_2O_2 by using reactive cysteine amino acids (Picazo & Molin, 2021). Under these conditions, all these individual factors will exert their cumulative catalytic action on hydrogen peroxide.

The reaction between wool keratin and oxidizing agents is extremely complex. It is known that most oxidizing agents act on the disulfide bonds in wool keratin, but other functional groups will also be oxidized, leading to an advanced degradation of the fiber (Wolfram et al., 1970).

Wool fibers exposed to alkaline solutions lead to the breaking of disulfide bonds (S-S) with the formation of sulfhydryl groups. Each temperature value corresponds to a pH value up to which the amount of reacted disulfide bonds is practically constant (Steinhardt et al., 1940). Thus, it is possible to treat wool in a basic environment at high temperatures, up to pH values where no appreciable destruction of the wool fiber is observed.

Based on the literature results, it can be concluded that the cystine present in keratin can be oxidized under different conditions, giving different products. The main product resulting from the reaction of hydrogen peroxide with wool fibers is cysteic acid (Alexander et al., 1950; 1951).

The action of hydrogen peroxide on wool leads to the oxidation of the disulfide group (S-S) in cystine and increases the possibility of damage to wool fibers in the alkaline medium. There is a functional relationship regarding the degree to which the wool fibers are oxidized, given by the decrease in the cystine content and the increase in their solubility in the alkaline medium (Smith & Harris, 1936a, 1936b).

Working at a pH lower than 11.75, the main species in the system will be the superoxide radical anion, along with the hydroperoxyl radical, which is found in smaller quantities. The superoxide radical anion (•O_2^-) (Asmus & Bonifačić, 2000; Wang et al., 2012) together with the conjugated acid radical present in the system (•OOH) will produce the superficial oxidation of raw wool fibers, initiating a chain reaction (Woodard & Curran, 2006). Following this treatment, negative charges will appear as a result of the oxidation of the S-S disulfide bonds of cystine into cysteic acid and the intermediate sulfoxides (Gacen & Cayuela, 2000; Reich, 2001; Cardamone et al., 2004; Cegarra et al., 2008; Ragelienė et al., 2009). So, during hydrogen peroxide oxidation of wool fibers, the functional group attacked is mainly the cystine disulfide bond present in the keratin of the fiber (Smith & Harris, 1936a).

INVOLVEMENT OF HYDROGEN PEROXIDE IN WOOL FIBERS' PROCESSING

In industrial practice, wool fibers collected from sheep are submitted to a series of operations in the presence of different chemical agents in order to either eliminate dirt and impurities or acquire certain properties specific to the final purpose for which they are to be used. Thus, the wool fibers are subjected to a washing process in the presence of detergents, followed by carbonization (to remove fat (lanolin) and vegetable matter) in the presence of H_2SO_4, $MgCl_2$, HCl (Czaplicki & Strzelecki, 2019), followed by neutralization with an alkaline solution (Na_2CO_3, NH_3, etc.). At this stage, processes of oxidation of the wool fibers may occur.

To bleach wool fibers, they are oxidized with various compounds (H_2O_2, peracetic acid, permolybdic and pertungstic acids, etc.) (Earland et al., 1955) in an acidic or basic medium. Alexander, Carter, & Earland (1951) treated wool fibers with peracetic acid, finding that all disulphide bonds in wool were completely oxidized to a final oxidation state of cysteic acid, as given by (Alexander et al., 1951):

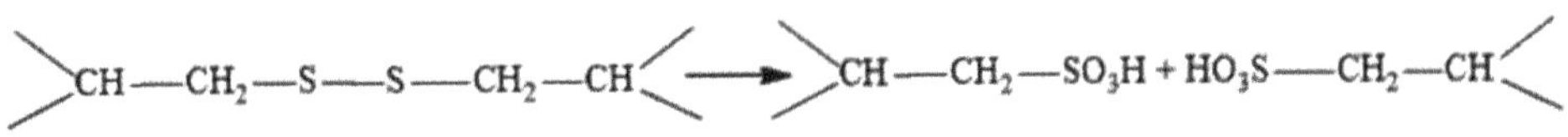

(9)

The acid chlorination process is used to prevent felting of wool fibers, but it results in large amounts of toxins. As a result of this process, their surface characteristics change due to the destruction of the scaly structure, which leads to the loss of the felting property (Pearson, 2008). The products obtained after acid chlorination include residues of cysteic acid (-CH2-SO_3^-) resulting from a series of partially oxidized intermediates and Bunte salt residues (-CH2-S-SO_3^-) (Maclaren, 1981; Woodhead et al., 1997).

In the textile industry, wastewater can come from any combination of washing, bleaching, mercerizing, dyeing, printing, finishing processes (Bisschops & Spanjers, 2003). Hydrogen peroxide is involved as active oxidation agent alone or in combination with other compounds/treatment methods in different steps of wool processing, such as the decontamination of the industrial waters resulted from the raw wool washing and fiber finishing processes or the bleaching processes.

Recovery of Wastewater Resulting From the Washing of Wool Fibers

Wool fibers contain various impurities, such as: oils, fats, waxes, including lanolin (a mixture of esters, fatty acids, alcohols and hydrocarbons (Bisschops & Spanjers, 2003)), minerals and vegetable matter. During the cleaning process, impurities are removed with water or solvents. Cleaning agents include detergents, soaps, alkalis, wetting agents, antifoams and lubricants. After the cleaning process, the wool fibers are rinsed to remove excess cleaning agents, and the generated wastewater contains substances with a toxic effect on the environment. The amounts of the different compounds found in the washing waters vary between 3000 – 20000 mg/L lanolin, 7000 – 15000 mg/L suint salts and 10000 – 30000 mg/L various residues that include soil, plant residues, etc. (Lapsirikul et al., 1994). This complex mixture consists of over 8000 organic and inorganic compounds, requiring a large amount of oxygen (chemical oxygen demand (COD) of 30.000 – 60.000 mg/L) to oxidize the organic matter present in the water. After chemical flocculation treatment, followed by a biological treatment under aerobic conditions, a reduction of COD value by more than 90% was registered, but the final effluent is colored (due to dissolved organic compounds that resist aerobic incubation) requiring a COD of 500 – 3000 mg/L (Bateup et al.,

1996; Poole, 2004). An extended incubation of more than three weeks with the addition of nutrients has no effect on color removal or COD reduction (Poole et al., 1999). Consequently, additional treatments must be carried out to reduce the COD to environmentally acceptable values, lower than 200 mg/L. Poole (2004) proposed the use of hydroxyl radicals resulting from the decomposition of hydrogen peroxide to oxidize stable organic compounds, showing that this reagent does not generate new amounts of residue after the treatment, so the contaminants can be removed without affecting the environment (Poole, 2004). It is important to find the right combination of biological, physical and chemical processes for the complete and most economical removal of contaminants present in the washing waters of wool fibers.

The hydroxyl radical (•OH) has a higher oxidation potential than other oxidizing agents (Plant et al., 1994), and the reaction rate constants are located in the range $10^7 - 10^{10}$ L/mol×s (Poole, 2004). During the degradation of organic compounds found in wool washing effluents by the attack of hydroxyl radicals, hydrogen extraction, OH addition or substitution and electron transfer can occur (Litter, 2005). The degradation of acidic compounds generally occurs through the extraction of hydrogen, as follows:

$$HO\cdot + RH \rightarrow R\cdot + H_2O \tag{10}$$

$$R\cdot + O_2 \rightarrow ROO\cdot \tag{11}$$

The degradation of aromatic compounds takes place by the hydroxylation of the aromatic, followed by the additional attack of the hydroxyl radical leading to the opening of the aromatic ring and the formation of open conjugated structures, according to the equations:

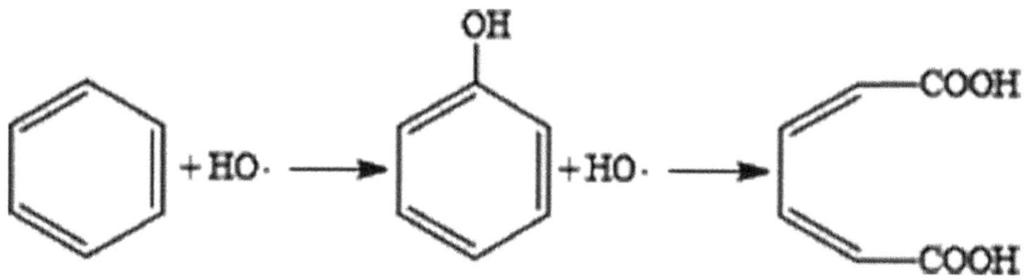

(12)

Hydroxyl radicals used for wastewater treatment are generated *in situ* by advanced oxidation processes (AOT), which include H_2O_2/UV, O_3/UV, Fenton's reagent, O_3/H_2O_2/UV, electron beam irradiation, photocatalysis, radiolysis, etc. (Litter, 2005; O'Shea & Dionysiou, 2012). Advanced oxidation methods can be used alone or in combination; they are useful for treating pollutants resistant to other (biological) treatments, without the formation of secondary products. Through advanced oxidation processes, effluents with a COD > 5 g/L cannot be properly treated, because they require large amounts of reagents and energy (Burrows et al., 2002). Advanced oxidation processes are useful for both pre-treatments, to convert pollutant into biodegradable compounds, or post-treatment to finishing wastewater before discharge to water (Litter, 2005).

In H_2O_2/UV oxidation process, the hydroxyl radical is generated by the homolytic cleavage of the O-O bond of hydrogen peroxide with the help of UV radiation. Radicals derived from the attack of hydroxyl radicals (•OH) on organic matter react with O_2 leading to the appearance of peroxyl radicals. The pH of the wastewater must be adjusted to have the best conditions for the degradation of organic compounds, but the H_2O_2 concentration must also be maintained at an optimal value (at too high H_2O_2 concentrations, water and oxygen result) (Schulte et al., 1995). Thus, Poole (2004) removed most of the organic compounds from wool washing waters by using hydroxyl radicals generated from hydrogen peroxide with the help of UV radiation and obtained a reduction of 75% of COD, 85% of total organic carbon (TOC) and a complete discoloration of the effluents.

We can say that the hydrogen peroxide used in the elimination of contaminants from wool washing effluents and not only, is a promising oxidant for future applications without leading to the generation of new harmful compounds.

Wool Bleaching

Natural wool exposed to different physical and chemical agents (light, humidity, temperature, alkalis) has a yellow color and should be bleached. The degree of yellowing of wool fibers is associated with their general degradation through exposure to light (UV rays), bacterial damage, the action of chemical agents, proteinoxidation and the increased pH of suint and processing (Winder et al., 1998; Simpson, 1999; Sumner et al., 2003; Jung & Sato, 2013). Moreover, the sensitivity of wool fibers to yellowing in the alkaline medium increases with the increase in hydrogen peroxide concentration (Cegarra et al., 2008).

The equation currently used to calculate the yellow index is derived from the Hunter equation (equation 1) (Hunter, 1942):

$$Y = (A - B)/G \tag{1}$$

where: A, B and G represent the color readings for red, blue and green.

The yellow index is calculated according to the formula (equation 2) (Arifoglu et al., 1990; ASTM E313; ASTM D1925):

$$YI = [100(1.28X - 1.06Z)]/Y \tag{2}$$

where: X, Y and Z are the measured tristimulus values.

If there is an absorption in the blue region of the spectrum, we can visually observe the yellowing of the wool fibers. The yellowness index is used to determine the degree to which wool fibers lack ideal whiteness (Jung & Sato, 2013; Millington, 2013).

For the purpose of bleaching, both oxidizing and reducing systems, or combined oxidation-reduction processes are used. The purpose of bleaching wool fibers is to remove the color of the fibers, which is closely related to the structure and reactivity of the melanin pigment that gives the color. Melanin granules are distributed in the cortex of the wool fiber. Therefore, during the bleaching process, the oxidation of keratin also takes place, as evidenced by the loss of cystine and the modification of the side chains of the amino acids that make up its composition.

Chemicals used for this purpose include sodium hypochlorite and hydrogen peroxide as optical brighteners (Bisschops & Spanjers, 2003). The bleaching of wool fibers using peroxides (H_2O_2) has been practiced for a very long time, for over five decades. The reaction of hydrogen peroxide with melanin has a maximum rate at a pH value below the pKa value of 11.75 of H_2O_2, which indicates that, in addition to the hydroperoxyl anion (HOO•), there are other reactive species (superoxide ($\bullet O_2^-$)) that will react with the pigment. Depending on the pH value, one of the two radicals predominates, as was shown many years before (Wolfram et al., 1970). For the same concentration of hydrogen peroxide, wool fibers become whiter in an alkaline medium (Smith & Harris, 1936b; Gacen & Cayuela, 2000).

The bleaching of wool fibers with hydrogen peroxide can be done under alkaline or acidic conditions. To stabilize hydrogen peroxide in an alkaline medium, sodium pyrophosphate/sodium oxalate or sodium metasilicate are used (Gacen & Cayuela, 2000). The process of bleaching wool fibers with hydrogen peroxide can also take place in an acidic medium in the presence of the Prestogen W activator (organic salt solution) which leads to the generation of peroxides, regulates the pH, inhibits the yellowing of wool and retains polyvalent metal ions. The advantages of wool bleaching in the presence of the Prestogen W activator consist in decreasing the bleaching time and the positive effects on fiber properties (the wool fibers become softer and more voluminous, the felting capacity), while the organic mass in the wastewater decreases (Bereck et al., 1997).

The degree of whiteness can be evaluated by the Berger index and the IWTO index as well as by Taube equations used in the Colorgard System, as defined by the equations (equation 3-5) (Berger, 1959; Cegarra et al., 1988; ASTM E313):

$$W = R_y + 3R_z - 3R_x \text{ – Berger index} \tag{3}$$

where: R_x, R_y and R_z are the reflectance obtained in a colorimeter with three filters.

$$W = [(100 - 0.94Y)^2 + (2.84X - 2.35Z)^2]^{1/2} \text{ – IWTO index} \tag{4}$$

where: X, Y and Z are the measured tristimulus values, which represent the amounts of red (X), green (Y) and blue (Z) in order to produce the color of the sample.

$$W = 3.387Z - 3Y \text{ – Taube equations} \tag{5}$$

In order to evaluate the degree of whiteness, the International Commission on Illumination (CIE) recommends using the following formula (equation 6) (Luo et al., 2005; Joiner, 2010; Westland, 2015; ASTM E313):

$$\mathrm{W = Y + 800(x_n - x) + 1700(y_n - y)} \tag{6}$$

where: Y is the luminance factor; x_n and y_n the chromaticity coordinates for the CIE standard illuminant and source used, respectively; x and y are the chromatic coordinates defined from the tristimulus values, as follows:

$$\mathrm{x=X/(X+Y+Z) \text{ and } y=Y/(X+Y+Z)} \tag{7}$$

Color evaluation is done by measuring the light reflected from a sample at a variety of wavelengths. Each color has its own fingerprint defined as the percentage of light reflected at a specific wavelength (Hurren, 2008).

A white surface reflects uniformly throughout the visible spectrum, while absorbing at a very low level. The higher the spectral reflectance, the whiter the wool fibers appear. So, a higher whiteness index means that wool fibers do not absorb in the blue-yellow region of the electromagnetic spectrum. Higher whiteness values can be obtained if the wool fibers are lighter or bluer than perfect white and lower whiteness values when the wool fibers are darker or more yellow than perfect white (Millington, 2013).

Several researchers have used hydrogen peroxide as an oxidative step combined with a reductive step in which thiourea, thiourea dioxide, or sodium hydroxymethanesulfinate were used, obtaining superior bleaching compared to conventional processes. The temperature, the concentration of the reactants and the duration of the treatment are the factors that influence the degree of whiteness (the whiter the fibers, the higher the cysteine content of the wool fibers) (Cegarra et al., 1988; Arifoglu et al., 1990; Gacen et al., 1991; Maiolo et al., 2007).

Other researchers have used sodium borohydride in the presence of sodium bisulphite solution for bleaching wool, obtaining similar results to that of hydrogen peroxide. Instead, a higher solubility in basic medium was obtained than bleaching with hydrogen peroxide (Yilmazer & Kanik, 2009), which means the increase of degradation of wool fibers.

Microwave irradiation has been used to reduce energy consumption and shorten the bleaching time of wool fibers with hydrogen peroxide, being an alternative to conventional heating. Bleaching wool with the help of microwave radiation led to an improved whiteness and a decrease in the degree of deterioration as compared to conventionally bleached wool (Abou El-Kheir, 2014).

TRANSFORMATION OF WOOL WASTE AND KERATIN INTO USEFUL MATERIALS

Wool wastes of livestock and industrial origins are usually thrown away, buried or incinerated, leading to an increase in greenhouse gas emissions. Wool waste management is a real problem as sheep are crossbred to improve meat production resulting in poor quality wool (Zoccola et al., 2015). Thus, these wastes are cheap sources of amino acids and proteins that can be used as fertilizers (Gousterova et al., 2003; Zoccola et al., 2015) or animal feed (Martínez-Alvarez et al., 2015).

Initially, keratin waste of animal origin, containing 20 – 25% protein was used as a protein additive in feed mixtures for domestic animals, but various diseases of unknown cause can be transmitted. This flour carries prions that cause animal diseases. In EC countries, feeding domestic animals with animal feed is strictly prohibited. The only way to stop the spread of diseases is to incinerate this keratin waste, which generates quite significant expenses. Although the natural degradation of keratin wastes thrown away is time-consuming, they represent a source of nitrogen for plants (Gousterova et al., 2003).

Raw wool is also rich in nutrients being released into the soil over a long period of time in a fluent manner. Wool contains carbon, nitrogen, sulfur, phosphorus that play an essential role for plants (Von Bergen, 1963). In natural conditions, wool fibers show high stability, thanks to intermolecular cystine bonds, peptide and hydrogen bonds, which give them resistance, rigidity and insolubility (Maclaren & Milligan, 1981).

Wool-Based Sorbents

The quality of surface waters is increasingly affected by pollution with hydrocarbons and heavy metal ions, generating great problems for the aquatic environment and beyond. In 2021, Romania owned approximately 12 million sheep and goats (INS, 2022), and wool production is approximately 23 thousand tons (INS, 2019). In this context, a waste resulting from the livestock sector, sheep wool of very coarse grade (74 ± 16 μm fiber width) from the Romanian Turcan breed, was transformed by simple methods into sorbents for heavy metal cations (Condurache et al., 2022a) and oil spills (Condurache et al., 2022b; 2022c). For both materials the unwoven wool fibers were superficially oxidized in solution of low concentration of H_2O_2 (0.3%) and ammonium hydroxide (pH, 11) resulting in increased oxidized cysteine species and nitrogen content (due to neutralization of acidic species by ammonium hydroxide), as proved by FTIR and SEM-EDX methods, respectively.

Oxidized Wool as Sorbent for Heavy Metal Cations

The reactions involved in the oxidation process of wool fibers by hydrogen peroxide in the presence of ammonium hydroxide are shown in Figure 1 (Condurache et al., 2022b). As proved by FTIR and SEM-EDX methods, respectively, the resulting fibers are characterized by increased oxidized cysteine species and nitrogen content (due to neutralization of acidic species by ammonium hydroxide).

Figure 1. Main chemical reactions involved in the oxidation of wool with H_2O_2/NH_4OH System

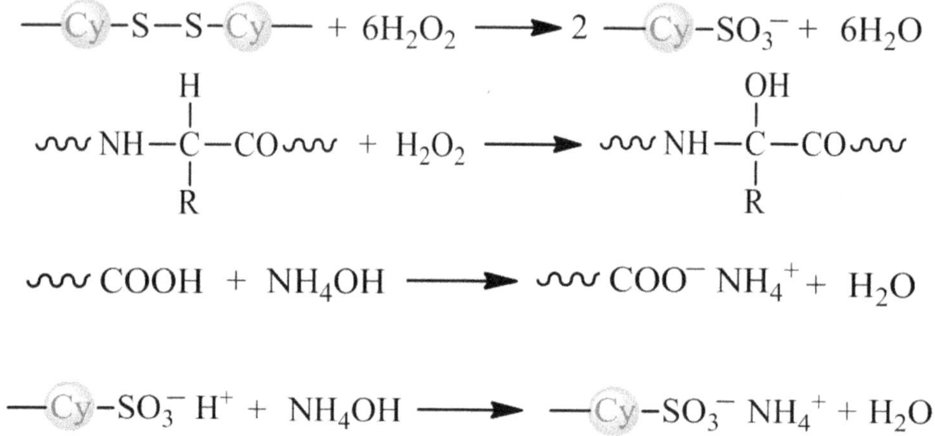

Similar results were obtained by Cegarra et al. (2008), who found that following the treatment with peroxide in a basic environment, the cysteic acid content increases, and the cystine content decreases (Cegarra et al., 2008).

As proved by the analysis of isothermal curves (Condurache et al., 2022a), the obtained materials were proved to absorb 9.41, 10.42, and 30.71 mg/g of Cu^{2+}, Cd^{2+} and Pb^{2+}, respectively, from their salt solutions in short time (less than 1 h), mainly through electrostatic interactions (physical sorption) between anionic groups of the sorbent and the metal cations, but coordination of cations to wool ligand groups (chemisorption) are also involved. Superior sorption capacities for heavy metal cations were achieved with wool-based sorbents prepared either by more energy consuming procedures such as powdering (Atef El-Sayed et al., 2014) or by chemical treatment with less environmentally friendly compounds such as Na_2S (Enkhzaya et al., 2017).

Superhydrophobic Wool-Magnetite Composite for Oil Spills Sorption From Contaminated Waters

For the preparation of oil composite sorbent, the oxidation reaction of wool was performed by hydrogen peroxide/ammonium hydroxide system in the presence iron sulfate (Fenton reagent), when a mixture of magnetite nanoparticles and wool/magnetite composite (W-Ma) were obtained. The composite was than submitted to a second step by reacting the OH groups of the wool with a poly(dimethyl-hydrogenmethyl) siloxane copolymer (Condurache et al., 2022b). A superhydrophobic (water contact angle, 152 ± 0.3 degrees) W-Ma-PSi composite was resulting, which possesses high enough magnetic properties (saturation magnetization, M_S=5.47emu/g; remanent magnetization, M_R=0.23emu/g) to be easily manipulated before and after oil sorption (Figure 1) The reactions involved in the preparation of the W-Ma-Psi composite are shown in Scheme 2.

(a) In Situ synthesis of Magnetite Nanoparticle through Haber-Weiss Mechanism, in the Presence of Wool (W)
(b) Hydrophobization of W-Ma Composite

According to the reaction mechanism, Fe^{2+} is oxidized by hydrogen peroxide (oxidizing agent) (Nayak & Rao, 2003) into Fe^{3+}, resulting in a radical and a hydroxyl ion. Then, Fe^{3+} is reduced back to Fe^{2+} by another hydrogen peroxide molecule or superoxide radical anion (reducing agent) (Rose & Waite, 2005), resulting in a hydroperoxyl radical and a proton. The hydroxyl radical can participate in reactions with hydrogen peroxide giving rise to other hydroperoxyl radicals, which will pass into superoxide radical anions at pH > 4.75 (Haber & Weiss, 1932; Duca, 2022), able to reduce Fe^{3+} to Fe^{2+}. The free radicals found in the reaction system will be engaged in secondary reactions, and the Fe^{2+} and Fe^{3+} cations will finally lead to magnetite. During the process, the ratio of the two (Fe^{2+} and Fe^{3+}) simultaneously present in solution is dependent on the value of the basic pH.

Figure 2. Manipulation of W-Ma-PSi sorbent: (a) W-Ma-PSi composite (left), top view of the oil layer colored with Sudan IV (left); (b) insertion of the composite on oil surface (right); (c), (d) manipulation of the sorbent before and after oil sorption; (e) exhausted sorbent (left), top view of cleaned water (right)

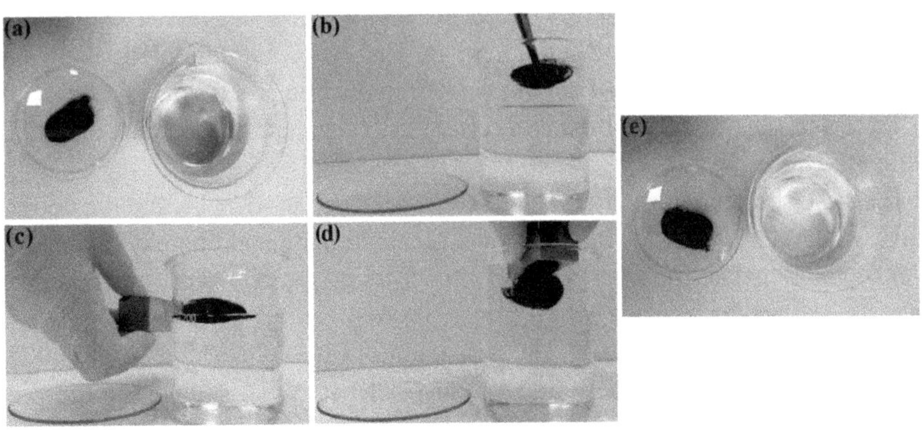

Figure 3. Preparation of W-Ma-PSi composite

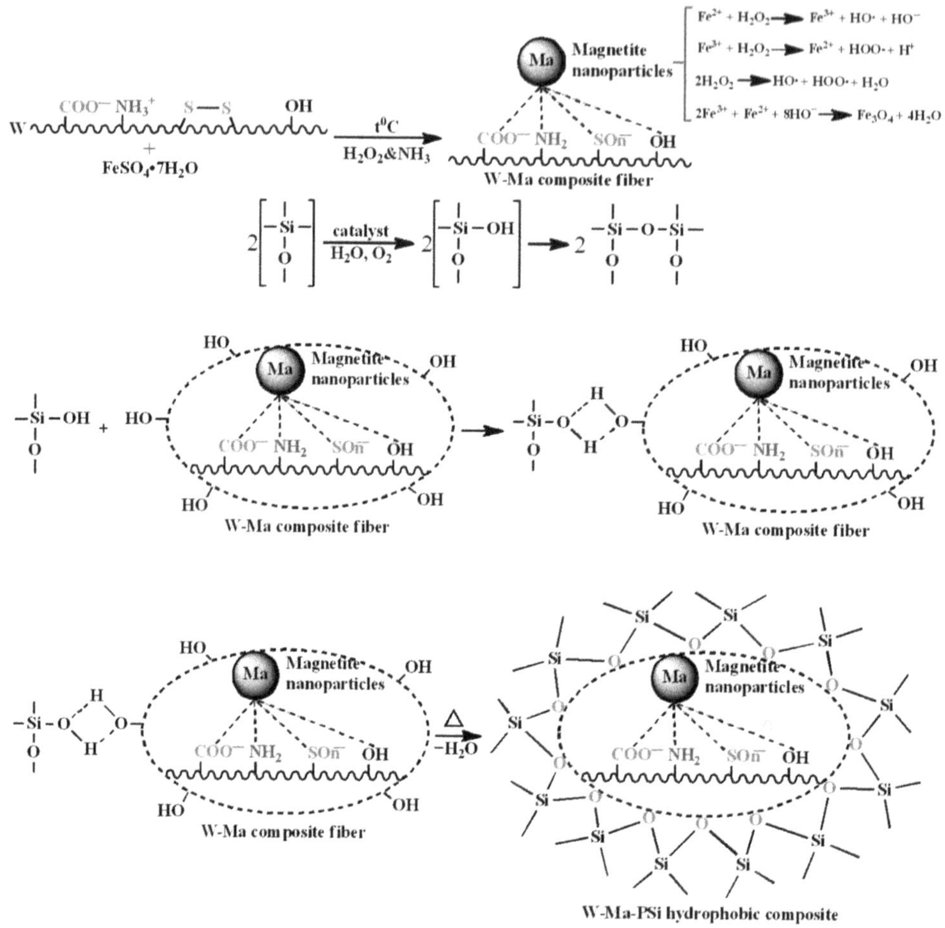

The pH value determines the reaction speed; at a low pH value the complexation of Fe^{2+} ions and the elimination of hydroxyl radicals due to the excess of protons in the solution occurs, leading to a decrease in the reaction speed (Gallard et al., 1998; Xu et al., 2009; Tang & Huang, 2016). A too high pH value leads to a decrease in the reaction speed, because it precipitates ferric hydroxide ($Fe(OH)_3$), having the effect of decreasing the concentration of the Fe^{3+} species in the solution (Pignatello, 1992; Cai et al., 2021). So, the kinetics of the Fenton oxidation-reduction process depend on the solubility of the iron ion species present in the solution at a certain pH value, and the reaction speed is influenced by the solubility of Fe^{3+} ions.

To eliminate these shortcomings, the studies were carried out at pH = 11, and the dominant species in the solution will be the superoxide radical anion, along with the hydroperoxyl radical. Under these conditions, the optimal molar ratio (Fe^{2+}: Fe^{3+} = 1:2) necessary for the formation of magnetite (Fe_3O_4) was reached.

The hydrophobization step occurs with the reticulation of polysiloxane change around the W-Ma fibers and the fixing of polysiloxane on solid wool surfaces with Si-O-Si dipoles aligned in the direction of the solid substrate (magnetic composite) and the methyl substituents ($-CH_3$) densely grouped and oriented outwards, leading to the formation of a superhydrophobic molecular layer (Cojocaru et al., 2017; Condurache et al., 2022b).

Table 1 shows the sorption capacities of wool, W-Ma magnetic composite and W-Ma-PSi superhydrophobic magnetic composite for oils and water.

As one may see from Table 1, the sorbtion capacities of both magnetic composites are only about one half from those of the raw wool (somewhat higher for the hydrophobic one). However, the sorbtion of water is less much less for hydrophobic W-Ma-PSi composite, this material representing a compromise of interest for the retention of oil products from contaminated waters.

Table 1. Sorbtion capacities for oils of wool and wool-based composites (Condurache et al., 2022b)

Sorbate Sorbent	MO-15W* (g/g)	MO-5W* (g/g)	Dodecan (g/g)	Water (g/g)
W	22,54	17,86	6,65	6,83
W-Ma	11,70	10,80	3,30	4,47
W-Ma-PSi	12,08	11,24	3,43	0,93

*MO-15W and MO-5W motor oils; viscosity (at 20 °C), 279 cP; respectively 133 cP

Bio-Oriented Keratin Materials

Keratin is obtained by the de-structuration of keratinous materials by different processes (Moore et al., 2016). It possesses a number of intrinsic characteristics, such as biocompatibility, biodegradability, receptivity to the specific biological environment and unique physicochemical properties. These properties were explored in order to produce nanocarriers of active principles for various biomedical and environmental applications (cancer treatment, tissue engineering, wound healing, and the removal of hazardous substances from wastewater: heavy metal ions etc.) (Jahangirian et al., 2017; Cañedo-Dorantes & Cañedo-Ayala, 2019; Gao et al., 2019; Varanko et al., 2020; Ferroni & Varchi, 2021; Lu et al., 2021a; Ranjit et al., 2022).

One of the most widely explored applications of keratin as a drug carrier involves the encapsulation of drugs in keratin nanoparticles based on its hydrophobic-hydrophilic behavior due to the presence of functional groups (-NH_2, -COOH, and -SH), allowing the physical incorporation or covalent attachment of different molecules (drugs, targeting moieties, dyes). Moreover, the presence of "Arg–Gly–Asp" and "Leu–Asp–Val" sequences on the protein backbone could ensure a certain degree of intrinsic tumor-targeting ability by specifically binding the vitronectin integrin receptors overexpressed in cancer cells (Ferroni & Varchi, 2021).

Cancer is one of the incurable diseases and one of the main causes of death worldwide. Despite the progress made in the use of nanotechnologies in drug delivery systems, due to insufficient absorption of antitumor substances, side effects and low bioavailability of drugs lead to a decrease in the effectiveness of the current treatments. Thus, by binding tumor-specific ligands on the surface of keratin nanoparticles, the selectivity of the tumor site can be improved, and the effectiveness of the applied treatment can be increased (Lu et al., 2021a; Lu et al., 2021b).

The presence of a large amount of cysteine can be efficiently exploited for the controlled release of drugs at the tumor site. Lu et al. (2021a) used keratin incorporating photosensitizers with specific ligands in the treatment of cancer by photodynamic therapy. The effectiveness of this method is limited by the hypoxic tumor microenvironment and the insufficient delivery of the photosensitizer (methylene blue) directed to the tumors, which can be amplified by the lack of a well-controlled nitric oxide release system at the target site. However, it was found that keratin generates nitric oxide, leading to increased efficiency of photodynamic therapy (Ferroni & Varchi, 2021; Lu et al., 2021a).

Keratin can be used in different forms (films, fibers, hydrogel, sponges) with applications in repairing damaged or diseased tissues, due to its mechanical and biological properties, self-assembly capacity, presence of cell-binding domains and bacteriostatic effects that distinguish keratin from many conventional biomaterials (Noishiki et al., 1982; O'Brien, 2011; Arslan et al., 2017; Ranjit et al., 2022).

Biofunctionalization and bacteriostatic properties of keratin (can be improved by adding natural or synthetic polymers or by introducing antimicrobial agents, while maintaining the biological compatibility (Tallawi et al., 2015; Khandaker et al., 2017; Ranjit et al., 2022).

CONCLUSION

The thermal decomposition of hydrogen peroxide is a complex reaction, and the degree to which wool fibers are oxidized by hydrogen peroxide solutions is influenced by concentration, temperature, pH, catalysts and duration of treatment. The presence of easily oxidizable substances and cysteine catalyzes the decomposition process of hydrogen peroxide, leading to the appearance of radical oxygen species.

The oxidation of wool fibers takes place by means of the superoxide radical anion ($\bullet O_2^-$), which predominates in the system under the designed experimental conditions. The appearance of the hydroperoxyl (HOO•) and superoxide ($\bullet O_2^-$), radicals, as well as their predominance in the solution, is closely related to the pH value at which the work is done.

The treatment with hydrogen peroxide results in the oxidation of the S-S disulfide bonds of cystine (present in wool keratin) to cysteic acid. Following the treatment with hydrogen peroxide in a basic medium, the cysteic acid content increases, but the cystine content decreases.

The practical applications of the oxidation of wool fibers consist of bleaching, carbonization, chlorination and modification of their morphological structure. Thus, new materials can be obtained that can be used to retain oils on the water surface or transition metal cations.

REFERENCES

Abou El-Kheir, A., Haggag, K., Mowafi, S. I., & El-Sayed, H. (2014). Microwave-assisted bleaching of wool fabrics. *Journal of Natural Fibers*, *12*(2), 97–107. doi:10.1080/15440478.2014.901202

Alexander, P., Carter, D., & Earland, C. (1951). The rôle of the disulphide bond in reactions which render wool non-felting. *Journal of the Society of Dyers and Colourists*, *67*(1), 23–27. doi:10.1111/j.1478-4408.1951.tb02670.x

Alexander, P., Fox, M., & Hudson, R. F. (1950). The reaction of oxidizing agents with wool. 5. The oxidation products of the disulphide bond and the formation of a sulphonamide in the peptide chain. *The Biochemical Journal, 49*(2), 129–138. doi:10.1042/bj0490129 PMID:14858297

Arifoglu, M., Marmer, W. N., & Carr, C. M. (1990). Sequential oxidative and reductive bleaching of wool in a single bath. *Textile Research Journal, 60*(6), 319–328. doi:10.1177/004051759006000602

Arslan, Y. E., Sezgin Arslan, T., Derkus, B., Emregul, E., & Emregul, K. C. (2017). Fabrication of human hair keratin/jellyfish collagen/eggshell-derived hydroxyapatite osteoinductive biocomposite scaffolds for bone tissue engineering: From waste to regenerative medicine products. *Colloids and Surfaces. B, Biointerfaces, 154*, 160–170. doi:10.1016/j.colsurfb.2017.03.034 PMID:28334693

Asmus, K.-D., & Bonifačić, M. (2000). Free radical chemistry. Handbook of Oxidants and Antioxidants in Exercise, 3–54. doi:10.1016/B978-044482650-3/50001-8

Atef El-Sayed, A., Salama, M., & Kantouch, A. A. M. (2014). Wool micro powder as a metal ion exchanger for the removal of copper and zinc. *Desalination and Water Treatment, 56*(4), 1010–1019. doi :10.1080/19443994.2014.941300

Bard, A. J., Parsons, R., & Jordan, J. (1985). Standard potentials in aqueous solution. International Union of Pure and Applied Chemistry, (1st ed.). M. Dekker. doi:10.1201/9780203738764

Bateup, B. O., Christoe, J. R., Jones, F. W., Poole, A. J., Skourtis, C., & Westmoreland, D. J. (1996). Effluent Management. In *Proceedings Top-Tech '96* (pp. 388–407). CSIRO Division of Wool Technology.

Behar, D., Czapski, G., Rabani, J., Dorfman, L. M., & Schwarz, H. A. (1970). Acid dissociation constant and decay kinetics of the perhydroxyl radical. *Journal of Physical Chemistry, 74*(17), 3209–3213. doi:10.1021/j100711a009

Bereck, A., Dillbohner, S., Weber, B., Riegel, D., Mosel, J., Reper, J. M., & Brakelmann, A. (1997). A simple method for the objective characterisation of fabric softness. Part 1: Influence of bleaching, dyeing and crosslinking of wool. *Journal of the Society of Dyers and Colourists, 113*(11), 322–326. doi:10.1111/j.1478-4408.1997.tb01853.x

Berger, A. (1959). Weissgradformeln und ihre praktische Bedeutung. *Farbe, 8*, 187.

Bielski, B. H. J., Cabelli, D. E., Arudi, R. L., & Ross, A. B. (1985). Reactivity of HO_2/O_2- radicals in aqueous solution. *Journal of Physical and Chemical Reference Data, 14*(4), 1041–1100. doi:10.1063/1.555739

Bisschops, I., & Spanjers, H. (2003). Literature review on textile wastewater characterisation. *Environmental Technology, 24*(11), 1399–1411. doi:10.1080/09593330309385684 PMID:14733393

Bradbury, J. H. (1976). The morphology and chemical structure of wool. *Pure and Applied Chemistry, 46*(2–4), 247–253. doi:10.1351/pac197646020247

Burrows, H. D., Canle, M. L., Santaballa, J. A., & Steenken, S. (2002). Reaction pathways and mechanisms of photodegradation of pesticides. *Journal of Photochemistry and Photobiology. B, Biology, 67*(2), 71–108. doi:10.1016/S1011-1344(02)00277-4 PMID:12031810

Cai, Q. Q., Lee, B. C. Y., Ong, S. L., & Hu, J. Y. (2021). Fluidized-bed Fenton technologies for recalcitrant industrial wastewater treatment–Recent advances, challenges and perspective. *Water Research, 190*, 116692. doi:10.1016/j.watres.2020.116692 PMID:33279748

Cañedo-Dorantes, L., & Cañedo-Ayala, M. (2019). Skin acute wound healing: A comprehensive review. *International Journal of Inflammation, 2019*, 1–15. doi:10.1155/2019/3706315 PMID:31275545

Cardamone, J. M., Nuñez, A., Garcia, R. A., & Aldema-Ramos, M. (2009). Characterizing wool keratin. *Research Letters in Materials Science, 1–5*, 1–5. doi:10.1155/2009/147175

Cardamone, J. M., Yao, J., & Nuñez, A. (2004). Controlling shrinkage in wool fabrics: Effective hydrogen peroxide systems. *Textile Research Journal, 74*(10), 887–898. doi:10.1177/004051750407401008

Cegarra, J., Gacen, J., Caro, M., & Pepio, M. (1988). Wool bleaching with thiourea dioxide. *Journal of the Society of Dyers and Colourists, 104*(7–8), 273–279. doi:10.1111/j.1478-4408.1988.tb01168.x

Cegarra, J., Ribé, J., & Gacén, J. (2008). Peroxide bleaching of different wools. *Journal of the Society of Dyers and Colourists, 85*(4), 147–149. doi:10.1111/j.1478-4408.1969.tb02890.x

Condurache, B., Cojocaru, C., Samoila, P., Cosmulescu, S. F., Predeanu, G., Enache, A., & Harabagiu, V. (2022a). Oxidized biomass and its usage as adsorbent for removal of heavy metal ions from aqueous solutions. *Molecules (Basel, Switzerland), 27*(6119), 1–19. doi:10.3390/molecules27186119 PMID:36144850

Condurache, B.-C., Cojocaru, C., Pascariu, P., Samoila, P., & Harabagiu, V. (2022b). Innovative nanostructured magnetite/wool/polysiloxane composite as magnetic adsorbent for oil spill removal. *Comptes Rendus. Chimie, 25*(S3), 245–260. doi:10.5802/crchim.168

Condurache, B.-C., Cojocaru, C., Samoila, P., Ignat, M., & Harabagiu, V. (2022c). Data-driven modeling and optimization of oil spill sorption by wool fibers: Retention kinetics and recovery by centrifugation. *International Journal of Environmental Science and Technology, 19*(1), 367–378. doi:10.100713762-021-03176-7

Czaplicki, Z., & Strzelecki, S. (2019). Wool carbonization with an energy-efficient drying process. *Journal of Natural Fibers, 17*(12), 1809–1818. doi:10.1080/15440478.2019.1599312

Davies, M. J. (2003). Singlet oxygen-mediated damage to proteins and its consequences. *Biochemical and Biophysical Research Communications, 305*(3), 761–770. doi:10.1016/S0006-291X(03)00817-9 PMID:12763058

Duca, G. (2022). Hydrogen peroxide in ecological and environmental chemistry. *Chemistry Journal of Moldova. General Industrial and Ecological Chemistry, 16*(2), 28–45. doi:10.19261/cjm.2021.918

Duca, G., & Travin, S. (2020). Reactions' mechanisms and applications of hydrogen peroxide. *American Journal of Physical Chemistry, 9*(2), 36–44. doi:10.11648/j.ajpc.20200902.13

Duca, G., & Vaseashta, A. (Eds.). (2020). *Handbook of Research on Emerging Developments and Environmental Impacts of Ecological Chemistry*. IGI Global. doi:10.4018/978-1-7998-1241-8

Dyer, J. M. (2018). Oxidative modification of trichocyte keratins. *The Hair Fibre: Proteins, Structure and Development,* 205–218. doi:10.1007/978-981-10-8195-8_14

Earland, C., MacRae, T. P., Weston, G. J., & Statham, K. (1955). The oxidation of wool by inorganic peracids. *Textile Research Journal*, *25*(11), 963–965. doi:10.1177/004051755502501109

Enkhzaya, S., Shiomori, K., & Oyuntsetseg, B. (2017). Removal of heavy metals from aqueous solution by adsorption using livestock biomass of Mongolia. *Journal of Environmental Science and Technology*, *10*(3), 107–119. doi:10.3923/jest.2017.107.119

Erdogan, U. H., Seki, Y., & Selli, F. (2020). Wool fibres. Handbook of Natural Fibres, 257–278. doi:10.1016/B978-0-12-818398-4.00011-6

Evans, M. G., & Uri, N. (1949). The dissociation constant of hydrogen peroxide and the electron affinity of the HO_2 radical. *Transactions of the Faraday Society*, *45*, 224–230. doi:10.1039/tf9494500224

Ferroni, C., & Varchi, G. (2021). Keratin-based nanoparticles as drug delivery carriers. *Applied Sciences (Basel, Switzerland)*, *11*(20), 9417. doi:10.3390/app11209417

Gacen, J., & Cayuela, D. (2000). Comparison of wool bleaching with hydrogen peroxide in alkaline and acidic media. *Coloration Technology*, *116*(1), 13–15. doi:10.1111/j.1478-4408.2000.tb00003.x

Gacen, J., Cegarra, J., & Caro, M. (1991). Wool bleaching with reducing agent in the presence of sodium lauryl sulphate. Part 3 - bleaching with thiourea dioxide. *Journal of Studies in Dynamics and Change*, *107*, 138–141.

Gallard, H., de Laat, J., & Legube, B. (1998). Influence du pH sur la vitesse d'oxydation de compose's organiques par FeII/H2O2. Me'canismes re'actionnels et mode'lisation. *New Journal of Chemistry*, *22*(3), 263–268. doi:10.1039/a708335a

Gao, F., Li, W., Deng, J., Kan, J., Guo, T., Wang, B., & Hao, S. (2019). Recombinant human hair keratin nanoparticles accelerate dermal wound healing. *ACS Applied Materials & Interfaces*, *11*(20), 18681–18690. doi:10.1021/acsami.9b01725 PMID:31038908

Ghiocel, I. E., Muresan, A., Irina, E. A., Ioanid, A., & Muresan, R. (2005). *Modification of the surface of textile materials*. PERFORMANTICA Publishing.

Giguere, P. A. (1975). *Complements au nouveau traite de chemie minerale – No. 4 Peroxyde d'Hydrogene et Polyoxydes d'Hydrogene*. Mason.

Gill, A. C., Ritchie, M. A., Hunt, L. G., Steane, S. E., Davies, K. G., Bocking, S. P., Rhie, A. G. O., Bennett, A. D., & Hope, J. (2000). Post-translational hydroxylation at the N-terminus of the prion protein reveals presence of PPII structure in vivo. *The EMBO Journal*, *19*(20), 5324–5331. doi:10.1093/emboj/19.20.5324 PMID:11032800

Goldstone, J. V., & Voelker, B. M. (2000). Chemistry of superoxide radical in seawater: Cdom associated sink of superoxide in coastal waters. *Environmental Science & Technology*, *34*(6), 1043–1048. doi:10.1021/es9905445

Gousterova, A., Nustorova, M., Goshev, I., Christov, P., Braikova, D., Tishinov, K., Haertlé, T., & Nedkov, P. (2003). Alkaline hydrolysate of waste sheep wool aimed as fertilizer. *Biotechnology, Biotechnological Equipment*, *17*(2), 140–145. doi:10.1080/13102818.2003.10817072

Gupta, P., Lakes, A., & Dziubla, T. (2016). A free radical primer. *Oxidative Stress and Biomaterials*, 1–33. doi:10.1016/B978-0-12-803269-5.00001-2

Haber, F., & Weiss, J. (1932). Über die katalyse des hydroperoxydes. *Naturwissenschaften*, *20*(51), 948–950. doi:10.1007/BF01504715

Haber, F., & Willstätter, R. (1931). Unpaarigkeit und radikalketten im reaktionsmechanismus organischer und enzymatischer vorgänge. [A and B Series]. *Berichte der Deutschen Chemischen Gesellschaft Abteilung B: Abhandlungen*, *64*(11), 2844–2856. doi:10.1002/cber.19310641118

Höcker, H. (2002). Fibre morphology. *Wool*, 60–79. doi:10.1533/9781855737648.60

Holder, T. M. (1974). A new technique for treating esophageal varices. *Journal of Pediatric Surgery*, *9*(3), 431. doi:10.1016/S0022-3468(74)80322-2

Hunter, R. S. (1942). Photoelectric tristimulus colorimetry with three filters. *Journal of the Optical Society of America*, *32*(9), 509–538. https://doi.org/doi:10.1364/josa.32.000509. doi:10.1364/JOSA.32.000509

Hurren, C. (2008). Dyeing and coloring tests for fabrics. *Fabric Testing*, 255–274. doi:10.1533/9781845695064.255

IWTO. (2022). *An annual compilation of the global wool market`s production statistics from farm to retail-world sheep numbers & wool production*. Prepared by Poimena Analysis & Delta Consultants, 28–43. https://iwto.org/resources/statistics

Jahangirian, H., Ghasemian lemraski, E., Webster, T. J., Rafiee-Moghaddam, R., & Abdollahi, Y. (2017). A review of drug delivery systems based on nanotechnology and green chemistry: Green nanomedicine. *International Journal of Nanomedicine*, *12*, 2957–2978. doi:10.2147/IJN.S127683 PMID:28442906

Joiner, A. (2010). Colorimetric evaluation of tooth color. *Color Measurement*, 343–371. doi:10.1533/9780857090195.2.343

Jones, L. N., & Rivett, D. E. (1997). The role of 18-methyleicosanoic acid in the structure and formation of mammalian hair fibres. *Micron (Oxford, England)*, *28*(6), 469–485. doi:10.1016/S0968-4328(97)00039-5 PMID:9519472

Jung, H., & Sato, T. (2013). Comparison between the color properties of whiteness index and yellowness index on the CIELAB. *Textile Coloration and Finishing.*, *25*(4), 241–246. doi:10.5764/TCF.2013.25.4.241

Kehrer, J. P. (2000). The Haber–Weiss reaction and mechanisms of toxicity. *Toxicology*, *149*(1), 43–50. doi:10.1016/S0300-483X(00)00231-6 PMID:10963860

Khadke, P., Sethuraman, P., Kandasamy, P., Parthasarathi, S., & Shukla, A. (2009). A self-supported direct borohydride-hydrogen peroxide fuel cell system. *Energies*, *2*(2), 190–201. doi:10.3390/en20200190

Khandaker, M., Riahinezhad, S., Sultana, F., Morris, T., Wolf, R., & Vaughan, M. (2017). Effect of collagen-polycaprolactone nanofibers matrix coating on the in vitro cytocompatibility and in vivo bone responses of titanium. *Journal of Medical and Biological Engineering*, *38*(2), 197–210. doi:10.100740846-017-0312-7 PMID:29861706

Koppenol, W. H. (1989). Generation and thermodynamic properties of oxyradicals. In C. Vigo-Pelfrey (Ed.), *Focus on Membrane Lipid Oxidation* (Vol. I, pp. 1–13). CRC Press.

Koppenol, W. H. (2001). The Haber-Weiss cycle – 70 years later. *Redox Report*, *6*(4), 229–234. doi:10.1179/135100001101536373 PMID:11642713

Korjenic, A., Klarić, S., Hadžić, A., & Korjenic, S. (2015). Sheep wool as a construction material for energy efficiency improvement. *Energies*, *8*(6), 5765–5781. doi:10.3390/en8065765

Lapsirikul, W., Cord-Ruwisch, R., & Ho, G. (1994). Treatment of wool scouring effluent by anaerobic bioflocculation. *Water Science and Technology*, *30*(12), 375–384. doi:10.2166/wst.1994.0637

Litter, M. I. (2005). Introduction to photochemical advanced oxidation processes for water treatment. *Environmental Photochemistry*, *M*(Part II), 325–366. doi:10.1007/b138188

Lu, T.-Y., Chiang, C.-Y., Fan, Y.-J., Jheng, P.-R., Quiñones, E. D., Liu, K.-T., Kuo, S.-H., Hsieh, H. Y., Tseng, C.-L., Yu, J., & Chuang, E.-Y. (2021b). Dual-targeting glycol chitosan/heparin-decorated polypyrrole nanoparticle for augmented photothermal thrombolytic therapy. *ACS Applied Materials & Interfaces*, *13*(8), 10287–10300. doi:10.1021/acsami.0c20940 PMID:33615773

Lu, T.-Y., Lu, W.-F., Wang, Y.-H., Liao, M.-Y., Wei, Y., Fan, Y.-J., Chuang, E. Y., & Yu, J. (2021a). Keratin-based nanoparticles with tumor-targeting and cascade catalytic capabilities for the combinational oxidation phototherapy of breast cancer. *ACS Applied Materials & Interfaces*, *13*(32), 38074–38089. doi:10.1021/acsami.1c10160 PMID:34351754

Luo, W., Westland, S., Ellwood, R., & Pretty, I. (2005). Evaluation of whiteness formulae for teeth. *Proceedings of the 10th Congress of the International Color Association, Granada, Spain., 3–7*. https://www.researchgate.net/publication/240633974

Maclaren, J. A., & Milligan, B. (1981). Wool Science: The chemical reactivity of the wool fibre, Science Press, Marrickville, 1–17, 53–75.

Maiolo, K., Marin, P. D., Bridges, T. E., & Heithersay, G. S. (2007). Evaluation of a combined thiourea and hydrogen peroxide regimen to bleach bloodstained teeth. *Australian Dental Journal*, *52*(1), 33–40. doi:10.1111/j.1834-7819.2007.tb00463.x PMID:17500162

Martínez-Alvarez, O., Chamorro, S., & Brenes, A. (2015). Protein hydrolysates from animal processing by-products as a source of bioactive molecules with interest in animal feeding : A review. *Food Research International*, *73*, 204–212. doi:10.1016/j.foodres.2015.04.005

Millington, K. R. (2013). Bleaching and whitening of wool: photostability of whites. *The Coloration of Wool and Other Keratin Fibres,* 131–155. doi:10.1002/9781118625118.ch5

Moore, K. E., Mangos, D. N., Slattery, A. D., Raston, C. L., & Boulos, R. A. (2016). Wool deconstruction using a benign eutectic melt. *RSC Advances*, *6*(24), 20095–20101. doi:10.1039/C5RA26516A

National Institute of Statistics. *Livestock numbers and animal production, in 2018*. (2019). 1–69. https://insse.ro/cms/ro/tags/efectivele-de-animale-si-productia-animala

National Institute of Statistics. *Existing animal numbers on December 1, 2021*. (2022). *122*, 1–4. https://insse.ro/cms/ro/tags/comunicat-efective-animale

Nayak, R., & Rao, J. R. (2003). Ferrous-ferric oxidation in pickle liquor. *Journal of Scientific and Industrial Research*, *62*, 1094–1097.

Nikiforova, T., Kozlov, V., & Islyaikin, M. (2019). Sorption of d-metal cations by keratin from aqueous solutions. *Journal of Environmental Chemical Engineering*, *7*(5), 103417. doi:10.1016/j.jece.2019.103417

Noishiki, Y., Ito, H., Miyamoto, T., & Inagaki, H. (1982). Application of denatured wool keratin derivatives to an antithrombogenic biomaterial-vascular graft coated with a heparinized keratin derivative. *Kobunshi Ronbunshu*, *39*(4), 221–227. doi:10.1295/koron.39.221

O'Brien, F. J. (2011). Biomaterials & scaffolds for tissue engineering. *Materials Today*, *14*(3), 88–95. doi:10.1016/S1369-7021(11)70058-X

O'Shea, K. E., & Dionysiou, D. D. (2012). Advanced oxidation processes for water treatment. *The Journal of Physical Chemistry Letters*, *3*(15), 2112–2113. doi:10.1021/jz300929x

Pearson, H. P. (2008). Chlorinated wool. *Journal of the Society of Dyers and Colourists*, *25*(3), 81–83. doi:10.1111/j.1478-4408.1909.tb00461.x

Pędziwiatr, P., Mikołajczyk, F., Zawadzki, D., Mikołajczyk, K., & Bedka, A. (2018). Decomposition of hydrogen perodixe - kinetics and review of chosen catalysts. *Acta Innovations*, *26*(26), 45–52. doi:10.32933/ActaInnovations.26.5

Picazo, C., & Molin, M. (2021). Impact of hydrogen peroxide on protein synthesis in yeast. *Antioxidants*, *10*(6), 952. doi:10.3390/antiox10060952 PMID:34204720

Pignatello, J. J. (1992). Dark and photoassisted iron (Fe^{3+})-catalyzed degradation of chlorophenoxy herbicides by hydrogen peroxide. *Environmental Science & Technology*, *26*(5), 944–951. doi:10.1021/es00029a012

Plant, L., Jeff, M., & Shelley, S. (1994). Hydrogen peroxide: A potent force to destroy organics in wastewater. *Chemical Engineering (Albany, N.Y.)*, 16–20.

Poole, A. J. (2004). Treatment of biorefractory organic compounds in wool scour effluent by hydroxyl radical oxidation. *Water Research*, *38*(14-15), 3458–3464. doi:10.1016/j.watres.2004.06.001 PMID:15276763

Poole, A. J., Cord-Ruwisch, R., & Jones, F. W. (1999). Biological treatment of chemically flocculated agro-industrial waste from the wool scouring industry by an aerobic process without sludge recycle. *Water Research*, *33*(9), 1981–1988. doi:10.1016/S0043-1354(98)00391-1

Popescu, C., & Wortmann, F.-J. (2010). Wool - Structure, mechanical properties and technical products based on animal fibres. *Industrial Applications of Natural Fibres*, 255–266. doi:10.1002/9780470660324.ch12

Rageliené, L., Treigiené, R., & Musnickas, J. (2009). Physical properties of human hair dyed using non-ionic surfactants. *Materials Science*, *15*(3), 277–282. https://www.researchgate.net/publication/228476027

Rajabinejad, H., Bucişcanu, I. I., & Maier, S. S. (2019). Current approaches for raw wool waste management and unconventional valorization: A review. *Environmental Engineering and Management Journal, 18*(7), 1439–1456. doi:10.30638/eemj.2019.136

Rajabinejad, H., Zoccola, M., Patrucco, A., Montarsolo, A., Rovero, G., & Tonin, C. (2018). Physicochemical properties of keratin extracted from wool by various methods. *Textile Research Journal, 88*(21), 2415–2424. doi:10.1177/0040517517723028

Ranjit, E., Hamlet, S., George, R., Sharma, A., & Love, R. M. (2022). Biofunctional approaches of wool-based keratin for tissue engineering, *Journal of Science: Advanced Materials and Devices, 7*(1), doi:10.1016/j.jsamd.2021.10.001

Rao, P. S., & Hayon, E. (1975). Redox potentials of free radicals. IV. Superoxide and hydroperoxy radicals O_2– and HO_2. *Journal of Physical Chemistry, 79*(4), 397–402. doi:10.1021/j100571a021

Rarata, G., & Surmacz, P. (2009). Hydrogen peroxide HTP class as a universal medium for propulsion and oxidiser. *Prace Instytutu Lotnictwa, 7*, 125–158.

Reich, C., & Su, D. T. (2001). Hair conditioners. Hair damage and cuticle surface. Handbook of Cosmetic Science and Technology. Marcel Dekker, Inc. 331–335

Robbins, C. R. (2012). Chemical and physical behavior of human hair. In Chemical and Physical Behavior of Human Hair. doi:10.1007/978-3-642-25611-0

Rose, A. L., & Waite, T. D. (2005). Reduction of organically complexed ferric iron by superoxide in a simulated natural water. *Environmental Science & Technology, 39*(8), 2645–2650. doi:10.1021/es048765k PMID:15884361

Schulte, P., Bayer, A., Kuhn, F., Luy, T., & Volkmer, M. (1995). H_2O_2/O_3, H_2O_2/UV and H_2O_2/Fe^{2+} processes for the oxidation of hazardous wastes. *Ozone Science and Engineering, 17*(2), 119–134. doi:10.1080/01919519508547541

Simpson, W. S. (1999) Physics and chemistry of wool yellowing. *WRONZ Report R217*. Wool Research Organisation of New Zealand. 36

Sinkiewicz, I., Śliwińska, A., Staroszczyk, H., & Kołodziejska, I. (2016). Alternative methods of preparation of soluble keratin from chicken feathers. *Waste and Biomass Valorization, 8*(4), 1043–1048. doi:10.100712649-016-9678-y

Smith, A. L., & Harris, M. (1936a). Oxidation of wool : Effect of hydrogen peroxide on wool. *Journal of Research of the National Bureau of Standards, 16*(4), 301–307. https://nvlpubs.nist.gov/nistpubs/jres/16/jresv16n4p301_a1b.pdf. doi:10.6028/jres.016.017

Smith, A. L., & Harris, M. (1936b). Oxidation Of Wool : The lead acetate test for hydrogen peroxide bleached wool. *Journal of Research of the National Bureau of Standards, 16*(4), 309–313. doi:10.6028/jres.016.018

St. Laurent, J. B., de Buzzaccarini, F., De Clerck, K., Demeyere, H., Labeque, R., Lodewick, R., & van Langenhove, L. (2007). Laundry cleaning of textiles. *Handbook for Cleaning/Decontamination of Surfaces*, 57–102. doi:10.1016/B978-0-444-51664-0.50003-6

Stadtman, E. R., & Levine, R. L. (2006). Chemical modification of proteins by reactive oxygen species. *Redox Proteomics,* 1–23. doi:10.1002/0471973122.ch1

Steinhardt, J., Fugitt, C. H., & Harris, M. (1940). Combination of wool protein with acid and base: The effect of tem perature on the titration curve. *Textile Research, 11*(2), 72–94. doi:10.1177/004051754001100203

Sulyman, M., Namiesnik, J., & Gierak, A. (2017). Greener cleaner : Sheep wool fiber as renewable sources for oil spill cleanup. *International Journal of Advances in Science, Engineering and Technology*, *5*(2), 84–92. http://iraj.in

Sumner, R. M. W., Young, S. R., & Upsdell, M. P. (2003). Wool yellowing and pH within Merino and Romney fleeces, *Proceedings of the New Zealand Society of Animal Production*, Queenstown, *63,* 155-159.

Swift, J. A., & Smith, J. R. (2001). Microscopical investigations on the epicuticle of mammalian keratin fibres. *Journal of Microscopy, 204*(3), 203–211. doi:10.1046/j.1365-2818.2001.00957.x PMID:11903797

Tallawi, M., Rosellini, E., Barbani, N., Cascone, M. G., Rai, R., Saint-Pierre, G., & Boccaccini, A. R. (2015). Strategies for the chemical and biological functionalization of scaffolds for cardiac tissue engineering: A review. *Journal of the Royal Society, Interface*, *12*(108), 20150254. doi:10.1098/rsif.2015.0254 PMID:26109634

Tang, W. Z., & Huang, C. P. (2016). 2,4-Dichlorophenol oxidation kinetics by Fenton's reagent. *Environmental Technology*, *17*(12), 1371–1378. doi:10.1080/09593330.1996.9618465

Varanko, A., Saha, S., & Chilkoti, A. (2020). Recent trends in protein and peptide-based biomaterials for advanced drug delivery. *Advanced Drug Delivery Reviews*, *156*, 133–187. Advance online publication. doi:10.1016/j.addr.2020.08.008 PMID:32871201

Vaseashta, A., Duca, G., & Travin, S. (Eds.). (2022). *Handbook of Research on Water Sciences and Society* (Vols. 1–2). IGI Global., doi:10.4018/978-1-7998-7356-3

Verlackt, C. C. W., Van Boxem, W., Dewaele, D., Lemière, F., Sobott, F., Benedikt, J., Neyts, E. C., & Bogaerts, A. (2017). Mechanisms of peptide oxidation by hydroxyl radicals: Insight at the molecular scale. *The Journal of Physical Chemistry C, 121*(10), 5787–5799. doi:10.1021/acs.jpcc.6b12278

Von Bergen, W. (1963). *Wool Handbook* (3rd ed., Vol. 1). John Wiley & Sons.

Wang, X., Shen, X., & Xu, W. (2012). Effect of hydrogen peroxide treatment on the properties of wool fabric. *Applied Surface Science*, *258*(24), 10012–10016. doi:10.1016/j.apsusc.2012.06.065

Westland, S. (2015). CIE Whiteness. Encyclopedia of Color Science and Technology, 1–5. doi:10.1007/978-3-642-27851-8_5-1

Weston, G. J. (1955). The infra-red spectrum of peracetic acid-treated wool. *Biochimica et Biophysica Acta*, *17*, 462–464. doi:10.1016/0006-3002(55)90407-8 PMID:13249992

Winder, L. M., Rea, A., Baronian, K., Webber, J., & Muller, B. (1998). Unravelling the causes of wool yellowing. Part II: Involvement of bacteria. *Proceedings of the New Zealand Society of Animal Production, 58*, 277–280.

Wojciechowska, E., Włochowicz, A., & Wesełucha-Birczyńska, A. (2000). Application of Fourier-transform infrared and Raman spectroscopy to the study of the influence of orthosilicic acid on the structure of wool fibre. *Journal of Molecular Structure, 555*(1-3), 397–406. doi:10.1016/S0022-2860(00)00626-8

Wolfram, L., Hall, K., & Hui, I. (1970). The mechanism of hair bleaching. *Journal of the Society of Cosmetic Chemists, 21*, 875–900.

Woodard & Curran, Inc. (2006). Methods for treating wastewater from industry. *Industrial Waste Treatment Handbook,* 149–334. doi:10.1016/B978-075067963-3/50009-6

Woodhead, A., Harrigan, F., & Church, J. (1997). Assessment of wool chlorination by infrared spectroscopy. *Vibrational Spectroscopy, 15*(2), 179–189. doi:10.1016/S0924-2031(97)00040-4

Xinhua, L., Shunbing, Z., Xizeng, Z., & Xiaoli, W. (2014). Hazard assessment of hydrogen peroxide with polyphosphonic acid by vent sizing package 2. *Procedia Engineering, 84*, 427–435. doi:10.1016/j.proeng.2014.10.453

Xu, X.-R., Li, X.-Y., Li, X.-Z., & Li, H.-B. (2009). Degradation of melatonin by UV, UV/H_2O_2, Fe^{2+}/H_2O_2 and UV/Fe^{2+}/H_2O_2 processes. *Separation and Purification Technology, 68*(2), 261–266. doi:10.1016/j.seppur.2009.05.013

Yilmazer, D., & Kanik, M. (2009). Bleaching of wool with sodium borohydride. *Journal of Engineered Fibers and Fabrics, 4*(3). doi:10.1177/155892500900400305

Zoccola, M., Montarsolo, A., Mossotti, R., Patrucco, A., & Tonin, C. (2015). Green hydrolysis is an emerging technology to turn wool waste into organic nitrogen fertilizer. *Waste and Biomass Valorization, 6*(5), 891–897. doi:10.100712649-015-9393-0

Chapter 18
Investigation of Redox Processes in Paliastomi Lake

Ketevan Kupatadze
https://orcid.org/0000-0003-4645-3374
Ilia State University, Georgia

ABSTRACT

Paliastomi lake is a small lake near the city Poti (Georgia), connected to the Black Sea by a narrow channel. Its surface area is 17.3 km2 and the depth is 2.6 m. The lake is included within the boundaries of the kolkheti National Park. Redox conditions in lakes generally are determined by the balance between the decomposition of organic matter and the supply of oxygen by circulation or vertical mixing of the water. The circulation in lakes is largely controlled by density differences, which are generally a consequence of temperature difference. That's why temperature and oxygen were also investigated. The chapter will present the results of a study of redox reactions taking place in the lake. The study is conducted by analyzing the results obtained by measuring special sensors. The chapter will discuss how this process affects the Black Sea coastline and the local flora and fauna of the lake.

INTRODUCTION

There are more than 26,000 rivers in Georgia, 99% of them are less than 25 km long. The number of average and large rivers is relatively small, however, their participation in the total volume of renewable water resources is significant. The volume of renewable surface water generated during the year is about 66 billion m^3. The rivers of Georgia belong to the basins of the Black and Caspian Seas. The Black Sea basin includes Engur-Rioni, Chorokhi-Achariskalli, and Kodori-Bzifi basin areas. Mtkvari, Alazan-Iori, and Khram-Debeda basin areas are part of the Caspian Sea basin. Rivers are unevenly distributed in the territory of Georgia. 75% of Georgia's water resources originate from Western Georgia. There are about 860 lakes in the country, the vast majority of which are small (Trapaidze, 2012; Abramia, 2022; Lomsadze, 2016; Kupatadze, 2022).

DOI: 10.4018/979-8-3693-0512-6.ch018

BACKGROUND

Study Site

Figure 1. Paliastomi Lake

Paliastomi Lake was selected as a research site (Figure 1). Paliastomi Lake is a flowing lake located on the Odish-Guria lowland near the city of Poti. The surface area of the lake is 18.2 km^2. Basin area - 547 km^2, maximum depth - 3.2 m, average depth 2.6 m. The volume of water is 52 mln. m^3. It is located 0.3 m below sea level. The Fichori River joins the lake, and the Kaparchina River flows out. The high-water level is characteristic in spring, summer, and autumn, and low in winter. Water temperature increases in July-August (25.1°C), and decreases in January (5.2°C). The lake is part of Kolkheti National Park (Aphkhazava et al., 1984).

According to the Ramsar Convention, Paliastomi Lake is a protected area of international importance and is part of the Kolkheti National Park. Ramsar site N° 893 "Wetlands of Central Kolkheti" consists of three different peat bog complexes (Anaklia-Churia, Nabada, and Fichora-Paliastomi), Paliastomi Lake, the surrounding sunken forests, the Black Sea coastal zone and the lower reaches of the Khobi (or Khobiskali) and Rion rivers. A total of 33,710 hectares (including the marine part of 55,500 hectares) in the central part of the Black Sea coastal plain, in Khobi and Lachkhuti administrative districts, and in the territory of Poti (Dassenakis et al., 2006).

Paliastomi Lake was considered one of the most highly productive reservoirs. The impact of environmental factors has led to the salinization of the lake with associated changes in water hydrology, and a sharp decrease in the biodiversity of the lake (the number of fish species has decreased from 40 to 24). The lake is fed by rainwater and tributaries.

Paliastomi Lake is separated from the sea by a 300-400 meter wide and 2-meter high sand section. The lake has a large amount of sediments and vegetation. Rivers (Fichori, Tkhorika, etc.) and lakes (Paliastomi, Imnati, Maltakva, etc.) receive water from large reservoirs, and the water level ranges from 0.2-0.7 m in swamps, 0.5-3 m in lakes, and 1- in rivers. 21 m. Until 1924, Paliastomi Lake was a freshwater lake: water flowed from the lake through the Kaparchina River, which originates near the northwestern shore of the lake. Leaving the lake, the river turns sharply to the south, runs almost parallel to the western shore, and flows into the sea 11 km from its source. Sea water freely flows into this strait during heavy storms and this has made the lake salty.

Purpose of Research

The aim of the study was to investigate the current redox potential of Paliastomi lake, its accompanying processes, and its impact on the local flora and fauna.

CHARACTERIZATION OF CONDUCTED STUDIES

Water quality is mainly determined by the environment, climatic conditions, and anthropogenic activities (Snoeyink et al., 2021; Cao et al., 2020). Water quality indicators used by environmental agencies include the following characteristics: turbidity; color; temperature; taste (for drinking water); smell; pH; conductivity; hardness; dissolved oxygen; dissolved inorganic anions (CO_3^{2-}, HCO_3^-, Cl^-, SO_4^{2-} and NO_3^-) and cations (Na^+, K^+, Mg^{2+}, Ca^{2+}, Al^{3+} and NH^{4+}); also dissolved organic substances (organic acids and phenol); presence or absence of microbiological organisms (bacteria); presence/absence of animals, algae, flora and fauna; presence/absence of garbage, sewage waste, oil;

Electrical conductivity is also considered one of the important characteristics of water. It depends on water temperature and mineralization (Smith, 2019; VanLoon, 2018). Natural waters are a mixture of strong and weak electrolytes. Water mineralization is mainly determined by: Na^+, K^+, Ca^{2+}, Cl^-, SO_4^{2-}, HCO_3^-. It is due to the presence of these ions that the conductivity of natural waters is determined. Electrical conductivity has no fixed norms, although e.g. 2000 μS/cm corresponds to about 1000 ppt salinity. Also, for example, we can mention that the conductivity of bi-distillate is 0.05 μS/cm, and 25 ^{0}C sea water is 50,000 μS/cm.

TDS or Totally Dissolved Particles provides additional information on the water hardness method (Smith, 2019; Mattson, 1993). 150-250 ppm is a good indicator of water clarity. That is, the particles are dissolved in a very small amount, and such an indicator is also useful for drinking water. More than 1200 ppm is already considered polluted. According to the received data, the data of Paliastom Lake sometimes exceeds the norm and this is mainly observed during high temperatures or hot seasons.

Amount of **Dissolved Oxygen (DO) -** Oxygen is transferred to the water as a result of contact with air, as well as released by the photosynthesis of algae. The amount of oxygen dissolved in water depends on temperature, turbidity index, and mineralization.

Measurement of Total Water Hardness

Environmental chemistry uses the hardness index to determine the concentration of Ca^{2+} and Mg^{2+} ions in water (USEPA. 1978b). From a chemical point of view, the hardness index is a combined concentration of Ca^{2+} and Mg^{2+} ions. The most common type of water hardness is temporary hardness. This is caused by calcium and magnesium compounds dissolved in water. The natural source of these salts are rocks containing calcium and magnesium salts - limestone ($CaCO_3$) and dolomite ($CaCO_3 \cdot MgCO_3$).

Calcium carbonate is insoluble in water, although it interacts with carbon dioxide (the latter dissolves well in water and is present in large quantities in the atmosphere). Water-soluble calcium bicarbonate is formed. The chemical reaction of this process is as follows:

$$CaCO_3(_s)+CO2(g)+H_20(_l)\leftrightarrow Ca(HCO_3)_2(_l) \qquad (1)$$

The resulting soluble salt causes water hardness. However, the process is reversible. Calcium and magnesium ions, forming a strong complex, interact with ethylenediaminetetraacetic acid (EDTA). Therefore, the hardness can be determined by titrating a standard EDTA solution. Eriochrome Black T-EBT solution is used as an indicator.

The structure of EDTA is as follows:

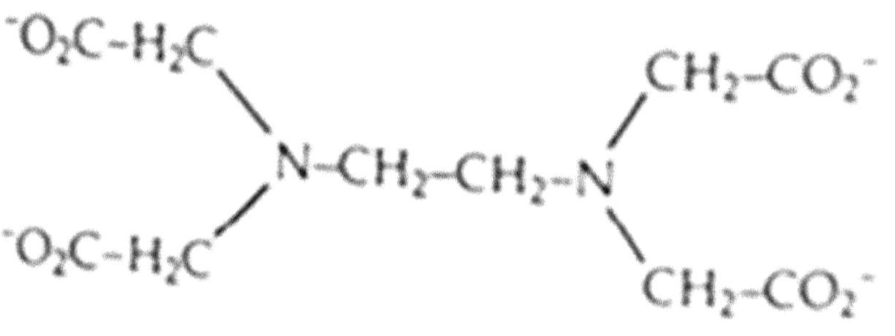

At the beginning of the reaction, the EBT indicator forms a complex with cations and the color of the solution changes. When we add EDTA during the titration, a new Ca^2+ and Mg^2+ complex is formed because the attraction between EDTA and cations is greater than between EBT and cations. Finally, all existing cations connect to EDTA, the color of the indicator changes from red to blue which is considered the end of the titration. Hardness is calculated by measuring the amount of EDTA consumed in the titration and corresponding to dissolved calcium and magnesium ions.

Total Alkalinity

Total alkalinity refers to the ability of water to neutralize an acidic environment (VanLoon et al., 2016). The measurement is based on titrating. The alkaline environment in most natural waters is due to the presence of (CO^{3-}), bicarbonate (HCO^{3-}), and hydroxyl anion (OH^-). However, borates, phosphates, and silicates also have their say in the formation of an alkaline environment. Alkalinity is usually associated with calcium carbonate ($CaCO_3$).

Relationship Between Alkalinity and Hardness

Alkalinity and hardness are linked by common ions. Alkalinity and hardness are measured by the same ions - bicarbonates and carbonates (USEPA, 1978a).

Measurement

Alkalinity measurement is based on the titration of a water sample with dilute sulfuric acid (0.1*N* or 0.02*N*). One ml 0.1*N* H_2SO_4 is equivalent to 5 mg $CaCO_3$; 1 ml 0.02*N* H_2SO_4 is equivalent to 1.00 mg $CaCO_3$ (APHA. 1998). A 1% solution of methyl orange is used as an indicator. 100 ml of water is poured for analysis into a conical flask with a capacity of 250-300 ml, 2-3 drops of each indicator are added and is titrated with 0.1N hydrochloric acid solution until the color of the solution changes from yellow to pink.

Formula: $\alpha \times K \times 1000/100 \times 100 = \alpha \times K/10 = 0.1 \times \alpha \times K$, where α is the amount of acid used for titration in ml, K is the correction coefficient.

General Acidity of Water

Required reagents: 0.1N NaOH; Methyl Red 1%.

100 ml of water is placed for analysis into a conical flask with a capacity of 250-300 ml, 2-3 drops of each indicator are added and titrated with 0.1N sodium alkali solution until the color of the solution changes from yellow to pink. Formula: $\alpha \times K \times 49$, where α is the amount of acid used for titration in ml, K-correction coefficient (APHA. 1998).

Redox Potential

Redox potential, in another way, is also called oxidation-reduction potential. Denoted by the symbol Eh, the dimension (mV) and pH, along with temperature and salinity, characterize the steady state of water. In lakes and surface waters, the indicator ranges from -0.5 to +0.7, although in some types of water, the indicator can be more than -0.6. At such a time, we can talk about the increased content of sulfur, hydrogen sulfide, and nutrients. Water with a redox indicator can be:

- Eh >+(more than 0.1-1.15) – in the water there is oxidizing environment, water contains dissolved oxygen and Fe^{3+}, Cu^{2+}, Pb^{2+}, Mo^{2+} ions.
- Eh from -0.0 to +0.1 - a transient oxidation-reduction environment, unstable
- Geochemical regime, variable amounts of oxygen and hydrogen sulfide, as well as weak oxidation-reduction reactions of various metals. It also indicates the presence of microorganisms and the process of eutrophication.
- Eh<0.0 - is a reduction environment. indicates the probable presence of hydrogen sulfide and Fe^{2+}, Mn^{2+}, Mo^{2+} metals in the water.

Along with the listed indicators, we also determined the concentration of chlorophyll. Its increased amount is related to redox reactions and the increased amount of nutrients. High concentrations of nitrates, nitrites, and phosphates in water bodies are the main cause of the intensive growth of algae, leading to the eutrophication of water bodies. Sulfates are considered to be less toxic compounds. Their high concentration affects the pH level of the water. Aquatic organisms are particularly sensitive to high concentrations of chlorides. Nitrates and nitrites, sulfates, phosphates, and chlorides are found in water mainly as a result of the discharge of nutrients from fertilizer-enriched agricultural lands, as well as the discharge of municipal and industrial wastewater (Smolders, 2006).

There is a classical method for determining the concentration of chlorophyll (Smith, 2019; Berner, 1971; Dévai, 1988). We used a mobile spectrophotometer, which allowed us to measure the amount of chlorophyll in situ. An OCEAN INSIDE miniature spectrophotometer was used, which measures pigment absorption through a supporting program. According to research (Schalles, 2006; Janelidze, 2021; Kupatadze, 2019), low chlorophyll concentration ($<2mg/m^3$) is well reflected in the blue part of the spectrum (400–500 nm), with increasing wavelength in the near-infrared spectrum (NIR, 700–800 nm), the reflectance decreases to 0; Chlorophyll concentrations between 2 and 30 mg/m^3 are reflected in the green (500-600 nm) and red bands (600-700 nm). At higher reflectance, the peak reflection in the green part of the spectrum and chlorophyll concentration above 300 mg/m^3 shows in the green part of the spectrum, the blue and red bands show little reflection in this case.

RESULTS AND THEIR DISCUSSION

The data presented in Table 1-11 were mainly measured on-site with the HANNA mobile spectrophotometer. Discovering reactions for nitrates and phosphates was carried out on the spot as well. We measured the concentration of chlorophyll with a mobile spectrophotometer. According to the obtained results, the electrical conductivity and, accordingly, the salinity index are always high. That is, there are large amounts of anions and cations, including nutrients, which make good conditions for eutrophication.

The redox index is also an indicator of various cations and anions, which includes the presence of phosphates and nitrates. The latter are nutrients and reasons for eutrophication. We also got positive results from their presence in the discovery reactions we conducted. For example, the redox maximum in August 2021 is 358.7 (mV). Accordingly, the measured chlorophyll concentration is also high: 6.812718379 on the shore of the lake and 11.02875784 in the inner part (Figure 2). In December 2021, the redox maximum is fixed at -388(mV). Chlorophyll concentration: 4.042831145 on the shore of the lake and 3.147683053 in the inner part. In this case, the higher concentration near the shore is probably related to anthropogenic factors. (Duca et al. 2021, Vaseashta et al, 2022). The redox state in lake waters is determined by the presence and decay of organic matter. The process is mainly related to the amount of dissolved oxygen (DO) in the water. Oxygen is a very important substance in this case. In water, oxygen is consumed rapidly by the oxidation of organic matter. So, the following processes might occur:

$$\{CH_2O\} + O_2 = CO_2 + H_2O$$

Also, oxygen in water may be consumed by the bio-oxidation of nitrogenous material:

$$NH_4^+ + 2O_2 = 2H^+ + NO_3^- + H_2O$$

Figure 2. Measured chlorophyll concentration at Paliastomi Lake

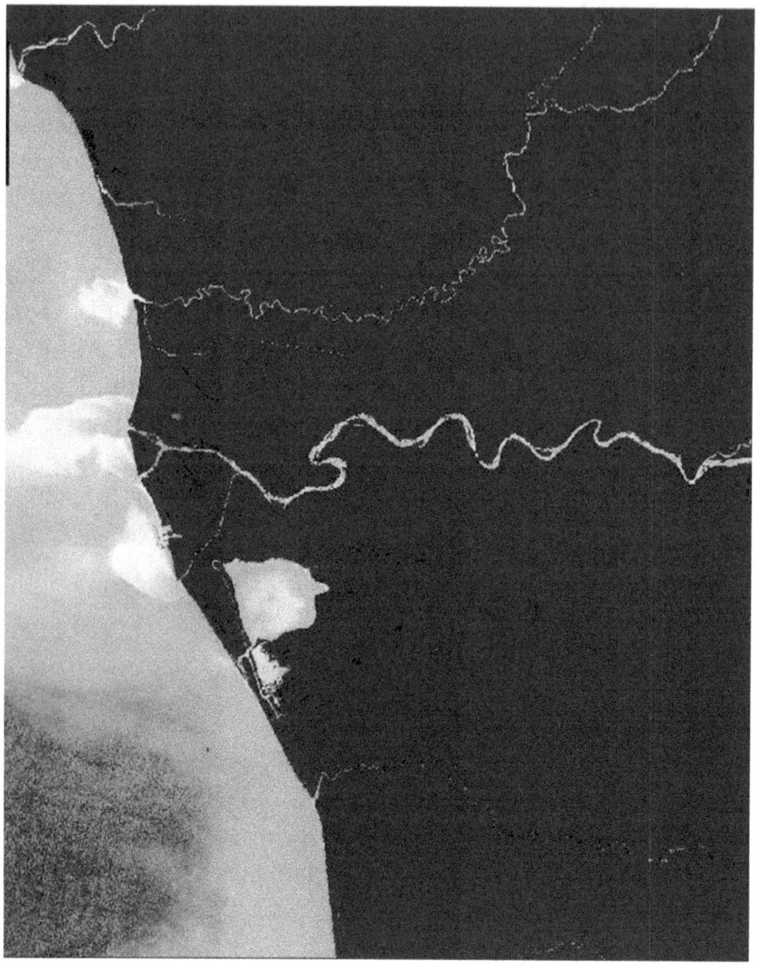

The amount of dissolved oxygen (DO) - 80-12-% DO is considered the norm in lakes and surface waters (Williams, 2014). According to the indicators obtained from the research, the excessive amount of DO in Paliastom Lake is found only in specific months (2021/08/09; 2022/02), and only in a certain period of time. So, it can be said that this indicator is within the norm.

The measured results show that the **TDS -Total Dissolved Particles** is quite high, which indicates a high level of dissolved oxygen in the water (which is confirmed by the measurements) and the presence of metal ions (metals were not measured and we can only assume). For example, Iron ions can be involved in the following reactions:

$$4Fe^{2+}+O_2+10H_2O=4Fe(OH)_3+8H^+$$

In some cases, the redox index is strongly negative, which indicates an increased rate of nutrients and eutrophication. e.g. In the month of September 2021, redox indicators range from -49.2 to 302. In the same period, the concentration of chlorophyll in Paliastomi Lake is 12.5 at the shore, 14.9 at depth. The data is interdependent. or, e.g. In November 2021, the eutrophication rate was 8.12 near the shore,

8.53 at depth. Redox is 72.04/262 (average of minimum and maximum taken). That is, it turns out that there is eutrophication, but not at a high level.

Electrical conductivity - electrical conductivity and, accordingly, the salinity index is always high. That is, there are large amounts of anions and cations, including nutrients, which create good conditions for redox reactions and eutrophication in Paliastomi Lake. The research results of water quality in Paliastomi Lake are presented in Tables 1-11.

Table 1. Research results- 2021-08.

	Min.	Max.
pH	7.33	8.84
Redox (mV)	36/-446	358.7
t (C)	24.02	30.06
Conductivity (μS/cm)	8220.29	12464.5
Salinity (ppt)	4.641	7.206
TDS (ppm)	4740.19	7272.56
DO (%)	4.53	139.3
Hardness	122	125
Turbidity(NTU)	12.49	131.8

Table 2. Research results - 2021-09.

	Min.	Max.
pH	7.597	8.62
Redox (mV)	-49.2	302
t (C)	17.71	24.39
Conductivity (μS/cm)	282.184	1646.07
Salinity (ppt)	0.18	1.389
TDS (ppm)	149	2236.01
DO (%)	26.97	140.02
Hardness	120	124
Turbidity(NTU)	19.5	20.89

Table 3. Research results - 2021-10

	Min.	Max.
pH	7.71	8.19
Redox (mV)	-87.3 /-442	285
t (C)	11.02	18.04
Conductivity (μS/cm)	457	1118
Salinity (ppt)	0.29	0.511
TDS (ppm)	209	804
DO (%)	42	94
Hardness	122	126
Turbidity (NTU)	1.92	15.42

Table 4. Research results - 2021-11.

	Min.	Max.
pH	7.79	8.19
Redox (mV)	72.04	262
t (C)	9.01	18.05
Conductivity (μS/cm)	432	4486
Salinity (ppt)	0.25	0.548
TDS (ppm)	235	3186
DO (%)	0.065	82
Hardness	125	130
Turbidity (NTU)	1.48	18.53

According to on-site research, the water of Paliastomi Lake has a weak alkaline reaction and the pH value is mainly in the range of 8-9. Carbonate hardness is also noted, which ranged from 125 mg/l to 130 mg/l to 122 mg/l during our measurements. General hardness and alkalinity are related to each other because with increased carbonate hardness, alkalinity can also increase. An increase in alkalinity initiates the process of eutrophication. The number of microplankton and phosphate ions increases.

The water of the researched lake can be considered as medium turbidity, although in some periods the increase of the turbidity indicator is recorded. Average and high turbidity indicates the presence of humic material and nutrients in the lake, which was confirmed by on-site research.

Table 5. Research results - 2021-12

	Min.	Max.
pH	7.71	8.09
Redox (mV)	-388	230
t (C)	7/2.54	12
Conductivity (μS/cm)	2722.02	6208.52
Salinity (ppt)	1.506	3.009
TDS (ppm)	1594	3597.99
DO (%)	0.076	25.22/100
Hardness	112	118
Turbidity(NTU)	0.778	1.714

Table 6. Research results - 2022-01.

	Min.	Max.
pH	7.87	8.31
Redox (mV)	153	319
t (C)	2	8
Conductivity (μS/cm)	1899	2315.89
Salinity (ppt)	1.302	1.505
TDS (ppm)	951	1406.98
DO (%)	71	108
Hardness	117	120
Turbidity(NTU)	41	514

Table 7. Research results - 2022-02

	Min.	Max.
pH	7.93	8.11
Redox (mV)	133	318
t (C)	4.94	10.05
Conductivity (μS/cm)	1246	2442
Salinity (ppt)	0.693	1.658
TDS (ppm)	771.682	1829.27
DO (%)	92.1	147.2
Hardness	117	119
Turbidity(NTU)	89	1127

Table 8. Research results - 2022-03

	Min.	Max.
pH	7.93	8.49
Redox (mV)	300	351
t (C)	4.88	11.65
Conductivity (μS/cm)	1873.01	4237.97
Salinity (ppt)	0.671	2.263
TDS (ppm)	1761.22	2441.23
DO (%)	87	120
Hardness	123	125
Turbidity(NTU)	59.43	1580

Table 9. Research results - 2022-04.

	Min.	Max.
pH	7.98	8.19
Redox (mV)	-239	179
t (C)	10.02	19.25
Conductivity (μS/cm)	2111.14	6410.71
Salinity (ppt)	2.761	3.547
TDS (ppm)	2994.27	3029.92
DO (%)	95.7	121.8
Hardness	120	124
Turbidity(NTU)	0.001	3.965

Table 10. Research results 2022-05.

	Min.	Max.
pH	7.28	8.67
Redox (mV)	-174.3	150
t (C)	16.02	20.17
Conductivity (μS/cm)	4823.1	5397.54
Salinity (ppt)	2.978	3.041
TDS (ppm)	2917.26	3099.36
DO (%)	68.36	92.34
Hardness	123	125
Turbidity(NTU)	87.3	328

Phosphate and nitrate ion detection reactions performed with qualitative analysis protocols showed positive samples.

Table 11. Research results - 2022-06.

	Min.	Max.
pH	6.55	8.03
Redox (mV)	-271.5	98.8
t (C)	22.12	28.34
Conductivity (μS/cm)	5537.73	6056.04
Salinity (ppt)	2.972	3.048
TDS (ppm)	1645	3347
DO (%)	15.03	77.28
Hardness	124	128
Turbidity (NTU)	69.12	270.7

The high data of the redox index allows us to say the same, which actually reflects the oxidation-reduction processes in the lake water. Eutrophication-promoting cations and anions (the same nutrients) participate in these processes. It also means the presence of heavy metal ions. Metals did not measured, but, when Eh >+(0.1-1.15 more) - dissolved oxygen and Fe^{3+}, Cu^{2+}, Pb^{2+}, Mo^{2+} ions are present in water. In addition, average and periodically increased turbidity indicates the presence of humic material, which binds heavy metal ions and becomes more stable in water. Therefore, due to turbidity and redox indicators, we can assume the presence of heavy metal ions along with nutrients. Which is also a good condition for eutrophication.

High levels of chlorophyll generally increase with increasing temperature. On-site research analyses showed that all conditions (pH, temperature, redox, turbidity) are present for a high eutrophication index. The rate of oxidation-reduction processes (redox) is also high, which indicates the presence of cations and anions (nutrients) and humic material. The latter can be naturally formed, e.g. By the decay of various plants, the metabolism of micro and zooplankton, or by the decomposition of the excrement of birds (a large number of them live in this area). However, some of them may be introduced by rivers polluted by various human industrial activities. Or, by poaching, if we assume that any chemicals or biological materials are used for fishing. No less contribution to the pollution process can be made by the so-called ferries tourist trips. These ferries are constantly anchored in the shoreline of the lake and that is why the chlorophyll level is higher near the shore than in the middle of the lake.

The increased amount of nutrients can also be related to the artificial fish farms, which are located in several locations in this region, and their waste flows into the sea through the two estuaries of the Iori River and may also end up in Paliastomi Lake.

CONCLUSION

Based on the conducted studies, it was established that oxidation-reduction reactions are actively taking place in Paliastomi Lake. This indicates the presence of various nutrients, metal ions and eutrophication index higher than average.

REFERENCES

Abramia, G., Gverdsiteli, L. (2022). *New challenges towards the ecological management of Paliastomi Lake will significantly improve the local environment with far-going positive global consequences.* Research Gate.

American Public Health Association (APHA), American Water Works Association (AWWA), and the Water Environmental Federation (WEF). (1998). *Standard methods for examinations of water and wastewater* (20th ed.). United Book Press, Inc.

Aphkhazava, I. (1984). *Georgian Encyclopedia*, *7*, 652.

Berner, R. (1971). *Principles of redox systems.* McGraw-Hill.

Cao, Z., Ma, R., Duan, H., Pahlevan, N., Melack, J., Shen, M., & Xue, K. (2020). A machine learning approach to estimate chlorophyll-a from Landsat-8 measurements in inland lakes. *Remote Sensing of Environment*, *248*, 111974. doi:10.1016/j.rse.2020.111974

Dassenakis, M., Botsou, F., Paraskevopoulou, V., Chikviladge, C., & Kachiasvili, K. (2006). Transport of pollutants in two estuarine systems on the coast of Georgia. *Chemistry and Ecology*, *22*(5), 379–393. doi:10.1080/02757540600917609

Dévai, I., Felföldy, L., Wittner, I., & Plósz, S. (1988). Detection of phosphine: New aspects of the phosphorus cycle in the hydrosphere. *Nature*, *333*(6171), 343–345. doi:10.1038/333343a0

Duca, G., & Vaseashta, A. (Eds.). (2020). *Handbook of Research on Emerging Developments and Environmental Impacts of Ecological Chemistry*. IGI Global. doi:10.4018/978-1-7998-1241-8

Janelidze, Z., Chikhradze, N., & Janelidze, G. (2021). Assessing the impact of anthropogenic factors on the ecosystems of the Paliastomi Lake region. *Georgian Geographical Journal*, *1*(1), 34–41.

Kupatadze, K. (2022). Surface water chemical composition monitoring in Several villages: Case study of Khakheti Region, Georgia. In: *Handbook of Research on Water Science and Society, IGI Global Publishing House, 26*, 587-598. https://www.igi-global.com/chapter/surface-water-chemical-composition-monitoring-in-several-villages/299900

Kupatadze, K., & Kiziloz, B. (2019). Researching some of the Parameters of Kura ("Mtkvari") river in two neighboring countries-Georgia and Turkey, in terms of environmental chemistry. *J. Tche Quimica*, *16*(31), 450–457. http://www.deboni.he.com.br/Periodico31.pdf

Lomsadze, Z., Makharadze, K., & Pirtskhalava, R. (2016). The Ecological Problems of Rivers of Georgia. *Annals of Agrarian Science*, *14*(3), 237–242. doi:10.1016/j.aasci.2016.08.009

Mattson, M., & Gene, E. (1993). Redox reactions of organic matter decomposition in a soft water lake. *Biogeochemistry*, *19*(3), 149–172. https://www.jstor.org/stable/1468581. doi:10.1007/BF00000876

Schalles, J. F. (2006). Optical remote sensing techniques to estimate phytoplankton chlorophyll concentrations in coastal. In L. L. Richardson & E. F. LeDrew (Eds.), *Remote sensing of aquatic coastal ecosystem processes* (pp. 27–79). Springer Netherlands. doi:10.1007/1-4020-3968-9_3

Smith, J. (2019). *General, organic, and biological chemistry*. McGraw Hill Education.

Smolders, J., Lamers, E., Lucassen, G., & Van Der Velde, G. (2006). Internal Eutrophication-a review. *Chemistry and Ecology*, *22*(2), 93–111. doi:10.1080/02757540600579730

Snoeyink, V. L., & Jenkins, D. (2015). *Water Chemistry*. John Wiley & Sons.

Trapaidze, V. (2012). *Water Recourse*. TSU. 8-12.

USEPA. (1978). EPA Method: 130.2: Hardness, total (mg/L as CaCO) (Titrimetric, EDTA). *Methods for the chemical analysis of water and wastes* (MCAWW) (EPA/600/4-79/020).

VanLoon, G. W., & Duffy, S. (2018). Environmental chemistry: A global perspective. Elsevier.

Vaseashta, A., Duca, G., & Travin, S. (Eds.). (2022). *Handbook of Research on Water Sciences and Society* (Vols. 1–2). IGI Global. doi:10.4018/978-1-7998-7356-3

Williams, I. (2014). Environmental chemistry, 157-159. Wiley.

Compilation of References

Abdol, R. S. M., & Mohammad, R. S. M. (2016). In-situ synthesis of highly stable CdS quantum dots and their application for photocatalytic degradation of dyes. *Spectrochimica Acta. Part A: Molecular and Biomolecular Spectroscopy, 152*, 156–164. doi:10.1016/j.saa.2015.07.052 PMID:26208270

Abdul, R. M., Mutnuri, L., Dattatreya, P. J., & Mohan, D. A. (2012). Assessment of drinking water quality using ICP-MS and microbiological methods in the Bholakpur area, Hyderabad, India. *Environmental Monitoring and Assessment, 184*(3), 1581–1592. doi:10.100710661-011-2062-2 PMID:21544503

Abdusamadzoda, D., Abdushukurov, D. A., Zinicovscaia, I., Duliu, O. G., & Vergel, K. N. (2020). Assessment of the ecological and geochemical conditions in surface sediments of the Varzob river, Tajikistan. *Microchemical Journal, 158*, 105173. doi:10.1016/j.microc.2020.105173

Abou El-Kheir, A., Haggag, K., Mowafi, S. I., & El-Sayed, H. (2014). Microwave-assisted bleaching of wool fabrics. *Journal of Natural Fibers, 12*(2), 97–107. doi:10.1080/15440478.2014.901202

Abramia, G., Gverdsiteli, L. (2022). *New challenges towards the ecological management of Paliastomi Lake will significantly improve the local environment with far-going positive global consequences.* Research Gate.

Abreu, T., Perestrelo, R., Bordiga, M., Locatelli, M., Coïsson, J. D., & Câmara, J. S. (2021). The flavor chemistry of fortified wines - a comprehensive approach. *Foods, 10*(6), 1239. doi:10.3390/foods10061239 PMID:34072391

Adekunle, K., & Okolie, J. (2015). A review of the biochemical process of anaerobic digestion. *Advances in Bioscience and Biotechnology, 6*(3), 205–212. doi:10.4236/abb.2015.63020

Ailer, Š., Jakabová, S., Benešová, L., & Ivanova-Petropulos, V. (2022). Wine Faults: State of Knowledge in Reductive Aromas, Oxidation and Atypical Aging, Prevention, and Correction Methods. *Molecules (Basel, Switzerland), 27*(11), 3535. doi:10.3390/molecules27113535 PMID:35684472

Akhter, M., Habib, G., & Qamar, S. U. (2018). Application of Electrodialysis in Wastewater Treatment and Impact of Fouling on Process Performance. *Journal of Membrane Science & Technology, 8*(2), 1000182. doi:10.4172/2155-9589.1000182

Akoh, C. C., & Min, D. B. (Eds.). (2002). Food lipids: Chemistry, nutrition, and biochemistry (2nd ed., rev.expanded). M. Dekker.

Alekhin, O. (1970). *Basics of hydrochemistry.* PC Leningrad.

Alexander, D., Ellerby, R., Hernandez, A., Wu, F., & Amarasiriwardena, D. (2017). Investigation of simultaneous adsorption properties of Cd, Cu, Pb and Zn by pristine rice husks using ICP-AES and LA-ICP-MS analysis. *Microchemical Journal, 135*, 129–139. doi:10.1016/j.microc.2017.08.001

Alexander, P., Carter, D., & Earland, C. (1951). The rôle of the disulphide bond in reactions which render wool non-felting. *Journal of the Society of Dyers and Colourists*, *67*(1), 23–27. doi:10.1111/j.1478-4408.1951.tb02670.x

Alexander, P., Fox, M., & Hudson, R. F. (1950). The reaction of oxidizing agents with wool. 5. The oxidation products of the disulphide bond and the formation of a sulphonamide in the peptide chain. *The Biochemical Journal*, *49*(2), 129–138. doi:10.1042/bj0490129 PMID:14858297

Almeida, C. A., Savio, M., González, P., Martinez, L. D., & Gil, R. A. (2013). Determination of chemical oxygen demand employed manganese as an environmental friendly oxidizing reagent by a low injection method base on microwave digestion an speciation coupled to ICP-OES. *Microchemical Journal*, *106*, 351–356. doi:10.1016/j.microc.2012.09.007

Alomary, A. A., & Belhadj, S. (2007). Determination of heavy metals (Cd, Cr, Cu, Fe, Ni, Pb, Zn) by ICP-OES and their speciation in Algerian Mediterranean Sea sediments after a five-stage sequential extraction procedure. *Environmental Monitoring and Assessment*, *135*(1–3), 265–280. doi:10.100710661-007-9648-8 PMID:17342430

Alom, M. M., Momin, M. A., Udi, M. J., & Islam, M. (2015). Structural and optical evaluation of CdS nanoparticles synthesized by chemical route. *J. Bangladesh Electron.*, *15*, 39–45.

Altemimi, A., Lakhssassi, N., Baharlouei, A., Watson, D., & Lightfoot, D. (2017). Phytochemicals: Extraction, isolation, and identification of bioactive compounds from plant extracts. *Plants*, *6*(4), 42. doi:10.3390/plants6040042 PMID:28937585

Al-Tohamy, R., Ali, S. S., Li, F., Okasha, K. M., Mahmoud, Y. A.-G., Elsamahy, T., Jiao, H., Fu, Y., & Sun, J. (2022). A critical review on the treatment of dye-containing wastewater: Ecotoxicological and health concerns of textile dyes and possible remediation approaches for environmental safety. *Ecotoxicology and Environmental Safety*, *231*, 113160. doi:10.1016/j.ecoenv.2021.113160 PMID:35026583

Al-Wadhav, Kh. A. (2012). Carbon carriers and the synthesis of palladium catalysts based on them. *Bulletin of MITHT*, *7*(1), 3–18.

Amaral, I. F., Granja, P. L., & Barbosa, M. A. (2005). Chemical modification of chitosan by phosphorylation: An XPS, FT-IR, and SEM study. *Journal of Biomaterials Science. Polymer Edition*, *6*(12), 1575–1593. doi:10.1163/156856205774576736 PMID:16366338

Amaro-Soriano, A., Hernández-Aldana, F., & Rivera, A. (2021). Photochemical treatments (UV/H_2O_2, UV/O_3 and UV/H_2O_2/O_3) and inverse osmosis in wastewater: Systematic review. *World Journal of Advanced Research and Reviews*, *10*(2), 229–240. doi:10.30574/wjarr.2021.10.2.0231

American Public Health Association (APHA), American Water Works Association (AWWA), and the Water Environmental Federation (WEF). (1998). *Standard methods for examinations of water and wastewater* (20th ed.). United Book Press, Inc.

American Public Health Organization. (2011). *APHA 5220, Standard methods for the examination of water and wastewater. Chemical oxygen demand.* COD.

American Society for Testing of Materials. (2012). ASTM D1252-06, Standard test methods for chemical oxygen demand (dichromate oxygen demand) of water, West Conshohocken, PA, USA.

Ames,, B.N., VcCann, J. & Jamasaki, E. (1975). Methods for detecting carmes mutagenesity tests. *Mutation Research*, *31*(3), 347–364. doi:10.1016/0165-1161(75)90046-1 PMID:768755

Andrea-Silva, J., Cosme, F., Ribeiro, L. F., Moreira, A. S. P., Malheiro, A. C., Coimbra, M., Domingues, R. M., & Nunes, F. M. (2014). Origin of the Pinking Phenomenon of White Wines. *Journal of Agricultural and Food Chemistry*, *62*(24), 5651–5659. doi:10.1021/jf500825h PMID:24857316

Aniipsitakis, G. P., & Dionysiou, D. D. (2006) Chemically Induced Redox Reactions in Water Treatment: A Summary of Advanced and Direct Technologies. *World Environmental and Water Resource Congress 2006*, (pp. 1-10). ASCE. 10.1061/40856(200)31

Anita, J., Tatjana, T., Damjana, D., Kristina, S., Polona, J., & Milenko, R. (2008). Biochemical biomarkers in chronically metal-stressed daphnids. *Comp Biochem Physiol C Toxicol Pharmacol. Jan., 147*(1), 61. doi:1. doi:0.1016/j.cbpc.2007.07.006

Aphkhazava, I. (1984). *Georgian Encyclopedia, 7*, 652.

Aranaz, I., Harris, R., & Heras, A. (2010). Chitosan amphiphilic derivatives. Chemistry and applications. *Current Organic Chemistry, 14*(3), 308–330. doi:10.2174/138527210790231919

Araujo, G.; Oliveira, S. (2021). Energy and environmental analysis of vinasse processing using internal circulation biodigesters and concentrators for different production ranges of ethanol and its potential impact on Brazil. *Environment, Development and Sustainability: A Multidisciplinary Approach to the Theory and Practice of Sustainable Development, 23*(3), 3130-3163.

Arcaño, Y. D., García, O. D., Mandelli, D., Carvalho, W. A., & Pontes, L. A. (2020). Xylitol: A review on the progress and challenges of its production by chemical route. *Catalysis Today, 344*, 2–14. doi:10.1016/j.cattod.2018.07.060

Ardila, N., Daigle, F., Heuzey, M. C., & Ajji, A. (2017). Effect of chitosan physical form on its antibacterial activity against pathogenic bacteria. *Journal of Food Science, 82*(3), 679–686. doi:10.1111/1750-3841.13635 PMID:28140469

Arias-Estevez, M., Lopez-Periago, E., Martinez-Carballo, E., Simal-Gandara, J., Mejuto, J. C., & Garcia-Rio, L. (2008). The mobility and degradation of pesticides in soils and the pollution of groundwater resources. *Agriculture, Ecosystems & Environment, 123*(4), 247–260. doi:10.1016/j.agee.2007.07.011

Arifoglu, M., Marmer, W. N., & Carr, C. M. (1990). Sequential oxidative and reductive bleaching of wool in a single bath. *Textile Research Journal, 60*(6), 319–328. doi:10.1177/004051759006000602

Arsanoglu, H., & Tumen, F. (2021). Potassium struvite (slow-release fertilizer) and activated carbon production: Resource recovery from vinasse and grape marc organic waste using thermal processing. *Process Safety and Environmental Protection, 147*, 1077–1087. doi:10.1016/j.psep.2021.01.025

Arslan, Y. E., Sezgin Arslan, T., Derkus, B., Emregul, E., & Emregul, K. C. (2017). Fabrication of human hair keratin/jellyfish collagen/eggshell-derived hydroxyapatite osteoinductive biocomposite scaffolds for bone tissue engineering: From waste to regenerative medicine products. *Colloids and Surfaces. B, Biointerfaces, 154*, 160–170. doi:10.1016/j.colsurfb.2017.03.034 PMID:28334693

Arvanitoyannis, I. S., Ladas, D., & Mavromatis, A. (2006). Potential uses and applications of treated wine waste: A review. *International Journal of Food Science & Technology, 41*(5), 475–487. doi:10.1111/j.1365-2621.2005.01111.x

Asam, Z. U. Z., Poulsen, T. G., Nzami, A.-S., Raxique, R., Kiely, G., & Murphy, J. D. (2011). How can we improve biomethane production per unit of feedstock in a biogas plant? *Applied Energy, 88*(6), 2013–2018. doi:10.1016/j.apenergy.2010.12.036

Asghar, A., Abdul Raman, A. A., & Wan Daud, W. M. A. (2015). Advanced oxidation processes for in-situ production of hydrogen peroxide/hydroxyl radical for textile wastewater treatment: A review. *Journal of Cleaner Production, 87*, 826–838. doi:10.1016/j.jclepro.2014.09.010

Ashrafi, M., Bayat, M., Mortazavi, P., Hashemi, S. J., & Meimandipour, A. (2020). Antimicrobial effect of chitosan–silver–copper nanocomposite on *Candida albicans*. *Journal of Nanostructure in Chemistry*, *10*(1), 87–95. doi:10.100740097-020-00331-3

Asmus, K.-D., & Bonifačić, M. (2000). Free radical chemistry. Handbook of Oxidants and Antioxidants in Exercise, 3–54. doi:10.1016/B978-044482650-3/50001-8

Asrar, J., & Ding, Y. (2003). *US Patent WO2003005816A1*. Monsanto Technology Lic.

Asrar, J., & Ding, Y. (2010). *US Patent 7771749 B2*. Monsanto Technology Lic.

Asrar, J., & Ding, Y. (2016). US *Patent 9,445,599 B2*. Monsanto Technology Lic.

Atef El-Sayed, A., Salama, M., & Kantouch, A. A. M. (2014). Wool micro powder as a metal ion exchanger for the removal of copper and zinc. *Desalination and Water Treatment*, *56*(4), 1010–1019. doi:10.1080/19443994.2014.941300

Atkinson, R. W., Kang, S., Anderson, H. R., Mills, I. C., & Walton, H. A. (2014). Epidemiological time series studies of PM2.5 and daily mortality and hospital admissions: A systematic review and meta-analysis. *Thorax*, *69*(7), 660–665. BMJ Publishing Group. doi:10.1136/thoraxjnl-2013-204492

Audemar, M., Ramdani, W., Junhui, T., Raluca Ifrim, A., Ungureanu, A., Jérôme, F., Royer, S., & Oliveira Vigier, K. (2020). Selective hydrogenation of xylose to xylitol over Co/SiO2 catalysts. *ChemCatChem*, *12*(7), 1973–1978. doi:10.1002/cctc.201901981

Azmir, J., Zaidul, I. S. M., Rahman, M. M., Sharif, K. M., Mohamed, A., Sahena, F., & Omar, A. K. M. (2013). Techniques for extraction of bioactive compounds from plant materials: A review. *Journal of Food Engineering*, *117*(4), 426–436. doi:10.1016/j.jfoodeng.2013.01.014

Baek, I. (2000). Preparation of semi-activated carbon fibers. *Korean Journal of Chemical Engineering*, *17*(5), 553–558. doi:10.1007/BF02707165

Bagheri-Khoulenjani, S., Taghizadeh, S. M., & Mirzadeh, H. (2009). An investigation on the short-term biodegradability of chitosan with various molecular weights and degrees of deacetylation. *Carbohydrate Polymers*, *78*(4), 773–778. doi:10.1016/j.carbpol.2009.06.020

Bajaj, P., & Dhawan, A. (1997). PAN-based activated carbon fibres: Production, characterization and applications. *Indian Journal of Fibre and Textile Research*, *22*, 222–235.

Balaz, P., Boldizarova, E., Godocikova, E., & Briancin, J. (2003). Mechanochemical route for sulphide nanoparticles preparation. *Materials Letters*, *57*(9-10), 1585–1589. doi:10.1016/S0167-577X(02)01037-6

Barandovski, L., Frontasyeva, M. V., Stafilov, T., Šajn, R., Pavlov, S., & Enimiteva, V. (2012). Trends of atmospheric deposition of trace elements in Macedonia studied by the moss biomonitoring technique. *Journal of Environmental Science and Health. Part A, Toxic/Hazardous Substances & Environmental Engineering*, *47*(13), 2000–2015. doi:10.1080/10934529.2012.695267 PMID:22870997

Barandovski, L., Stafilov, T., Šajn, R., Frontasyeva, M. V., & Bačeva, K. (2013). Air pollution study in macedonia using a moss biomonitoring technique, ICP-AES and AAS. *Macedonian Journal of Chemistry and Chemical Engineering*, *32*(1), 89–107. doi:10.20450/mjcce.2013.137

Barbateskovic, M., Schjørring, O. L., Krauss, S. R., Jakobsen, J. C., Meyhoff, C. S., Dahl, R. M., Rasmussen, B. S., Perner, A., & Wetterslev, J. (2021). Higher versus lower fraction of inspired oxygen or targets of arterial oxygenation for adults admitted to the intensive care unit. *The Cochrane Library*, *33*(4). Advance online publication. doi:10.1002/14651858.CD012631.pub2 PMID:34251144

Bard, A. J., Parsons, R., & Jordan, J. (1985). Standard potentials in aqueous solution. International Union of Pure and Applied Chemistry, (1st ed.). M. Dekker. doi:10.1201/9780203738764

Barril, C., Rutledge, D. N., Scollary, G. R., & Clarc, A. C. (2016). Ascorbic Acid and White wine production: a review of beneficial versus detrimental impacts, 169-181. Australian Society of Viticulture and Oenology Inc.

Bartkovský, M., Semjon, B., Marcinčák, S., Turek, P., & Baričičová, V. (2020). Effect of wine maturing on the colour and chemical properties of chardonnay wine. *Czech Journal of Food Sciences*, *38*(4), 223–228. doi:10.17221/139/2019-CJFS

Bassani, A., Alberici, N., Fiorentini, C., Giuberti, G., Dordoni, R., & Spigno, G. (2020). Hydrothermal treatment of grape skins for sugars, antioxidants and soluble fibers production. *Chemical Engineering (Albany, N.Y.)*, 79.

Bateup, B. O., Christoe, J. R., Jones, F. W., Poole, A. J., Skourtis, C., & Westmoreland, D. J. (1996). Effluent Management. In *Proceedings Top-Tech '96* (pp. 388–407). CSIRO Division of Wool Technology.

Beecher, G. R. (2003). Overview of dietary flavonoids: Nomenclature, occurrence, and intake. *The Journal of Nutrition*, *133*(10), 3248S–3254S. doi:10.1093/jn/133.10.3248S PMID:14519822

Behar, D., Czapski, G., Rabani, J., Dorfman, L. M., & Schwarz, H. A. (1970). Acid dissociation constant and decay kinetics of the perhydroxyl radical. *Journal of Physical Chemistry*, *74*(17), 3209–3213. doi:10.1021/j100711a009

Bekhterev, A. N., & Zolotarev, V. M. (2006). IR spectroscopic study of vibrational states in low-dimensional nanocrystalline carbon with a changing structure. *Scientific and technical bulletin of information technologies, mechanics and optics*, (34), 91-96.

Bekker, M. Z., Day, M. P., & Smith, P. A. (2019). Changes in metal ion concentrations in a Chardonnay wine related to oxygen exposure during vinification. *Molecules (Basel, Switzerland)*, *24*(8), 1523. doi:10.3390/molecules24081523 PMID:30999713

Bekker, M. Z., Smith, M. E., Smith, P. A., & Wilkes, E. N. (2016). Formation of hydrogen sulfide in wine: Interactions between copper and sulfur dioxide. *Molecules (Basel, Switzerland)*, *21*(9), 1214. doi:10.3390/molecules21091214 PMID:27626394

Bektaşoğlu, B., Esin Çelik, S., Özyürek, M., Güçlü, K., & Apak, R. (2006). Novel hydroxyl radical scavenging antioxidant activity assay for water-soluble antioxidants using a modified CUPRAC method. *Biochemical and Biophysical Research Communications*, *345*(3), 1194–1200. doi:10.1016/j.bbrc.2006.05.038 PMID:16716257

Bereck, A., Dillbohner, S., Weber, B., Riegel, D., Mosel, J., Reper, J. M., & Brakelmann, A. (1997). A simple method for the objective characterisation of fabric softness. Part 1: Influence of bleaching, dyeing and crosslinking of wool. *Journal of the Society of Dyers and Colourists*, *113*(11), 322–326. doi:10.1111/j.1478-4408.1997.tb01853.x

Berger, A. (1959). Weissgradformeln und ihre praktische Bedeutung. *Farbe*, *8*, 187.

Berner, R. (1971). *Principles of redox systems*. McGraw-Hill.

Berovič, M., & Nemanič, J. (2019). The role of redox potential measurement in oak barrel wine maturation. *Chemical and Biochemical Engineering Quarterly*, *33*(1), 153–160. Advance online publication. doi:10.15255/CABEQ.2018.1469

Bhadwal, A. S., Tripathi, R. M., Gupta, R. K., Kumar, N., Singh, R. P., & Shrivastav, A. (2014). Biogenic synthesis and photocatalytic activity of CdS nanoparticles. *RSC Advances*, *4*(19), 9484–9490. doi:10.1039/c3ra46221h

Bhat, K. V., Kumari, R., Pathak, N., & Rai, A. K. (2014). Value addition in sesame: A perspective on bioactive components for enhancing utility and profitability. *Pharmacognosy Reviews*, *8*(16), 147. doi:10.4103/0973-7847.134249 PMID:25125886

Bhattarai, N., Gunn, J., & Zhang, M. (2010). Chitosan-based hydrogels for controlled, localized drug delivery. *Advanced Drug Delivery Reviews, 62*(1), 83–99. doi:10.1016/j.addr.2009.07.019 PMID:19799949

Bielski, B. H. J., Cabelli, D. E., Arudi, R. L., & Ross, A. B. (1985). Reactivity of HO_2/O_2- radicals in aqueous solution. *Journal of Physical and Chemical Reference Data, 14*(4), 1041–1100. doi:10.1063/1.555739

Biglar, M., Khanavi, M., Hajimahmoodi, M., Hassani, S., Moghaddam, G., Sadeghi, N., & Oveisi, M. R. (2012). Tocopherol content and fatty acid profile of different Iranian date seed oils. *Iranian Journal of Pharmaceutical Research: IJPR, 11*(3), 873–878. PMID:24250514

Bilanovic, D., Loewenthal, R. E., Avnimelech, Y., & Green, M. (1997). Potentiometric measurement of chemical oxygen demand. *Water S.A., 23*, 301.

Binder, J., & Raines, R. T. (2010). Fermentable sugars by chemical hydrolysis of biomass. *Proceedings of the National Academy of Sciences of the USA (PNAS), 107*(10), 4516-4521. 10.1073/pnas.0912073107

Bindoli, A., Fukuto, J. M., & Forman, H. J. (2008). Thiol chemistry in peroxidase catalysis and redox signaling. *Antioxidants & Redox Signaling, 10*(9), 1549–1564. doi:10.1089/ars.2008.2063 PMID:18479206

Bisschops, I., & Spanjers, H. (2003). Literature review on textile wastewater characterisation. *Environmental Technology, 24*(11), 1399–1411. doi:10.1080/09593330309385684 PMID:14733393

Blizzard, T. A., Mrozik, H., & Fisher, M. H. (1988). Chemical degradation of avermectin B1. *Tetrahedron Letters, 29*(26), 3163–3166. doi:10.1016/0040-4039(88)85111-6

Blonschi, V. (2021). *Photochemical transformations of some thiol substances and their participation in chemical self-purification processes of natural waters. (No. 57760).* [Doctoral dissertation, Moldova State University]. The National Council on Accreditation and Attestation.

Blonschi, V., Gladchi, V., & Bunduchi, E. (2022a). Monitoring of the thiol substances content in some surface waters. In V. Gladchi & A. Aricu (Eds.), *Ecological Chemistry: History and Achievements: Academician Gheorghe Duca, 70th Anniversary: Monograph* (pp. 141–159). Publishing Center of Moldova State University.

Blonschi, V., Gladchi, V., & Duca, Gh. (2022b). *Participation of thiol substances in chemical self-purification processes of natural waters: Monograph.* Publishing Center of Moldova State University.

Bobu, E., Raluca, N., Lupei, M., Ciolacu, F. L., & Desbri'eres, J. (2011). Synthesis and characterization of N-alkyl chitosan for papermaking applications. *Cellulose Chemistry and Technology, 45*(9-10), 619–625.

Bogacheva, S. Yu., Kapitanov, A. F., & Izmailov, B. A. (2009). Patent N° 2370583. *A method for imparting hydrophilic properties to fibrous materials.* Russia.

Borodaev, R. I. (2008). *Redox potential of the Dniester water as an indicator of its biological valorization.* In Proceeding of International Conference, Managementul bazinul transfrontalier al fluviului Nistru şi Directiva - cadru a apelor a Uniunii Europene (pp. 38-41). Chisinau, Republic of Moldova.

Borodaev, R. I., & Duca, G. G. (2002). *Characteristics of the redox conditions of the lower Dniester.* In Collection of abstracts of reports of the All-Russian scientific youth symposium, Biosphere Safety–2001/2002 (p. 2). Ekaterinburg, Russia.

Borodaev, R. I., Bunduchi, E. G., & Goreacheva, N. V. (2019). *On the mechanism of "Quasi Toxicity" of the photic layer of the waters from the Dniester basin.* In International scientific conference, Achievements in studies of marginal effect in water ecosystems and their practical significance, (p. 14). Odessa, Ukraine.

Borodaev, R. I. (2002). *Redox conditions of the lower Dniester and the state of dissolved forms of iron and copper. Analele ştiinţifice ale Universităţii de Stat din Moldova.* Chemical-Biological Sciences.

Bradbury, J. H. (1976). The morphology and chemical structure of wool. *Pure and Applied Chemistry, 46*(2–4), 247–253. doi:10.1351/pac197646020247

Brazinha, C., Cadima, M., & Crespo, J. G. (2014). Optimization of extraction of bioactive compounds from different types of grape pomace produced at wineries and distilleries. *Journal of Food Science, 79*(6), E1142–E1149. doi:10.1111/1750-3841.12476 PMID:24891032

Brezonik, P. L., & Fulkerson-Brekken, J. (1998). Nitrate-induced photolysis in natural waters: Controls on concentrations of hydroxyl radical photo-intermediates by natural scavenging agents. *Environmental Science & Technology, 32*(19), 3004–3010. doi:10.1021/es9802908

Briedis, V., Povilaitytė, V., Kazlauskas, S., & Venskutonis, P. R. (2003). *Polifenolių ir antocianinų kiekis vynuogėse, vynuogių sultyse ir raudonuose vynuose bei jų antioksidacinio aktyvumo įvertinimas.*

Brugnari, T., Braga, D. M., dos Santos, C. S. A., Torres, B. H. C., Modkovski, T. A., Haminiuk, C. W. I., & Maciel, G. M. (2021). Laccases as green and versatile biocatalysts: From lab to enzyme market—an overview. *Bioresources and Bioprocessing, 8*(1), 131. doi:10.118640643-021-00484-1

Bruk, L., Titov, D., Ustyugov, A., Zubavichus, Y., Chernikova, V., Tkachenko, O., Kustov, L., Murzin, V., Oshanina, I., & Temkin, O. (2018). The Mechanism of Low-Temperature Oxidation of Carbon Monoxide by Oxygen over the $PdCl_2$–CuCl2/γ-Al2O3 Nanocatalyst. *Nanomaterials (Basel, Switzerland), 8*(4), 217. doi:10.3390/nano8040217 PMID:29614029

Bueno, M., Marrufo-Curtido, A., Carrascón, V., Fernández-Zurbano, P., Escudero, A., & Ferreira, V. (2018). Formation and Accumulation of Acetaldehyde and Strecker Aldehydes during Red Wine Oxidation. *Frontiers in Chemistry, 6*, 20. doi:10.3389/fchem.2018.00020 PMID:29492401

Bu, L., Zhu, N., Li, C., Huang, Y., Kong, M., Duan, X., & Dionysiou, D. D. (2020). Susceptibility of atrazine photo-degradation in the presence of nitrate: Impact of wavelengths and significant role of reactive nitrogen species. *Journal of Hazardous Materials, 388*, 121760. doi:10.1016/j.jhazmat.2019.121760 PMID:31810806

Bulska, E., & Wagner, B. (2016). Quantitative aspects of inductively coupled plasma mass spectrometry. *Philosophical Transactions of the Royal Society A: Mathematical, Physical and Engineering Sciences, 374*(2079). Royal Society of London. doi:10.1098/rsta.2015.0369

Bulushev, D. A., Kiwi-Minsker, L., Yuranov, I., Suvorova, E. I., Buffat, P. A., & Renken, A. (2002). Structured Au/FeOx/C catalysts for low-temperature CO oxidation. *Journal of Catalysis, 210*(1), 149–159. doi:10.1006/jcat.2002.3632

Bulushev, D. A., Yuranov, I., Suvorova, E. I., Buffat, P. A., & Kiwi-Minsker, L. (2004). Highly dispersed gold on activated carbon fibers for low-temperature CO oxidation. *Journal of Catalysis, 224*(1), 8–17. doi:10.1016/j.jcat.2004.02.014

Bunduchi, E., Gladchi, V., & Duca, Gh. (2019). Natural water auto purification capacity through the redox process. *Achievements and perspectives of modern chemistry.* Chisinau: ASM.

Bunduchi, E., Duca, Gh., Gladchi, V., Goreaceva, N., & Mardari, I. (2006). *Assessment of kinetic parameters in the water of the Nistru course in the section Naslavcea-Dubasari.* Chemistry Journal of Moldova., doi:10.19261/cjm.2006.01(1).08

Burgos, G., Amoros, W., Muñoa, L., Sosa, P., Cayhualla, E., Sanchez, C., Diaz, C., & Bonierbale, M. (2013). Total phenolic, total anthocyanin and phenolic acid concentrations and antioxidant activity of purple-fleshed potatoes as affected by boiling. *Journal of Food Composition and Analysis, 30*(1), 6–12. doi:10.1016/j.jfca.2012.12.001

Burri, L., Hoem, N., Banni, S., & Berge, K. (2012). Marine omega-3 phospholipids: Metabolism and biological activities. *International Journal of Molecular Sciences*, *13*(12), 15401–15419. doi:10.3390/ijms131115401 PMID:23203133

Burrows, H. D., Canle, M. L., Santaballa, J. A., & Steenken, S. (2002). Reaction pathways and mechanisms of photodegradation of pesticides. *Journal of Photochemistry and Photobiology. B, Biology*, *67*(2), 71–108. doi:10.1016/S1011-1344(02)00277-4 PMID:12031810

Burtch, N. C., Jasuja, H., & Walton, K. S. (2014). Water stability and adsorption in metal–organic frameworks. *Chemical Reviews*, *114*(20), 10575–10612. doi:10.1021/cr5002589 PMID:25264821

Bušić, A., Marđetko, N., Kundas, S., Morzak, G., Belskaya, H., Ivančić Šantek, M., Komes, D., Novak, S., & Šantek, B. (2018). Bioethanol production from renewable raw materials and its separation and purification: A review. *Food Technology and Biotechnology*, *56*(3), 289–311. doi:10.17113/ftb.56.03.18.5546 PMID:30510474

Bustamante, M. A., Moral, R., Paredes, C., Pérez-Espinosa, A., Moreno-Caselles, J., & Pérez-Murcia, M. D. (2008). Agrochemical characterization of the solid by-products and residues from the winery and distillery industry. *Waste Management (New York, N.Y.)*, *28*(2), 372–380. doi:10.1016/j.wasman.2007.01.013 PMID:17433659

Bybik, E. E., Buzunov, O. V., Gribanov, N. M., & Lavrov, I. S. (1979). Study of the kinetics of colliodal magnetite particles (Rus.). *Journal of Applied Chemistry*, *7*, 1631–1632.

Bybik, E. E., Buzunov, O. V., & Skobochkin, V. E. (1980). Magnetometric method of fine ferromagnetics study. *Industrial Laboratory*, *46*(7), 618–619.

Byrne, A. R. (1993). Review of neutron activation analysis in the standardization and study of reference materials, including its application to radionuclide reference materials. *Fresenius' Journal of Analytical Chemistry*, *345*(2–4), 144–151. doi:10.1007/BF00322575

Cai, Q. Q., Lee, B. C. Y., Ong, S. L., & Hu, J. Y. (2021). Fluidized-bed Fenton technologies for recalcitrant industrial wastewater treatment–Recent advances, challenges and perspective. *Water Research*, *190*, 116692. doi:10.1016/j.watres.2020.116692 PMID:33279748

Cameron, W. M., & Moore, T. B. (1957). The influence of chloride on the dichromate-value test. *Analyst (London)*, *82*(979), 677–682. doi:10.1039/an9578200677

Campos, E. V. R., de Oliveira, J. L., & Fraceto, L. F. (2014). Applications of controlled release systems for fungicides, herbicides, acaricides, nutrients, and plant growth hormones: A review. *Advanced Science, Engineering and Medicine*, *6*(4), 373–387. doi:10.1166/asem.2014.1538

Cañedo-Dorantes, L., & Cañedo-Ayala, M. (2019). Skin acute wound healing: A comprehensive review. *International Journal of Inflammation*, *2019*, 1–15. doi:10.1155/2019/3706315 PMID:31275545

Cano-López, M., Pardo-Mínguez, F., Schmauch, G., Saucier, C., Teissedre, P. L., López-Roca, J. M., & Gómez-Plaza, E. (2008). Effect of micro-oxygenation on color and anthocyanin-related compounds of wines with different phenolic contents. *Journal of Agricultural and Food Chemistry*, *56*(14), 5932–5941. doi:10.1021/jf8006147 PMID:18558704

Cao, Z., Ma, R., Duan, H., Pahlevan, N., Melack, J., Shen, M., & Xue, K. (2020). A machine learning approach to estimate chlorophyll-a from Landsat-8 measurements in inland lakes. *Remote Sensing of Environment*, *248*, 111974. doi:10.1016/j.rse.2020.111974

Carbonell-Capella, J. M., Buniowska, M., Barba, F. J., Esteve, M. J., & Frígola, A. (2014). Analytical methods for determining bioavailability and bioaccessibility of bioactive compounds from fruits and vegetables: A review: Bioavailability of bioactive compounds…. *Comprehensive Reviews in Food Science and Food Safety, 13*(2), 155–171. doi:10.1111/1541-4337.12049 PMID:33412647

Cardamone, J. M., Nuñez, A., Garcia, R. A., & Aldema-Ramos, M. (2009). Characterizing wool keratin. *Research Letters in Materials Science, 1–5*, 1–5. doi:10.1155/2009/147175

Cardamone, J. M., Yao, J., & Nuñez, A. (2004). Controlling shrinkage in wool fabrics: Effective hydrogen peroxide systems. *Textile Research Journal, 74*(10), 887–898. doi:10.1177/004051750407401008

Carpena, M., Pereira, A. G., Prieto, M. A., & Simal-Gandara, J. (2020). Wine aging technology: Fundamental role of wood barrels. *Foods, 9*(9), 1160. doi:10.3390/foods9091160 PMID:32842468

Carr, M. E., Doane, W. M., Wing, R. E., & Bagley, E. B. (1993). *Patent US 5183690.* The United States of America as represented by the Secretary of Agriculture.

Caruso, G., de Pasquale, F., Mita, D. G., & Micale, V. (2016). Digestive enzymatic patterns as possible biomarkers of endocrine disruption in the red mullet (Mullus barbatus): A preliminary investigation. *Marine Pollution Bulletin, 105*(1), 37–42. doi:10.1016/j.marpolbul.2016.03.005 PMID:26971230

Carvajal, A. K., Rustad, T., Mozuraityte, R., & Storrø, I. (2009). Kinetic studies of lipid oxidation induced by hemoglobin measured by consumption of dissolved oxygen in a liposome model system. *Journal of Agricultural and Food Chemistry, 57*(17), 7826–7833. doi:10.1021/jf9013394 PMID:19691337

Casbeer, E., Sharma, V. K., & Li, X. (2012). Synthesis and photocatalytic activity of ferrites under visible light: A review. *Separation and Purification Technology, 87*, 1–14. doi:10.1016/j.seppur.2011.11.034

Catherine, J. M. (2002). Optical Sensing with Quantum Dots. *Analytical Chemistry, 74*(19), 520–526. doi:10.1021/ac022124v PMID:12380801

Cegarra, J., Gacen, J., Caro, M., & Pepio, M. (1988). Wool bleaching with thiourea dioxide. *Journal of the Society of Dyers and Colourists, 104*(7–8), 273–279. doi:10.1111/j.1478-4408.1988.tb01168.x

Cegarra, J., Ribé, J., & Gacén, J. (2008). Peroxide bleaching of different wools. *Journal of the Society of Dyers and Colourists, 85*(4), 147–149. doi:10.1111/j.1478-4408.1969.tb02890.x

Cejudo-Bastante, M. J., Pérez-Coello, M. S., & Hermosín-Gutiérrez, I. (2011). Effect of wine micro-oxygenation treatment and storage period on colour-related phenolics, volatile composition and sensory characteristics. *Lebensmittel-Wissenschaft + Technologie, 44*(4), 866–874. doi:10.1016/j.lwt.2010.10.015

Celotti, E., Lazaridis, G., Figelj, J., Scutaru, Y., & Natolino, A. (2022). Comparison of a Rapid Light-Induced and Forced Test to Study the Oxidative Stability of White Wines. *Molecules (Basel, Switzerland), 27*(1), 326. doi:10.3390/molecules27010326 PMID:35011556

Centre for Inland Waters. (1988). In B. J. Dutka, & the Dept of Environment Rivers Research Branch National Waters Research Institute Canada (Ed.), Method for microbiological and toxicological analysis of waters, wastewaters and sediments (p. 334). Dept of Environment Rivers Research Branch National Waters Research Institute Canada.

Chakka, A. K., & Babu, A. S. (2022). Bioactive compounds of winery by-products: Extraction techniques and their potential health benefits. *Applied Food Research, 2*(1), 100058. doi:10.1016/j.afres.2022.100058

Chaligava, O., Shetekauri, S., Badawy, W. M., Frontasyeva, M. V., Zinicovscaia, I., Shetekauri, T., Kvlividze, A., Vergel, K., & Yushin, N. (2021). Characterization of trace elements in atmospheric deposition studied by moss biomonitoring in Georgia. *Archives of Environmental Contamination and Toxicology*, *80*(2), 350–367. doi:10.100700244-020-00788-x PMID:33236186

Chandran, J., Nayana, N., Roshini, N., & Nisha, P. (2017). Oxidative stability, thermal stability and acceptability of coconut oil flavored with essential oils from black pepper and ginger. *Journal of Food Science and Technology*, *54*(1), 144–152. doi:10.100713197-016-2446-y PMID:28242912

Chang, C., Li, J., Li, X., Wang, C., Zhou, B., Su, Y., & Yang, Y. (2017). Effect of protein microparticle and pectin on properties of light mayonnaise. *Lebensmittel-Wissenschaft + Technologie*, *82*, 8–14. doi:10.1016/j.lwt.2017.04.013

Chaturvedi, P., Giri, B. S., Shukla, P., & Gupta, P. (2021). Recent advancement in remediation of synthetic organic antibiotics from environmental matrices: Challenges and perspective. *Bioresource Technology*, *319*, 124161. doi:10.1016/j.biortech.2020.124161 PMID:33007697

Chauhan, S., Darvishzadeh, R., Lu, Y., Stroppiana, M., Pepe, B.M., & Nelson, A. (2019) *Wheat lodging assessment using multispectral UAV Data.*

Chauhan, R., Kumar, A., & Chaudhary, R. P. (2013). Visible-light photocatalytic degradation of methylene blue with Fe doped CdS nanoparticles. *Applied Surface Science*, *270*, 655–660. doi:10.1016/j.apsusc.2013.01.110

Chen, B., McClements, D. J., & Decker, E. A. (2011). Minor components in food oils: A critical review of their roles on lipid oxidation chemistry in bulk oils and emulsions. *Critical Reviews in Food Science and Nutrition*, *51*(10), 901–916. doi:10.1080/10408398.2011.606379 PMID:21955091

Cheng, X., Wang, S., Huang, W., Wang, F., Fang, S., Ge, R., Zhang, Q., Zhang, L., Du, W., Fang, F., Feng, Q., Cao, J., & Luo, J. (2022). Current status of hypochlorite technology on the wastewater treatment and sludge disposal: Performance, principals and prospects. *The Science of the Total Environment*, *803*, 150085. doi:10.1016/j.scitotenv.2021.150085 PMID:34525771

Chen, H., Cui, S., Zhao, Y., Wang, B., Zhang, S., & Peng, X. (2012). O-Alkylation of chitosan for gene delivery by using ionic liquid in an in-situ reactor. *Engineering (London)*, *04*(10, 10B), 114–117. doi:10.4236/eng.2012.410B029

Chen, J. L., & Zhu, C. Q. (2005). Functionalized cadmium sulfide quantum dots as fluorescence probe for silver ion determination. *Analytica Chimica Acta*, *546*(2), 147–153. doi:10.1016/j.aca.2005.05.006

Chen, L., Lin, Y., Zhou, H., Hao, L., Chen, H., & Zhou, X. (2021). A stable polyamine-modified zein-based nanoformulation with high foliar affinity and lowered toxicity for sustained avermectin release. *Pest Management Science*, *77*(7), 3300–3312. doi:10.1002/ps.6374 PMID:33763979

Chen, W., Luo, Y., Ran, G., & Li, Q. (2020). Microwave-induced persulfate-hydrogen peroxide binary oxidant process for the treatment of dinitrodiazophenol industrial wastewater. *Chemical Engineering Journal*, *382*, 122803. doi:10.1016/j.cej.2019.122803

Che, P., Zhao, Z.-Y., Glassman, K., Dolde, D., Hu, T. X., Jones, T. J., Gruis, D. F., Obukosia, S., Wambugu, F., & Albertsen, M. C. (2016). Elevated vitamin E content improves all- *trans* β-carotene accumulation and stability in biofortified sorghum. *Proceedings of the National Academy of Sciences of the United States of America*, *113*(39), 11040–11045. doi:10.1073/pnas.1605689113 PMID:27621466

Cheremisinoff, N. P. (1996). *Elemental and structural characterization tests. Polymer Characterization, 43–81*. Elsevier. doi:10.1016/B978-081551403-9.50006-6

Cheremnykh E.G., Ivanov P.A., Faktor M.I., Karpova N.S., Vasiljeva E.F., Gusev K.V. & Brusov O.S. (2015). A new method to assess functional activity of serum complement system. *Medical Immunology*, *17*(5), 479-488. . doi:10.15789/1563-0625-2015-5-479-488

Cheremnykh, E. G., Pokataev, A. S., & Gridunova, V. N. (2008). *Device for biological research*. Patent RU 2361913 Publ.27.04.2008 Bull. No. 12. doi:10.17816/phf46438/2713-153X-2020-3-2-64-73

Chochorek, A., Bobrowski, A., Kiralyova, Z., & Mocak, J. (2010). ICP-OES determination of select metals in surface water - A metrological study. *Polish Journal of Environmental Studies*, *19*(1), 59–64. http://www.pjoes.com/ICP-OES-Determination-of-Select-Metals-r-nin-Surface-Water-a-Metrological-Study,88356,0,2.html

Choe, E., & Min, D. B. (2006). Mechanisms and factors for edible oil oxidation. *Comprehensive Reviews in Food Science and Food Safety*, *5*(4), 169–186. doi:10.1111/j.1541-4337.2006.00009.x

Choeisai, P., Jitkam, N., Silapanoraset, K., Yubolsai, C., Yoochatchava, W., Yamaguchi, T., Onodera, T., & Syutsubo, K. (2014). Sugarcane molasses-based bio-ethanol wastewater treatment by two-phase multi-staged up-flow anaerobic sludge blanket (UASB) combination with up-flow UASB and down-flow hanging sponge. *Water Science and Technology*, *69*(6), 1174–1180. doi:10.2166/wst.2013.794 PMID:24647181

Chokshi, N. P., & Ruparelia, J. P. (2021). Synthesis of nano Ag-La-Co composite metal oxide for degradation of RB 5 dye using catalytic ozonation process. *Ozone Science and Engineering*, *44*(2), 182–195. doi:10.1080/01919512.2021.1901070

Christofoletti, C. A., Escher, J. P., Correia, J. E., Marinho, J. F. U., & Fontanetti, C. S. (2013). Sugarcane vinasse: Environmental implications of its use. *Waste Management (New York, N.Y.)*, *33*(12), 2752–2761. doi:10.1016/j.wasman.2013.09.005 PMID:24084103

Chu, C., Stamatelatos, D., & Mcneill, K. (2017). Aquatic indirect photochemical transformations of natural peptidic thiols: Impact of thiol properties, solution pH, solution salinity and metal ions. *Environmental Science and Pollution Research International*, *19*(12), 1518–1527. doi:10.1039/C7EM00324B PMID:29090717

Chu, L., Wang, J., Dong, J., Liu, H., & Sun, X. (2012). Treatment of coking wastewater by an advanced Fenton oxidation process using iron powder and hydrogen peroxide. *Chemosphere*, *86*(4), 409–414. doi:10.1016/j.chemosphere.2011.09.007 PMID:22014660

Coelho, C., Heim, B., Foerster, S., Brosinsky, A., & De Araújo, J. C. (2017). In situ and satellite observation of CDOM and chlorophyll-a dynamics in small water surface reservoirs in the Brazilian semiarid region. *Water (Basel)*, *9*(12), 913. doi:10.3390/w9120913

Collin, F. (2019). Chemical basis of reactive oxygen species reactivity and involvement in neurodegenerative diseases. *International Journal of Molecular Sciences*, *20*(10), 2407. doi:10.3390/ijms20102407 PMID:31096608

Condurache, B.-C., Cojocaru, C., Pascariu, P., Samoila, P., & Harabagiu, V. (2022b). Innovative nanostructured magnetite/wool/polysiloxane composite as magnetic adsorbent for oil spill removal. *Comptes Rendus. Chimie*, *25*(S3), 245–260. doi:10.5802/crchim.168

Condurache, B.-C., Cojocaru, C., Samoila, P., Ignat, M., & Harabagiu, V. (2022c). Data-driven modeling and optimization of oil spill sorption by wool fibers: Retention kinetics and recovery by centrifugation. *International Journal of Environmental Science and Technology*, *19*(1), 367–378. doi:10.100713762-021-03176-7

Condurache, B., Cojocaru, C., Samoila, P., Cosmulescu, S. F., Predeanu, G., Enache, A., & Harabagiu, V. (2022a). Oxidized biomass and its usage as adsorbent for removal of heavy metal ions from aqueous solutions. *Molecules (Basel, Switzerland)*, *27*(6119), 1–19. doi:10.3390/molecules27186119 PMID:36144850

Corobov, R., Trombitchi, I., Siroedov, G., & Andreev, A. (2014). *Vulnerability to climate change. Moldovan part of the Dniester basin.* Monograph. Elan Poligraf.

Covaliov, V., & Covaliova, O. (2002). *MD Patent Nr. 2427.* Reactor for magnetic treatment of liquids.

Covaliov, V.V., Caraghiaur, V.G. (1985). Intensification of electrocoagulation wastewater treatment/ Electromagnetite treatment. *Chisinau, VINITI. 85*(18).

Covaliova, O. (2008). *MD Patent nr. 3912.* Method of electrochemical purification of waste waters from heavy metal ions.

Covaliova, O. (2010). Role of iron compounds in photocatalytic destruction of 2-aminobenzothiazole in water environment. *Studia Universitatis Moldaviae. Nature and Science, 1*(31), 199–206.

Covaliova, O., & Covaliov, V. (2004). *Fundamentals of intensification of electrochemical processes of wastewaters treatment and sludge ferritization in plating industry. Ecologie-Acoperiri Metalice.* ICTCM.

Covaliov, V. (2015). Improving the biogas production systems. *Academos, 4*(39), 38–44.

Covaliov, V., & Covaliova, O. (2000). Autocatalysis in the processes of wastewaters treatment. *Intellectus, 4*, 55–60.

Covaliov, V., & Covaliova, O. (2004). *Theoretical and practical aspects of electrochemical water treatment.* CEP USM.

Covaliov, V., Malina, J. F., Duca, Gh., Covaliova, O., & Bobeica, V. (2014). Application of Bioactive Substances in the Biogas Technology. In G. Duca (Ed.), *Management of Water Quality in Moldova.* doi:10.1007/978-3-319-02708-1_12

Covaliov, V., Senicovscaia, I., Nenno, V., Bobeica, V., & Covaliova, O. (2015). Effects of amaranths' seeds on dehydrogenase activity and gases emission in methanogenic bioreactors. *Studia Universitatis Moldaviae. Real and Natural Sciences Ser., 1*(81), 230–235.

Covinich, L. G., Bengoechea, D. I., Fenoglio, R. J., & Area, M. C. (2014). Advanced Oxidation Processes for Wastewater Treatment in the Pulp and Paper Industry: A Review. *American Journal of Environmental Engineering, 4*, 56–70. doi:10.5923/j.ajee.20140403.03

Cretescu, I., Borodaev, R. I, Lutic, D., Soreanu, G., Lazar, L., Gitlestad, J., & Zubcov, E. (2022b). Redox state assessment for the Danube water in the Sulina channel using a mobile water quality monitoring station. *Annals of "Dunarea de Jos" University of Galati, Mathematics, Physics, Theoretical Mechanics*, Fascicle II, Year XIV (XLV), 2, 122-125. University of Galati.

Cretescu, I., Soreanu, G., & Dragoi, E. (2022c). *Patent RO135089/2022.* Surface water quality monitoring system located on a passenger ship.

Cretescu, I., Kovacs, Z., Lazar, L., Burada, A., Sbarcea, M., Teodorof, L., Padure, D., & Soreanu, G. (2022a). Danube Delta: Water Management on the Sulina Channel in the Frame of Environmental Sustainability. Chapter. In A. J. Manning (Ed.), *River Deltas Research.* IntechOpen Limited. doi:10.5772/intechopen.97877

Crouch, L. S., Feely, W. F., Arison, B. H., VandenHeuvel, W. J. A., Colwell, L. F., Stearns, R. A., Kline, W. F., & Wislocki, P. G. (1991). Photodegradation of avermectin B1a thin films on glass. *Journal of Agricultural and Food Chemistry, 39*(7), 1310–1319. doi:10.1021/jf00007a024

Cruz-Salomón, A., Ríos-Valdovinos, E., Pola, F., Meza Gordillo, R., Lagunas Rivera, S., & Ruíz-Valdiviezo, V. (2017). Anaerobic treatment of agro-industrial wastewaters for COD removal in expanded granular sludge bed bioreactor. *Biofuel Research Journal, 16*(4), 715–720. doi:10.18331/BRJ2017.4.4.3

Cubadda, F. (2007). Inductively coupled plasma mass spectrometry. In *Food Toxicants Analysis* (pp. 697–751). Elsevier. doi:10.1016/B978-044452843-8/50020-1

Cui, J., Mo, D., Jiang, Y., Gan, C., Li, W., Wu, A., Li, X.-Y., Xiao, J.-A., Hu, Q., Yuan, H.-Y., Lu, R., & Huang, Y. (2019). Fabrication, Characterization, and Insecticidal Activity Evaluation of Emamectin Benzoate-Sodium Lignosulfonate Nanoformulation with pH-responsive. *Industrial & Engineering Chemistry Research, 58*(43), 19741–19751. doi:10.1021/acs.iecr.9b03171

Cvetković, S., Purenović, J. M., & Jovićević, J. N. (2008). Change of water redox potential, pH and rH in contact with magnesium-enriched kaolinite–bentonite ceramics. *Applied Clay Science, 38*(3-4), 268–278. doi:10.1016/j.clay.2007.03.009

Czaplicki, Z., & Strzelecki, S. (2019). Wool carbonization with an energy-efficient drying process. *Journal of Natural Fibers, 17*(12), 1809–1818. doi:10.1080/15440478.2019.1599312

D'Angelo, A., Tedesco, M., Cipollina, A., Galia, A., Micale, G., & Scialdone, O. (2017). Reverse electrodialysis performed at pilot plant scale: Evaluation of redox processes and simultaneous generation of electric energy and treatment of wastewater. *Water Research, 125*, 123–131. doi:10.1016/j.watres.2017.08.008 PMID:28843152

Da Silva, V. B., Barreira, J. C. M., & Oliveira, M. B. P. P. (2016). Naturalphytochemicals and probiotics as bioactive ingredients for functional foods: Extraction, biochemistry and protected-delivery technologies. *Trends in Food Science & Technology, 50*, 144–158. doi:10.1016/j.tifs.2015.12.007

Dai, X., Chen, L., Zhou, H., & Zhou, X. (2022). Development of stable and efficient zein based nanopesticides through green and simple ways for improving utilization efficiency. *Journal of Asia-Pacific Entomology, 25*(2), 101906. doi:10.1016/j.aspen.2022.101906

Dalpaz, R., Konrad, O., Da Silva Cyrne, C., Barzotto, H. P., Hasan, C., & Guerini Filho, M. (2020). Using biogas for energy cogeneration: An analysis of electric and thermal energy generation from agro-industrial waste. *Sustainable Energy Technologies and Assessments, 40*, 100774. doi:10.1016/j.seta.2020.100774

Danilewicz, J. C. (2012). Review of oxidative processes in wine and value of reduction potentials in enology. *American Journal of Enology and Viticulture, 63*(1), 1–10. doi:10.5344/ajev.2011.11046

Danilewicz, J. C. (2016). Reaction of oxygen and sulfite in wine. *American Journal of Enology and Viticulture, 67*(1), 13–17. doi:10.5344/ajev.2015.15069

Darwish, A. S., Attia, S. K., & Osman, D. I. (2022). Accelerated activation of H_2O_2 and persulfate by Sm-doped ZnO@ highly- defective layered yttria nanocomposite under visible-light irradiation for dyeing wastewater treatment: Comprehensive dominance of oxygen vacancies in photocatalytic advanced oxidation processes. *Journal of Alloys and Compounds, 925*, 166742. doi:10.1016/j.jallcom.2022.166742

Das, A., Adak, M. K., Mahata, N., & Biswas, B. (2021). Wastewater treatment with the advent of TiO_2 endowed photocatalysts and their reaction kinetics with scavenger effect. *Journal of Molecular Liquids, 338*, 116479. doi:10.1016/j.molliq.2021.116479

Dassenakis, M., Botsou, F., Paraskevopoulou, V., Chikviladge, C., & Kachiasvili, K. (2006). Transport of pollutants in two estuarine systems on the coast of Georgia. *Chemistry and Ecology, 22*(5), 379–393. doi:10.1080/02757540600917609

Davies, M. J. (2003). Singlet oxygen-mediated damage to proteins and its consequences. *Biochemical and Biophysical Research Communications, 305*(3), 761–770. doi:10.1016/S0006-291X(03)00817-9 PMID:12763058

Davini, P. (2003). Flue gas desulphurization by activated carbon fibers obtained from polyacrylonitrile by-product. *Carbon, 41*(2), 277–284. doi:10.1016/S0008-6223(02)00282-8

De Godoi, L. A. G., Camiloti, P. R., Bernardes, A. N., Sanchez, B. L. S., Torres, A. P. R., da Conceição Gomes, A., & Botta, L. S. (2019). Seasonal variation of the organic and inorganic composition of sugarcane vinasse: Main implications for its environmental uses. *Environmental Science and Pollution Research International*, *26*(28), 29267–29282. doi:10.100711356-019-06019-8 PMID:31396875

de la Guardia, M., & Armenta, S. (2011). Multianalyte determination versus one-at-a-time methodologies. In *Comprehensive Analytical Chemistry, (57)*. Elsevier. doi:10.1016/B978-0-444-53709-6.00006-9

de Oliveira, J. L., Campos, E. V. R., Pereira, A. E. S., Pasquoto, T., Lima, R., Grillo, R., Andrade, D. J., Santos, F. A., & Fraceto, L. F. (2018). Zein nanoparticles as eco-friendly carrier systems for botanical repellents aiming sustainable agriculture. *Journal of Agricultural and Food Chemistry*, *66*(6), 1130–1340. doi:10.1021/acs.jafc.7b05552 PMID:29345934

De Oliveira, M. C., da Silva, D. M., & de Almeida Faleiros, C. (2013). The effect of including liquid vinasse in the diet of rabbits on growth performance. *Revista Brasileira de Zootecnia*, *42*(4), 259–263. doi:10.1590/S1516-35982013000400005

De Satour, C. (1973). Magnetic separation in water pollution control. *IEEE Transactions on Magnetics*, *3*(3), 314–316. doi:10.1109/TMAG.1973.1067685

de Simone, N., Pace, B., Grieco, F., Chimienti, M., Tyibilika, V., Santoro, V., Capozzi, V., Colelli, G., Spano, G., & Russo, P. (2020). Botrytis cinerea and table grapes: A review of the main physical, chemical, and bio-based control treatments in post-harvest. *Foods*, *9*(9), 1138. doi:10.3390/foods9091138 PMID:32824971

Decker, G., Hönicke, P., & Hall, D. (2015). COD analysis and REACH authorization. *International Environmental Technology, Water/Wastewater*, 56.

Decker, E. A., McClements, D. J., Bourlieu-Lacanal, C., Durand, E., Figueroa-Espinoza, M. C., Lecomte, J., & Villeneuve, P. (2017). Hurdles in predicting antioxidant efficacy in oil-in-water emulsions. *Trends in Food Science & Technology*, *67*, 183–194. doi:10.1016/j.tifs.2017.07.001

Deiana, M., Rosa, A., Cao, C. F., Pirisi, F. M., Bandino, G., & Dessì, M. A. (2002). Novel approach to study oxidative stability of extra virgin olive oils: Importance of α-tocopherol concentration. *Journal of Agricultural and Food Chemistry*, *50*(15), 4342–4346. doi:10.1021/jf020033t PMID:12105968

Dellicolli, H. T., Dilling, P., & Falkehag, S. I. (1980). *US Patent 4,3184,866*. Westvaco.

Demirel, D., Yenigün, O., & Bekbölet, M. (1999). Removal of Cu, Ni and Zn from wastewater by the ferrite process. *Environmental Technology*, *20*(9), 963–970. doi:10.1080/09593332008616891

Deng, Y., Chen, N., Feng, C., Chen, F., Wang, H., Kuang, P., Feng, Z., Liu, T., Gao, Y., & Hu, W. (2019). Treatment of organic wastewater containing nitrogen and chlorine by combinatorial electrochemical system: Taking biologically treated landfill leachate treatment as an example. *Chemical Engineering Journal*, *364*, 349–360. doi:10.1016/j.cej.2019.01.176

Derjaguin, B., & Landau, L. D. (1941). Theory of the stability of strongly charged lyophobic sols and of the adhesion of strongly charged particles in solutions of electrolytes. *Acta Phys. Chim.*, *14*, 633–662.

Desai, K. R., Pathan, A. A., & Bhasin, C. P. (2017). Synthesis, characterization of cadmium sulphide nanoparticles and its application as photocatalytic degradation of Congored. *Int. J. Nanomater. Chem.*, *3*(2), 39–43. doi:10.18576/ijnc/030204

Deublein, D., & Steinhauser, A. (2008). *Biogas from Waste and Renewable Resources: An Introduction*. Wiley-VCH. doi:10.1002/9783527621705

Deutsches Institut für Normung. (1980). *DIN 38409 Teil 41, Deutsche Einheitsverfahren zur Wasser-, Abwasser- und Schlammuntersuchung Summarische Wirkungs- und Stoffkenngrößen (Gruppe H) Bestimmung des Chemischen Sauerstoffbedarfs (CSB) im Bereich über 15 mg/l (H 41)*. Beuth.

Dévai, I., Felföldy, L., Wittner, I., & Plósz, S. (1988). Detection of phosphine: New aspects of the phosphorus cycle in the hydrosphere. *Nature*, *333*(6171), 343–345. doi:10.1038/333343a0

Devesa-Rey, R. X., Vecino, J. L., Varela-Alende, J. L., Barral, M. T., Cruz, J. M., & Moldes, A. B. (2011). Valorization of winery waste vs. the costs of not recycling. *Waste Management (New York, N.Y.)*, *31*(11), 2327–2335. doi:10.1016/j.wasman.2011.06.001 PMID:21752623

Dhaka, S., Kumar, R., Lee, S., Kurade, M. B., & Jeon, B.-H. (2018). Degradation of ethyl paraben in aqueous medium using advanced oxidation processes: Efficiency evaluation of UVC supported oxidants. *Journal of Cleaner Production*, *180*, 505–513. doi:10.1016/j.jclepro.2018.01.197

Diamante, M. S., Vanz Borges, C., Minatel, I. O., Jacomino, A. P., Basílio, L. S. P., Monteiro, G. C., & Pace Pereira Lima, G. (2021). Domestic coking practices influence the carotenoid and tocopherol content in colored cauliflower. *Food Chemistry*, *340*, 127901. doi:10.1016/j.foodchem.2020.127901 PMID:32889204

Diehl, A., & Levin, Y. (2006). Smoluchowski equation and the colloidal charge reversal. *The Journal of Chemical Physics*, *125*(5), 054902. doi:10.1063/1.2222372 PMID:16942253

Dilag, J., Kobus, H., & Ellis, A. V. (2009). Cadmium sulfide quantum dot/chitosan nanocomposites for latent fingermark detection. *Forensic Science International*, *187*(1-3), 97–102. doi:10.1016/j.forsciint.2009.03.006 PMID:19356872

Ding, K., Shi, L., Zhang, L., Zeng, T., Yin, Y., & Yi, Y. (2016). Synthesis of photoresponsive polymeric propesticide micelles based on PEG for the controlled release of a herbicide. *Polymer Chemistry*, *7*(4), 899–904. doi:10.1039/C5PY01690H

Doane, W. M., Maiti, S., & Wing, R. E. (1990). *US Patent 4,911,952 A*. The United States of America as represented by the Secretary of Agriculture.

Domínguez, R., Pateiro, M., Gagaoua, M., Barba, F. J., Zhang, W., & Lorenzo, J. M. (2019). A Comprehensive review on lipid oxidation in meat and meat products. *Antioxidants*, *8*(10), 10. doi:10.3390/antiox8100429 PMID:31557858

Domini, C. E., Vidal, L., & Canals, A. (2009). Trivalent manganese as an environmentally friendly oxidizing reagent for microwave- and ultrasound-assisted chemical oxygen demand determination. *Ultrasonics Sonochemistry*, *008*(5), 686–691. doi:10.1016/j.ultsonch.2009.01.008 PMID:19250852

Dong, J., Chen, W., Qin, D., Chen, Y., Li, J., Wang, C., Yu, Y., Feng, J., & Du, X. (2021a). Cyclodextrin polymer-valved MoS_2-embedded mesoporous silica nanopesticides toward hierarchical targets via multidimensional stimuli of biological and natural environments. *Journal of Hazardous Materials*, *419*, 126404. doi:10.1016/j.jhazmat.2021.126404 PMID:34153613

Dong, J., Liu, X., Chen, Y., Yang, W., & Du, X. (2021b). User-safe and efficient chitosan-gated porous carbon nanopesticides and nanoherbicides. *Journal of Colloid and Interface Science*, *594*, 20–34. doi:10.1016/j.jcis.2021.03.001 PMID:33744730

Drasch, G., Horvat, M., & Stoeppler, M. (2004). Mercury. In M. Anke, M. Ihnat, & M. Stoeppler (Eds.), *Elements and their compounds in the environment* (2nd ed., p. 931). Wiley-VCH. doi:10.1002/9783527619634.ch38

du Toit, W. J., Marais, J., Pretorius, I. S., & du Toit, M. (2017). Oxygen in Must and Wine: A review. *South African Journal of Enology and Viticulture*, *27*(1). doi:10.21548/27-1-1610

Duca, Gh. (2012). Heterogeneous catalysis with metal complexes: fundamentals and applications. Springer. doi:10.1007/978-3-642-24629-6

Duca, Gh., Blonschi, V., Gladchi, V., Travin, S. (2020). Dynamics of different sulfur forms in natural waters and their influence on the redox state. *American Journal of Physical Chemistry*. . doi:10.11648/j.ajpc.20200903.12

Duca, Gh., Gladchi, V., & Romanciuc, L. (2002). Processes of pollution and self-purification of natural waters. Chisinau: EC MSU, 145.

Duca, Gh., Gladchi, V., & Romnaciuc, L. (2002). *Processes of pollution and self-purification of natural waters.* Chisinau, Republic of Moldova: CE USM.

Duca, Gh., Skurlatov, Iu., Misiti, A., Macoveanu, M., & Surpăţeanu, M. (2003). *Ecological Chemistry.* Chisinau, Republic of Moldova: CE USM.

Duca, G. (2021). Hydrogen peroxide in ecological and environmental chemistry, *Chemistry Journal of Moldova. General. Industrial and Ecological Chemistry, 16*, 28–45.

Duca, G. (2022). Hydrogen peroxide in ecological and environmental chemistry. *Chemistry Journal of Moldova. General Industrial and Ecological Chemistry, 16*(2), 28–45. doi:10.19261/cjm.2021.918

Duca, G., Bunduchi, E., Gladchi, V., Romanciuc, L., & Goreaceva, N. (2008). Estimation of the natural water self-purification capacity from the kinetic standpoint. *Chemistry Journal of Moldova, 3*(1), 10–12. doi:10.19261/cjm.2008.03(1).16

Duca, G., & Travin, S. (2020). Reactions' mechanisms and applications of hydrogen peroxide. *American Journal of Physical Chemistry, 9*(2), 36–44. doi:10.11648/j.ajpc.20200902.13

Duca, G., & Vaseashta, A. (Eds.). (2020). *Handbook of Research on Emerging Developments and Environmental Impacts of Ecological Chemistry.* IGI Global., doi:10.4018/978-1-7998-1241-8

Duca, Gh. (2017). Ecological and environmental chemistry. *Chemistry Journal of Moldova. General. Industrial and Ecological Chemistry., 12*(1), 9–19.

Duca, Gh., Covaliov, V., & Covaliova, O. (2020). Intensive biochemical processes of wastewater treatment with high caloricity biogas production. In *Emerging Developments and Environmental Impacts of Ecological Chemistry* (pp. 291–306). IGI Global Publ. House. doi:10.4018/978-1-7998-1241-8.ch013

Duca, Gh., Gladchi, V., & Goreacheva, N. (2007). *Practical works on the chemistry of natural waters.* Publishing Center of Moldova State University.

Duca, Gh., Gladchi, V., & Romanciuc, L. (2002). *Processes of pollution and self-purification of natural waters.* CE USM.

Duca, Gh., Skurlatov, Iu., Misiti, A., Macoveanu, M., & Surpăţeanu, M. (2003). *Ecological chemistry.* CE USM.

Duca, Gh., Skurlatov, Yu., & Sychev, A. (2002). *Redox catalysis and ecological chemistry.* PC MSU.

Duchaniya, R. K. (2014). Optical studies of chemically synthesized CdS nanoparticles. *International Journal of Minerals Metallurgy and Materials, 2*, 54–56.

Duka G. & Skurlatov Yu. (2002). *Ecological Chemistry*, 289. Publishing Center M.S.U. Chisinau.

Du, X., Li, H. Y., Yu, J., Xiao, X., Shi, Z., Mao, D., & Lu, G. (2015). Realization of a highly effective Pd–Cu–Cl_x/Al_2O_3 catalyst for low-temperature CO oxidation by pre-synthesizing the active copper phase of $Cu_2Cl(OH)_3$. *Catalysis Science & Technology, 5*(8), 3970–3979. doi:10.1039/C5CY00545K

Dyer, J. M. (2018). Oxidative modification of trichocyte keratins. *The Hair Fibre: Proteins, Structure and Development,* 205–218. doi:10.1007/978-981-10-8195-8_14

Dzyazko, Y. S., Volfkovich, Y. M., & Chaban, M. O. (2020). Composites containing inorganic ion exchangers and graphene oxide: hydrophilic–hydrophobic and sorption properties. In *Nanomaterials and Nanocomposites, Nanostructure Surfaces, and Their Applications* (pp. 93–110). Springer.

Earland, C., MacRae, T. P., Weston, G. J., & Statham, K. (1955). The oxidation of wool by inorganic peracids. *Textile Research Journal, 25*(11), 963–965. doi:10.1177/004051755502501109

EBPI. (n.d.). Algae Toxicity Tests. *Biotoxicity.* https://www.biotoxicity.com/index.php/microbiotests/algae-toxicity-tests/fresh-water

Echave, J., Barral, M., Fraga-Corral, M., Prieto, M. A., & Simal-Gandara, J. (2021). Bottle aging and storage of wines: A review. *Molecules (Basel, Switzerland), 26*(3), 713. doi:10.3390/molecules26030713 PMID:33573099

Edori, O. S., Festus, C., & Edori, E. S. (2014). Comparative effects of petrol and diesel on enzyme activity in Tympanotonus fuscatus after sublethal exposure. *Pakistan Journal of Biological Sciences, 17*(4), 545–549. doi:10.3923/pjbs.2014.545.549 PMID:25911844

Eggersdorfer, M., & Wyss, A. (2018). Carotenoids in human nutrition and health. *Archives of Biochemistry and Biophysics, 652*, 18–26. doi:10.1016/j.abb.2018.06.001 PMID:29885291

Eidelman, N., & Simon, C. G. (2004). Characterization of combinatorial polymer blend composition gradients by FTIR microspectroscopy. *Journal of Research of the National Institute of Standards and Technology, 09*(2), 219–231. doi:10.6028/jres.109.014 PMID:27366606

El Brahmi, A., & Abderafi, S. (2021). Hydrogen sulfide removal from wastewater using hydrogen peroxide insitu treatment: Case study of Moroccan urban sewers. *Materials Today: Proceedings, 45*, 7424–7427. doi:10.1016/j.matpr.2021.01.641

Ellman, G. L. (1959). Tissue sulfhydryl groups. *Archives of Biochemistry and Biophysics, 82*(1), 70–77. doi:10.1016/0003-9861(59)90090-6 PMID:13650640

Elsevier. (2010). *Managing Wine Quality: Viticulture and Wine Quality.* Woodhead Publishing Series in Food Science, Technology and Nutrition.

Enaru, B., Dreţcanu, G., Pop, T. D., Stǎnilǎ, A., & Diaconeasa, Z. (2021). Anthocyanins: Factors affecting their stability and degradation. *Antioxidants, 10*(12), 1967. doi:10.3390/antiox10121967 PMID:34943070

Ene, A., Zubcov, E., Bulat, D., Bulat, D., Cristea, V., & Florea, L. Gh. (2021). Pollutants and biodiversity in the Prut River. HydroEcoNex Project Report, Public information seminar, Vladesti, Galati county. *Romania.* Advance online publication. doi:10.13140/RG.2.2.28146.02247

Engin, A. B., Bukan, N., Kurukahvecioglu, O., Memis, L., & Engin, A. (2011). Effect of butylated hydroxytoluene (E321) pretreatment versus l-arginine on liver injury after sub-lethal dose of endotoxin administration. *Environmental Toxicology and Pharmacology, 32*(3), 457–464. doi:10.1016/j.etap.2011.08.014 PMID:22004966

Englezos, V., Cravero, F., Torchio, F., Rantsiou, K., Ortiz-Julien, A., Lambri, M., Gerbi, V., Rolle, L., & Cocolin, L. (2018). Oxygen availability and strain combination modulate yeast growth dynamics in mixed culture fermentations of grape must with Starmerella bacillaris and Saccharomyces cerevisiae. *Food Microbiology, 69*, 179–188. doi:10.1016/j.fm.2017.08.007 PMID:28941899

Engwa, G. A., Nweke, F. N., & Nkeh-Chungag, B. N. (2022). Free radicals, oxidative stress-related diseases, and antioxidant supplementation. *Alternative Therapies in Health and Medicine, 28*(1), 144–128. PMID:32827401

Enkhzaya, S., Shiomori, K., & Oyuntsetseg, B. (2017). Removal of heavy metals from aqueous solution by adsorption using livestock biomass of Mongolia. *Journal of Environmental Science and Technology, 10*(3), 107–119. doi:10.3923/jest.2017.107.119

Erdogan, U. H., Seki, Y., & Selli, F. (2020). Wool fibres. Handbook of Natural Fibres, 257–278. doi:10.1016/B978-0-12-818398-4.00011-6

Esimbekova, E. N., Torgashina, I. G., Kalyabina, V. P., & Kratasyuk, V. A. (2021). Enzymatic biotesting: Scientific basis and application. [in Russian]. *Siberian Journal of Ecology*, *3*(3), 364–382. doi:10.15372/SEJ20210308

España-Gamboa, E., Mijangos-Cortes, J., Barahona-Perez, L., Dominguez-Maldonado, J., Hernández-Zarate, G., & Alzate-Gaviri, L. (2011). Vinasse: Characterization and treatments. *Waste Management & Research*, *29*(12), 1235–1250. doi:10.1177/0734242X10387313 PMID:21242176

Esparza, I., Martínez-Inda, B., Cimminelli, M. J., Jimeno-Mendoza, M. C., Moler, J. A., Jiménez-Moreno, N., & Ancín-Azpilicueta, C. (2020). Reducing SO_2 doses in red wines by using grape stem extracts as antioxidants. *Biomolecules*, *10*(10), 1369. doi:10.3390/biom10101369 PMID:32992977

Estela-Escalante, W., Rychtera, M., Melzoch, K., Hatta-Sakoda, B. (2012). Effect of aeration on the fermentative activity of saccharomyces cerevisiae cultured in apple juice. *Revista Mexicana de Ingeniera Quimica, 1*(2).

European Chemicals Agency. (2010). Member state committee support document for identification of potassium dichromate as a substance of very high concern because of its CMR properties, adopted on 4 June 2010.

European Chemicals Agency. (2016). Annex XVII to REACH. *Substances restricted under REACH.* https://echa.europa.eu/substances-restricted-under-reach

European Union (2011). *Directive 2011/65/EU on the restriction of hazardous substances in electrical and electronic equipment (RoHS2)*. EU.

European Union. (2008). *Directive 2008/105/EC of the Europeans Parliament and of the Council of 16 December 2008, on environmental quality standards in the field of water policy*. EU.

European Union. (2013). Commission Regulation 2013/348/EU amending Annex XIV to Regulation (EC) No 1907/2006 of the European Parliament and of the Council on the Registration, Evaluation, Authorisation and Restriction of Chemicals (REACH).

Evans, M. G., & Uri, N. (1949). The dissociation constant of hydrogen peroxide and the electron affinity of the HO_2 radical. *Transactions of the Faraday Society*, *45*, 224–230. doi:10.1039/tf9494500224

Ewuzie, U., Saliu, O. D., Dulta, K., Ogunniyi, S., Bajeh, A. O., Iwuozori, K. O., & Ighalo, J. O. (2022). A review on treatment technologies for printing and dyeing wastewater (PDW). *Journal of Water Process Engineering*, *50*, 103273. doi:10.1016/j.jwpe.2022.103273

Ferna'ndez-Pe'rez, M., Flores-Ce'spedes, F., Daza-Fernández, I., Vidal-Peña, F., & Villafranca-Sánchez, M. (2014). Lignin and lignosulfonate-based formulations to protect pyrethrins against photodegradation and volatilization. *Industrial & Engineering Chemistry Research*, *53*(35), 13557–13564. doi:10.1021/ie500186e

Fernandes, C., Encarnacao, I., Gaspar, A., Garrido, J., Borges, F., & Garrido, E. M. (2014). Influence of hydroxypropyl-β-cyclodextrin on the photostability of fungicide pyrimethanil. *International Journal of Photoenergy*, *2014*, 1–8. doi:10.1155/2014/489873

Fernandez-Ibanez, P., McMichael, S., Cabanillas, A. R., Alkharabsheh, S., Moranchel, A. T., & Abyrne, J. (2021). New trends on photoelectrocatalysis (PEC): Nanomaterials, wastewater treatment and hydrogen generation. *Current Opinion in Chemical Engineering*, *34*, 100725. doi:10.1016/j.coche.2021.100725

Ferrarello, C. N., Fernández de la Campa, M. D. R., Muñiz, C. S., & Sanz-Medel, A. (2000). Metal distribution patterns in the mussel Mytilus edulis cytosols using size-exclusion chromatography and double focusing ICP-MS detection. *Analyst (London)*, *125*(12), 2223–2229. doi:10.1039/b008181g PMID:11219056

Ferraresi-de Arraujo, J. G., & Nino-Castillo, I. N. (2021). Vinasse: Current concept, challenges, and opportunities for sustainability. *ECORFAN-Journal-Republic of Nicaragua.*, *7*(13), 1–9. doi:10.35429/EJRN.2021.13.7.1.9

Ferreira, L. F., Aguiar, M., Pompeu, G., Messias, T., & Monteiro, R. T. (2010). Selection of vinasse degrading microorganisms. *World Journal of Microbiology & Biotechnology*, *26*(9), 1613–1621. doi:10.100711274-010-0337-3

Ferri, M., Vannini, M., Ehrnell, M., Eliasson, L., Xanthakis, E., Monari, S., & Tassoni, A. (2020). From winery waste to bioactive compounds and new polymeric biocomposites: A contribution to the circular economy concept. *Journal of Advanced Research*, *24*, 1–11. doi:10.1016/j.jare.2020.02.015 PMID:32181012

Ferroni, C., & Varchi, G. (2021). Keratin-based nanoparticles as drug delivery carriers. *Applied Sciences (Basel, Switzerland)*, *11*(20), 9417. doi:10.3390/app11209417

Fitzer, E. (1988). *Carbon fibers and carbon composites.* Moscow: Mir.

Flieger, J., Flieger, W., Baj, J., & Maciejewski, R. (2021). Antioxidants: Classification, natural sources, activity/capacity measurements, and usefulness for the synthesis of nanoparticles. *Materials (Basel)*, *14*(15), 4135. doi:10.3390/ma14154135 PMID:34361329

Florêncio, M. H., Pires, E., Castro, A. L., Nunes, M. R., Borges, C., & Costa, F. M. (2004). Photodegradation of Diquat and Paraquat in aqueous solutions by titanium dioxide: Evolution of degradation reactions and characterization of intermediates. *Chemosphere*, *55*(3), 345–355. doi:10.1016/j.chemosphere.2003.11.013 PMID:14987933

Food and Agriculture Organization of the United Nations (FAO). (2018). *FAOSTAT – Pesticide use (Food and Agriculture Organization of the United Nations: Rome, Italy).* FAO. http://www.fao.org/faostat/en/#data/RP

Foyer, C. H., & Allen, J. F. (2003). Lessons from redox signaling in plants. *antioxidants & redox signaling*, *5*(1), 3–5. doi:10.1089/152308603321223487

Francesca, N., Sannino, C., Settanni, L., Corona, O., Barone, E., & Moschetti, G. (2014). Microbiological and chemical monitoring of Marsala base wine obtained by spontaneous fermentation during large-scale production. *Annals of Microbiology*, *64*(4), 1643–1657. doi:10.100713213-014-0808-0

Frankel, E. N. (2012). Hydroperoxide decomposition. In *Lipid Oxidation* (pp. 67–98). Elsevier. doi:10.1533/9780857097927.67

Fratianni, A., D'Agostino, A., Niro, S., Bufano, A., Paura, B., & Panfili, G. (2021). Loss or gain of lipophilic bioactive compounds in vegetables after domestic cooking? Effect of steaming and boiling. *Foods*, *10*(5), 960. doi:10.3390/foods10050960 PMID:33924885

Frois, S. R., Tadeu Grassi, M., De Campos, M. S., & Abate, G. (2012). Determination of Cr(VI) in water samples by ICP-OES after separation of Cr(III) by montmorillonite. *Analytical Methods*, *4*(12), 4389–4394. doi:10.1039/c2ay26125a

Frontasyeva M., Harmens H., Uzhinskiy A., & Chaligava, O. (2020). Mosses as biomonitors of air pollution: 2015/2016 survey on heavy metals, nitrogen and POPs in Europe and beyond. *Report of the ICP Vegetation Moss Survey Coordination Centre.* Joint Institute for Nuclear Research, Dubna, Russian Federation. https://www.researchgate.net/publication/342804243_Mosses_as_biomonitors_of_air_pollution_20152016_survey_on_heavy_metals_nitrogen_and_POPs_in_Europe_and_beyond

Fujii, M., & Otani, E. (2017). Photochemical generation and decay kinetics of superoxide and hydrogen peroxide in the presence of standard humic and fulvic acids. *Water Research*, *123*, 642–654. doi:10.1016/j.watres.2017.07.015 PMID:28710981

Gacen, J., & Cayuela, D. (2000). Comparison of wool bleaching with hydrogen peroxide in alkaline and acidic media. *Coloration Technology*, *116*(1), 13–15. doi:10.1111/j.1478-4408.2000.tb00003.x

Gacen, J., Cegarra, J., & Caro, M. (1991). Wool bleaching with reducing agent in the presence of sodium lauryl sulphate. Part 3 - bleaching with thiourea dioxide. *Journal of Studies in Dynamics and Change*, *107*, 138–141.

Gallard, H., de Laat, J., & Legube, B. (1998). Influence du pH sur la vitesse d'oxydation de compose's organiques par FeII/H2O2. Me'canismes re'actionnels et mode'lisation. *New Journal of Chemistry*, *22*(3), 263–268. doi:10.1039/a708335a

Gallejones, P., Pardo, G., Aizpurua, A., & Del Prado, A. (2014). Life cycle assessment of first-generation biofuels using a nitrogen crop model. *The Science of the Total Environment*, *505C*, 1191–1201. PMID:25461117

Gan, P., Sun, Y., Li, Y., Liu, W., Ye, J., Tong, M., & Liang, J. (2023). The degradation of municipal solid waste incineration leachate by UV/ persulfate and UV/H_2O_2 processes: The different selectivity of $SO_4 \bullet^-$ and •OH. *Chemosphere*, *311*, 137009. doi:10.1016/j.chemosphere.2022.137009 PMID:36326516

Gao, F., Li, W., Deng, J., Kan, J., Guo, T., Wang, B., & Hao, S. (2019). Recombinant human hair keratin nanoparticles accelerate dermal wound healing. *ACS Applied Materials & Interfaces*, *11*(20), 18681–18690. doi:10.1021/acsami.9b01725 PMID:31038908

Garcia, P. E., Gerea, M., & Diéguez, M. (2020). Natural levels of hydrogen peroxide (H_2O_2) in deep clear South temperate lakes: Field and laboratory evidence of photo - and biotic production. *The Science of the Total Environment*, *727*, 138641. Advance online publication. doi:10.1016/j.scitotenv.2020.138641 PMID:32334222

García, P. E., Queimaliños, C., & Diéguez, M. C. (2018). Natural levels and photo-production rates of hydrogen peroxide (H2O2) in Andean Patagonian aquatic systems: Influence of the dissolved organic matter pool. *Chemosphere*. doi:10.1016/j.chemosphere.2018.10.179 PMID:30445400

Gawrysiak-Witulska, M., Rudzińska, M., Wawrzyniak, J., & Siger, A. (2012). The effect of temperature and moisture content of stored rapeseed on the phytosterol degradation rate. *Journal of the American Oil Chemists' Society*, *89*(9), 1673–1679. doi:10.100711746-012-2064-4 PMID:22923815

Gęgotek, A., & Skrzydlewska, E. (2022). Antioxidative and anti-inflammatory activity of ascorbic acid. *Antioxidants*, *11*(10), 1993. doi:10.3390/antiox11101993 PMID:36290716

Ge, L., Zhang, P., Halsall, C., Li, Y., Chen, C. E., Li, J., & Yao, Z. (2019). The importance of reactive oxygen species on the aqueous phototransformation of sulfonamide antibiotics: Kinetics, pathways, and comparisons with direct photolysis. *Water (Basel)*, *149*, 243–250. doi:10.1016/j.watres.2018.11.009 PMID:30448736

Gerbens-Leenes, W., & Hoekstra, A. Y. (2012). The water footprint of sweeteners and bioethanol. *Environment International*, *40*, 202–211. doi:10.1016/j.envint.2011.06.006 PMID:21802146

Gerringa, L. J. A., Rijkenberg, M. J. A., Timmermans, R., & Buma, A. G. J. (2004). The influence of solar ultraviolet radiation on the photochemical production of H2O2 in the equatorial Atlantic Ocean. *Journal of Sea Research*, *51*(1), 3–10. doi:10.1016/j.seares.2003.03.002

Gharby, S., Harhar, H., Matthäus, B., Bouzoubaa, Z., & Charrouf, Z. (2016). The chemical parameters and oxidative resistance to heat treatment of refined and extra virgin Moroccan Picholine olive oil. *Journal of Taibah University for Science : JTUSCI*, *10*(1), 100–106. doi:10.1016/j.jtusci.2015.05.004

Ghazal, H., Koumaki, E., Hoslett, J., Malamis, S., Katsou, E., Barcelo, D., & Jouhara, H. (2022). Insights into current physical, chemical and hybrid technologies used for the treatment of wastewater contaminated with pharmaceuticals. *Journal of Cleaner Production*, *361*, 132079. doi:10.1016/j.jclepro.2022.132079

Ghendov-Mosanu, A., Cojocari, D., Balan, G., Patras, A., Lung, I., Soran, M.-L., Opris, O., Cristea, E., & Sturza, R. (2022). Chemometric optimization of biologically active compounds extraction from grape marc: Composition and antimicrobial activity. *Molecules (Basel, Switzerland), 27*(5), 1610. doi:10.3390/molecules27051610 PMID:35268711

Ghiocel, I. E., Muresan, A., Irina, E. A., Ioanid, A., & Muresan, R. (2005). *Modification of the surface of textile materials.* PERFORMANTICA Publishing.

Gibson, L. T. (2004). Archaeometry and Antique Analysis - Metallic and Ceramic Objects. In *Encyclopedia of Analytical Science* (2nd ed., pp. 117–123). Elsevier. doi:10.1016/B0-12-369397-7/00020-0

Giguere, P. A. (1975). *Complements au nouveau traite de chemie minerale – No. 4 Peroxyde d'Hydrogene et Polyoxydes d'Hydrogene.* Mason.

Gil T. A., Balayan A. E. & Stom D. I. (1988). *Method of biotesting for quenching the luminescence of luminous bacteria. Methods of biotesting of waters.* Chernogolovka: Publishing house of OIKhFI, 17-18.

Gill, A. C., Ritchie, M. A., Hunt, L. G., Steane, S. E., Davies, K. G., Bocking, S. P., Rhie, A. G. O., Bennett, A. D., & Hope, J. (2000). Post-translational hydroxylation at the N-terminus of the prion protein reveals presence of PPII structure in vivo. *The EMBO Journal, 19*(20), 5324–5331. doi:10.1093/emboj/19.20.5324 PMID:11032800

Gimaeva, A. R., Valinurova, E. R., Igdavletova, D. K. & Kudasheva, F. C. (2011). Sorption of heavy metal ions from water by activated carbon adsorbents. *Sorption and chromatographic processes, 11*(3), 350-356.

Gladchi, V., Bunduchi, E., Blonschi, V., & Romanciuc, L. (2022a). Ecological chemistry of natural waters and self-purification processes of aquatic systems. History and development at the State University of Moldova. In: V. Gladchi, A. Aricu (Eds.), Ecological Chemistry: History and Achievements: Academician Gheorghe Duca, 70th Anniversary: Monograph (pp. 37-85). Publishing Center of Moldova State University.

Gladchi, V., Goreaceva, N., Duca, Gh., Bunduchi, E., Borodaev, R., Şurîghina, O., & Lis, A. (2014). Chemical composition of right bank tributaries of Dniester River and their general impact. Chapter in Gh. Duca (Ed.), Management of Water Quality in Moldova (pp. 81-96). Water Sciences and Technology Library, vol. 69, Springer, Cham. https://doi.org/ doi:10.1007/978-3-319-02708-1_6

Gladchi, V., Goreaceva, N., Duca, Gh., Bunduchi, E., Borodaev, R., Şurîghina, O., & Lis, A. (2014). Chemical compozition of right bank tributaries of Nistru river and their general impact. Management of Water Quality in Moldova. Springer.

Gladchi, V. (2018). *Catalytic transformations and the redox state of the environment.* CEP USM.

Gladchi, V., Bunduchi, E., Blonschi, V., & Romanciuc, L. (2022b). Ecological chemistry of natural waters and self-purification processes of aquatic systems. History and development at the State University of Moldova. In V. Gladchi & A. Aricu (Eds.), *Ecological Chemistry: History and Achievements: Academician Gheorghe Duca, 70th Anniversary: Monograph* (pp. 37–85). Publishing Center of Moldova State University.

Gladchi, V., Bunduchi, E., Blonschi, V., & Romanciuc, L. (2022b). Hydrochemistry of polluted surface water: case study of Moldova. In A. Vaseashta, G. Duca, & S. Travin (Eds.), *Handbook of Research on Water Sciences and Society* (pp. 501–529). IGI Global. doi:10.4018/978-1-7998-7356-3.ch022

Gladchi, V., Bunduchi, E., & Romanciuc, L. (2020). Ecological Chemistry of the natural water. In I. G. I. Global (Ed.), *Handbook of Research on Emerging Developments and Environmental Impacts of Ecological Chemistry* (pp. 196–211). doi:10.4018/978-1-7998-1241-8.ch009

Gladchi, V., Duca, Gh., Goreaceva, N., Bunduchi, E., & Lis, A. (2013). Chemical composition from the Dniester river tributaries. *Chemistry Journal of Moldova, General. Industrial and Ecological Chemistry, 8*(1), 23–32.

Gladchi, V., Goreaceva, N., Duca, G., Bunduchi, E., Romanciuc, L., Mardari, I., & Borodaev, R. (2008). The study of redox conditions in the Dniester river. *Chemistry Journal of Moldova*, *3*(1), 70–76. doi:10.19261/cjm.2008.03(1).14

Gmelin Institut. (1963). *Gmelins Handbook of Inorganic Chemistry, Sulfur* (Vol. 8). Verlag Chemie. (in German)

Gogos, A., Knauer, K., & Bucheli, T. D. (2012). Nanomaterials in plant protection and fertilization: Current state, foreseen applications, and research priorities. *Journal of Agricultural and Food Chemistry*, *60*(39), 9781–9792. doi:10.1021/jf302154y PMID:22963545

Goldstone, J. V., & Voelker, B. M. (2000). Chemistry of superoxide radical in seawater: Cdom associated sink of superoxide in coastal waters. *Environmental Science & Technology*, *34*(6), 1043–1048. doi:10.1021/es9905445

Goncharuk, V. V., Bagrii, V. A., Melnik, L. A., Chebotareva, R. D., & Bashtan, S. Y. (2010). The Use of Redox Potential in Water Treatment Processes. *Journal of Water Chemistry and Technology*, *32*(1), 1–9. doi:10.3103/S1063455X10010017

Gong, S., Ding, Ch., Liu, J., Fu, K., Pan, Y., Shi, J., & Deng, H. (2021). Degradation of naproxen by UV-irradiation in the presence of nitrate: Efficiency, mechanism, products, and toxicity change. *Chemical Engineering Journal, 133016.* doi:10.1016/j.cej.2021.133016

Gopalan, P. R., Annaselvi, A. G., & Subramania, P. (2013). Synthesis and characterization of β-Cyclodextrin capped silver nanoparticles. *Int. J. Nanomater. Biostruct.*, *3*, 26–30.

Goreacheva, N., & Duca, Gh. (2004). *Hydrochemistry of small Rivers of the Republic of Moldova.* Publishing Center of Moldova State University.

Gorina, V. A., & Cheblakova, E. G. (2016). Influence of activation modes on the specific surface and development of a microporous structure of viscose-based carbon fibers. *Russian Journal of Non-Ferrous Metals*, *57*(7), 734–738. doi:10.3103/S1067821216070087

Goullé, J. P., Mahieu, L., Castermant, J., Neveu, N., Bonneau, L., Lainé, G., Bouige, D., & Lacroix, C. (2005). Metal and metalloid multi-elementary ICP-MS validation in whole blood, plasma, urine and hair: Reference values. *Forensic Science International*, *153*(1), 39–44. doi:10.1016/j.forsciint.2005.04.020 PMID:15979835

Gousterova, A., Nustorova, M., Goshev, I., Christov, P., Braikova, D., Tishinov, K., Haertlé, T., & Nedkov, P. (2003). Alkaline hydrolysate of waste sheep wool aimed as fertilizer. *Biotechnology, Biotechnological Equipment*, *17*(2), 140–145. doi:10.1080/13102818.2003.10817072

Grabner, G., & Richard, C. (2005). Mechanisms of direct photolysis of biocides based on halogenated phenols and anilines. *Environmental Photochemistry*, *M*(Part II), 161–192. doi:10.1007/b138183

Greenberg, R. R., Bode, P., & De Nadai Fernandes, E. A. (2011). Neutron activation analysis: A primary method of measurement. *Spectrochimica Acta - Part B Atomic Spectroscopy, 66*(3–4), 193–241. Elsevier B.V. doi:10.1016/j.sab.2010.12.011

Greer, J. B., Magnuson, J. T., Hester, K., Giroux, M., Pope, C., Anderson, T., Liu, J., Dang, V., Denslow, N. D., & Schlenk, D. (2019). Effects of chlorpyrifos on cholinesterase and serine li pase activities and li pid metabolism in brains of rainbow trout (Oncorhynchus mykiss). *Toxicological Sciences*, *172*(1), 146–154. doi:10.1093/toxsci/kfz167 PMID:31359069

Gregg, S., & Sing, K. (1982). *Adsorption, Surface Area, and Porosity* (2nd ed.). Academic.

Gregorin, C., Albarano, L., Somma, E., Costantini, M., & Zupo, V. (2021). Assessing the ecotoxicity of copper and polycyclic aromatic hydrocarbons: Comparison of effects on *Paracentrotus lividus* and *Botryllus schlosseri*, as alternative bioassay methods. *Water (Basel)*, *13*(5), 711. doi:10.3390/w13050711

Grieve, M. (2021). *Field Measurement of Oxidation Reduction Potential*. U.S. Environmental Protection Agency, Laboratory Services and Applied Science Division.

Grossi-Botelho, R., Tornisielo, V. L., Alves de Olinda, R., Maranho, L. A., & Machado-Neto, L. (2012). Acute toxicity of sugarcane vinasse to aquatic organisms before and after pH adjustment. *Toxicological and Environmental Chemistry, 94*(10), 2035–2045. doi:10.1080/02772248.2012.738516

Guan, C., Guo, Q., Wang, Z., Wei, X., Han, B., Luo, X., Pan, H., & Jiang, J. (2022). Bisulfite activated permanganate for oxidative water decontamination. *Water Research, 216*, 118331. doi:10.1016/j.watres.2022.118331 PMID:35358879

Guerreiro, L. F., Rodrigues, C. S., Duda, R. M., de Oliveira, R. A., Boaventura, R. A., & Madeira, L. M. (2016). Treatment of sugarcane vinasse by a combination of coagulation/flocculation and Fenton's oxidation. *Journal of Environmental Management, 181*, 237–248. doi:10.1016/j.jenvman.2016.06.027 PMID:27353374

Guillén, M. D., & Cabo, N. (2002). Fourier transform infrared spectra data versus peroxide and anisidine values to determine oxidative stability of edible oils. *Food Chemistry, 77*(4), 503–510. doi:10.1016/S0308-8146(01)00371-5

Gul, H., Raza, W., Lee, J., Azam, M., Ashraf, M., & Kim, K.-H. (2021). Progress in microbial fuel cell technology for wastewater treatment and energy harvesting. *Chemosphere, 281*, 130828. doi:10.1016/j.chemosphere.2021.130828 PMID:34023759

Gunaseelan, V. N. (1997). Anaerobic digestion of biomass for methane production: A review. *Biomass and Bioenergy, 13*(1-2), 83–144. doi:10.1016/S0961-9534(97)00020-2

Gupta, P., Lakes, A., & Dziubla, T. (2016). A free radical primer. *Oxidative Stress and Biomaterials,* 1–33. doi:10.1016/B978-0-12-803269-5.00001-2

Gupta, K., Rachna, Singh, B., Mungray, A. K., Bharti, R., Nema, A. K., Pant, K. K., & Mulla, S. I. (2022). Bioelectrochemical technologies for removal of xenobiotics from wastewater. *Sustainable Energy Technologies and Assessments, 49*, 101652. doi:10.1016/j.seta.2021.101652

Gutiérrez-Escobar, R., Aliaño-González, M. J., & Cantos-Villar, E. (2021). Wine polyphenol content and its influence on wine quality and properties: A review. *Molecules (Basel, Switzerland), 26*(3), 718. doi:10.3390/molecules26030718 PMID:33573150

Haber, F., & Weiss, J. (1932). Über die katalyse des hydroperoxydes. *Naturwissenschaften, 20*(51), 948–950. doi:10.1007/BF01504715

Haber, F., & Willstätter, R. (1931). Unpaarigkeit und radikalketten im reaktionsmechanismus organischer und enzymatischer vorgänge. [A and B Series]. *Berichte der Deutschen Chemischen Gesellschaft Abteilung B: Abhandlungen, 64*(11), 2844–2856. doi:10.1002/cber.19310641118

Hach. (2014). *Oxygen Demand Chemical, Manganese III reactor digestion method 190067 without chloride removal 30 to 1000 mg/L COD Mn.* Hach Company, DOC316.53.01102.

Häkkinen, P. J., Anesio, A. M., & Granéli, W. (2004). Hydrogen peroxide distribution, production, and decay in boreal lakes. *Canadian Journal of Fisheries and Aquatic Sciences, 61*(8), 1520–1527. doi:10.1139/f04-098

Halliwell, B., & Gutteridge, J. M. C. (2015). *Free Radicals in Biology and Medicine*. Oxford University Press. doi:10.1093/acprof:oso/9780198717478.001.0001

Hani, Y. M. I., Turies, C., Palluel, O., Delahaut, L., Gaillet, V., Bado-nilles, A., Porcher, J.-M., Geffard, A., & Dedourgegeffard, O. (2018). Effects of chronic exposure to cadmium and temperature, alone or combined, on the threespine stickleback (Gasterosteus aculeatus): Interest of digestive enzymes as biomarkers. *Aquatic Toxicology (Amsterdam, Netherlands), 199*, 252–262. doi:10.1016/j.aquatox.2018.04.006 PMID:29677587

Hansen, S., Pedersen-Bjergaard, S., & Rasmussen, K. (2012). *Introduction to pharmaceutical chemical analysis. Redox Titrations.* John Wiley & Sons., doi:10.1002/9781119953647

Hao, L., Gong, L., Chen, L., Guan, M., Zhou, H., Qiu, S., Wen, H., Chen, H., Zhou, X., & Akbulut, M. (2020). Composite pesticide nanocarriers involving functionalized boron nitride nanoplatelets for pH-responsive release and enhanced UV stability. *Chemical Engineering Journal, 396*, 125233. doi:10.1016/j.cej.2020.125233

Hao, L., Lin, G., Lian, J., Chen, L., Zhou, H., Chen, H., Hu, H., & Zhou, X. (2019). Carboxymethyl cellulose capsulated zein as pesticide nano-delivery system for improving adhesion and anti-UV properties. *Carbohydrate Polymers, 231*, 115725. doi:10.1016/j.carbpol.2019.115725 PMID:31888813

He, M., Huang, L., Zhao, B., Chen, B., & Hu, B. (2017). Advanced functional materials in solid phase extraction for ICP-MS determination of trace elements and their species - A review. *Analytica Chimica Acta, 973,* 1–24. Elsevier B.V. doi:10.1016/j.aca.2017.03.047

Helmke, P. A. (2018). Neutron activation analysis. In *Methods of Soil Analysis, Part 3: Chemical Methods* (pp. 141–159). Springer US. doi:10.2136ssabookser5.3.c6

Hernanz, D., Recamales, Á. F., González-Miret, M. L., Gómez-Míguez, M. J., Vicario, I. M., & Heredia, F. J. (2007). Phenolic composition of white wines with a prefermentative maceration at experimental and industrial scale. *Journal of Food Engineering, 80*(1), 327–335. doi:10.1016/j.jfoodeng.2006.06.006

He, S., Zhang, W., Li, D., Li, P., Zhu, Y., Ao, M., Ki, J., & Cao, Y. (2013). Preparation and characterization of double-shelled avermectin microcapsules based on copolymer matrix of silica–glutaraldehyde–chitosan. *Journal of Materials Chemistry. B, Materials for Biology and Medicine, 1*(9), 1270–1278. doi:10.1039/c2tb00234e PMID:32260800

Heymann, T., Heinz, P., & Glomb, M. A. (2015). Lycopene inhibits the isomerization of β-carotene during quenching of singlet oxygen and free radicals. *Journal of Agricultural and Food Chemistry, 63*(12), 3279–3287. doi:10.1021/acs.jafc.5b00377 PMID:25803572

Hidalgo, D., Martin-Marroquin, J. M., & Corona, F. (2019). A multi-waste management concept as a basis for a circular economy model. *Renewable & Sustainable Energy Reviews, 111*, 481–489. doi:10.1016/j.rser.2019.05.048

Hira, S., Saleem, U., Anwar, F., Sohail, M. F., Raza, Z., & Ahmad, B. (2019). β-Carotene: A natural compound improves cognitive impairment and oxidative stress in a mouse model of streptozotocin-induced Alzheimer's disease. *Biomolecules, 9*(9), 441. doi:10.3390/biom9090441 PMID:31480727

Hoang, H., & Quang, P. ThiDinh, N., Ngoc, T., Thi, T., Tran, L., Bao, N., Yuvakkumar, R., & Nguyen, V.-H. (2022) Fe2+, Fe3+, Co2+ as highly efficient cocatalysts in the homogeneous electro-Fenton process for enhanced treatment of real pharmaceutical wastewater. *Journal of Water Process Engineering, 46*, 102635.

Höcker, H. (2002). Fibre morphology. *Wool,* 60–79. doi:10.1533/9781855737648.60

Hofmann, T., Lowry, G. V., Ghoshal, S., Tufenkji, N., Brambilla, D., Dutcher, J. R., Gilbertson, L. M., Giraldo, J. P., Kinsella, J. M., Landry, M. P., & Lowell, W. (2020). Technology readiness and overcoming barriers to sustainably implement nanotechnology-enabled plant agriculture. *Nature Food, 1*(7), 416–425. doi:10.103843016-020-0110-1

Holder, T. M. (1974). A new technique for treating esophageal varices. *Journal of Pediatric Surgery*, *9*(3), 431. doi:10.1016/S0022-3468(74)80322-2

Holt, E. A., & Miller, S. W. (2010). Bioindicators: Using Organisms to Measure Environmental Impacts. *Nature Education Knowledge*, *3*(10), 8.

Holubová, Z., Moos, M., & Sommer, L. (2012). Simultaneous determination of metal traces by ICP-MS in environmental waters using SPE preconcentration on different polymeric sorbents. *Chemical Papers*, *66*(10), 899–906. doi:10.247811696-012-0185-8

Hong, J., Cho, K.-H., Presser, V., & Su, X. (2022). Recent advances in wastewater treatment using semiconductor photocatalysts. *Current Opinion in Green and Sustainable Chemistry*, *36*, 100644. doi:10.1016/j.cogsc.2022.100644

Hopfer, H., Ebeler, S. E., & Heymann, H. (2012). The combined effects of storage temperature and packaging type on the sensory and chemical properties of chardonnay. *Journal of Agricultural and Food Chemistry*, *60*(43), 10743–10754. doi:10.1021/jf302910f PMID:23035911

Hopwood, M. J., Rapp, I., Schlosser, C., & Achterberg, E. P. (2017). Hydrogen peroxide in deep waters from the Mediterranean Sea, South Atlantic and South Pacific Oceans. *Scientific Reports*, *7*(1), 43436. doi:10.1038rep43436 PMID:28266529

Horváth, G., & Kawazoe, K. (1983). Method for the calculation of effective pore size distribution in molecular sieve carbon. *Journal of Chemical Engineering of Japan*, *16*(6), 470–475. doi:10.1252/jcej.16.470

Hristozova, G., Marinova, S., Motyka, O., Svozilík, V., & Zinicovscaia, I. (2020). Multivariate assessment of atmospheric deposition studies in Bulgaria based on moss biomonitors: Trends between the 2005/2006 and 2015/2016 surveys. *Environmental Science and Pollution Research International*, *27*(31), 39330–39342. doi:10.100711356-020-10005-w PMID:32648219

Hsu, E., & Parthasarathy, S. (2017). Anti-inflammatory and antioxidant effects of sesame oil on atherosclerosis: A descriptive literature review. *Cureus*. doi:10.7759/cureus.1438 PMID:28924525

Hua, M., Zhang, S. J., Pan, B. C., Zhang, W. M., Lv, L., & Zhang, Q. X. (2012). Heavy metal removal from water/wastewater by nanosized metal oxides: A review. *Journal of Hazardous Materials*, *211–212*, 317–331. doi:10.1016/j.jhazmat.2011.10.016 PMID:22018872

Huang, G., Qian, G., Yan, Y., Xu, D., Xu, Ch., Fu, L., & Lin, B. (2019). A super long-acting and anti-photolysis pesticide release platform through self-assembled natural polymer-based polyelectrolyte. *Reactive & Functional Polymers*, *146*, 104429. doi:10.1016/j.reactfunctpolym.2019.104429

Huang, Y., Li, X., Xiong, Q., Chen, Y., Peng, Z., Chen, J., Li, J., Zhang, Y., & Cui, J. (2022). Preparation and insecticidal activity evaluation of emamectin–lignin sulfonic acid conjugate with antiphotolysis property. *ACS Omega*, *7*(33), 29046–29053. doi:10.1021/acsomega.2c02883 PMID:36033669

Hunter, R. S. (1942). Photoelectric tristimulus colorimetry with three filters. *Journal of the Optical Society of America*, *32*(9), 509–538. https://doi.org/doi:10.1364/josa.32.000509. doi:10.1364/JOSA.32.000509

Hurren, C. (2008). Dyeing and coloring tests for fabrics. *Fabric Testing*, 255–274. doi:10.1533/9781845695064.255

International Organization for Standards Group. (1987). ISO 6341-82. *Water quality. Determination of inhibition of Daphnia magna Straus (Cladocera, Crustacea) mobility, 58,* 13. - M.: Publishing house of standards.

International Standard Organization (2002). ISO 15705, Water quality – Determination of the chemical oxygen demand index (ST-COD – Small-scale sealed tube method, Genf, Schweiz. Janicke, W. (1983). *Chemical oxidability of water constituents*. WaBoLu Berichte 1/1983, Dietrich Reimer Verlag GmbH.

International Standard Organization. (1989). ISO 6060, Water quality - Determination of the chemical oxygen demand. Genf, Schweiz.

Issakhov, A., Alimbek, A., & Zhandaulet, Y. (2021). The assessment of water pollution by chemical reaction products from the activities of industrial facilities: Numerical study. *Journal of Cleaner Production*, *282*, 125239. doi:10.1016/j.jclepro.2020.125239

Istomina, A., Chelomin, V., Kukla, S., Zvyagintsev, A., Karpenko, A., Slinko, E., Dovzhenko, N., Slobodskova, V., & Kolosova, L. (2019). Copper effect on the biomarker state of the Mizuhopecten yessoensis tissues in the prespawning period. *Environmental Toxicology and Pharmacology*, *70*, 103189. doi:10.1016/j.etap.2019.04.010 PMID:31103490

Iti, S. (2020). ICP-OES: An Advance Tool in Biological Research. *Open Journal of Environmental Biology*, 027–033. doi:10.17352/ojeb.000018

Ito, S., Sugumaran, M., & Wakamatsu, K. (2020). Chemical Reactivities of *ortho*-Quinones Produced in Living Organisms: Fate of Quinonoid Products Formed by Tyrosinase and Phenoloxidase Action on Phenols and Catechols. *International Journal of Molecular Sciences*, *21*(17), 6080. doi:10.3390/ijms21176080 PMID:32846902

IWTO. (2022). *An annual compilation of the global wool market`s production statistics from farm to retail-world sheep numbers & wool production.* Prepared by Poimena Analysis & Delta Consultants, 28–43. https://iwto.org/resources/statistics

Jahangirian, H., Ghasemian lemraski, E., Webster, T. J., Rafiee-Moghaddam, R., & Abdollahi, Y. (2017). A review of drug delivery systems based on nanotechnology and green chemistry: Green nanomedicine. *International Journal of Nanomedicine*, *12*, 2957–2978. doi:10.2147/IJN.S127683 PMID:28442906

Jalali, S., Ardjmand, M., Ramavandi, B., & Nosratinia, F. (2021). Removal of amoxicillin from wastewater in the presence of H_2O_2 using modified zeolite Y- MgO catalyst: An optimization study. *Chemosphere*, *274*, 129844. doi:10.1016/j.chemosphere.2021.129844 PMID:33582537

Janelidze, Z., Chikhradze, N., & Janelidze, G. (2021). Assessing the impact of anthropogenic factors on the ecosystems of the Paliastomi Lake region. *Georgian Geographical Journal*, *1*(1), 34–41.

Jankowski, J., Kieber, D., & Mopper, K. (1999). Nitrate and nitrite ultraviolete actinometers. *Photochemistry and Photobiology*, *70*(3), 319–326. doi:10.1111/j.1751-1097.1999.tb08143.x

Javidipour, I., Erinç, H., Baştürk, A., & Tekin, A. (2017). Oxidative changes in hazelnut, olive, soybean, and sunflower oils during microwave heating. *International Journal of Food Properties*, *20*(7), 1582–1592. doi:10.1080/10942912.2016.1214963

Jayakumar, R., Nwe, N., Tokura, S., & Tamura, H. (2007). Sulfated chitin and chitosan as novel biomaterials. *International Journal of Biological Macromolecules*, *40*(3), 175–181. doi:10.1016/j.ijbiomac.2006.06.021 PMID:16893564

Jayakumar, R., Prabaharan, M., Reis, R. L., & Mano, J. F. (2005). Graft copolymerized chitosan - present status and applications. *Carbohydrate Polymers*, *62*(2), 142–158. http://repositorium.sdum.uminho.pt/bitstream/1822/20220/1/file.pdf. doi:10.1016/j.carbpol.2005.07.017

Jerzykiewicz, M., Ćwieląg-Piasecka, I., & Jezierski, A. (2013). Pro- and antioxidative effect of α-tocopherol on edible oils, triglycerides and fatty acids. *Journal of the American Oil Chemists' Society*, *90*(6), 803–811. doi:10.100711746-013-2227-y PMID:23710072

Jiang, H., Wang, Y., Tan, Z., Hu, L., Shi, J., Liu, G., Yin, Y., Cai, Y., & Jiang, G. (2022). Dissolved metal ion removal by online hollow fiber ultrafiltration for enhanced size characterization of metal-containing nanoparticles with single-particle ICP-MS. *Journal of Environmental Sciences (China)*, *126*, 494–505. doi:10.1016/j.jes.2022.05.034 PMID:36503776

Jiexiang, Y. (2009). *Synthesis and characterization of cyclodextrin capped Au and Ag nanoparticles*. NUS.

Ji, M., Zhang, J., Li, S., Mu, G., & Di, H. (2019). Evaluating CDOM sources using excitation-emission matrix fluorescence and parallel factor analysis, and their links to water quality in highly polluted rivers in China. *Polish Journal of Environmental Studies*, *28*(3), 1203–1214. doi:10.15244/pjoes/85951

Joiner, A. (2010). Colorimetric evaluation of tooth color. *Color Measurement*, 343–371. doi:10.1533/9780857090195.2.343

Jones, L. N., & Rivett, D. E. (1997). The role of 18-methyleicosanoic acid in the structure and formation of mammalian hair fibres. *Micron (Oxford, England)*, *28*(6), 469–485. doi:10.1016/S0968-4328(97)00039-5 PMID:9519472

Joo, M. S., Tourillon, G., Sayers, D. E., & Theil, E. C. (1990). Rapid reduction of iron in horse spleen ferritin by thioglycolic acid measured by dispersive X-ray absorption spectroscopy. *Biology of Metals*, *3*(3-4), 171–175. doi:10.1007/BF01140575 PMID:2073457

Jung, H., & Sato, T. (2013). Comparison between the color properties of whiteness index and yellowness index on the CIELAB. *Textile Coloration and Finishing.*, *25*(4), 241–246. doi:10.5764/TCF.2013.25.4.241

Jurminskaia, O., Ene, A., & Zubcov, E. (2021). *Monitoring of aquatic ecosystems based on biological parameters*. Conference: International Conference, Bender, Republic of Moldova.

Jusufi, K., Stafilov, T., Vasjari, M., Korca, B., Halili, J., & Berisha, A. (2016). Determination of heavy metals by ICP-AES in the agricultural soils surrounding Kosovo's power plants. *Fresenius Environmental Bulletin*, *5*(25), 1313–1321. https://hero.epa.gov/hero/index.cfm/reference/details/reference_id/3462260

Kah, M., Beulke, S., Tiede, K., & Hofmann, T. (2013). Nanopesticides: State of knowledge, environmental fate, and exposure modeling. *Critical Reviews in Environmental Science and Technology*, *43*(16), 1823–1867. doi:10.1080/10643389.2012.671750

Kah, M., & Hofmann, T. (2014). Nanopesticide research: Current trends and future priorities. *Environment International*, *63*, 224–235. doi:10.1016/j.envint.2013.11.015 PMID:24333990

Kah, M., Kookana, R. S., Gogos, A., & Bucheli, T. D. (2018). A critical evaluation of nanopesticides and nanofertilizers against their conventional analogues. *Nature Nanotechnology*, *13*(8), 677–684. doi:10.103841565-018-0131-1 PMID:29736032

Kah, M., Weniger, A., & Hofmann, T. (2016). Impacts of (nano)-formulations on the fate of an insecticide in soil and consequences for environmental exposure assessment. *Environmental Science & Technology*, *50*(20), 10960–10967. doi:10.1021/acs.est.6b02477 PMID:27648740

Kamaraj, M., Ramachandran, K. K., & Aravind, J. (2020). Biohydrogen production from waste materials: Benefits and challenges. *International Journal of Environmental Science and Technology*, *17*(1), 559–576. doi:10.100713762-019-02577-z

Kanavouras, A., Karanika, E., Frank, A. C., Kotseridis, Y., & Kallithraka, S. (2019). Preliminary study of flavor compounds as oxidation markers in bottled white wines of Greek origin. *OENO One*, *53*(3). doi:10.20870/oeno-one.2019.53.3.2439

Karapinar, K. I., & Kargi, F. (2006). Bio-hydrogen production from waste materials. *Enzyme and Microbial Technology*, *38*(5), 569–582. doi:10.1016/j.enzmictec.2005.09.015

Karimi, K., & Taherzadeh, M. J. (2016). A critical review on analysis in pretreatment of lignocelluloses: Degree of polymerization, adsorption/desorption, and accessibility. *Bioresource Technology*, *203*, 348–356. doi:10.1016/j.biortech.2015.12.035 PMID:26778166

Kashyap, A., Ramasamy, E., Ramalingam, V., & Pattabiraman, M. (2021). Supramolecular control of singlet oxygen generation. *Molecules (Basel, Switzerland)*, *26*(9), 2673. doi:10.3390/molecules26092673 PMID:34063309

Kaur, A., Kumar, P., Kaur, L., Sharma, R., & Kush, P. (2020). Thiolated chitosan nanoparticles for augmented oral bioavailability of gemcitabine: Preparation, optimization, in vitro and in vivo study. *Journal of Drug Delivery Science and Technology*, *61*, 102169. doi:10.1016/j.jddst.2020.102169

Kaur, P., Thakur, R., Bernela, M., Chopra, M., Manuja, A., & Chaudhury, A. (2014). Synthesis, characterization and in vitro evaluation of cytotoxicity and antimicrobial activity of chitosan-metal nanocomposites: Antimicrobial & biocompatible chitosan-metal nanocomposites. *Journal of Chemical Technology and Biotechnology*, *90*(5), 867–873. doi:10.1002/jctb.4383

Kaushik, A., & Aradhana Singh, M. (2020). Metal removal and recovery using bioelectrochemical technology: The major determinants and opportunities for synchronic wastewater treatment and energy production. *Journal of Environmental Management*, *270*, 110826. doi:10.1016/j.jenvman.2020.110826 PMID:32721300

Kaziem, A. E., Gao, Y., Zhang, Y., Qin, X., Xiao, Y., Zhang, Y., You, H., Li, J., & He, S. (2018). α-Amylase triggered carriers based on cyclodextrin anchored hollow mesoporous silica for enhancing insecticidal activity of avermectin against *Plutella xylostella*. *Journal of Hazardous Materials*, *359*, 213–221. doi:10.1016/j.jhazmat.2018.07.059 PMID:30036751

Kefeni, R. R., Mamba, B. B., & Msagati, T. A. M. (2017). Application of spinel ferrite nanoparticles in water and wastewater treatment: A review. *Separation and Purification Technology*, *188*, 399–422. doi:10.1016/j.seppur.2017.07.015

Kehrer, J. P. (2000). The Haber–Weiss reaction and mechanisms of toxicity. *Toxicology*, *149*(1), 43–50. doi:10.1016/S0300-483X(00)00231-6 PMID:10963860

Kępczyńsqi, M., Czosnyka, A., & Nowakowska, M. (2004). Photooxidation of phenol in aqueous nanodispersion of humic acid. *Journal of Photochemistry and Photobiology*, *79*(3), 259–264.

Khadke, P., Sethuraman, P., Kandasamy, P., Parthasarathi, S., & Shukla, A. (2009). A self-supported direct borohydride-hydrogen peroxide fuel cell system. *Energies*, *2*(2), 190–201. doi:10.3390/en20200190

Khandaker, M., Riahinezhad, S., Sultana, F., Morris, T., Wolf, R., & Vaughan, M. (2017). Effect of collagen-polycaprolactone nanofibers matrix coating on the in vitro cytocompatibility and in vivo bone responses of titanium. *Journal of Medical and Biological Engineering*, *38*(2), 197–210. doi:10.100740846-017-0312-7 PMID:29861706

Khare, A., Chhawani, N., & Kumari, K. (2019). Glutathione reductase and catalase as potential biomarkers for synergistic intoxication of pesticides in fish. *Biomarkers*, *24*(7), 666–676. doi:10.1080/1354750X.2019.1651902 PMID:31368361

Kholodkevich, S. V., Kuznetsova, T. V., Kirin, M. P., Smirnov, I. S., Rudakova, O. A., Lyubimtsev, V. A., Manvelova, A. B., Susloparova, O. N., Perelygin, V. V., & Sakharova, O. A. (2020). Bioindication of the ecological state (health) of coastal waters based on the use of automated bioelectronic systems. *Pharmacy Formulas*, *2*(3), 64–73. doi:10.17816/phf46438

Khor, E., & Lim, L. Y. (2003). Implantable applications of chitin and chitosan. *Biomaterials*, *24*(13), 2339–2349. doi:10.1016/S0142-9612(03)00026-7 PMID:12699672

Khurshid, C. A., Mahdi, K., Ahmed, O. I., Osman, R., Rahman, M., & Ritsema, C. (2022). Assessment of potentially toxic elements in the urban soil and plants of Kirkuk city in Iraq. *Sustainability (Basel)*, *14*(9), 5655. doi:10.3390u14095655

Kilic, M. Y., Yona, T., & Mert, B. K. (2014). Landfill leachate treatment by Fenton and Fenton-like oxidation processes. *Clean (Weinheim)*, *42*(5), 586–593. Advance online publication. doi:10.1002/clen.201200714

Killeen, D. J., Boulton, R., & Knoesen, A. (2018). Advanced monitoring and control of redox potential in wine fermentation. *American Journal of Enology and Viticulture*, *69*(4), 394–399. Advance online publication. doi:10.5344/ajev.2018.17063

Kim, I. Y., Seo, S. J., Moon, H. S., Yoo, M. K., Park, I. Y., Kim, B. C., & Cho, C. S. (2008). Chitosan and its derivatives for tissue engineering applications. *Biotechnology Advances*, *26*(1), 1–21. doi:10.1016/j.biotechadv.2007.07.009 PMID:17884325

Kim, J. Y., Choi, D. S., & Jung, M. Y. (2003). Antiphoto-oxidative activity of sesamol in methylene blue- and chlorophyll-sensitized photo-oxidation of oil. *Journal of Agricultural and Food Chemistry*, *51*(11), 3460–3465. doi:10.1021/jf026056p PMID:12744684

Kim, N., & Choe, E. (2012). Effects of monoacylglycerols on the oil oxidation of acidic water/perilla oil emulsion under light in the presence of chlorophyll. *Food Science and Biotechnology*, *21*(1), 183–189. doi:10.100710068-012-0023-3

Kleszczewska, E. (1999). The spectrophotometry determination of chelate complex: L-ascorbic acid with cuprum (II) and mercury (II) in alkaline solution. *Polish Journal of Environmental Studies*, *8*(5), 313–318.

Klimuk, A. I., Obvintseva, L. A., Kuchaev, V. L., Shepelev, A. D., Sadovskaya, N. V., Tomashpolsky, Yu. Ya., Kozlova, N. V., & Avetisov, A. K. (2008). Interaction of ozone with microfiber materials. *Journal of the Russian chemical about them D. I. Mendeleev*, *LII*(5), 102–111.

Knidri, H. E., Laajeb, A., & Lahsini, A. (2020). Chitin and chitosan: Chemistry, solubility, fiber formation, and their potential applications. *Handb. Chitin Chitosan*, *1*, 35–57. doi:10.1016/B978-0-12-817970-3.00002-X

Kokkinos, P., Venieri, D., & Mantzavinos, D. (2021). Advanced Oxidation Processes for Water and Wastewater Viral Disinfection. A Systematic Review. *Food and Environmental Virology*, *13*(3), 283–302. doi:10.100712560-021-09481-1 PMID:34125359

Kolb, M., Bahadir, M., & Teichgräber, B. (2017). Determination of chemical oxygen demand (COD) using an alternative wet chemical method free of mercury and dichromate. *Water Research*, *122*, 645–654. doi:10.1016/j.watres.2017.06.034 PMID:28649029

Kolesnik, I. V., & Sapoletova, N. A. (2011). *Infrared spectroscopy*. Moscow state university.

Kookana, R. S., Boxall, A. B. A., Reeves, P. T., Ashauer, R., Beulke, S., Chaudhry, Q., Cornelis, G., Fernandes, T. F., Gan, J., Kah, M., Lynch, I., Ranville, J., Sinclair, C., Spurgeon, D., Tiede, K., & Van den Brink, P. J. (2014). Nanopesticides: Guiding principles for regulatory evaluation of environmental risks. *Journal of Agricultural and Food Chemistry*, *62*(19), 4227–4240. doi:10.1021/jf500232f PMID:24754346

Koppenol, W. H. (1989). Generation and thermodynamic properties of oxyradicals. In C. Vigo-Pelfrey (Ed.), *Focus on Membrane Lipid Oxidation* (Vol. I, pp. 1–13). CRC Press.

Koppenol, W. H. (2001). The Haber-Weiss cycle–70 years later. *Redox Report*, *6*(4), 229–234. doi:10.1179/135100001101536373 PMID:11642713

Korenaga, T., Zhou, X., Okada, K., Moriwake, T., & Shinoda, S. (1993). Determination of chemical oxygen demand by a flow-injection method using cerium (IV) sulfphate as oxidizing agent. *Analytica Chimica Acta*, *272*(2), 237–244. doi:10.1016/0003-2670(93)80575-6

Korenika, A. M. J., Biloš, J., Kozina, B., Tomaz, I., Preiner, D., & Jeromel, A. (2020). Effect of different reducing agents on aromatic compounds, antioxidant and chromatic properties of sauvignon blanc wine. *Foods*, *9*(8), 996. Advance online publication. doi:10.3390/foods9080996 PMID:32722317

Korjenic, A., Klarić, S., Hadžić, A., & Korjenic, S. (2015). Sheep wool as a construction material for energy efficiency improvement. *Energies*, *8*(6), 5765–5781. doi:10.3390/en8065765

Köşker, A. R., Gündoğdu, S., Ayas, D., & Bakan, M. (2022). Metal levels of processed ready-to-eat stuffed mussels sold in Turkey: Health risk estimation. *Journal of Food Composition and Analysis, 106*, 104326. doi:10.1016/j.jfca.2021.104326

Kowalczyk, P., Terzyk, A. P., Gauden, P. A., & Rychlicki, G. (2002). Numerical Analysis of the Horvath–Kawazoe Equation – The Adsorption of Nitrogen, Argon, Benzene, Carbon Tetrachloride and Sulphur Hexafluoride. *Adsorption Science and Technology, 20*(3), 295–305. doi:10.1260/026361702760254478

Krakovská, A. S., Svozilík, V., Zinicovscaia, I., Vergel, K., & Jančík, P. (2020). Analysis of spatial data from moss biomonitoring in czech-polish border. *Atmosphere (Basel), 11*(11), 1237. doi:10.3390/atmos11111237

Kritzinger, E., Bauer, F., & du Toit, W. (2012). Role of Glutathione in Winemaking: A Review. *Journal of Agricultural and Food Chemistry, 61*(2), 269–277. doi:10.1021/jf303665z PMID:23240621

Krumova, K., & Cosa, G. (2016). Overview of reactive oxygen species. In S. Nonell & C. Flors (Eds.), *Comprehensive Series in Photochemical & Photobiological Sciences* (Vol. 1, pp. 1–21). Royal Society of Chemistry. doi:10.1039/9781782622208-00001

Ksibi, M. (2006). Chemical oxidation with hydrogen peroxide for domestic wastewater treatment. *Chemical Engineering Journal, 119*(2-3), 161–165. doi:10.1016/j.cej.2006.03.022

Ksibi, M., Zemzemi, A., & Bouqchina, R. (2003). Photocatalytic degradability of substituted phenols over UV irradiated TiO_2. *Journal of Photochemistry and Photobiology A Chemistry, 159*(1), 61–70. doi:10.1016/S1010-6030(03)00114-X

Kučera, J., Granja, C., Leroy, C., & Stekl, I. (2007). The use of neutron activation analysis in environmental and biomedical research. *AIP Conference Proceedings, 958*, 28–32. doi:10.1063/1.2825821

Kuijpers, T. F. M., Gruppen, H., Sforza, S., van Berkel, W. J. H., & Vincken, J. P. (2013). The antibrowning agent sulfite inactivates Agaricus bisporus tyrosinase through covalent modification of the copper-B site. *The FEBS Journal, 280*(23), 6184–6195. doi:10.1111/febs.12539 PMID:24112416

Kukec, A., Berovic, M., Celan, S., & Wondra, M. (2002). The role of on-line redox potential measurement in Sauvignon blanc fermentation. *Food Technology and Biotechnology, 40*(1).

Kumari, S., Singh, R. P., Chavan, N. N., & Annamalai, P. K. (2016). Chitosan-based bionanocomposites for biomedical application. *Bioinspired Biomim. Nanobiomater, 7*(4), 219–227. doi:10.1680/jbibn.15.00015

Kupatadze, K. (2022). Surface water chemical composition monitoring in Several villages: Case study of Khakheti Region, Georgia. In: *Handbook of Research on Water Science and Society, IGI Global Publishing House, 26*, 587-598. https://www.igi-global.com/chapter/surface-water-chemical-composition-monitoring-in-several-villages/299900

Kupatadze, K., & Kiziloz, B. (2019). Researching some of the Parameters of Kura ("Mtkvari") river in two neighboring countries-Georgia and Turkey, in terms of environmental chemistry. *J. Tche Quimica, 16*(31), 450–457. http://www.deboni.he.com.br/Periodico31.pdf

Lam, M., & Mabury, S. (2005). Photodegradation of the farmaceuticals atorvastatin, carbamazepine, levofloxacin, and sulfamethoxazole in natural waters. *Aquatic Sciences, 67*(2), 177–188. doi:10.100700027-004-0768-8

Lamuela Raventos, R. M., Huix-Blanquera, M., & Waterhouse, A. (2001). Treatments for Pinking Alteration in White Wines. *American Journal of Enology and Viticulture, 52*(2), 156–158. doi:10.5344/ajev.2001.52.2.156

Lapsirikul, W., Cord-Ruwisch, R., & Ho, G. (1994). Treatment of wool scouring effluent by anaerobic bioflocculation. *Water Science and Technology, 30*(12), 375–384. doi:10.2166/wst.1994.0637

Larson, R. A., & Mccord, J. M. (1977). Environmental chemistry of reactive oxygen species. *CRC Critical Reviews in Environmental Control*, *8*(1-4), 197–246. doi:10.1080/10643387709381663

Lawless, H. T., & Heymann, H. (2010). *Sensory Evaluation of Food: Principles and Practices*. Springer New York. doi:10.1007/978-1-4419-6488-5

Lazo, P., Kika, A., Qarri, F., Bekteshi, L., Allajbeu, S., & Stafilov, T. (2022). Air quality assessment by moss biomonitoring and trace metals atmospheric deposition. *Aerosol and Air Quality Research*, *22*(5), 220008. doi:10.4209/aaqr.220008

LeBaron, T. W., & Sharpe, R. (2022). ORP should not be used to estimate or compare concentrations of aqueous H_2: An in-silico analysis and narrative synopsis. *Front. Food. Sci. Technol.*, *2*, 1007001. doi:10.3389/frfst.2022.1007001

Lee, S., Choi, Y., Jeong, H. S., Lee, J., & Sung, J. (2018). Effect of different coking methods on the content of vitamins and true retention in selected vegetables. *Food Science and Biotechnology*, *27*(2), 333–342. PMID:30263756

Lee, Y. (2004). Chemical composition and oxidative stability of safflower oil prepared from safflower seed roasted with different temperatures. *Food Chemistry*, *84*(1), 1–6. doi:10.1016/S0308-8146(03)00158-4

Lesica, M., & Košmerl, T. (2009). Microoxygenation of red wines. *Acta Agriculturae Slovenica*, *93*(3).

Liang, Y., Gao, Y., Wang, W., Dong, H., Tang, R., Yang, J., Niu, J., Zhou, Z., Jiang, N., & Cao, Y. (2020). Fabrication of smart stimuli-responsive mesoporous organosilica nano-vehicles for targeted pesticide delivery. *Journal of Hazardous Materials*, *389*, 122075. doi:10.1016/j.jhazmat.2020.122075 PMID:31972522

Liang, Y., Wang, S., Jia, H., Yao, Y., Song, J., Dong, H., Cao, Y., Zhu, F., & Huo, Z. (2022). Pectin functionalized metal-organic frameworks as dual-stimuli-responsive carriers to improve the pesticide targeting and reduce environmental risks. *Colloids and Surfaces. B, Biointerfaces*, *219*, 112796. doi:10.1016/j.colsurfb.2022.112796 PMID:36063717

Li, J., Li, F., Liu, Q., & Zhang, Y. (2014). Trace metal in surface water and groundwater and its transfer in a Yellow River alluvial fan: Evidence from isotopes and hydrochemistry. *The Science of the Total Environment*, *472*, 979–988. doi:10.1016/j.scitotenv.2013.11.120 PMID:24355394

Li, K. G., Chen, J. T., Bai, S. S., Wena, X., Song, S. Y., Yua, Q., & Wanga, Y. Q. (2009). Intracellular oxidative stress and cadmium ions release induce cytotoxicity of unmodified cadmium sulfide quantum dots. *Toxicology In Vitro*, *23*(6), 1007–1013. doi:10.1016/j.tiv.2009.06.020 PMID:19540911

Li, K., Ling, L., Lu, C., Liu, Z., Liu, L., & Mochida, I. (2001a). Influence of CO-evolving groups on the activity of activated carbon fiber for SO_2 removal. *Fuel Processing Technology*, *70*(3), 151–158. doi:10.1016/S0378-3820(01)00175-8

Li, K., Ling, L., Lu, C., Qiao, W., Liu, Z., Liu, L., & Mochida, I. (2001b). Catalytic removal of SO_2 over ammonia-activated carbon fibers. *Carbon*, *39*(12), 1803–1808. doi:10.1016/S0008-6223(00)00320-1

Lima, L. B. D., de Morais, P. B., de Andrade, R. L. T., Mattos, L. V., & Moron, S. E. (2018). Use of biomarkers to evaluate the ecological risk of xenobiotics associated with agriculture. *Environmental Pollution*, *237*, 611–624. doi:10.1016/j.envpol.2018.02.011 PMID:29525628

Lindeboom, R. E. F., Fermoso, F. G., Weijma, J., Zaqt, K., & van Lier, J. B. (2011). Autogenerative high pressure digestion: Anaerobic digestion and biogas upgrading in a single step reactor system. *Water Science and Technology*, *64*(3), 647–653. doi:10.2166/wst.2011.664 PMID:22097043

Lin, N., Gong, Y., Wang, R., Yin Wang, Y., & Zhang, X. (2022). Critical review of perovskite-based materials in advanced *oxidation system for wastewater treatment: Design, applications and mechanisms. Journal of Hazardous Materials*, *424*, 127637. doi:10.1016/j.jhazmat.2021.127637 PMID:34753649

Lin, R., Li, Y., Yong, T., Cao, W., Wu, J., & Shen, Y. (2022). Synergistic effects of oxidation, coagulation and adsorption in the integrated Fenton-based process for wastewater treatment: A review. *Journal of Environmental Management, 306*, 114460. doi:10.1016/j.jenvman.2022.114460 PMID:35026715

Li, R., Speed, D., Siriwardena, D., Fernando, S., Thagard, S. M., & Holsen, T. M. (2021). Comparison of hydrogen peroxide-based advanced oxidation processes for the treatment of azole-containing industrial wastewater. *Chemical Engineering Journal, 425*, 131785. doi:10.1016/j.cej.2021.131785

Lis, A. (2016). Direct photolysis of cysteine and thiourea in aqueous solutions. *Studia Universitatis Moldaviae, 6*(96), 133–141.

Lis, A., Gladchi, V., Duca, G., & Travin, S. (2021). Sensitized photolysis of thioglycolic acid in aquatic environment. *Chemistry Journal of Moldova., 16*(1), 46–59. doi:10.19261/cjm.2021.796

Lise Halvorsen, B., & Blomhoff, R. (2011). Determination of lipid oxidation products in vegetable oils and marine omega-3 supplements. *Food & Nutrition Research, 55*(1), 5792. doi:10.3402/fnr.v55i0.5792 PMID:21691461

Li, T., Yang, X.-L., Song, H.-L., Xu, H., & Chen, Q.-L. (2022). Quinones contained in wastewater as redox mediators for the synergistic removal of azo dye in microbial fuel cells. *Journal of Environmental Management, 301*, 113924. doi:10.1016/j.jenvman.2021.113924 PMID:34731947

Litter, M. I. (2005). Introduction to photochemical advanced oxidation processes for water treatment. *Environmental Photochemistry, M*(Part II), 325–366. doi:10.1007/b138188

Liu, X. (2018). Progress in the mechanism and kinetics of Fenton reaction. *MOJ Ecology & Environmental Sciences, 3*(1). doi:10.15406/mojes.2018.03.00060

Liu, C.-G., Qin, J.-C., & Lin, Y.-H. (2017). *Fermentation and Redox Potential*. Fermentation Processes. doi:10.5772/64640

Liu, H., Gao, Y., Wang, J., Pan, J., Gao, B., & Yue, Q. (2022). Catalytic ozonation performance and mechanism ofMn-CeOx@γ-Al2O3/O3 in the treatment of sulfate-containing hypersaline antibiotic wastewater. *The Science of the Total Environment, 807*, 150867. doi:10.1016/j.scitotenv.2021.150867 PMID:34627913

Liu, H., Ren, L., Zhuo, H., & Fu, S. (2019). Water footprint and water pinch analysis in ethanol industrial production for water management. *Water (Basel), 11*(3), 518. doi:10.3390/w11030518

Liu, L., Chen, Z., Zhang, J., Shan, D., Wu, Y., Bai, L., & Wang, B. (2021). Treatment of industrial dye wastewater and pharmaceutical residue wastewater by advanced oxidation processes and its combination with nanocatalysts: A review. *Journal of Water Process Engineering, 42*, 102122. doi:10.1016/j.jwpe.2021.102122

Liu, W. G., Zhu, X. Y., Yin, M. M., & Chen, F. L. (2012). Photodegradation of three avermectin liquid formulations. *Journal of Environmental Sciences (China), 31*, 1906–1912.

Liu, X., Chen, Z., Du, W., Liu, P., Zhang, L., & Shi, F. (2022). Treatment of wastewater containing methyl orange dye by fluidized three-dimensional electrochemical oxidation process integrated with chemical oxidation and adsorption. *Journal of Environmental Management, 311*, 114775. doi:10.1016/j.jenvman.2022.114775 PMID:35245840

Liu, Y., Guo, H., Zhang, Y., Tang, W., Cheng, X., & Li, W. (2018). Heterogeneous activation of peroxymonosulfate by sillenite Bi 25 FeO 40: Singlet oxygen generation and degradation for aquatic levofloxacin. *Chemical Engineering Journal, 343*, 128–137. doi:10.1016/j.cej.2018.02.125

Li, W., Xu, X., Lyu, B., Tang, Y., Zhang, Y., Chen, F., & Korshin, G. (2019). Degradation of typical macrolide antibiotic roxithromycin by hydroxyl radical: Kinetics, products, and toxicity assessment. [online]. *Environmental Science and Pollution Research International, 26*(14), 14570–14582. doi:10.100711356-019-04713-1 PMID:30877533

Li, Z. H., Li, P., & Shi, Z.-C. (2014). Molecular responses in digestive tract of juvenile common carp after chronic exposure to sublethal tributyltin. *Ecotoxicology and Environmental Safety*, *109*, 10–14. doi:10.1016/j.ecoenv.2014.07.031 PMID:25129219

Li, Z. H., Zlabek, V., Grabic, R., Li, P., Machova, J., Velisek, J., & Randak, T. (2010). Effects of exposure to sublethal propicon azole on intestine-related biochemical responses in rainbow trout, Oncorhynchus mykiss. *Chemico-Biological Interactions*, *185*(3), 241–246. doi:10.1016/j.cbi.2010.02.040 PMID:20193671

Li, Z.-Z., Xu, S.-A., Wen, L.-X., Liu, F., Liu, A.-Q., Wang, Q., Sun, H.-Y., Yu, W., & Chen, J.-F. (2006). Controlled release of avermectin from porous hollow silica nanoparticles: Influence of shell thickness on loading efficiency, UV-shielding property and release. *Journal of Controlled Release*, *111*(1-2), 81–88. doi:10.1016/j.jconrel.2005.10.020 PMID:16388871

Lo Conte, M., & Carroll, K.S. (2013). The chemistry of thiol oxidation and detection. *Oxidative Stress and Redox Regulation*, 1-42. . doi:10.1007/978-94-007-5787-5_1

Lobato, L. C. S., Chernicharo, C. A. L., Pujatti, F. J. P., Martins, O. M., Melo, G. C. B., & Recio, A. A. R. (2013). Use of biogas for cogeneration of heat and electricity for local application: Performance evaluation of an engine power generator and a sludge thermal dryer. *Water Science and Technology*, *67*(1), 159–167. doi:10.2166/wst.2012.549 PMID:23128634

Lomsadze, Z., Makharadze, K., & Pirtskhalava, R. (2016). The Ecological Problems of Rivers of Georgia. *Annals of Agrarian Science*, *14*(3), 237–242. doi:10.1016/j.aasci.2016.08.009

Lopes, R. M., Filho, M. V. S., de Salles, J. B., Bastos, V. L. F. C., & Bastos, J. C. (2014). Cholinesterase activity of muscle tissue from freshwater fishes: Characterization and sensitivity analysis to the organophosphate methyl-paraoxon. *Environmental Toxicology and Chemistry*, *33*(6), 1331–1336. doi:10.1002/etc.2556 PMID:24648156

López-Campos, O., Bodas, R., Prieto, N., Frutos, P., Andrés, S., & Giráldez, F. (2010). Vinasse added to the concentrate for fattening lambs: Intake, animal performance, and carcass and meat characteristics. *Journal of Animal Science*, *89*(4), 1153–1162. doi:10.2527/jas.2010-2977 PMID:21148788

Lou, J.-C., Huang, Yu.-J., & Han, J.-Yu. (2009). Treatment of printed circuit board industrial wastewater by ferrite process combined with Fenton method. *Journal of Hazardous Materials*, *170*(2-3), 620–626. doi:10.1016/j.jhazmat.2009.05.020 PMID:19493613

Lu, S., Zhang, H., Du, W., & Deng, J. (2020). Determination of metal elements in workplace air by ICP-MS. *MATEC Web of Conferences, 319*, 02005. 10.1051/matecconf/202031902005

Lu, A. H., & Zheng, J. T. (2001). Study of microstructure of high-surface-area polyacrylonitrile activated carbon fibers. *Journal of Colloid and Interface Science*, *236*(2), 369–374. doi:10.1006/jcis.2000.7425 PMID:11401385

Lu, C., Yu, J., Wang, C., Wang, J., & Chu, F. (2018). Fabrication of UV-absorbent cellulose-rosin based thermoplastic elastomer via "graft from" ATRP. *Carbohydrate Polymers*, *188*, 128–135. doi:10.1016/j.carbpol.2018.01.062 PMID:29525148

Lueder, U., Jorgensen, B. B., Kappler, A., & Schmidt, C. (2020). Photochemistry of iron in aquatic environments. *Environmental Science. Processes & Impacts*, *22*(1), 12–24. Advance online publication. doi:10.1039/C9EM00415G PMID:31904051

Luo, W., Westland, S., Ellwood, R., & Pretty, I. (2005). Evaluation of whiteness formulae for teeth. *Proceedings of the 10th Congress of the International Color Association, Granada, Spain., 3–7.* https://www.researchgate.net/publication/240633974

Luo, Y., & Wang, Q. (2014). Recent development of chitosan-based polyelectrolyte complexes with natural polysaccharides for drug delivery. *International Journal of Biological Macromolecules*, *64*, 353–367. doi:10.1016/j.ijbiomac.2013.12.017 PMID:24360899

Lu, T.-Y., Chiang, C.-Y., Fan, Y.-J., Jheng, P.-R., Quiñones, E. D., Liu, K.-T., Kuo, S.-H., Hsieh, H. Y., Tseng, C.-L., Yu, J., & Chuang, E.-Y. (2021b). Dual-targeting glycol chitosan/heparin-decorated polypyrrole nanoparticle for augmented photothermal thrombolytic therapy. *ACS Applied Materials & Interfaces*, *13*(8), 10287–10300. doi:10.1021/acsami.0c20940 PMID:33615773

Lu, T.-Y., Lu, W.-F., Wang, Y.-H., Liao, M.-Y., Wei, Y., Fan, Y.-J., Chuang, E. Y., & Yu, J. (2021a). Keratin-based nanoparticles with tumor-targeting and cascade catalytic capabilities for the combinational oxidation phototherapy of breast cancer. *ACS Applied Materials & Interfaces*, *13*(32), 38074–38089. doi:10.1021/acsami.1c10160 PMID:34351754

Macaev, F., Boldescu, V., Geronikaki, A., & Sucman, N. (2013). Recent advances in the use of cyclodextrins in antifungal formulations. *Current Topics in Medicinal Chemistry*, *13*(21), 2677–2683. doi:10.2174/15680266113136660194 PMID:24083792

Mack, J., & Bolton, R. (1999). Photochemistry and nitrite and nitrate in aqueous solution. *Journal of Photochemistry and Photobiology A Chemistry*, *128*(1-3), 1–13. doi:10.1016/S1010-6030(99)00155-0

Maclaren, J. A., & Milligan, B. (1981). Wool Science: The chemical reactivity of the wool fibre, Science Press, Marrickville, 1–17, 53–75.

Madhusudhan, R. A., Karthikeyan, R., Vejandla, R. S., Divya, G., & Babu, P. S. (2017). Controlled release matrix drug delivery system – a review. *Int. J. Allied Med. Sci.Clin. Res.*, *5*(2), 384–398.

Mahler, C. F., Svierzoski, N. D. S., & Bernardino, C. A. R. (2021). Chemical characteristics of humic substances in nature. In A. Makan (Ed.), *Humic Substances* (pp. 1–13). IntechOpen.

Maifadi, S., Mhlanga, S. D., Nxumalo, E. N., Motsa, M. M., & Kuvarega, A. T. (2022). Treatment of salon wastewater by peroxydisulfate based advanced oxidation process (PDS-AOP) under solar light: Synergy through integrated technologies. *Journal of Water Process Engineering*, *49*, 103062. doi:10.1016/j.jwpe.2022.103062

Maiolo, K., Marin, P. D., Bridges, T. E., & Heithersay, G. S. (2007). Evaluation of a combined thiourea and hydrogen peroxide regimen to bleach bloodstained teeth. *Australian Dental Journal*, *52*(1), 33–40. doi:10.1111/j.1834-7819.2007.tb00463.x PMID:17500162

Makarov, V.M., Kalaeva, S.Z., Zakharova, I.N., Nevzorov, I.A., Maltseva, M.S., Shipilin, A.M., Krzhizhanovskaya, M.G. (2015). Magnetic and X-ray studies of nanodispersed magnetite synthesized from chrome containing galvanic sludge. *Journal of Nano- and Electronic Physics*, *7*(4), 04010.

Ma, L., & Waterhouse, A. L. (2018). Flavanols react preferentially with quinones through an electron transfer reaction, stimulating rather than preventing wine browning. *Analytica Chimica Acta*, *18*(1039), 162–171. doi:10.1016/j.aca.2018.07.013 PMID:30322547

Malakof, D., & Stokstad, M. (2013). Infographic – Pesticide planet. *Science*, *341*(6147), 730–731. doi:10.1126cience.341.6147.730 PMID:23950524

Malik, S. N., Ghosh, P. C., Vaidya, A. N., & Mudliar, S. N. (2018). Catalytic ozone pretreatment of complex textile effluent using Fe2+and zero-valent iron nanoparticles. *Journal of Hazardous Materials*, *357*, 363–375. doi:10.1016/j.jhazmat.2018.05.070 PMID:29909170

Manisalidis, I., Stavropoulou, E., Stavropoulos, A., & Bezirtzoglou, E. (2020). Environmental and health impacts of air pollution: A review. In Frontiers in Public Health, 8, 14. Frontiers Media SA. doi:10.3389/fpubh.2020.00014

Marinoni, A., Parazols, M., Brigante, M., Deguillaume, L., Amato, P., Delort, A. M., Laj, P., & Mailhot, G. (2011). Hydrogen peroxide in natural cloud water: Sources and photoreactivity. *Atmospheric Research, 101*(1-2), 256–263. doi:10.1016/j.atmosres.2011.02.013

Márquez, K., Contreras, D., Salgado, P., & Mardones, C. (2019). Production of hydroxyl radicals and their relationship with phenolic compounds in white wines. *Food Chemistry, 271*, 80–86. doi:10.1016/j.foodchem.2018.07.165 PMID:30236744

Martin, C., Perrard, A., Joly, J. P., Gaillard, F., & Delecroix, V. (2002). Dynamic adsorption on activated carbons of SO_2 traces in air: I. Adsorption capacities. *Carbon, 40*(12), 2235–2246. doi:10.1016/S0008-6223(02)00108-2

Martínez-Alvarez, O., Chamorro, S., & Brenes, A. (2015). Protein hydrolysates from animal processing by-products as a source of bioactive molecules with interest in animal feeding : A review. *Food Research International, 73*, 204–212. doi:10.1016/j.foodres.2015.04.005

Martínez, S. B., Pérez-Parra, J., & Suay, R. (2011). Use of Ozone in Wastewater Treatment to Produce Water Suitable for Irrigation. *Water Resources Management, 25*(9), 2109–2124. doi:10.100711269-011-9798-x

Martín-Gullón, I., Andrews, R., Jagtoyen, M., & Derbyshire, F. (2001). PAN-based activated carbon fiber composites for sulfur dioxide conversion: Influence of fiber activation method. *Fuel, 80*(7), 969–977. doi:10.1016/S0016-2361(00)00186-1

Martín-Polvillo, M., Márquez-Ruiz, G., & Dobarganes, M. C. (2004). Oxidative stability of sunflower oils differing in unsaturation degree during long-term storage at room temperature. *Journal of the American Oil Chemists' Society, 81*(6), 577–583. doi:10.100711746-006-0944-1

Masunga, N., Kelebogile, O., Kefeni, K. K., & Mamba, B. B. (2019). Recent advances in copper ferrite nanoparticles and nanocomposites synthesis, magnetic properties and application in water treatment [Review]. *Journal of Environmental Chemical Engineering, 7*(3), 103179. doi:10.1016/j.jece.2019.103179

Matkovsky, P.E.; Yarullin, R.S.; Startseva, G.P.; Sedov, I. (2010). Bioethanol: technologies of its production from renewable raw materials and application areas. *Alternative power engineering and ecology, 6*(86), 95-105.

Mattson, M., & Gene, E. (1993). Redox reactions of organic matter decomposition in a soft water lake. *Biogeochemistry, 19*(3), 149–172. https://www.jstor.org/stable/1468581. doi:10.1007/BF00000876

Ma, Z., & Boye, J. I. (2013). Advances in the design and production of reduced-fat and reduced-cholesterol salad dressing and mayonnaise: A Review. *Food and Bioprocess Technology, 6*(3), 3. doi:10.100711947-012-1000-9

McClements, D. J. (2016). Food emulsions: Principles, practices, and techniques (Third edition). CRC Press, Taylor & Francis Group.

Meijide, J., Lama, G., Pazos, M., Sanroman, M. A., & Dunlop, P. S. M. (2022). Ultraviolet-based heterogeneous advanced oxidation processes as technologies to remove pharmaceuticals from wastewater: An overview. *Journal of Environmental Chemical Engineering, 10*(3), 107630. doi:10.1016/j.jece.2022.107630

Mekoue Nguela, J., Vernhet, A., Julien-Ortiz, A., Sieczkowski, N., & Mouret, J. R. (2019). Effect of grape must polyphenols on yeast metabolism during alcoholic fermentation. *Food Research International, 121*, 161–175. doi:10.1016/j.foodres.2019.03.038 PMID:31108737

Mierzwa, J.C., Rodrigues, R., & Teixeira, A.C. (2018) UV-hydrogen peroxide processes. *Advanced oxidation processes for wastewater treatment, 3*, 13–48.

Mignogna, R., Fratianni, A., Niro, S., & Panfili, G. (2015). Tocopherol and tocotrienol analysis as a tool to discriminate different fat ingredients in bakery products. *Food Control, 54*, 31–38. doi:10.1016/j.foodcont.2015.01.032

Mikova, N., Ivanov I., Chesnokov N., & Kargin V. (2014). Effect of oxidative treatment on the properties of carbon materials from microcrystalline cellulose. *Chemistry of plant raw material, 2*, 51-59.

Mikucka, W., & Zielińska, M. (2020). Distillery stillage: Characteristics, treatment, and valorization. *Applied Biochemistry and Biotechnology, 192*(3), 770–793. doi:10.100712010-020-03343-5 PMID:32557233

Mikulski, D., & Kłosowski, G. (2022). Integration of first- and second-generation bioethanol production from beet molasses and distillery stillage after dilute sulfuric acid pretreatment. *BioEnergy Research, 15*(6), 454–465. doi:10.100712155-021-10260-w

Miller, D. G., Brayton, S. V., & Boyles, W. T. (2001). Chemical oxygen demand analysis of wastewater using trivalent manganese oxidant with chloride removal by sodium bismuthate pretreatment. *Water Environment Research, 73*(1), 63–71. doi:10.2175/106143001X138705 PMID:11558305

Millington, K. R. (2013). Bleaching and whitening of wool: photostability of whites. *The Coloration of Wool and Other Keratin Fibres,* 131–155. doi:10.1002/9781118625118.ch5

Mill, T., Hendry, D., & Richardson, H. (1980). Tree radical oxidants in natural waters. *Science, 207*(4433), 886–887. doi:10.1126cience.207.4433.886 PMID:17729869

Min, D. B., Callison, A. L., & Lee, H. O. (2003). Singlet oxygen oxidation for 2-pentylfuran and 2-pentenyfuran formation in soybean oil. *Journal of Food Science, 68*(4), 1175–1178. doi:10.1111/j.1365-2621.2003.tb09620.x

Minero, Cl., Chiron, S., Falletti, G., Maurino, V., Pelizzetti, E., Ajassa, R., Carlotti, M. E., & Vione, D. (2007). Photechemical processes involving nitrite in surface water samples. *Aquatic Sciences, 69*(1), 71–85. doi:10.100700027-007-0881-6

Ministry of Natural Resources. (2002). *Guidelines for determining the toxicity of water, bottom sediments, pollutants and drilling fluids by biotesting method.* M.: REFIA, NIA Priroda.

Miret, C., Chazara, P., Montastruc, L., Negny, S., & Domenech, S. (2015). Design of bioethanol green supply chain: Comparison between first and second generation biomass concerning economic, environmental and social criteria. *Computers & Chemical Engineering*, 85.

Mishra, S. K., Srivastava, R. K., Prakash, S. G., Yadav, R. S., & Panday, A. C. (2011). Structural, photoconductivity and photoluminescence characterization of cadmium sulfide quantum dots prepared by a co-precipitation method. *Electronic Materials Letters, 7*(1), 31–38. doi:10.100713391-011-0305-6

Moate, P. J., Williams, S. R. O., Torok, M. C., Hannah, B. E., Ribaux, M. H., Tavendale, M. H., Eckard, R. J., Jacobs, J. L., Auldist, M. J., & Wales, W. J. (2014). Grape marc reduces methane emissions when fed to dairy cows. *Journal of Dairy Science, 97*(8), 5073–5087. doi:10.3168/jds.2013-7588 PMID:24952778

Modiri-Gharehveran, M., Hain, E., Blaney, L., & Shah, A. (2020). Influence of dissolved organic matter on carbonyl sulfide and carbon disulfide formation from cysteine during sunlight photolysis. *Environmental Science. Processes & Impacts, 22*(9), 1852–1864. doi:10.1039/D0EM00219D PMID:32966465

Mohammed, H. K., Mamoori, A., Mahdi, D. K., & Alshrefi, S. M. (2018). Synthesis and spectroscopic study of CdS nanoparticle using hydrothermal method. *AIP Conference Proceedings*, 1968.

Mohyudin, S., Farooq, R., Jubeen, F., Rasheed, T., Fatima, M., & Sher, F. (2022). Microbial fuel cells a state-of-the-art technology for wastewater treatment and bioelectricity generation. *Environmental Research, 204*, 112387. doi:10.1016/j.envres.2021.112387 PMID:34785206

Mollica, F., Lucarini, M., Passerini, C., Carati, C., Pavoni, S., Bonoldi, L., & Amorati, R. (2020). Effect of antioxidants on high-temperature stability of renewable bio-oils revealed by an innovative method for the determination of kinetic parameters of oxidative reactions. *Antioxidants*, *9*(5), 399. doi:10.3390/antiox9050399 PMID:32397271

Moore, K. E., Mangos, D. N., Slattery, A. D., Raston, C. L., & Boulos, R. A. (2016). Wool deconstruction using a benign eutectic melt. *RSC Advances*, *6*(24), 20095–20101. doi:10.1039/C5RA26516A

Mopper, K., & Zhou, X. (1990). Photochemical formation of hidroxyl radical by constituents of natural waters. *Science*, *25*, 661. doi:10.1126cience.250.4981.661 PMID:17810867

Moraes, B. S., Junqueira, T. L., Pavanello, L. G., Cavaletti, O., Mantelatto, P. E., & Bonomi, A. (2014). Anaerobic digestion of vinasse from sugarcane biorefineries in Brazil from energy, environmental, and economic perspectives: Profit or expense? *Applied Energy*, *113*, 825–835. doi:10.1016/j.apenergy.2013.07.018

Morales, P., Rojas, V., Quirós, M., & Gonzalez, R. (2015). The impact of oxygen on the final alcohol content of wine fermented by a mixed starter culture. *Applied Microbiology and Biotechnology*, *99*(9), 3993–4003. doi:10.100700253-014-6321-3 PMID:25582558

Moran-Salazar, R. G., Sanchez-Lizarraga, A. L., Rodriguez-Campos, J., Davila-Vazquez, G., Marino-Marmolejo, E. N., Dendooven, L., & Contreras-Ramos, S. M. (2016). Utilization of vinasses as soil amendment: Consequences and perspectives. *SpringerPlus*, *5*(1), 1007. doi:10.118640064-016-2410-3 PMID:27441131

Mordi, R. C., Ademosun, O. T., Ajanaku, C. O., Olanrewaju, I. O., & Walton, J. C. (2020). Free radical mediated oxidative degradation of carotenes and xanthophylls. *Molecules (Basel, Switzerland)*, *25*(5), 1038. doi:10.3390/molecules25051038 PMID:32110916

Morelló, J.-R., Motilva, M.-J., Tovar, M.-J., & Romero, M.-P. (2004). Changes in commercial virgin olive oil (cv Arbequina) during storage, with special emphasis on the phenolic fraction. *Food Chemistry*, *85*(3), 357–364. doi:10.1016/j.foodchem.2003.07.012

Moreno-Arribas, M. V., & Polo, M. C. (2005). Winemaking biochemistry and microbiology: Current knowledge and future trends. *Critical Reviews in Food Science and Nutrition*, *45*(4), 265–286. doi:10.1080/10408690490478118 PMID:16047495

Morillo, E. (2006). Application of cyclodextrins in agrochemistry. In: cyclodextrins and their complexes. In H. Dodziuk (Ed.), *Chemistry, analytical methods, applications* (pp. 459–467). Wiley-VCH.

Morris, J., Rose, A., & Lu, Z. (2022). Reactive oxygen species in the world ocean and their impacts on marine ecosystems. Elsevier. *Redox Biology*, *52*, 102285. doi:10.1016/j.redox.2022.102285 PMID:35364435

Mostofa, K. M. G., Liu, C. Q., Sakugawa, H., Vione, D., Minakata, D., Saquib, M., & Mottaleb, M. A. (2013). Photoinduced generation of hydroxyl radical in natural waters. In K. Mostofa, T. Yoshioka, A. Mottaleb, & D. Vione (Eds.), *Photobiogeochemistry of Organic Matter. Environmental Science and Engineering* (pp. 209–272). Springer. doi:10.1007/978-3-642-32223-5_3

Mourya, V. K., Inamdar, N. N., & Tiwari, A. (2010). Carboxymethyl chitosan and its applications. *Advanced Materials Letters*, *1*(1), 11–33. doi:10.5185/amlett.2010.3108

Müller, N., & Rauhut, D. (2018). Recent developments on the origin and nature of reductive sulfurous off-odours in wine. *Fermentation (Basel, Switzerland)*, *4*(3), 62. doi:10.3390/fermentation4030062

Muniz, J., Herrero, J. E., & Fuertes, A. B. (1998). Treatments to enhance the SO_2 capture by activated carbon fibres. *Applied Catalysis B: Environmental*, *18*(1-2), 171–179. doi:10.1016/S0926-3373(98)00039-3

Munné-Bosch, S., & Alegre, L. (2002). The function of tocopherols and tocotrienols in plants. *Critical Reviews in Plant Sciences*, *21*(1), 31–57. doi:10.1080/0735-260291044179

Munteanu, C., Dumitrascu, M., & Iliuta, R. A. (2011). *Ecology and environmental quality protection.*

Mureseanu, M., Chivu, V., Osiac, M., Ciobanu, M., Bucur, C., Parvulescu, V., & Cioatera, N. (2021). New photoactive mesoporous Ce-modified TiO_2 for simultaneous wastewater treatment and electric power generation. *Catalysis Today*, *366*, 164–176. doi:10.1016/j.cattod.2020.09.035

Musteata, G., Balanuta, A., Restica, A., Filimon, R. V., Baetu, M. M., & Patras, A. (2021). Capitalization of secondary wine products – an opportunity for the wine sector of Republic of Moldova and Romania. *Journal of Social Sciences*, *IV*(2), 117–127. doi:10.52326/jss.utm.2021.4(2).12

Nageswaran, G., Choudhary, Y. S., & Jagannathan, S. (2017). Inductively Coupled Plasma Mass Spectrometry. In *Spectroscopic Methods for Nanomaterials Characterization, 2, 163–194.* Elsevier. doi:10.1016/B978-0-323-46140-5.00008-X

Naik, R. M., Agarwal, A., Prasad, S., & Verma, A. K. (2009). Trace determination of thiosulphate and thioglycolic acid using novel inhibitory kinetic spectrophotometric method. *Microchemical Journal*, *93*(1), 43–48. doi:10.1016/j.microc.2009.04.006

Nair, A., Hykkerud, A., & Ratnaweera, H. (2022). Estimating Phosphorus and COD Concentrations Using a Hybrid Soft Sensor: A Case Study in a Norwegian Municipal Wastewater Treatment Plant. *Water (Basel)*, *14*(3), 332. doi:10.3390/w14030332

Narayanan, M., Loganathan, S., Valapa, R. B., Thomas, S., & Varghese, T. O. (2017). UV protective poly(lactic acid)/rosin films for sustainable packaging. *International Journal of Biological Macromolecules*, *99*, 37–45. doi:10.1016/j.ijbiomac.2017.01.152 PMID:28212931

Narra, M. R., Begum, G., Rajender, K., & Rao, J. V. (2011). *In vivo* impact of monocrotophos on biochemical parameters of a freshwater fish during subacute toxicity and following cessation of exposure to the insecticide. *Zeitschrift für Naturforschung. C, A Journal of Biosciences*, *66*(9-10), 507–514. doi:10.1515/znc-2011-9-1011 PMID:22191217

National Institute of Statistics. *Existing animal numbers on December 1, 2021.* (2022). *122,* 1–4. https://insse.ro/cms/ro/tags/comunicat-efective-animale

National Institute of Statistics. *Livestock numbers and animal production, in 2018.* (2019). 1–69. https://insse.ro/cms/ro/tags/efectivele-de-animale-si-productia-animala

Nayak, R., & Rao, J. R. (2003). Ferrous-ferric oxidation in pickle liquor. *Journal of Scientific and Industrial Research*, *62*, 1094–1097.

Ndungu, L. K., Steele, J. H., Hancock, T. L., Bartleson, R. D., Milbrandt, E. C., Parsons, M. L., & Urakawa, H. (2019). Hydrogen peroxide measurements in subtropical aquatic systems and their implications for cyanobacterial blooms. *Ecological Engineering*, *138*, 444–453. Advance online publication. doi:10.1016/j.ecoleng.2019.07.011

Nederlands Normalisatie Instituut. (2006). *NEN 6633, Water en (zuiverings) slib - Bepaling van het chemisch zuurstofverbruik (CZV).* Niederlande.

Negash, Y. A., Amare, D. E., Bitew, B. D., & Dagne, H. (2019). Assessment of quality of edible vegetable oils accessed in Gondar City, Northwest Ethiopia. *BMC Research Notes*, *12*(1), 793. doi:10.118613104-019-4831-x PMID:31801628

Negoescu, D., Atkinson, I., Gherendi, M., Culita, D. C., Baran, A., Petrescu, S., Trica, B., Pelinescu, D., Ionescu, R., Bratan, V., & Parvulescu, V. (2023). Brij 58–activated carbon assisted synthesis of Ag/Ag_2O/TiO_2-AC photocatalysts for efficient organic pollutants degradation. *Journal of Alloys and Compounds*, *931*, 167528. doi:10.1016/j.jallcom.2022.167528

Negro, C., Tommasi, L., & Miceli, A. (2003). Phenolic compounds and antioxidant activity from red grape marc extracts. *Bioresource Technology*, *87*(1), 41–44. doi:10.1016/S0960-8524(02)00202-X PMID:12733573

Nekhoroshkov, P., Bezuidenhout, J., Zinicovscaia, I., Yushin, N., Vergel, K., & Frontasyeva, M. (2021). Levels of elements in typical mussels from the southern coast of africa (Namibia, south africa, mozambique): Safety aspect. *Water (Basel)*, *13*(22), 3238. doi:10.3390/w13223238

Nel, A. P., du Toit, W. J., & van Jaarsveld, F. P. (2020). Pinking in White Wines -A Review. *South African Journal of Enology and Viticulture*, *41*(2), 151. 10.21548/41-2-3952

Nerantzis, E., & Tataridis, P. (2006). Integrated enology utilization of winery by-products into high added value products. *E-. Journal of Science and Technology*, *1*, 79–89.

Neri-Badang, M. C., & Chakraborty, S. (2019). Carbohydrate polymers as controlled release devices for pesticides. *Journal of Carbohydrate Chemistry*, *38*(2), 1–19. doi:10.1080/07328303.2019.1568449

Neubauer, M. P., Poehlmann, M., & Fery, A. (2014). Microcapsule mechanics: From stability to function. *Advances in Colloid and Interface Science*, *207*, 65–80. doi:10.1016/j.cis.2013.11.016 PMID:24345731

Nevers, L., Ribeiro, R., Oliveira, R., & Alves, M. M. (2006). Enhancement of Methane Production from Barley Waste. *Biomass and Bioenergy*, *30*(6), 599–560. doi:10.1016/j.biombioe.2005.12.003

Nguyen, H. M., Hwang, I.-C., Park, J.-W., & Park, H.-J. (2012). Photoprotection for deltamethrin using chitosan-coated beeswax solid lipid nanoparticles. *Pest Management Science*, *68*(7), 1062–1068. doi:10.1002/ps.3268 PMID:22383397

Nguyen, T. H., & Waterhouse, A. L. (2021). Redox Cycling of Iron: Effects of Chemical Composition on 2 Reaction Rates with Phenols and Oxygen in Model Wine. [AJEV]. *American Journal of Enology and Viticulture*, *72*(3), 209–216. doi:10.5344/ajev.2021.20024

Nicanorov, A. (1989). *Hydrochemistry*. Hidrometeoizdat.

Nidheesh, P. V., Divyapriya, G., Titchou, F. E., & Hamdani, M. (2022). Treatment of textile wastewater by sulfate radical based advanced oxidation processes. *Separation and Purification Technology*, *293*, 121115. doi:10.1016/j.seppur.2022.121115

Nikiforova, T., Kozlov, V., & Islyaikin, M. (2019). Sorption of d-metal cations by keratin from aqueous solutions. *Journal of Environmental Chemical Engineering*, *7*(5), 103417. doi:10.1016/j.jece.2019.103417

Nikolantonaki, M., Magiatis, P., & Waterhouse, A. L. (2014). Measuring protection of aromatic wine thiols from oxidation by competitive reactions vs wine preservatives with ortho-quinones. *Food Chemistry*, *163*, 61–67. doi:10.1016/j.foodchem.2014.04.079 PMID:24912696

Noishiki, Y., Ito, H., Miyamoto, T., & Inagaki, H. (1982). Application of denatured wool keratin derivatives to an antithrombogenic biomaterial-vascular graft coated with a heparinized keratin derivative. *Kobunshi Ronbunshu*, *39*(4), 221–227. doi:10.1295/koron.39.221

Nugraha, W. D., Sarminingsih, A., & Alfisya, B. (2020). The study of self purification capacity based on Biological Oxygen Demand (BOD) and Dissolved Oxygen (DO) parameters. *IOP Conference Series. Earth and Environmental Science*, *448*(1), 012105. doi:10.1088/1755-1315/448/1/012105

Nuruzzaman, M., Rahman, M. M., Liu, Y. J., & Naidu, R. (2016). Nanoencapsulation, nano-guard for pesticides: Anew window for safe application. *Journal of Agricultural and Food Chemistry*, *64*(7), 1447–1483. doi:10.1021/acs.jafc.5b05214 PMID:26730488

O'Brien, F. J. (2011). Biomaterials & scaffolds for tissue engineering. *Materials Today, 14*(3), 88–95. doi:10.1016/S1369-7021(11)70058-X

O'Shea, K. E., & Dionysiou, D. D. (2012). Advanced oxidation processes for water treatment. *The Journal of Physical Chemistry Letters, 3*(15), 2112–2113. doi:10.1021/jz300929x

Oae, S. (1991). *Organic sulfur chemistry.* CRC Press.

Oliveira, C. M., Ferreira, A. C. S., De Freitas, V., & Silva, A. M. S. (2011). Oxidation mechanisms occurring in wines. *Food Research International, 44*(5), 1115–1126. Advance online publication. doi:10.1016/j.foodres.2011.03.050

Onwudiwe, D. C., & Ajibade, P. A. (2011). ZnS, CdS and HgS nanoparticles via alkyl-phenyl dithiocarbomate complexed as single source precursors. *International Journal of Molecular Sciences, 12*(9), 5538–5551. doi:10.3390/ijms12095538 PMID:22016607

Ordaz-Diaz, L. A., & Bailon-Salas, A. M. (2020). Molecular identification of microbial communities in the methane production from vinasse: A review. *BioResources, 15*(2), 4528–4552. doi:10.15376/biores.15.2.Diaz

Osborne, H., & Akoh, C. C. (2003). Effects of natural antioxidants on iron-catalyzed lipid oxidation of structured lipid-based emulsions. *Journal of the American Oil Chemists' Society, 80*(9), 847–852. doi:10.100711746-003-0784-z

Ossola, R., Clerc, B., & Mcneill, K. (2020). Mechanistic insights into dissolved organic sulfur photomineralization through the study of cysteine sulfinic acid. *Environmental Science & Technology, 54*(20), 13066–13076. doi:10.1021/acs.est.0c04340 PMID:32936630

Ossola, R., Jönsson, O. M., Moor, K., & Mcneill, K. (2021). Singlet oxygen quantum yields in environmental waters. *Chemical Reviews, 121*(7), 4100–4146. Advance online publication. doi:10.1021/acs.chemrev.0c00781 PMID:33683861

Ostafiychuk, B., Budzulyak, I., Ivanichok, N., Lisovsky, R., & Rachiy, B. (2013). Injection of thermal modification on the development of the microporous structure of the carbonaceous material. *Physics of solid body, 14*(4), 891-896.

Otero, R., Fernandez, J. M., Ulibarri, M. A., Celis, R., & Bruna, F. (2012). Adsorption of nonionic pesticide S-Metolachlor on layered double hydroxides intercalated with dodecylsulfate and tetradecanedioate. *Applied Clay Science, 65–66*, 72–79. doi:10.1016/j.clay.2012.04.022

Oturan, M. A., & Aaron, J.-J. (2014). Advanced oxidation processes in water/wastewater treatment: Principles and applications. A review. *Critical Reviews in Environmental Science and Technology, 44*(23), 2577–2641. doi:10.1080/10643389.2013.829765

Ouyang, Z., Yang, C., He, J., Yao, Q., Zhang, B., Wang, H., & Dang, Z. (2020). Homogeneous photocatalytic degradation of sulfamethazine induced by Fe(III)-carboxylate complexes: Kinetics, mechanism and products. *Chemical Engineering Journal, 126122*, 126122. doi:10.1016/j.cej.2020.126122

Ozcan, A., & Ogun, M. (2014), Biochemistry of reactive oxygen and nitrogen species. Chapter in Sivakumar Joghi Thatha Gowder (Ed.). *Basic Principles and Clinical Significance of Oxidative Stress.* IntechOpen, DOI:, https://www.intechopen.com/chapters/48975 doi:10.5772/61193

Packer, J., Jeffrey, J., Latch, D., McNeill, K., & Arnold, W. (2003). Photochemical fate of pharmaceuticals in the environment: Naproxen, diclofenac, clofibric acid, and ibuprofen. *Aquatic Sciences, 65*(4), 342–351. doi:10.100700027-003-0671-8

Paiva, R. P., Munita, C. J. S., Cunha, I. I. L., Romano, J., & Alonso, C. D. (1990). Determination of trace elements in aerosol samples by instrumental neutron activation analysis. *Biological Trace Element Research, 26–27*(1), 231–242. doi:10.1007/BF02992676 PMID:1704723

Pande, S., Ghosh, S. K., Praharaj, S., Panigrahi, S., Basu, S., Jana, S., Pal, A., Tsukuda, T., & Pal, T. (2017). Synthesis of Normal and inverted gold-silver core-shell architectures in â-cyclodextrin and their applications in SERS. *The Journal of Physical Chemistry C*, *111*(29), 10806–10813. doi:10.1021/jp0702393

Pandey, A. K., Kumar, R. R., Kalidasan, B., Laghari, I. A., Samykano, M., Kothari, R., Abusorrah, A. M., Sharma, K., & Tyagi, V. V. (2021). Utilization of solar energy for wastewater treatment: Challenges and progressive research trends. *Journal of Environmental Management*, *297*, 113300. doi:10.1016/j.jenvman.2021.113300 PMID:34293672

Panfili, G., Fratianni, A., & Irano, M. (2003). Normal Phase High-Performance Liquid Chromatography Method for the Determination of Tocopherols and Tocotrienols in Cereals. *Journal of Agricultural and Food Chemistry*, *51*(14), 3940–3944. doi:10.1021/jf030009v PMID:12822927

Panfili, G., Niro, S., Bufano, A., D'Agostino, A., Fratianni, A., Paura, B., Falasca, L., & Cinquanta, L. (2020). Bioactive compounds in wild Asteraceae edible plants consumed in the Mediterranean diet. *Plant Foods for Human Nutrition (Dordrecht, Netherlands)*, *75*(4), 540–546. doi:10.100711130-020-00842-y PMID:32803719

Pang, L., Gao, Z., Feng, H., Wang, S., & Wang, Q. (2019). Cellulose based materials for controlled release formulations of agrochemicals: A review of modifications and applications. *Journal of Controlled Release*, *316*, 105–115. doi:10.1016/j.jconrel.2019.11.004 PMID:31704109

Panigrahi, S., Kundu, S., Ghosh, S. K., Nath, S., & Pal, T. (2005). Sugar assisted evolution of mono- and bimetallic nanoparticles. *Colloids and Surfaces. A, Physicochemical and Engineering Aspects*, *264*(1-3), 133–138. doi:10.1016/j.colsurfa.2005.04.017

Pariente, M. I., Segura, Y., Álvarez-Torrellas, S., Casas, J. A., de Pedro, Z. M., Diaz, E., García, J., López-Muñoz, M. J., Marugán, J., Mohedano, A. F., Molina, R., Munoz, M., Pablos, C., Perdigón-Meló, J. A., Petre, A. L., Rodríguez, J. J., Tobajas, M., & Martínez, F. (2022). Critical review of technologies for the on-site treatment of hospital wastewater: From conventional to combined advanced processes. *Journal of Environmental Management*, *320*, 115769. doi:10.1016/j.jenvman.2022.115769 PMID:35944316

Park, E. D., Choi, S. H., & Lee, J. S. (2000). Active states of Pd and Cu in carbon-supported Wacker-type catalysts for low-temperature CO oxidation. *The Journal of Physical Chemistry B*, *104*(23), 5586–5594. doi:10.1021/jp000583z

Park, S. Y., Jun, S. T., & Marsh, K. S. (2001). Physical properties of PVOH/chitosan-blended films cast from different solvents. *Food Hydrocolloids*, *15*(4-6), 499–502. doi:10.1016/S0268-005X(01)00055-8

Parmar, T. K., Rawtani, D., & Agrawal, Y. K. (2016). Bioindicators: The natural indicator of environmental pollution. *Frontiers in Life Science*, *9*(2), 110–118. doi:10.1080/21553769.2016.1162753

Parsaee, M., Kiani, M. K. D., & Karimi, K. (2019). A review of biogas production from sugarcane vinasse. *Biomass and Bioenergy*, *122*, 117–125. doi:10.1016/j.biombioe.2019.01.034

Patai, S. (Ed.). (1974). *The chemistry of the thiol group* (Vol. 2). Wiley. doi:10.1002/9780470771327

Patel, M., Kumar, R., Kishor, K., Mlsna, T., Pittman, C. U. Jr, & Mohan, D. (2019). Pharmaceuticals of emerging concern in aquatic systems: Chemistry, occurrence, effects, and removal methods. *Chemical Reviews*, *119*(6), 3510–3673. doi:10.1021/acs.chemrev.8b00299 PMID:30830758

Pavanello, A., Miranda, M. A., & Marin, M. L. (2022). Organic photoredox catalysts for wastewater remediation: Beyond the established advanced oxidation processes. *Chemical Engineering Journal Advances*, *11*, 100296. doi:10.1016/j.ceja.2022.100296

Pavlíková, I., Hladký, D., Motyka, O., Vergel, K. N., Strelkova, L. P., & Shvetsova, M. S. (2021). Characterization of pm10 sampled on the top of a former mining tower by the high-volume wind direction-dependent sampler using inna. *Atmosphere (Basel)*, *12*(1), 1–19. doi:10.3390/atmos12010029

Pavlíková, I., Motyka, O., Plášek, V., & Bitta, J. (2021). Monitoring of heavy metals and nitrogen concentrations in mosses in the vicinity of an integrated iron and steel plant: Case study in Czechia. *Applied Sciences (Basel, Switzerland)*, *11*(17), 8262. doi:10.3390/app11178262

Pearson, H. P. (2008). Chlorinated wool. *Journal of the Society of Dyers and Colourists*, *25*(3), 81–83. doi:10.1111/j.1478-4408.1909.tb00461.x

Pędziwiatr, P., Mikołajczyk, F., Zawadzki, D., Mikołajczyk, K., & Bedka, A. (2018). Decomposition of hydrogen perodixe - kinetics and review of chosen catalysts. *Acta Innovations*, *26*(26), 45–52. doi:10.32933/ActaInnovations.26.5

Peng, R., Yang, D., Qiu, X., Qin, Y., & Zhou, M. (2020). Preparation of self-dispersed lignin-based drug-loaded material and its application in avermectin nano-formulation. *International Journal of Biological Macromolecules*, *151*, 421–427. doi:10.1016/j.ijbiomac.2020.02.114 PMID:32061696

Penicaud, C., Achir, N., Dhuique-Mayer, C., Dornier, M., & Bohuon, P. (2011). Degradation of β-carotene during fruit and vegetable processing or storage: Reaction mechanisms and kinetic aspects: A review. *Fruits*, *66*(6), 417–440. doi:10.1051/fruits/2011058

Pérez-Navarro, J., Izquierdo-Cañas, P. M., Mena-Morales, A., Chacón-Vozmediano, J. L., Martínez-Gascueña, J., García-Romero, E., Hermosín-Gutiérrez, I., & Gómez-Alonso, S. (2020). Comprehensive chemical and sensory assessment of wines made from white grapes of vitis vinifera cultivars albillo dorado and montonera del casar: A comparative study with airén. *Foods*, *9*(9), 1282. doi:10.3390/foods9091282 PMID:32932602

Perlatti, B., de Souza Bergo, P. L., Silva, M. F. G. F. d., Fernandes, J. B., & Forim, M. R. (2013). Polymeric Nanoparticle-Based Insecticides: A Controlled Release Purpose for Agrochemicals. In Insecticides - Development of Safer and More Effective Technologies. IntechOpen. https://doi.org/ doi:10.5772/53355

Perron, N. R., & Brumaghim, J. L. (2009). A review of the antioxidant mechanisms of polyphenol compounds related to iron binding. *Cell Biochemistry and Biophysics*, *53*(2), 75–100. doi:10.100712013-009-9043-x PMID:19184542

Petcu, G., Anghel, E. M., Buixaderas, E., Atkinson, I., Somacescu, S., Baran, A., Culita, D. C., Trica, B., Bradu, C., Ciobanu, M., & Parvulescu, V. (2022). Au/Ti Synergistically Modified Supports Based on SiO_2 with Different Pore Geometries and Architectures. *Catalysts*, *12*(10), 1129. doi:10.3390/catal12101129

Petrou, A., Petrou, P., Ntanos, T., & Liapis, A. (2018). A possible role for singlet oxygen in the degradation of various antioxidants. A meta-analysis and review of literature Data. *Antioxidants*, *7*(3), 35. doi:10.3390/antiox7030035 PMID:29495515

Petrozziello, M., Torchio, F., Piano, F., Giacosa, S., Ugliano, M., Bosso, A., & Rolle, L. (2018). Impact of increasing levels of oxygen consumption on the evolution of color, phenolic, and volatile compounds of Nebbiolo wines. *Frontiers in Chemistry*, *6*(APR), 137. doi:10.3389/fchem.2018.00137 PMID:29755971

Petterson, C. O., & Meyers, J. (1973). Photosintetic production of hydrogen peroxide by Anacystis nidulans. *Plant Physiology*, *51*(1), 104–109. doi:10.1104/pp.51.1.104 PMID:16658269

Phan Quang, H. H., Dinh, N. T., Tran Thi, T. N., Ngoc Bao, L. T., Yuvakkumar, R., & Nguyen, V.-H. (2022). Fe^{2+}, Fe^{3+}, Co^{2+}as highly efficient cocatalysts in the homogeneous electro-Fenton process for enhanced treatment of real pharmaceutical wastewater. *Journal of Water Process Engineering*, *46*, 102635. doi:10.1016/j.jwpe.2022.102635

Picazo, C., & Molin, M. (2021). Impact of hydrogen peroxide on protein synthesis in yeast. *Antioxidants, 10*(6), 952. doi:10.3390/antiox10060952 PMID:34204720

Pignatello, J. J. (1992). Dark and photoassisted iron (Fe^{3+})-catalyzed degradation of chlorophenoxy herbicides by hydrogen peroxide. *Environmental Science & Technology, 26*(5), 944–951. doi:10.1021/es00029a012

Pilarski, K., Pilarska, A. A., Boniecki, P., Niedbała, G., Witaszek, K., Piekutowska, M., Idzior-Haufa, M., & Wawrzyniak, A. (2021). Degree of biomass conversion in the integrated production of bioethanol and biogas. *Energies, 14*(22), 7763. doi:10.3390/en14227763

Pinelo, M., & Rodriguez, R. (2007). Supercritical fluid and solid-liquid extraction of phenolic antioxidants from grape pomace: A comparative study. *European Food Research and Technology, 226*(1), 199–205. doi:10.100700217-006-0526-3

Pinte, G., Veado, M. A. R., Oliveira, A. H., Khalis, M., Ayrault, S., & Revel, G. (1998). Comparison of neutron activation analysis and ICP-MS used for river water pollution control. *Hydrobiologia, 373–374*(0), 61–73. doi:10.1023/A:1017004329179

Piva, G. G., Casalta, E., Legras, J. L., Tesnière, C., Sablayrolles, J. M., Ferreira, D., Ortiz-Julien, A., Galeote, V., & Mouret, J. R. (2022). Characterization and Role of Sterols in Saccharomyces cerevisiae during White Wine Alcoholic Fermentation. *Fermentation (Basel, Switzerland), 8*(2), 90. doi:10.3390/fermentation8020090

Plant, L., Jeff, M., & Shelley, S. (1994). Hydrogen peroxide: A potent force to destroy organics in wastewater. *Chemical Engineering (Albany, N.Y.)*, 16–20.

Poljsak, B., Kovač, V., & Milisav, I. (2021). Antioxidants, food processing and health. *Antioxidants, 10*(3), 433. doi:10.3390/antiox10030433 PMID:33799844

Polovetskaya, O. S., Arsenieva, T. A., & Polovetsky, M. D. (2015). Study of the organic matter of peat by IR-Fourier spectroscopy. *Privolzhsky Nauch. West., 7*(47), 12–18.

Poole, A. J. (2004). Treatment of biorefractory organic compounds in wool scour effluent by hydroxyl radical oxidation. *Water Research, 38*(14-15), 3458–3464. doi:10.1016/j.watres.2004.06.001 PMID:15276763

Poole, A. J., Cord-Ruwisch, R., & Jones, F. W. (1999). Biological treatment of chemically flocculated agro-industrial waste from the wool scouring industry by an aerobic process without sludge recycle. *Water Research, 33*(9), 1981–1988. doi:10.1016/S0043-1354(98)00391-1

Poole, L. B. (2015). The basics of thiols and cysteines in redox biology and chemistry. *Free Radical Biology & Medicine, 80*, 148–157. doi:10.1016/j.freeradbiomed.2014.11.013 PMID:25433365

Popescu, C., & Wortmann, F.-J. (2010). Wool - Structure, mechanical properties and technical products based on animal fibres. *Industrial Applications of Natural Fibres,* 255–266. doi:10.1002/9780470660324.ch12

Pravst, I., Žmitek, K., & Žmitek, J. (2010). Coenzyme Q10 contents in foods and fortification strategies. *Critical Reviews in Food Science and Nutrition, 50*(4), 269–280. doi:10.1080/10408390902773037 PMID:20301015

Prescha, A., Grajzer, M., Dedyk, M., & Grajeta, H. (2014). The antioxidant activity and oxidative stability of cold-pressed oils. *Journal of the American Oil Chemists' Society, 91*(8), 1291–1301. doi:10.100711746-014-2479-1 PMID:25076788

Prya, M., & Giridnar, M. (2006). Kinetics of photocatalytic degradation of phenols with multiple substituent groups. *Journal of Photochemistry and Photobiology A Chemistry, 179*(3), 256–262. doi:10.1016/j.jphotochem.2005.08.022

Pynaev, A. V. (2006). Study of heavy metal ions leaching from the ferritized plating sludge under the field conditions (Rus.). *Proceedings of Universities from North-Caucasus Region. Technical Sciences., 4*, 46–48.

Pysarenko, A. E., Pospelova, K. A., & Yakovlev, A. G. (1964). The course of colloidal chemistry. M: Vysshaia Shkola (High School).

Qiao, J., & Xiong, Y. (2021). Electrochemical oxidation technology: A review of its application in high-efficiency treatment of wastewater containing persistent organic pollutants. *Journal of Water Process Engineering*, *44*, 102308. doi:10.1016/j.jwpe.2021.102308

Qiao, R., Tian, W., Bai, J., Wang, L., Zhao, J., Du, Zh., & Gong, X. (2019). Application of magnetic adsorbents based on iron oxide nanoparticles for oil spill remediation: A review. *Journal of the Taiwan Institute of Chemical Engineers*, *97*, 227–236. doi:10.1016/j.jtice.2019.01.029

Quan, X., Luo, D., Wu, J., Li, R., Cheng, W., & Ge. (2017). Ozonation of acid red 18 wastewater using O3/Ca (OH) 2 system in a micro bubble gas-liquid reactor. *Journal of Environmental Chemical Engineering*, *5*(1), 283–291. doi:10.1016/j.jece.2016.12.007

Qutub, N., & Sabir, S. (2013). Characterization of cadmium sulphide nanoparticles synthesized by chemical precipitation method. *Advanced Science, Engineering and Medicine*, *5*(8), 1–6. doi:10.1166/asem.2013.1350

Rabadiya, B., Thakkar, V., & Rabadiya, P. (2013). Drug excipients interaction and solubility enhancement study of simvastatin. *IJPRBS*, *2*, 168–185.

Radkevich, V. Z., Wilson, K., Khaminets, S. G., & Sen'ko, T. L. (2014). Effect of preparation conditions on the formation of the active phase of carbon fiber catalytic systems for the low-temperature oxidation of carbon monoxide. *Kinetics and Catalysis*, *55*(2), 252–267. doi:10.1134/S0023158414020086

Radkevich, V.Z., Senko, T.L., & Khaminets, S.G., S. G., Wilson, K., & Egiazarov, & Yu. G. (2008). Catalytic systems based on carbon carriers for low-temperature CO oxidation. *Kinetics and Catalysis*, *49*(4), 570–576. doi:10.1134/S0023158408040149

Ragelienė, L., Treigienė, R., & Musnickas, J. (2009). Physical properties of human hair dyed using nonionic surfactants. *Materials Science*, *15*(3), 277–282. https://www.researchgate.net/publication/228476027

Ragno, G., Risoli, A., Ioele, G., & Luca, M. D. (2006). Photo- and Thermal-Stability Studies on Benzimidazole Anthelmintics by HPLC and GC-MS. *Chemical & Pharmaceutical Bulletin*, *54*(6), 802–806. doi:10.1248/cpb.54.802 PMID:16755047

Rai, P. K., Kumar, V., Lee, S., Raza, N., Kim, K.-H., Ok, Y. S., & Tsang, D. C. W. (2018). Nanoparticle-plant interaction: Implications in energy, environment, and agriculture. *Environment International*, *119*, 1–19. doi:10.1016/j.envint.2018.06.012 PMID:29909166

Rajabinejad, H., Bucişcanu, I. I., & Maier, S. S. (2019). Current approaches for raw wool waste management and unconventional valorization: A review. *Environmental Engineering and Management Journal*, *18*(7), 1439–1456. doi:10.30638/eemj.2019.136

Rajabinejad, H., Zoccola, M., Patrucco, A., Montarsolo, A., Rovero, G., & Tonin, C. (2018). Physicochemical properties of keratin extracted from wool by various methods. *Textile Research Journal*, *88*(21), 2415–2424. doi:10.1177/0040517517723028

Rakitskaya, T. L., Kiose, T. A., Vasylechko, V. O., Volkova, V. Y., & Gryshchouk, G. V. (2011). Adsorption-desorption properties of clinoptilolites and the catalytic activity of surface Cu(II)–Pd(II) complexes in the reaction of carbon monoxide oxidation with oxygen. *Chemistry of metals and alloys*, *4*(3-4), 213-218.

Rakitskaya, T. L., Vasylechko, V. O., Kiose, T. A., Gryshchouk, G. V., Dzhiga, A. M., & Volkova, V. Y. (2015). Catalytic activity of Pd(II) and Cu(II) complexes anchored to natural and pre-modified bentonite in the oxidation of carbon monoxide. *Chemistry of metals and alloys, 8*(1-2), 32-38.

Rakitskaya, T.L., Truba, A.S., Kiose, T.A., & Raskola, L.A. (2015). Mechanisms of formation of d metal complexes on porous carriers and their catalytic activity in redox reactions. *Odesa National University Herald. Chemistry, 20*(54), 27-48.

Rakitskaya, T. L., Bandurko, A. Y., Ennan, A. A., Paina, V. Y., & Litvinskaya, V. V. (2000). Low-temperature catalytic decomposition of ozone microconcentrations by carbon fibrous materials. *Advances in Environmental Research, 3*(4), 472–487.

Rakitskaya, T. L., Bandurko, A. Y., Ennan, A. A., Paina, V. Y., & Rakitskiy, A. S. (2001). Carbon-fibrous-material-supported base catalysts of ozone decomposition. *Microporous and Mesoporous Materials, 43*(2), 153–160. doi:10.1016/S1387-1811(00)00358-9

Rakitskaya, T. L., Dzhyga, G. M., Kiose, T. A., Oleksenko, L. P., & Volkova, V. Y. (2019). Pd(II), Cu(II), and pillared clay based nanocatalysts for low-temperature CO oxidation. *SN Applied Sciences, 1*(4), 1–8. doi:10.100742452-019-0314-x

Rakitskaya, T. L., & Ennan, A. A. (2012). *Phosphine. Physicochemical Properties and Practical Aspects of Elimination.* Astroprint.

Rakitskaya, T. L., Kiose, T. A., Ennan, A. A., Golubchik, K. O., Oleksenko, L. P., & Gerasiova, V. G. (2016). The Influence of Conditions of Acid-Thermal Modification of Clinoptilolite on Catalytic Properties of Palladium-Copper Complexes Anchored on it in the Reaction of Carbon Monoxide Oxidation. *Russian Journal of Physical Chemistry A. Focus on Chemistry, 90*(6), 1128–1135. doi:10.1134/S0036024416060182

Rakitskaya, T. L., Kiose, T. A., Golubchik, K. O., Ennan, A. A., & Volkova, V. Y. (2017). Acid-modified clinoptilolite as a support for palladium–copper complexes catalyzing carbon monoxide oxidation with air oxygen. *Chemistry Central Journal, 11*(1), 1–10. doi:10.118613065-017-0256-6 PMID:29086820

Rakitskaya, T. L., Kiose, T. A., Truba, A. S., & Ennan, A. A. A. (2022). Effect of Water on Activity and Protective Properties of Catalysts Used in Respiratory Protective Equipment. In *Handbook of Research on Water Sciences and Society* (pp. 469–499). IGI Global. doi:10.4018/978-1-7998-7356-3.ch021

Rakitskaya, T. L., Kiose, T. A., Voloshchuk, A. G., Oleksenko, L. P., Volkova, V. Y., & Reznik, L. I. (2009). Effect exerted by acid modification of bazalt tuff on catalytic activity of fixed acido complexes of palladium (II) and copper (II) in the reaction of carbon (II) oxide oxidation with air oxygen. *Russian Journal of Applied Chemistry, 82*(2), 204–208. doi:10.1134/S1070427209020074

Rakitskaya, T. L., Kiose, T. A., Zryutina, A. M., Gladyshevskii, R. E., Truba, A. S., Vasylechko, V. O., Demchenko, P. Y., Gryschouk, G. V., & Volkova, V. Y. (2013). Solid-state catalysts based on bentonites and Pd (II) Cu (II) complexes for low-temperature carbon monoxide oxidation. [). Trans Tech Publications Ltd.]. *Diffusion and Defect Data, Solid State Data. Part B, Solid State Phenomena, 200*, 299–304. doi:10.4028/www.scientific.net/SSP.200.299

Rakitskaya, T. L., Litvinskaya, V. V., Abramova, N. N., Redko, T. D., & Popova, N. A. (1987). Influence of adsorption-structural characteristics of carbon fiber material on the catalyst activity of phosphine oxidation. *Journal of applied chemistry, 60*(6), 1415–1417.

Rakitskaya, T. L., Truba, A. S., & Ennan, A. A. (2019). *Ozone. Physical and chemical properties, and catalytic methods of decomposition.* Astroprint.

Rakitskaya, T., Dzhyga, G., Kiose, T., & Volkova, V. (2020). Natural Nanobentonites as Supports in Palladium(II)–Copper(II) Catalysts for Carbon Monoxide Oxidation with Air Oxygen. In *Nanooptics and Photonics, Nanochemistry and Nanobiotechnology, and Their Applications* (pp. 141–157). Springer. doi:10.1007/978-3-030-52268-1_11

Rakotondramavo, A., Ribourg, L., Meynier, A., Guyon, C., de Lamballerie, M., & Pottier, L. (2019). Monitoring oxidation during the storage of pressure-treated cooked ham and impact on technological attributes. *Heliyon*, *5*(8), e02285. doi:10.1016/j.heliyon.2019.e02285 PMID:31485513

Ramana, K. V., Reddy, A. B. M., Majeti, N. V. R. K., & Singhal, S. S. (2018). Therapeutic potential of natural antioxidants. *Oxidative Medicine and Cellular Longevity*, *1–3*, 1–3. doi:10.1155/2018/9471051 PMID:30186551

Rangan, C., & Barceloux, D. G. (2009). Food additives and sensitivities. *Disease-a-Month*, *55*(5), 292–311. doi:10.1016/j.disamonth.2009.01.004 PMID:19362177

Ranjit, E., Hamlet, S., George, R., Sharma, A., & Love, R. M. (2022). Biofunctional approaches of wool-based keratin for tissue engineering, *Journal of Science: Advanced Materials and Devices,* *7*(1), doi:10.1016/j.jsamd.2021.10.001

Rao, B. S., Rajeshkumar, B., Rajagopalreddy, V., & Subbarao, T. (2011). Preparation and characterization of CdS nanoparticles by chemical co-precipitation technique. *Chalcogenide Letters*, *8*, 177–185.

Rao, P. S., & Hayon, E. (1975). Redox potentials of free radicals. IV. Superoxide and hydroperoxy radicals O_2– and HO_2. *Journal of Physical Chemistry*, *79*(4), 397–402. doi:10.1021/j100571a021

Rarata, G., & Surmacz, P. (2009). Hydrogen peroxide HTP class as a universal medium for propulsion and oxidiser. *Prace Instytutu Lotnictwa*, *7*, 125–158.

Rastogi, A., Yadav, D. K., Szymańska, R., Kruk, J., Sedlářová, M., & Pospíšil, P. (2014). Singlet oxygen scavenging activity of tocopherol and plastochromanol in *Arabidopsis thaliana*: Relevance to photooxidative stress: $^1O_2$ scavenging by tocopherol and plastochromanol. *Plant, Cell & Environment*, *37*(2), 392–401. doi:10.1111/pce.12161 PMID:23848570

Regulation (EU) No 1169/2011 of the European Parliament and of the Council of 25 October 2011 on the provision of food information to consumers, Official Journal of the European Union (L 304/18, Luxembourg) 2011.

Reich, C., & Su, D. T. (2001). Hair conditioners. Hair damage and cuticle surface. Handbook of Cosmetic Science and Technology. Marcel Dekker, Inc. 331–335

Rekhate, C. V., & Srivastava, J. K. (2020). Recent advances in ozone-based advanced oxidation processes for treatment of wastewater- A review. *Chemical Engineering Journal Advances*, *3*, 100031. doi:10.1016/j.ceja.2020.100031

Rekhate, C. V., & Srivastava, J. K. (2021). Effectiveness of O3/Fe2+/H2O2 process for detoxification of heavy metals in municipal wastewater by using RSM. *Chemical Engineering and Processing*, *165*, 108442. doi:10.1016/j.cep.2021.108442

Ribéreau-Gayon, P., Glories, Y., Maujean, A., & Dubourdieu, D. (2006). The chemistry of wine stabilization and treatments. In: Handbook of Enology, (2).

Rice, C., Young, R. J., Zan, R., Bangert, U., Wolverson, D., Georgiou, T., Jalil, R., & Novoselov, K. S. (2013). Raman-scattering measurements and first-principles calculations of strain-induced phonon shifts in monolayer MoS2. *Physical Review B: Condensed Matter and Materials Physics*, *87*(8), 081307. doi:10.1103/PhysRevB.87.081307

Richard, L., Peake, B., Rusak, S., Cooper, W., & Burritt, D. (2007). Production and decomposition dynamics of hydrogen peroxide in freshwater. *Environmental Chemistry*, *4*(1), 49–54. doi:10.1071/EN06068

Riddles, P. W., Blakeley, R. L., & Zerner, B. (1983). Reassessment of Ellman's reagent. *Methods in Enzymology*, *91*, 49–60. doi:10.1016/S0076-6879(83)91010-8 PMID:6855597

Rihak, Z., Prusova, B., Kumsta, M., & Baron, M. (2022). Effect of must hyperoxygenation on sensory expression and chemical composition of the resulting wines. *Molecules (Basel, Switzerland)*, *27*(1), 235. doi:10.3390/molecules27010235 PMID:35011467

Rivas, F. J., Beltran, F. J., Gimeno, O., & Frades, J. (2001). Treatment of olive oil mill wastewater by Fenton's Reagent. *Journal of Agricultural and Food Chemistry*, *49*(4), 1873–1880. doi:10.1021/jf001223b PMID:11308339

Robbins, C. R. (2012). Chemical and physical behavior of human hair. In Chemical and Physical Behavior of Human Hair. doi:10.1007/978-3-642-25611-0

Romanenko, A. V., & Simonov, P. A. (2007). Carbon materials and their physical and chemical properties: textbook, 128. M.: Kalvis.

Rondeau, P., Gambier, F., Jolibert, F., & Brosse, N. (2013). Compositions and chemical variability of grape pomaces from French vineyard. *Industrial Crops and Products*, *43*, 251–254. doi:10.1016/j.indcrop.2012.06.053

Rose, A. L., & Waite, T. D. (2005). Reduction of organically complexed ferric iron by superoxide in a simulated natural water. *Environmental Science & Technology*, *39*(8), 2645–2650. doi:10.1021/es048765k PMID:15884361

Ruiz-Rodriguez, A., Marín, F. R., Ocaña, A., & Soler-Rivas, C. (2008). Effect of domestic processing on bioactive compounds. *Phytochemistry Reviews*, *7*(2), 345–384. doi:10.100711101-007-9073-1

Russian Federation State Committee on hydrometeorology (1991) *Method of determination of toxicity of natural and waste waters on the reaction of lipid peroxidation of liposomes*. RD 52. 18. Moscow, 14.

Ryzo, E. G., Liubman, I. A., & Dmitriev, S. N. (2006). Universal technology of heavy metal ions sedimentation from wastewaters – ferritization technology. *Water and Ecology: Problems and Solutions*, *3*, 32–37.

Saad, B., Sing, Y., Nawi, M., Hashim, N., Mohamedali, A., Saleh, M., Sulaiman, S., Talib, K., & Ahmad, K. (2007). Determination of synthetic phenolic antioxidants in food items using reversed-phase HPLC. *Food Chemistry*, *105*(1), 389–394. doi:10.1016/j.foodchem.2006.12.025

Saa, P. A., Moenne, M. I., Pérez-Correa, J. R., & Agosin, E. (2012). Modeling oxygen dissolution and biological uptake during pulse oxygen additions in oenological fermentations. *Bioprocess and Biosystems Engineering*, *35*(7), 1167–1178. doi:10.100700449-012-0703-7 PMID:22349928

Saerens, S. M. G., Delvaux, F. R., Verstrepen, K. J., & Thevelein, J. M. (2010). Production and biological function of volatile esters in Saccharomyces cerevisiae. *Microbial Biotechnology*, *3*(2), 165–177. doi:10.1111/j.1751-7915.2009.00106.x PMID:21255318

Sahu, S., Rani Sahoo, P., Patel, S., & Mishra, B. K. (2011). Oxidation of thiourea and substituted thioureas: A review. *Journal of Sulfur Chemistry*, *32*(2), 171–197. doi:10.1080/17415993.2010.550294

Samal, K., Mahapatra, S., & Ali, M. H. (2022). Pharmaceutical wastewater as Emerging Contaminants (EC): Treatment technologies, impact on environment and human health. *Energy Nexus*, *6*, 100076. doi:10.1016/j.nexus.2022.100076

Samuel, B., Mathew, S., Anand, V. R., Correya, A. A., Nampoori, V. P. N., & Mujeeb, A. (2018). Surface defect assisted broad spectra emission from CdSe quantum dots for white LED application. *Materials Research Express*, *5*(2), 025009. doi:10.1088/2053-1591/aaaa83

Sandmann, G. (1985). Photosynthetic and respiratory electron transport in Cu^{2+}-deficient Dunaliella. *Physiologia Plantarum*, *65*(4), 481–486. doi:10.1111/j.1399-3054.1985.tb08677.x

Saravanan, A., Deivayanai, V. C., Senthil Kumar, P., Rangasamy, G., Hemavathy, R. V., Harshana, T., Gayathri, N., & Alagumalai, K. (2022). A detailed review on advanced oxidation process in treatment of wastewater: Mechanism, challenges and future outlook. *Chemosphere*, *308*, 136524. doi:10.1016/j.chemosphere.2022.136524 PMID:36165838

Schalles, J. F. (2006). Optical remote sensing techniques to estimate phytoplankton chlorophyll concentrations in coastal. In L. L. Richardson & E. F. LeDrew (Eds.), *Remote sensing of aquatic coastal ecosystem processes* (pp. 27–79). Springer Netherlands. doi:10.1007/1-4020-3968-9_3

Schmidtko, S., Stramma, L., & Visbeck, M. (2017). Decline in global oceanic oxygen content during the past five decades. *Nature*, *542*(7641), 335–339. doi:10.1038/nature21399 PMID:28202958

Schneider, V. (1998). Must hyperoxidation: A review. *American Journal of Enology and Viticulture*, *49*(1), 65–73. doi:10.5344/ajev.1998.49.1.65

Schröder, W., & Nickel, S. (2019). Spatial structures of heavy metals and nitrogen accumulation in moss specimens sampled between 1990 and 2015 throughout Germany. *Environmental Sciences Europe*, *31*(1), 33. doi:10.118612302-019-0216-y

Schulte, P., Bayer, A., Kuhn, F., Luy, T., & Volkmer, M. (1995). H_2O_2/O_3, H_2O_2/UV and H_2O_2/Fe^{2+} processes for the oxidation of hazardous wastes. *Ozone Science and Engineering*, *17*(2), 119–134. doi:10.1080/01919519508547541

Scull, I., Savón, L., Gutiérrez, O., Valiño, E., Orta, I., Mora, P. O., Orta, H., Ramos, Y., Molineda, A., Coto, G., & Noda, A. (2012). Physico-chemical composition of concentrated vinasse for their assessment in animal diets. Cuban. *Journal of Agricultural Science*, *46*(4), 385.

Scutaru, I., Balanuta, A., & Zgardan, D. (2018). The determination of oxidation behavior of white wines produced from local and European grape varieties using spectrophotometric method. *Journal of Engineering Science*, *25*(4), 82–93.

Scutaru, I., Sturza, R., & Gherdelescu, L. (2022). Oxidative-reducing processes in winemaking. *Modern Technologies, in the Food Industry – 2022: Proc. of the Intern. Conf.* (pp. 84–85). UTM.

Sébille, B. (1987). Cyclodextrin derivatives. In D. Duchêne (Ed.), *Cyclo- dextrins and their industrial uses* (pp. 351–393). Edition de Santé.

Seeprasert, P., Anurakpongsatorn, P., Laoharojanaphand, S., & Busamongkol, A. (2017). Instrumental neutron activation analysis to determine inorganic elements in paddy soil and rice and evaluate bioconcentration factors in rice. *Agriculture and Natural Resources (Bangkok)*, *51*(3), 154–157. doi:10.1016/j.anres.2016.09.005

Seitkalieva, A. V., Menzorova, N. I., & Rasskazov, V. A. (2016). Application of different enzyme assays and biomarkers for pollution monitoring of the marine environment. *Environmental Monitoring and Assessment*, *188*(1), 70. doi:10.100710661-015-5066-5 PMID:26721566

Semenov, V. V., Varlamova, S. I., & Klimov, E. S. (2005). Utilization of plating sludge by ferritization method. *University News*, *48*(2), 111–112.

Senthamilselvi, S., Saravanakumar, K., Jabena Begum, N., Anandhi, R., Ravichandran, A. T., Sakthivel, B., & Ravichandran, K. (2012). Photovoltaic properties of nanocrystalline CdS films deposited by SILAR and CBD techniques—A comparative study. *Journal of Materials Science Materials in Electronics*, *23*(1), 302–308. doi:10.100710854-011-0409-7

Septembre-Malaterre, A., Remize, F., & Poucheret, P. (2018). Fruits and vegetables, as a source of nutritional compounds and phytochemicals: Changes in bioactive compounds during lactic fermentation. *Food Research International*, *104*, 86–99. doi:10.1016/j.foodres.2017.09.031 PMID:29433787

Seraghni, N., Ghoul, I., Dekkiche, B. A., Bouaziz, C., Debbache, N., & Sehili, T. (2021). Effect of Fe(III)-bicarboxylic complexes in removal pollutant under UV and sunlight in aqueous solutions. *Environmental Technology*. doi:10.1080/09593330.2021.1891142 PMID:33583355

Servili, M., & Montedoro, G. F. (2002). Contribution of phenolic compounds to virgin olive oil quality. *European Journal of Lipid Science and Technology, 104*(9-10), 602–613. doi:10.1002/1438-9312(200210)104:9/10<602::AID-EJLT602>3.0.CO;2-X

Servos, M., Carey, J., & Fergusson, M. (1992). Impact of a modern bleached kraft mill with secondary treatment on white suckers. *Water Pollution Research Journal of Canada, 273*, 423–437.

Severini, C., Giuliani, R., De Filippis, A., Derossi, A., & De Pilli, T. (2016). Influence of different blanching methods on colour, ascorbic acid and phenolics content of broccoli. *Journal of Food Science and Technology, 53*(1), 501–510. doi:10.100713197-015-1878-0 PMID:26787969

Shabir, M., Yasin, M., Hussain, M., Shafiq, I., Akhter, P., Nizami, A.-S., Jeon, B.-H., & Park, Y.-K. (2022). A review on recent advances in the treatment of dye-polluted wastewater. *Journal of Industrial and Engineering Chemistry, 112*, 1–19. doi:10.1016/j.jiec.2022.05.013

Shahidi, F., & De Camargo, A. C. (2016). Tocopherols and tocotrienols in common and emerging dietary sources: Occurrence, applications, and health benefits. *International Journal of Molecular Sciences, 17*(10), 1745. doi:10.3390/ijms17101745 PMID:27775605

Shahidi, F., & Shukla, V. K. S. (1996). Nontriacylglycerol constituents of fats, oils. INFORM - *International News on Fats. Oils and Related Materials, 7*(11), 1227–1232.

Shao, Y., Chen, Y., Liu, J., Wu, P., Ying, Y., & Xie, J. (2016). ICP-MS Determination of potential toxic elements in soil and rice (Oryza sativa L.) and related health risk. *Food Analytical Methods, 9*(12), 3501–3508. doi:10.100712161-016-0536-0

Shariatinia, Z. (2019). Pharmaceutical applications of chitosan. *Advances in Colloid and Interface Science, 263*, 131–194. doi:10.1016/j.cis.2018.11.008 PMID:30530176

Sharma, S., Mothudi, B. M., & Dhlamini, M. S. (2016). Observed coexistence of memristive, memcapacitive and meminductive characteristic in polyvinyl alchol/cadmium sulphide nanocomposites. *Journal of Materials Science Materials in Electronics, 27*(5), 4551–4558. doi:10.100710854-016-4330-y

Sharma, V., Singh, H., Guleria, S., Bhardwaj, N., Puri, S., Arya, S. K., & Khatri, M. (2022). Application of superparamagnetic iron oxide nanoparticles (SPIONs) for heavy metal adsorption: A 10-year meta-analysis. *Environmental Nanotechnology, Monitoring & Management, 18*, 100716. doi:10.1016/j.enmm.2022.100716

Sharpless, Ch., Seibold, D., & Linden, K. (2003). Nitrate photosensitized degradation of atrazine during UV water treatment. *Aquatic Sciences, 65*(4), 359–366. doi:10.100700027-003-0674-5

Shcherban, N. D., & Filonenko, S. M. (2017). Synthesis, power, and determination of the functionalization of carbon nanostructures. *Nanosystems, nanomaterials, nanotechnologies, 15*(1), 113-131.

Shen, C., Li, H., Yu, J., Wu, G., Mao, D., & Lu, G. (2013). A First Principles DFT Study on the Active Sites of Pd-Cu-Cl_x/Al_2O_3 Catalyst for Low-Temperature CO Oxidation. *ChemCatChem, 5*(10), 2813–2817. doi:10.1002/cctc.201300356

Shibasaki-Kitakawa, N., Kato, H., & Yonemoto, T. (2004). Oxidation kinetics of β—Carotene in oleic acid solvent with addition of an antioxidant, α-tocopherol. *Journal of the American Oil Chemists' Society, 81*(4), 389–394. doi:10.100711746-004-0911-x

Shih, P.-Y., Liao, Y.-T., Tseng, Y.-K., Deng, F.-S., & Lin, C.-H. (2019). A potential antifungal effect of chitosan against *Candida albicans* is mediated via the inhibition of SAGA complex component expression and the subsequent alteration of cell surface integrity. *Frontiers in Microbiology, 10*, 602. doi:10.3389/fmicb.2019.00602 PMID:30972050

Shiota, M., Uchida, T., Oda, T., & Kawakami, H. (2006). Utilization of lactoferrin as an iron-stabilizer for soybean and fish oil. *Journal of Food Science, 71*(3), C120–C123. doi:10.1111/j.1365-2621.2006.tb15606.x

Shrestha, S., Kognou, A. L. M., Zhang, J., & Qin, W. (2021). Different Facets of Lignocellulosic Biomass Including Pectin and its Perspectives. *Waste and Biomass Valorization, 12*(9), 4805–4823. doi:10.100712649-020-01305-w

Shtamm, E. V. (1988). Redox state of water fish reproduction Proceedings of the first all-union workshop on ecological chemistry of natural waters, 279-295. Centre for International Projects, GKNT, Moscow.

Shtamm, E. V., & Skurlatov, Yu. I. (1993). *Toxicity caused by disbalance of redox processes in natural waters.* 20-th Annual aquatic toxicity workshop. Quebec, Canada.

Shtamm, E. V., Alexandrova, E. V., Kozlova, N. B., Shishkina, L. N., Semenova, I. V., Skurlatov, Yu. I., & Dedenchuk, I. V. (2005). *Lipid peroxidation of liposomes – sensitive and express method of biotesting of the water quality and biologically active substances.* Third Moscow Intern. Congress Biotechnology: State of the Art and Prospects of Development. Congress Proceedings, Part 2, (pp. 68 – 69). Moscow, Russia.

Shtamm, E. V., Kozlova, N. B., Alexandrova, E. V., Skurlatov, Yu. I., Frog, B. N., & Frog, D. B. (2003). Microalgae for control of water quality of fishery and drinking reservoirs undergoing the influence of industrial enterprises sewage. *Abstr. Biotechnology of microalgae, 5th European Workshop*, (p. 63).

Shtamm, E. V., Skurlatov, Yu. I., Kozlova, N. B., Zaitseva, N. I., & Alexandrova, E. V. (2011). Bioassaying in the Assessment of Wastewater Treatment Process Efficiency. *Water Resources, 38*(2), 237–243. Pleiades Publishing, Ltd.

Shtamm, E. V., Frog, B. N., Skurlatov, Yu. I., Kozlova, N. B., Alexandrova, E. V., & Shishkina, L. N. (2002). Possible role of redused sulfur compounds in toxic properties of pulp and paper mill effluents. *Acta Hydrochimica et Hydrobiologica, 30*(5-6), 256–265. doi:10.1002/aheh.200390007

Shtamm, E. V., Kozlova, N. B., Alexandrova, E. V., Razumovsky, S. D., Konstantinova, M. L., Emnéus, J., & Winther-Nielsen, M. (2004). In G. E. Zaikov (Ed.), *Dinamics of surfactant destruction in complex photoozonation + microbial degradation. Biotechnology and the environment including biogeotechnology* (pp. 35–42). Nova Science Publishers Inc.

Shtamm, E. V., Purmal, A. P., & Skurlatov, Y. I. (1991). The role of hydrogen peroxide in natural aquatic media. *Russian Chemical Reviews, 60*(11), 1228–1248. http://mi.mathnet.ru/eng/rcr982. doi:10.1070/RC1991v060n11ABEH001141

Shtamm, E. V., Skurlatov, Y. I., Ernestova, L. S., Baikova, I. S., & Dyubanov, M. V. (2017). Chemical safety of aquatic ecosystems: Biological methods of control. *Russian Journal of Physical Chemistry B. Focus on Physics, 11*(4), 635–642. doi:10.1134/S1990793117040236

Shtamm, E. V., Skurlatov, Yu. I., Kozlova, N. B., Zaitseva, N. I., & Aleksandrova, E. V. (2011). Bioassaying in the Assessment of Wastewater Treatment Process Efficiency. *Water Resources, 38*(2), 237–243. doi:10.1134/S0097807811010155

Shukla, S. K., Mishra, A. K., Arotiba, O. A., & Mamba, B. B. (2013). Chitosan-based nanomaterials: A state-of-the-art review. *International Journal of Biological Macromolecules, 59*, 46–58. doi:10.1016/j.ijbiomac.2013.04.043 PMID:23608103

Sibhatu, A. K., Weldegebrieal, G. K., Sagadevan, S., Tran, N. N., & Hessel, V. (2022). Photocatalytic activity of CuO nanoparticles for organic and inorganic pollutants removal in wastewater remediation. *Chemosphere, 300*, 134623. doi:10.1016/j.chemosphere.2022.134623 PMID:35439489

Siciliano, A., Limonti, C., & Curcio, C. M. (2022). Performance evaluation of pressurized anaerobic digestion (PDA) of raw compost leachate. *Fermentation (Basel, Switzerland)*, *8*(1), 15. doi:10.3390/fermentation8010015

Siddique, N., Jawad, M., & Waheed, S. (2013). Instrumental neutron activation analysis for the study of size-fractionated airborne particulate matter samples. *Radiochimica Acta*, *101*(4), 273–277. doi:10.1524/ract.2013.2020

Simanova, S. A., Shchukarev, A. V., & Lysenko, A. A. (2008). Adsorption of chloro complexes of palladium, platinum, and gold by carbon fibers of different structures. *Journal "Chemical fibers"*, *4*, 1-7.

Simonic, M. (2021). Reverse Osmosis Treatment of Wastewater for Reuse as Process Water—A Case Study. *Membranes*, *11*(12), 976. doi:10.3390/membranes11120976 PMID:34940477

Simonov, P. A., Romanenko, A. V., Prosvirin, I. P., Moroz, E. M., Boronin, A. I., Chuvilin, A. L., & Likholobov, V. A. (1997). On the nature of the interaction of H2PdCl4 with the surface of graphite-like carbon materials. *Carbon*, *35*(1), 73–82. doi:10.1016/S0008-6223(96)00129-7

Simpson, W. S. (1999) Physics and chemistry of wool yellowing. *WRONZ Report R217*. Wool Research Organisation of New Zealand. 36

Sinelnikov, V., Demina, V. (1974). About the origin of hydrogen peroxide contained in open water bodies. *Hydrochemical Materials, X.*

Singh, A., Dhiman, N., Kar, A. K., Singh, D., Purohit, M. P., Ghosh, D., & Patnaik, S. (2020). Advances in controlled release pesticide formulations: Prospects to safer integrated pest management and sustainable agriculture. *Journal of Hazardous Materials*, *385*, 121525. doi:10.1016/j.jhazmat.2019.121525 PMID:31740313

Singh, H., Bhardwaj, N., Arya, S. K., & Khatri, M. (2020). Environmental impacts of oil spills and their remediation by magnetic nanomaterials. *Environmental Nanotechnology, Monitoring & Management*, *14*, 100305. doi:10.1016/j.enmm.2020.100305

Singh, V., & Chauhan, P. (2009). Structural and optical characterization of CdS nanoparticles prepared by chemical precipitation method. *Journal of Physics and Chemistry of Solids*, *70*(7), 1074–1079. doi:10.1016/j.jpcs.2009.05.024

Sinkiewicz, I., Śliwińska, A., Staroszczyk, H., & Kołodziejska, I. (2016). Alternative methods of preparation of soluble keratin from chicken feathers. *Waste and Biomass Valorization*, *8*(4), 1043–1048. doi:10.100712649-016-9678-y

Skurlatov, Y. I., Shtamm, E. V., Travin, S. O., Shvydkiy, V. O., & Semenyak, L. V. (2020). Role of Oxidation-Reduction Processes in Formation of Toxic Properties of Natural Aqueous Environment. *Handbook of Research on Emerging Developments and Environmental Impacts of Ecological Chemistry,* 149-163. IGI Global. https://www.igi-global.com/book/emerging-developments-environmental-impacts-ecological/231905

Skurlatov, Yu. I., Ernestova, L. S., Shtamm, E. V., Shpotova, T. V., & Kalinin, V. B. (1984). Redox state and seasonal toxicity of natural waters. [in Russian]. *Doklady of the USSR Academy of Sciences*, *276*(4), 1014–1016.

Skurlatov, Yu. I., Shtamm, E. V., & Semenyak, L. V. (1991). Role of algae in aquatic environment redox-state formation. Interfacial phenomena in the Environment. *Internat. Conf. Switzerland*, (October), 103–104.

Skurlatov, Yu. I., Zepp, R. G., & Pierce, J. T. (1987). Algal-induced decay and formation of hydrogen peroxide in water. Its possible role in oxidation of anilenes by algae. *Ann. Chem. Soc. Symp. Ser.*, *327*, 215–224.

Smith, A. L., & Harris, M. (1936a). Oxidation of wool : Effect of hydrogen peroxide on wool. *Journal of Research of the National Bureau of Standards*, *16*(4), 301–307. https://nvlpubs.nist.gov/nistpubs/jres/16/jresv16n4p301_a1b.pdf. doi:10.6028/jres.016.017

Smith, A. L., & Harris, M. (1936b). Oxidation Of Wool : The lead acetate test for hydrogen peroxide bleached wool. *Journal of Research of the National Bureau of Standards, 16*(4), 309–313. doi:10.6028/jres.016.018

Smith, J. (2019). *General, organic, and biological chemistry*. McGraw Hill Education.

Smolders, J., Lamers, E., Lucassen, G., & Van Der Velde, G. (2006). Internal Eutrophication-a review. *Chemistry and Ecology, 22*(2), 93–111. doi:10.1080/02757540600579730

Smotkin, E. S., Lee, C., Brad, A. J., Campion, A., Fox, M. A., Mallouk, T. E., Webber, S. E., & White, J. M. (1988). Size quantization effects in cadmium sulfide layers formed by a langmiur-blodgett technique. *Chemical Physics Letters, 152*(2-3), 265–268. doi:10.1016/0009-2614(88)87365-2

Snoeyink, V. L., & Jenkins, D. (2015). *Water Chemistry*. John Wiley & Sons.

Sommer, B. A., & Margerum, D. W. (1970). Kinetic study of the hidroxoiron(III) dimmer. *Inorganic Chemistry, 9*(11), 2517–2521. doi:10.1021/ic50093a028

Søndergaard, J., Asmund, G., & Larsen, M. M. (2015). Trace elements determination in seawater by ICP-MS with on-line pre-concentration on a Chelex-100 column using a "standard" instrument setup. *MethodsX, 2*, 323–330. doi:10.1016/j.mex.2015.06.003 PMID:26258050

Song, C., Liu, H.Y., Guo, S., & Wang, S.G. (2020). Photolysis mechanisms of tetracycline under UV irradiation in simulated aquatic environment surrounding limestone. *Chemosphere*, 125582. doi:. 2019.125582. doi:10.1016/j.chemosphere

Song, C., Liu, H. Y., Guo, S., & Wang, S. G. (2020). Photolysis mechanisms of tetracycline under UV irradiation in simulated aquatic environment surrounding limestone. *Chemosphere, 244*, 125582. doi:10.1016/j.chemosphere.2019.125582 PMID:32050352

Song, H., Wignall, P. B., Song, H., Dai, X., & Chu, D. (2019). Seawater temperature and dissolved oxygen over the past 500 million years. *Journal of Earth Science, 30*(2), 236–243. doi:10.100712583-018-1002-2

Song, S., Wang, Y., Xie, J., Sun, B., Zhou, N.-L., Shen, H., & Shen, J. (2019). Carboxymethyl chitosan modified carbon nanoparticle for controlled emamectin benzoate delivery: Improved solubility, ph-responsive release, and sustainable pest control. *ACS Applied Materials & Interfaces, 11*(37), 34258–34267. doi:10.1021/acsami.9b12564 PMID:31461267

Son, M. H., Gong, J., Seo, S., Yoon, H., & Chang, Y. S. (2019). Photosensitized diastereoisomer-specific degradation of hexabromocyclododecane (HBCD) in the presence of humic acid in aquatic systems. *Journal of Hazardous Materials, 369*, 171–179. doi:10.1016/j.jhazmat.2019.02.035 PMID:30776600

Sousa, B. C., Pitt, A. R., & Spickett, C. M. (2017). Chemistry and analysis of HNE and other prominent carbonyl-containing lipid oxidation compounds. *Free Radical Biology & Medicine, 111*, 294–308. doi:10.1016/j.freeradbiomed.2017.02.003 PMID:28192230

South African National Standard. (2010). SANS 6048, Water – Chemical oxygen demand. SABS standards division, Pretoria, South Africa.

Southworth, B. A., & Voelker, B. M. (2003). Hydroxyl radical production via the photo-fenton reaction in the presence of fulvic acid. *Environmental Science & Technology, 37*(6), 1130–1136. doi:10.1021/es0207571 PMID:12680665

Sova, M., & Saso, L. (2020). Natural sources, pharmacokinetics, biological activities and health benefits of hydroxycinnamic acids and their metabolites. *Nutrients, 12*(8), 2190. doi:10.3390/nu12082190 PMID:32717940

Spinei, M., & Oroian, M. (2021). The potential of grape pomace varieties as a dietary source of pectic substances. *Foods, 10*(4), 867. doi:10.3390/foods10040867 PMID:33921097

Špirić, Z., Vučković, I., Stafilov, T., Kušan, V., & Frontasyeva, M. (2013). Air pollution study in croatia using moss biomonitoring and ICP-AES and AAS analytical techniques. *Archives of Environmental Contamination and Toxicology*, *65*(1), 33–46. doi:10.100700244-013-9884-6 PMID:23467710

Spiropoulos, A., & Bisson, L. F. (2000). MET17 and hydrogen sulfide formation in Saccharomyces cerevisiae. *Applied and Environmental Microbiology*, *66*(10), 4421–4426. doi:10.1128/AEM.66.10.4421-4426.2000 PMID:11010893

Sprynsky, M., Lebedynets, M., Namieśnik, J., & Buszewski, B. (2007). Phenolics occurrence in surface water of the Dniester river basin (West Ukraine): Natural background and industrial pollution. *Environmental Geology (Berlin)*, *53*(1), 67–75. doi:10.100700254-006-0619-0

St. Laurent, J. B., de Buzzaccarini, F., De Clerck, K., Demeyere, H., Labeque, R., Lodewick, R., & van Langenhove, L. (2007). Laundry cleaning of textiles. *Handbook for Cleaning/Decontamination of Surfaces*, 57–102. doi:10.1016/B978-0-444-51664-0.50003-6

Stadtman, E. R., & Levine, R. L. (2006). Chemical modification of proteins by reactive oxygen species. *Redox Proteomics*, 1–23. doi:10.1002/0471973122.ch1

Steenson, D. F., Lee, J. H., & Min, D. B. (2002). Solid phase microextraction of volatile soybean oil and corn oil compounds. *Journal of Food Science*, *67*(1), 71–76. doi:10.1111/j.1365-2621.2002.tb11361.x

Steinhardt, J., Fugitt, C. H., & Harris, M. (1940). Combination of wool protein with acid and base: The effect of tem perature on the titration curve. *Textile Research*, *11*(2), 72–94. doi:10.1177/004051754001100203

Stemme, K., Gerdes, B., Harms, A., & Kamphues, J. (2005). Beet-vinasse (condensed molasses solubles) as an ingredient in diets for cattle and pigs - Nutritive value and limitations. *Journal of Animal Physiology and Animal Nutrition*, *89*(3-6), 179–183. doi:10.1111/j.1439-0396.2005.00554.x PMID:15787991

Stihi, C., Popescu, I. V., Frontasyeva, M., Radulescu, C., Ene, A., Culicov, O., Zinicovscaia, I., Dulama, I. D., Cucu-Man, S., Todoran, R., Gheboianu, A. I., Bucurica, A., Bancuta, I., & Dima, G. (2017). Characterization of heavy metal air pollution in Romania using moss biomonitoring, neutron activation analysis, and atomic absorption spectrometry. *Analytical Letters*, *50*(17), 2851–2858. doi:10.1080/00032719.2016.1275661

Stoecker, B. (2004). Chromium. In Anke M., Ihnat M., Stoeppler M. (eds.) Elements and their compounds in the environment. 2nd Edition. Wiley-VCH Verlag, Weinheim, pp. 709.

Stubenrauch, J., Ekardt, F., Heyl, K., Garske, B., Schott, V. L., & Ober, S. (2021). How to legally overcome the distinction between organic and conventional farming—Governance approaches for sustainable farming on 100% of the land. *Sustainable Production and Consumption*, *28*, 716–725. doi:10.1016/j.spc.2021.06.006

Sturza, R., Scutaru, I., Duca, G. (2021). Quality Management of Wines and Redox Processes. *Advances in Intelligent Systems and Computing*. AISC. . doi:10.1007/978-3-030-49889-4_45

Subramanian, G., & Madras, G. (2017). Remarkable enhancement of Fenton degradation at wide pH range promoted by thioglycolic acid. *Chemical Communications*, *53*(6), 1136–1139. doi:10.1039/C6CC09962A PMID:28054055

Sulyman, M., Namiesnik, J., & Gierak, A. (2017). Greener cleaner : Sheep wool fiber as renewable sources for oil spill cleanup. *International Journal of Advances in Science, Engineering and Technology*, *5*(2), 84–92. http://iraj.in

Sumner, R. M. W., Young, S. R., & Upsdell, M. P. (2003). Wool yellowing and pH within Merino and Romney fleeces, *Proceedings of the New Zealand Society of Animal Production*, Queenstown, *63*, 155-159.

Sun, B., Neves, A. C., Fernandes, T. A., Fernandes, A. L., Mateus, N., de Freitas, V., Leandro, C., & Spranger, M. I. (2011). Evolution of phenolic composition of red wine during vinification and storage and its contribution to wine sensory properties and antioxidant activity. *Journal of Agricultural and Food Chemistry, 59*(12), 6550–6557. doi:10.1021/jf201383e PMID:21561162

Sun, B., Zhang, Y., Gong, Z., Zhang, J., & Zhang, J. (2021). Reducing substances-enhanced degradation of pollutants by permanganate: The role of in situ formed colloidal MnO_2. *Chemosphere, 276*, 130203. doi:10.1016/j.chemosphere.2021.130203 PMID:33725625

Sun, C., Chen, C., Ma, W., & Zhao, J. (2011). Photodegradation of organic pollutants catalyzed by iron species under visible light irradiation. *Physical Chemistry Chemical Physics, 13*(6), 1957–1969. doi:10.1039/C0CP01203C PMID:21082142

Suresh, R., Saravanan, R., Ponnusamy, S. K., Vo, D.-V., & Cornejo-Ponce, L. (2021). Recent advancements of spinel ferrite based binary nanocomposite photocatalysts in wastewater treatment. *Chemosphere, 274*, 129734. doi:10.1016/j.chemosphere.2021.129734 PMID:33548641

Swift, J. A., & Smith, J. R. (2001). Microscopical investigations on the epicuticle of mammalian keratin fibres. *Journal of Microscopy, 204*(3), 203–211. doi:10.1046/j.1365-2818.2001.00957.x PMID:11903797

Sychev, A., & Duca, G. G. (2002). *Fundamental and practical aspects of homogeneous catalysis of metal complexes.* Stiinta I.E.P.

Sychev, A., Travin, S. O., Duca, G. G., & Skurlatov, Iu. (1983). *Catalytical reactions and environmental protection.* Stiinta I.E.P.

Szejtli, J. (2004). Past, present, and future of cyclodextrin research. *Pure and Applied Chemistry, 76*(10), 1825–1845. doi:10.1351/pac200476101825

Szente, L., & Szejtli, J. (1996). Cyclodextrins in pesticides. In J. Szejtli & T. Osa (Eds.), *Comprehensive supramolecular chemistry* (Vol. 3, pp. 503–514). Pergamon.

Takeda, K., Takedoi, H., Yamaji, S., Ohta, K., & Sakugawa, H. (2004). Determination of hydroxil radical photoproduction rates in natural waters. *Analytical Sciences, 67*(1), 153–158. doi:10.2116/analsci.20.153 PMID:14753275

Tallawi, M., Rosellini, E., Barbani, N., Cascone, M. G., Rai, R., Saint-Pierre, G., & Boccaccini, A. R. (2015). Strategies for the chemical and biological functionalization of scaffolds for cardiac tissue engineering: A review. *Journal of the Royal Society, Interface, 12*(108), 20150254. doi:10.1098/rsif.2015.0254 PMID:26109634

Tamarkina, Yu.V., Bovan, L. A., Tsyba, N. N., & Kucherenko, V. A. (2008). Development of brown coal porosity during thermolysis with alkalis of different nature. *Chemistry, physics and surface technology, 14*, 224-229.

Tan, C. P., Che Man, Y. B., Selamat, J., & Yusoff, M. S. A. (2002). Comparative studies of oxidative stability of edible oils by differential scanning calorimetry and oxidative stability index methods. *Food Chemistry, 76*(3), 385–389. doi:10.1016/S0308-8146(01)00272-2

Tang, D., Hung, C.-C., Warnken, K. W., & Santschi, P. H. (2000). The distribution of biogenic thiols in surface waters of Galveston Bay. *Limnology and Oceanography, 45*(6), 1289–1297. doi:10.4319/lo.2000.45.6.1289

Tang, T., Teng, T., Hou, Y., Wang, Y., Tan, F., Qu, S., & Wang, Z. (2010). Optical properties and electrical bistability of CdS nanoparticles synthesized in dodecanethiol. *Applied Physics Letters, 96*(16), 163112–163113. doi:10.1063/1.3402770

Tang, W. Z., & Huang, C. P. (1995). Inhibitory effect of thioacetamide on CdS dissolution during photocatalytic oxidation of 2,4-dichlorophenol. *Chemosphere, 30*(7), 1385–1399. doi:10.1016/0045-6535(95)00029-8

Tang, W. Z., & Huang, C. P. (2016). 2,4-Dichlorophenol oxidation kinetics by Fenton's reagent. *Environmental Technology, 17*(12), 1371–1378. doi:10.1080/09593330.1996.9618465

Taranto, F., Pasqualone, A., Mangini, G., Tripodi, P., Miazzi, M. M., Pavan, S., & Montemurro, C. (2017). Polyphenol oxidases in crops: Biochemical, physiological and genetic aspects. *International Journal of Molecular Sciences, 18*(2), 377. doi:10.3390/ijms18020377 PMID:28208645

Tardy, B. L., Mattos, B. D., Otoni, C. G., Beaumont, M., Majoinen, J., Kämäräinen, T., & Rojas, O. J. (2021). Deconstruction and reassembly of renewable polymers and biocolloids into next generation structured materials. *Chemical Reviews, 121*(22), 14088–14188. doi:10.1021/acs.chemrev.0c01333 PMID:34415732

Tarko, T., Duda-Chodak, A., Sroka, P., & Siuta, M. (2020). The Impact of Oxygen at Various Stages of Vinification on the Chemical Composition and the Antioxidant and Sensory Properties of White and Red Wines. *International Journal of Food Sciences, 2020*, 1–11. doi:10.1155/2020/7902974 PMID:32309422

Tchounwou, P. B., Yedjou, C. G., Patlolla, A. K., & Sutton, D. J. (2012). *Heavy metal toxicity and the environment. EXS, 101, 133–164.* NIH Public. doi:10.1007/978-3-7643-8340-4_6

Tedesco, F., Siesto, G., Pietrafesa, R., Romano, P., Salvia, R., Scieuzo, C., Falabella, P., & Capece, A. (2022). Chemical Methods for Microbiological Control of Winemaking: An Overview of Current and Future Applications. *Beverages, 8*(3), 58. doi:10.3390/beverages8030058

Teixeira, A., Baenas, N., Dominguez-Perles, R., Barros, A., Rosa, E., Moreno, D., & Garcia-Viguera, C. (2014). Natural bioactive compounds from winery by-products as health promoters: A review. *International Journal of Molecular Sciences, 15*(9), 15638–15678. doi:10.3390/ijms150915638 PMID:25192288

Tejada, M., & Gonzalez, J. L. (2006). Effects of two beet vinasse forms on soil physical properties and soil loss. *Catena, 68*(1), 41–50. doi:10.1016/j.catena.2006.04.025

Thakur, A., Singh, S., Dulta, K., Singh, N., Ali, B., Hafeez, A., & Marc, R. A. (2022). Nutritional evaluation, phytochemical makeup, antibacterial and antioxidant properties of wild plants utilized as food by the gaddis-a tribal tribe in the western himalayas. *Frontiers in Agronomy, 4*, 1010309. doi:10.3389/fagro.2022.1010309

Titov, D. N., Ustyugov, A. V., Tkachenko, O. P., Kustov, L. M., Zubavichus, Y. V., Veligzhanin, A. A., & Temkin, O. N. (2012). State of active components on the surface of the $PdCl_2$-$CuCl_2/\gamma$-Al_2O_3 catalyst for the low-temperature oxidation of carbon monoxide. *Kinetics and Catalysis, 53*(2), 262–273. doi:10.1134/S0023158412020140

Tleuova, A. B., Wielogorska, E., Talluri, V. S. S. L. P., Štěpánek, F., Elliott, C. T., & Grigoriev, D. O. (2020). Recent advances and remaining barriers to producing novel formulations of fungicides for safe and sustainable agriculture. *Journal of Controlled Release, 326*, 468–481. doi:10.1016/j.jconrel.2020.07.035 PMID:32721524

Tofan, L., Paduraru, C., Sluser, B., & Teodosiu, C. (2019). *A small encyclopedia of water*. Ed. EcoZone. (in Romanian)

Torres, E., Marín, V., Aburto, J., Beltr´an, H.I., Shirai, K., Villanueva, S., & Sandoval, G. (2012) Enzymatic modification of chitosan with quercetin and its application as antioxidant edible films. *Appl. Biochem. Microbiol., 48*(12, 2), 175-182. . doi:10.1134/S0003683812020123

Trapaidze, V. (2012). *Water Recourse*. TSU. 8-12.

Travin, S., Duca, Gh., & Gladchi, V. (2019). *Self-purification of aquatic media from hexachlorocyclohexane in a radical process*. Chemistry Journal of Moldova. doi:10.19261/cjm.2018.537

Trincherini, P. R., Domergue, C., Manteca, I., Nesta, A., & Quarati, P. (2009). The identification of lead ingots from the Roman mines of Cartagena: The role of lead isotope analysis. *Journal of Roman Archaeology*, *22*(1), 123–145. doi:10.1017/S1047759400020626

Trubetskoi, O. A., Patsaeva, S. V., & Trubetskaya, O. E. (2019). Photochemical degradation of organic pollutants in solutions of soil humic acids. *Eurasian Soil Science*, *52*(9), 1075–1080. doi:10.1134/S1064229319090102

Trueman, C. (2004). Geochemistry - inorganic. In *Encyclopedia of Analytical Science* (2nd ed., pp. 171–181). Elsevier Inc. doi:10.1016/B0-12-369397-7/00240-5

Tsuneda, T. (2020). Fenton reaction mechanism generating no OH radicals in Nafion membrane decomposition. *Scientific Reports*, *10*(1), 18144. doi:10.103841598-020-74646-0 PMID:33097759

Tufail, A., Price, W. E., Mohseni, M., Pramanik, B. K., & Hai, F. I. (2020). A critical review of advanced oxidation processes for emerging trace organic contaminant degradation: Mechanisms, factors, degradation products, and effluent toxicity. *Journal of Water Process Engineering*, *40*, 101778. doi:10.1016/j.jwpe.2020.101778

Turchi, C. S., & Ollis, D. F. (1990). Photocatalytic Degradation of Organic Water Contaminants: Mechanisms Involving Hydroxyl Radical Attack. *Journal of Catalysis*, *122*(1), 178–192. doi:10.1016/0021-9517(90)90269-P

Tyagi, C., Sharma, A., & Kurchania, R. (2014). Synthesis of CdS Quantum dots using wet chemical co-precipitation method. *Journal of Non-Oxide Glasses*, *6*, 23–26.

Tynek, M., Pawłowicz, R., Gromadzka, J., Tylingo, R., Wardencki, W., & Karlovits, G. (2012). Virgin rapeseed oils obtained from different rape varieties by cold pressed method—Their characteristics, properties, and differences. *European Journal of Lipid Science and Technology*, *114*(3), 357–366. doi:10.1002/ejlt.201100296

Ulitzur, S., Lahav, T., & Ulitzur, N. (2002). A novel and sensitive test for rapid determination of water toxicity. *Environmental Toxicology*, *17*(3), 291–296. doi:10.1002/tox.10060 PMID:12112638

Ullah, J. M. H. T. A. M. A., &. M. Z. (2003). Effect of light, natural and synthetic antioxidants on stability of edible oil and fats. *Asian Journal of Plant Sciences*, *2*(17), 1192–1194. doi:10.3923/ajps.2003.1192.1194

Ünal, M. Ü., & Şener, A. (2016). Correlation between browning degree and composition of important Turkish white wine grape varieties. *Turkish Journal of Agriculture and Forestry*, *40*(1), 62–67. doi:10.3906/tar-1412-54

USEPA. (1978). EPA Method: 130.2: Hardness, total (mg/L as CaCO) (Titrimetric, EDTA). *Methods for the chemical analysis of water and wastes* (MCAWW) (EPA/600/4-79/020).

Vahterova, Y. V., Shtamm, E. V., Shvydky, V. O., & Saratovskikh, E. A. (2020). Investigation of the toxicity of the products of photochemical destruction of glyphosate by changing the intensity of bioluminescence of the bacterial test system "ecolum". *Vth International conference actual scientific & technical issues of chemical safety (ASTICS-2020)*. KAZAN.

Van Lier, J. B., Tilche, A., Ahring, B. K., Macaria, H., Moletta, R., Dohanyos, M., Hulstoff Pol, L. W., Lens, P., & Versaete, W. (2001). New perspectives in anaerobic digestion. *Water Science and Technology*, *43*(1), 1–18. doi:10.2166/wst.2001.0001 PMID:11379079

VanLoon, G. W., & Duffy, S. (2018). Environmental chemistry: A global perspective. Elsevier.

Varanko, A., Saha, S., & Chilkoti, A. (2020). Recent trends in protein and peptide-based biomaterials for advanced drug delivery. *Advanced Drug Delivery Reviews*, *156*, 133–187. Advance online publication. doi:10.1016/j.addr.2020.08.008 PMID:32871201

Vaseashta, A., Duca, G., Covaliova, O., & Romanciuc, L. (2022). Water Safety, Security, and Sustainability: Emerging Trends and Future Pathways. In A. Vaseashta, G. Duca, & S. Travin (Eds.), *Handbook of Research on Water Sciences and Society* (pp. 1–36). IGI Global. doi:10.4018/978-1-7998-7356-3.ch001

Vaseashta, A., Duca, G., & Travin, S. (Eds.). (2022). *Handbook of Research on Water Sciences and Society* (Vols. 1–2). IGI Global. doi:10.4018/978-1-7998-7356-3

Vaseashta, A., Gevorgyan, G., Kavaz, D., Ivanov, O., Jawaid, M., & Vasović, D. (2021). Exposome, Biomonitoring, Assessment and Data Analytics to Quantify Universal Water Quality. In A. Vaseashta & C. Maftei (Eds.), *Water Safety, Security and Sustainability. Advanced Sciences and Technologies for Security Applications*. Springer. doi:10.1007/978-3-030-76008-3_4

Vaughan, P., & Blough, N. (1998). Hydroxil radical photoproduction in the sea and its potential impact on marine processes. *Environmental Science & Technology, 32*, 2947. doi:10.1021/es9710417

Velasco, J., Andersen, M. L., & Skibsted, L. H. (2004). Evaluation of oxidative stability of vegetable oils by monitoring the tendency to radical formation. A comparison of electron spin resonance spectroscopy with the Rancimat method and differential scanning calorimetry. *Food Chemistry, 85*(4), 623–632. doi:10.1016/j.foodchem.2003.07.020

Velásquez-Riaño, M., Meneses-Sánchez, J. S., & Carvajal Arias, C. E. (2019). Evaluation of acute toxicity of vinasse by means of *Daphnia magna* and *Aliivibrio fischeri*: A comparative study. *Interdisciplinary Toxicology, 2*(3), 143–148. doi:10.2478/intox-2019-0017 PMID:32210703

Vergel, K., Zinicovscaia, I., Yushin, N., Chaligava, O., Nekhoroshkov, P., & Grozdov, D. (2022). Moss biomonitoring of atmospheric pollution with trace elements in the Moscow region, Russia. *Toxics, 10*(2), 66. Advance online publication. doi:10.3390/toxics10020066 PMID:35202252

Vergel, K., Zinicovscaia, I., Yushin, N., & Gundorina, S. (2020). Assessment of atmospheric deposition in Central Russia using moss biomonitors, neutron activation analysis and GIS technologies. *Journal of Radioanalytical and Nuclear Chemistry, 325*(3), 807–816. doi:10.100710967-020-07234-1

Verlackt, C. C. W., Van Boxem, W., Dewaele, D., Lemière, F., Sobott, F., Benedikt, J., Neyts, E. C., & Bogaerts, A. (2017). Mechanisms of peptide oxidation by hydroxyl radicals: Insight at the molecular scale. *The Journal of Physical Chemistry C, 121*(10), 5787–5799. doi:10.1021/acs.jpcc.6b12278

Verlee, A., Mincke, S., & Stevens, C. V. (2017). Recent developments in antibacterial and antifungal chitosan and its derivatives. *Carbohydrate Polymers, 164*, 268–283. doi:10.1016/j.carbpol.2017.02.001 PMID:28325326

Vignault, A., Pascual, O., Jourdes, M., Moine, V., Fermaud, M., Roudet, J., Canals, J. M., Teissedre, P. L., & Zamora, F. (2019). Impact of enological tannins on laccase activity. *OENO One, 53*(1). doi:10.20870/oeno-one.2019.53.1.2361

Vincoli, J. W. (1997). *Risk Management for Hazardous Chemicals, 2*. CRC Press.

Vione, D. (2016). Photochemical reactions in sunlit surface waters. *Applied Photochemistry, 92*, 343–376. doi:10.1007/978-3-319-31671-0_7

Vione, D., Maurino, V., & Minero, C. (2006). Sources and sinks of hydroxil radicals upon irradiation of natural water samples. *Environmental Science & Technology, 40*(12), 3775–3781. doi:10.1021/es052206b PMID:16830541

Vione, D., & Scozzaro, A. (2019). Photochemistry of surface fresh waters in the framework of climate change. *Environmental Science & Technology, 53*(14), 7945–7963. doi:10.1021/acs.est.9b00968 PMID:31241909

Visioli, F., Panaite, S. A., & Tomé-Carneiro, J. (2020). Wine's phenolic compounds and health: A pythagorean view. *Molecules (Basel, Switzerland), 25*(18), 4105. doi:10.3390/molecules25184105 PMID:32911765

Von Bergen, W. (1963). *Wool Handbook* (3rd ed., Vol. 1). John Wiley & Sons.

Wagner, R. (1976). The COD method in Wastewater Charges Act (AbwAG). [in German]. *Vom Wasser, 76*, 139–159.

Walker, G. W., Kookana, R. S., Smith, N. E., Kah, M., Doolette, C. L., Reeves, P. T., Lovell, W., Anderson, D. J., Turney, T. W., & Navarro, D. A. (2018). Ecological risk assessment of nano-enabled pesticides: A perspective on problem formulation. *Journal of Agricultural and Food Chemistry, 66*(26), 6480–6486. doi:10.1021/acs.jafc.7b02373 PMID:28812885

Wang, F. (2020). *Hydrogen peroxide and antioxidant enzymes moderate interaction of the carbon and oxygen cycles at the redoxcline* [Doctoral dissertation, University of South Carolina]. https://scholarcommons.sc.edu/etd/5982

Wang, H., Zhou, H., Ma, J., Nie, J., Yan, S., & Song, W. (2020). Triplet photochemistry of dissolved black carbon and its effects on the photochemical formation of reactive oxygen species. *Environmental Science & Technology, 54*(8), 4903–4911. doi:10.1021/acs.est.0c00061 PMID:32227921

Wang, L., Chen, Y. C., & Rojanasakul, Y. (2014). Protein nanoparticles as drug delivery carriers for cancer therapy. *BioMed Research International, 2014*, 180549. PMID:24772414

Wang, L., Luo, Z., Hong, Y., Chelme-Ayala, P., Meng, L., Wu, Z., & El-Din, M. G. (2022). The treatment of electroplating wastewater using an integrated approach of interior micro electrolysis and Fenton combined with recycle ferrite. *Chemosphere, 286*(1), 131543. doi:10.1016/j.chemosphere.2021.131543 PMID:34303045

Wang, M., Xiang, X., Zuo, Y., Peng, J., Lu, K., Dempsey, C., & Gao, S. (2020). Singlet oxygen production abilities of oxidated aromatic compounds in natural water. *Chemosphere, 258*, 127308. doi:10.1016/j.chemosphere.2020.127308 PMID:32535450

Wang, N., Ji, Y., Zhu, Y., Wu, X., Mei, L., Zhang, H., Deng, J., & Wang, S. (2020). Antibacterial effect of chitosan and its derivative on *enterococcus faecalis* associated with endodontic infection. *Experimental and Therapeutic Medicine, 19*, 3805–3813. doi:10.3892/etm.2020.8656 PMID:32346445

Wang, S., Chen, Z. H., Ma, W. J., & Ma, Q. S. (2006). Influence of heat treatment on physical–chemical properties of PAN-based carbon fiber. *Ceramics International, 32*(3), 291–295. doi:10.1016/j.ceramint.2005.02.014

Wang, T., Wang, S.-P., Zhong, X.-Z., Sun, Z.-Y., Huang, Y.-L., Tan, L., Tang, Y.-Q., & Kida, K. (2017). Converting digested residue eluted from dry anaerobic digestion of distilled grain waste into value-added fertilizer by aerobic composting. *Journal of Cleaner Production, 166*, 530–536. doi:10.1016/j.jclepro.2017.08.075

Wang, W., Gao, Y., Wang, W., Zhang, J., Yin, J., Le, T., Xue, J., Engelhardt, U. H., & Jiang, H. (2022). Kojic Acid Showed Consistent Inhibitory Activity on Tyrosinase from Mushroom and in Cultured B16F10 Cells Compared with Arbutins. *Antioxidants, 11*(3), 502. doi:10.3390/antiox11030502 PMID:35326152

Wang, X., Jing, J., Zhou, M., & Dewil, R. (2022). (in press). Recent advances in H_2O_2-based advanced oxidation processes for removal of antibiotics from wastewater. *Chinese Chemical Letters*. doi:10.1016/j.cclet.2022.06.044

Wang, X., Shen, X., & Xu, W. (2012). Effect of hydrogen peroxide treatment on the properties of wool fabric. *Applied Surface Science, 258*(24), 10012–10016. doi:10.1016/j.apsusc.2012.06.065

Wang, Y., Zheng, K., Guo, H., Tian, L., He, Y., Wang, X., Zhu, T., Sun, P., & Liu, Y. (2023). Potassium permanganate-based advanced oxidation processes for wastewater decontamination and sludge treatment: A review. *Chemical Engineering Journal, 452*, 139529. doi:10.1016/j.cej.2022.139529

Wąsowicz, E., Gramza, A., Hęś, M., Jeleń, H. H., Korczak, J., Małecka, M., Mildner-Szkudlarz, S., Rudzińska, M., Samotyja, U., & Zawirska-Wojtasiak, R. (2004). Oxidation of lipids in food. *Polish Journal of Food and Nutrition Sciences, 54*(1s), 87–100.

Waterhouse, A. L., & Laurie, V. F. (2006). Oxidation of wine phenolics: A critical evaluation and hypotheses. *American Journal of Enology and Viticulture*, *57*(3), 306–313. doi:10.5344/ajev.2006.57.3.306

Wei, W., Guo, K., Kang, X., Zhang, J., Li, C., & Fang, J. (2021). Fang, J., Complete Removal of Organoarsenic by the UV/Permanganate Process via HO• Oxidation and in Situ- Formed Manganese Dioxide Adsorption. *ACS ES&T Engineering*, *1*(4), 794–803. doi:10.1021/acsestengg.1c00004

Weng, H., Yang, Y., Zhang, C., Cheng, M., Wang, W., Song, B., Luo, H., Qin, D., Huang, C., Qin, F., & Li, K. (2023). Insight into FeOOH-mediated advanced oxidation processes for the treatment of organic polluted wastewater. *Chemical Engineering Journal*, *453*, 139812. doi:10.1016/j.cej.2022.139812

Westland, S. (2015). CIE Whiteness. Encyclopedia of Color Science and Technology, 1–5. doi:10.1007/978-3-642-27851-8_5-1

Weston, G. J. (1955). The infra-red spectrum of peracetic acid-treated wool. *Biochimica et Biophysica Acta*, *17*, 462–464. doi:10.1016/0006-3002(55)90407-8 PMID:13249992

Williams, I. (2014). Environmental chemistry, 157-159. Wiley.

Wilschefski, S. C., & Baxter, M. R. (2019). Inductively coupled plasma mass spectrometry: Introduction to analytical aspects. *The Clinical Biochemist. Reviews / Australian Association of Clinical Biochemists*, *40*(3), 115–133. doi:10.33176/AACB-19-00024 PMID:31530963

Winder, L. M., Rea, A., Baronian, K., Webber, J., & Muller, B. (1998). Unravelling the causes of wool yellowing. Part II: Involvement of bacteria. *Proceedings of the New Zealand Society of Animal Production*, *58*, 277–280.

Witkowska, E., Szczepaniak, K., & Biziuk, M. (2005). Some applications of neutron activation analysis: A review. *Journal of Radioanalytical and Nuclear Chemistry*, *265*(1), 141–150. doi:10.100710967-005-0799-1

Wojciechowska, E., Włochowicz, A., & Wesełucha-Birczyńska, A. (2000). Application of Fourier-transform infrared and Raman spectroscopy to the study of the influence of orthosilicic acid on the structure of wool fibre. *Journal of Molecular Structure*, *555*(1-3), 397–406. doi:10.1016/S0022-2860(00)00626-8

Wolfram, L., Hall, K., & Hui, I. (1970). The mechanism of hair bleaching. *Journal of the Society of Cosmetic Chemists*, *21*, 875–900.

Wong, G. T. F., Dunstan, W. M., & Kim, D. B. (2003). The decomposition of hydrogen peroxide by marine phytoplankton. *Oceanologica Acta*, *26*(2), 191–198. doi:10.1016/S0399-1784(02)00006-3

Woodard & Curran, Inc. (2006). Methods for treating wastewater from industry. *Industrial Waste Treatment Handbook*, 149–334. doi:10.1016/B978-075067963-3/50009-6

Woodhead, A., Harrigan, F., & Church, J. (1997). Assessment of wool chlorination by infrared spectroscopy. *Vibrational Spectroscopy*, *15*(2), 179–189. doi:10.1016/S0924-2031(97)00040-4

Worrall, E. A., Hamid, A., Mody, K. T., Mitter, N., & Pappu, H. R. (2018). Nanotechnology for Plant Disease Management. *Agronomy (Basel)*, *8*(12), 285. doi:10.3390/agronomy8120285

Wszelaki, A. L., Delwiche, J. F., Walker, S. D., Liggett, R. E., Miller, S. A., & Kleinhenz, M. D. (2005). Consumer liking and descriptive analysis of six varieties of organically grown edamame-type soybean. *Food Quality and Preference*, *16*(8), 651–658. doi:10.1016/j.foodqual.2005.02.001

Wu, L., Pan, H., Huang, W., Wang, M., Hu, Z., & Zhang, F. (2022). Self-assembled degradable iron-doped mesoporous silica nanoparticles for the smart delivery of prochloraz to improve plant protection and reduce environmental impact. *Environmental Technology & Innovation*, *28*, 102890. doi:10.1016/j.eti.2022.102890

Wu, M., Mintz, M., Wang, M., & Arora, S. (2009). Water consumption in the production of ethanol and petroleum gasoline. *Environmental Management*, *44*(5), 981–997. doi:10.100700267-009-9370-0 PMID:19774326

Wu, Z.-Y., Zhu, W.-P., Liu, Y., Zhou, L.-L., Liu, P.-X. J., & Xu, J. (2021). An integrated biological-electrocatalytic process for highly efficient treatment of coking wastewater. *Bioresource Technology*, *339*, 125584. doi:10.1016/j.biortech.2021.125584 PMID:34303099

Xinhua, L., Shunbing, Z., Xizeng, Z., & Xiaoli, W. (2014). Hazard assessment of hydrogen peroxide with polyphosphonic acid by vent sizing package 2. *Procedia Engineering*, *84*, 427–435. doi:10.1016/j.proeng.2014.10.453

Xu, L., Yu, X., Li, M., Chen, J., & Wang, X. (2017). Monitoring oxidative stability and changes in key volatile compounds in edible oils during ambient storage through HS-SPME/GC–MS. *International Journal of Food Properties*, *20*(sup3), S2926–S2938. doi:10.1080/10942912.2017.1382510

Xu, C., Cao, L., Bilal, M., Cao, C., Zhao, P., Zhang, H., & Huang, Q. (2021). Multifunctional manganese-based carboxymethyl chitosan hydrogels for pH-triggered pesticide release and enhanced fungicidal activity. *Carbohydrate Polymers*, *262*, 117933. doi:10.1016/j.carbpol.2021.117933 PMID:33838810

Xu, H., Li, Y., Liu, J., Du, H., Du, Y., Su, Y., & Jiang, H. (2020). Photogeneration and steady-state concentration of hydroxyl radical in river and lake waters along middle-lower Yangtze region, China. *Water Research*, *176*, 115774. . doi:10.1016/j.watres.2020.115774 PMID:32272319

Xu, K., Dong, H., Li, M., & Qiang, Z. (2018). Quinone group enhances the degradation of levofloxacin by aqueous permanganate: Kinetics and mechanism. *Water Research*, *143*, 109–116. doi:10.1016/j.watres.2018.06.026 PMID:29940356

Xu, X.-R., Li, X.-Y., Li, X.-Z., & Li, H.-B. (2009). Degradation of melatonin by UV, UV/H_2O_2, Fe^{2+}/H_2O_2 and UV/Fe^{2+}/H_2O_2 processes. *Separation and Purification Technology*, *68*(2), 261–266. doi:10.1016/j.seppur.2009.05.013

Yakovlev, S.V., Volkov, L.S., Voronov, Yu.V. (1999). *Treatment and utilization of sludge from industrial waste water.* (Rus.). M.: Chemistry.

Yang, H., Huang, C., Li, X., Shi, R., & Zhang, K. (2005). Luminescent and photocatalytic properties of cadmium sulfide nanoparticles synthesized via microwave irradiation. *Materials Chemistry and Physics*, *90*(1), 155–158. doi:10.1016/j.matchemphys.2004.10.028

Yan, Y., Hou, H., Ren, T., Xu, Y., Wang, Q., & Xu, W. (2013). Utilization of environmental waste cyanobacteria as a pesticide carrier: Studies on controlled release and photostability of avermectin. *Colloids and Surfaces. B, Biointerfaces*, *102*, 341–347. doi:10.1016/j.colsurfb.2012.07.043 PMID:23006573

Yayintas, O. T., Yılmaz, S., Turkoglu, M., & Dilgin, Y. (2007). Determination of heavy metal pollution with environmental physicochemical parameters in waste water of Kocabas Stream (Biga, Canakkale, Turkey) by ICP-AES. *Environmental Monitoring and Assessment*, *127*(1–3), 389–397. doi:10.100710661-006-9288-4 PMID:16897498

Yilmazer, D., & Kanik, M. (2009). Bleaching of wool with sodium borohydride. *Journal of Engineered Fibers and Fabrics*, *4*(3). doi:10.1177/155892500900400305

Yin, R., Chen, Y., Hu, J., Lu, G., Zeng, L., Choi, W., & Zhu, M. (2021). Complexes of Fe(III)-organic pollutants that directly activate Fenton-like processes under visible light. *Applied Catalysis B: Environmental*, *283*, 119663. doi:10.1016/j.apcatb.2020.119663

Yoon, H., Kim, H. C., & Kim, S. (2021). Long-term seasonal and temporal changes of hydrogen peroxide from cyanobacterial blooms in fresh waters. *Journal of Environmental Management, 298*, 113515. doi:10.1016/j.jenvman.2021.113515 PMID:34403920

Younes, I., & Rinaudo, M. (2015). Chitin and chitosan preparation from marine sources. structure, properties and applications. *Marine Drugs, 13*(3), 1133–1174. doi:10.3390/md13031133 PMID:25738328

Young, A., & Lowe, G. (2018). Carotenoids—Antioxidant Properties. *Antioxidants, 7*(2), 28. doi:10.3390/antiox7020028 PMID:29439455

Yuan, S., Liu, X., Liao, W., Zhang, P., Wang, X., & Tong, M. (2018). Mechanisms of electron transfer from structrual Fe(II) in reduced nontronite to oxygen for production of hydroxyl radicals. *Geochimica et Cosmochimica Acta, 223*, 422–436. doi:10.1016/j.gca.2017.12.025

Yucel, E., Yucel, Y. Durak, M. (2016). *Process optimization for window material CdS thin films grown by a successive ionic layer adsorption and reaction method using response surface methodology.* Research Gate, 530-537.

Yu, J., & Ahmedna, M. (2013). Functional components of grape pomace: Their composition, biological properties and potential applications. *International Journal of Food Science & Technology, 48*(2), 221–237. doi:10.1111/j.1365-2621.2012.03197.x

Yu, M., Yao, J., Liang, J., Zeng, Z., Cui, B., Zhao, X., Sun, Ch., Wang, Y., Liu, G., & Cui, H. (2017). Development of functionalized abamectin poly(lactic acid) nanoparticles with regulatable adhesion to enhance foliar retention. *RSC Advances, 7*(19), 11271–11280. doi:10.1039/C6RA27345A

Yun, J.-M., & Surh, J.-H. (2012). Fatty acid composition as a predictor for the oxidation stability of Korean vegetable oils with or without induced oxidative stress. *Preventive Nutrition and Food Science, 17*(2), 158–165. doi:10.3746/pnf.2012.17.2.158 PMID:24471078

Zaim, N., Dogan, C., & Camtakan, Z. (2016). Neutron activation analysis of soil samples from different parts of edirne in Turkey. *Journal of Applied Spectroscopy, 83*(2), 271–276. doi:10.100710812-016-0280-7

Zaitseva, N. I., Baikova, I. S., Shvydkii, V. O., Borodulin, R. R., Shtamm, E. V., & Skurlatov, Yu. I. (2009). Kinetics of the Oxidation of Model Toxicants during Photolysis of Nitrites and Nitrates in an Aqueous Medium. ISSN 1990-7931, *Russian Journal of Physical Chemistry B, 3*(2), 185–190. doi:10.1134/S1990793109020031

Zakharov V.M. and Clark D.M. (1993). *Biotest: integral assessment of ecosystems and individual species health.* Moscow department of the International Foundation “Biotest.”

Zebib, H., Bultosa, G., & Abera, S. (2015). Physico-chemical Properties of Sesame (<i>Sesamum indicum</i> L.) Varieties Grown in Northern Area, Ethiopia. *Agricultural Sciences, 06*(02), 238–246. doi:10.4236/as.2015.62024

Zehiroglu, C., & Sarikaya, S. B. (2019). The importance of antioxidants and place in today’s scientific and technological studies. *Journal of Food Science and Technology, 56*(11), 4757–4774. doi:10.100713197-019-03952-x PMID:31741500

Zemskova, L. A., & Sheveleva, I. V. (2004). Modified sorption-active carbon fiber materials. *Russian chemical journal, 48*(5), 53-57.

Zepp, R.G., Cline, D.M. (1977). Rates of direct photolysis in aquatic environments. *Environ. Sci. Technol., V*(11), 359.

Zhang, L., Peng, X., Zhong, L., Chua, W., Xiang, Z., & Sun, R. (2019). Lignocellulosic biomass derived functional materials: Synthesis and applications in biomedical engineering. *Current Medicinal Chemistry, 26*(14), 2456–2474. doi:10.2174/0929867324666170918122125 PMID:28925867

Zhang, X., Tang, X., Zhao, C., Yuan, Z., Zhang, D., Zhao, H., Yang, N., Guo, K., He, Y., He, Y., Hu, J., He, L., He, L., & Qian, K. (2022). A pH-responsive MOF for site-specific delivery of fungicide to control citrus disease of *Botrytis cinerea*. *Chemical Engineering Journal*, *431*, 133351. doi:10.1016/j.cej.2021.133351

Zhang, X., Xu, Z., Wu, M., Qian, X., Lin, D., Zhang, H., Tang, J., Zeng, T., Yao, W., Filser, J., Li, L., & Sharma, V. K. (2019). Potential environmental risks of nanopesticides: Application of $Cu(OH)_2$ nanopesticides to soil mitigates the degradation of neonicotinoid thiacloprid. *Environment International*, *129*, 42–50. doi:10.1016/j.envint.2019.05.022 PMID:31108392

Zhao, M., Chen, Z., Hao, L., Chen, H., Zhou, X., & Zhou, H. (2023). CMC based microcapsules for smart delivery of pesticides with reduced risks to the environment. *Carbohydrate Polymers*, *300*, 120260, doi:10.1016/j.carbpol.2021.117749 PMID:33673979

Zhao, P., Cao, L., Ma, D., Zhou, Z., Huang, Q., & Pan, C. (2018). Translocation, distribution and degradation of prochloraz-loaded mesoporous silica nanoparticles in cucumber plants. *Nanoscale*, *10*(4), 1798–1806. doi:10.1039/C7NR08107C PMID:29308814

Zhao, R., Torley, P., & Halley, P. J. (2008). Emerging biodegradable materials: Starch- and proteinbased bio-nanocomposites. *Journal of Materials Science*, *43*(9), 3058–3071. doi:10.100710853-007-2434-8

Zhou, X., Chen, Q., Liu, C., & Fang, Y. (2017). Using moss to assess airborne heavy metal pollution in Taizhou, China. *International Journal of Environmental Research and Public Health*, *14*(4), 430. Advance online publication. doi:10.3390/ijerph14040430 PMID:28420186

Zhu, J., He, Y., Gao, M., Zhou, W., Hu, Y., Shen, J., & Zhu, Y. Ch. (2011). Photodegradation of emamectin benzoate and its influence on efficacy against the rice stem borer, *Chilo suppressalis*. *Crop Protection (Guildford, Surrey)*, *30*(10), 1356–1362. doi:10.1016/j.cropro.2011.06.006

Zielinski, Z. A. M., & Pratt, D. A. (2017). Lipid Peroxidation: Kinetics, Mechanisms, and Products. *The Journal of Organic Chemistry*, *82*(6), 2817–2825. doi:10.1021/acs.joc.7b00152 PMID:28248497

Ziembowicz, S., & Kida, M. (2022). Limitations and future directions of application of the Fenton-like process in micro-pollutants degradation in water and wastewater treatment: A critical review. *Chemosphere*, *296*, 134041. doi:10.1016/j.chemosphere.2022.134041 PMID:35189198

Zinicovscaia, I., Cepoi, L., Povar, I., Chiriac, T., Rodlovskaya, E., & Culicov, O. A. (2018). Metal Uptake from complex industrial effluent by *Cyanobacteria Arthrospira* platensis. *Water, Air, and Soil Pollution*, *229*(7), 1–10. doi:10.100711270-018-3873-3

Zinicovscaia, I., Chaligava, O., Yushin, N., Grozdov, D., Vergel, K., & Hramco, C. (2022). Moss biomonitoring of atmospheric trace element pollution in the Republic of Moldova. *Archives of Environmental Contamination and Toxicology*, *82*(3), 355–366. doi:10.100700244-022-00918-7 PMID:35266044

Zinicovscaia, I., Hramco, C., Chaligava, O., Yushin, N., Grozdov, D., Vergel, K., & Duca, G. (2021). Accumulation of potentially toxic elements in mosses collected in the Republic of Moldova. *Plants*, *10*(3), 1–13. doi:10.3390/plants10030471 PMID:33801412

Zinicovscaia, I., Sturza, R., Duliu, O., Grozdov, D., Gundorina, S., Ghendov-Mosanu, A., & Duca, G. (2020). Major and trace elements in moldavian orchard soil and fruits: Assessment of anthropogenic contamination. *International Journal of Environmental Research and Public Health*, *17*(19), 1–19. doi:10.3390/ijerph17197112 PMID:32998367

Zinicovscaia, I., Yushin, N., Grozdov, D., Vergel, K., Nekhoroshkov, P., & Rodlovskaya, E. (2021). Treatment of rhenium-containing effluents using environmentally friendly sorbent, saccharomyces cerevisiae biomass. *Materials (Basel), 14*(16), 4763. Advance online publication. doi:10.3390/ma14164763 PMID:34443283

Zinicovscaia, I., Yushin, N., Grozdov, D., Vergel, K., Popova, N., Artemiev, G., & Safonov, A. (2020). Metal removal from nickel-containing effluents using mineral–organic hybrid adsorbent. *Materials (Basel), 13*(19), 1–21. doi:10.3390/ma13194462 PMID:33050087

Zinicovscaia, I., Yushin, N., Humelnicu, D., Grozdov, D., Ignat, M., Demcak, S., & Humelnicu, I. (2021). Sorption of ce(Iii) by silica sba-15 and titanosilicate ets-10 from aqueous solution. *Water (Basel), 13*(22), 3263. doi:10.3390/w13223263

Zironi, R., Comuzzo, P., Tat, L., Scobioala, S. (2010). Oxygen and wine. *Internet Journal of Viticulture and Enology, 3*(2).

Zoccola, M., Montarsolo, A., Mossotti, R., Patrucco, A., & Tonin, C. (2015). Green hydrolysis is an emerging technology to turn wool waste into organic nitrogen fertilizer. *Waste and Biomass Valorization, 6*(5), 891–897. doi:10.100712649-015-9393-0

Zubaidah, T., Karnaningroem, N., & Slamet, A. (2019). The self-purification ability in the rivers of Banjarmasin, Indonesia. *Journal of Ecological Engineering, 20*(2), 177–182. doi:10.12911/22998993/97286

About the Contributors

Ashok Vaseashta is CEO/CTO and Executive Director for research with the International Clean Water Institute in Virginia, USA. He also serves in several honorary positions, such as a Professor at the Transylvania University of Brasov; Chaired Professor of Nanotechnology at the Ghitu Institute of Electrical Engineering and Nanotechnologies, Academy of Sciences of Moldova; Academician at the Euro-Mediterranean Academy of Arts and Sciences, member of CIRET, France, and Senior Strategic Research Advisor for several organizations. Inspired by nature and guided by societal necessities, he strives for technological innovations to address the global challenges of the 21st century. His research interests include critical infrastructure safety, security, and environmental sustainability - all using the nexus of advanced technological solution platforms. He is a scholar, visionary, strategist, and dedicated futurist providing strategic leadership to promote and advance research initiatives and priorities using data-driven decisions. He received Ph.D. from the Virginia Polytechnic Institute and State University, Blacksburg, VA in 1990 followed by Kobe's post-doctoral fellowship. Following his Ph.D., he served as a professor and researcher at Virginia Tech and Marshall University. He also served as the Director of Research at the Institute for Advanced Sciences Convergence and International Clean Water Institute for Norwich University Applied Research Institutes, Vice-Provost (Rector) for Research in South Carolina, and Executive Director and Chair of the Institutional Review Board at a State University in New Jersey. His earlier honorary positions included a Visiting Professor at the Riga Technical University, Latvia, 3 Nano-SAE Research Centre, University of Bucharest, Romania, and Visiting Scientist at the Weizmann Institute of Science, Israel. He served the U.S. Department of State in two rotations, as a strategic S&T advisor in the Bureau of International Security and Nonproliferation, Office of Weapons of Mass Destruction and Terrorism, and U.S. diplomat. His research interests span foresight, nanotechnology, environmental/ecological science, and critical infrastructure safety and security. He is the author/editor of 14 books and has published over 350 articles in scientific journals, book chapters, and conferences. He serves on the editorial boards of several international journals and is an active member of various professional organizations.

* * *

Müfit Bahadir (1947), Univ.-Prof. Dr. rer. nat. Dr. agr. habil. Dr. h.c., studied chemistry at the Free University Berlin and Bonn University in Germany, received his PhD 1975 from Bonn University and his Assoc. Prof. from Munich Technical University. He became a full Professor of Environmental and Sustainable Chemistry at the Braunschweig Technical University in 1989. Since 1997, he is an honorary doctor and visiting professor at the Selcuk and Necmetting Erbakan Universities, Konya, Turkey. His research fields cover environmental chemistry and analyses, environmental pollution through industrial processes and products, pesticide chemistry and metabolism in soil and water, ecotoxicology, sustainable chemistry, renewable feed stocks, biodiesel and bio-lubricants. Since 2010, his research focused on R&D of sustainable water and wastewater management in developing countries. Müfit Bahadir chaired several international networks like MESAEP (Mediterranean Scientific Association for Environmental Protection), the Transformation Partnership Project "EMPOWER Tunisia – Emerging Pollutants in Water and Wastewater in Tunisia", and the worldwide network for capacity building "EXCEED – Excellence Center for Development Cooperation – Sustainable Water Management in Developing Countries" that he founded with 40 partner universities and research centers from 20 countries on 4 continents. He was also visiting or guest professor at several universities: Selçuk University, Konya/Turkey; Al-Farabi University, Almaty/Kazakhstan; Gadjah Mada University, Yogyakarta/ Indonesia; Jordan University, Amman/Jordan; Universidad Federal de Santa Maria, Santa Maria RS/Brazil; Cairo University, Giza/Egypt. Müfit Bahadir holds several international awards for his scientific merits, i. a., Honorary Doctorate of Selçuk University Konya/Turkey; Honorary Medal of Mediterranean Scientific Association of Environmental Protection; Election to Full Member of Braunschweig Society of Sciences (Academy); UNESCO Award for Education for Sustainable Development 2006/2007 and 2011/2012; International Award of Applied Sciences, Turkey; Environmental Sciences Award, Akdeniz University Antalya/Turkey; Honorary Citizen Award, Konya/Turkey; and Honorary Member of Turkish Academy of Sciences (TÜBA), Ankara/Turkey. Müfit Bahadir retired 2016 from the Technical University Braunschweig, Braunschweig/Germany.

Vladislav Blonschi, PhD, university lecturer, Department of Industrial and Ecological Chemistry "Academician Gheorghe Duca", Faculty of Chemistry and Chemical Technology, State University of Moldova. The doctoral thesis in chemical sciences, the specialty Ecological chemistry of water was supported in 2021. Fields of scientific interest: ecological chemistry of waters, photochemical processes in the environment, processes of self-purification of the environment, the hydrochemical circuit of the elements.

Ruslan Borodaey began in 1995 both scientific research and teaching activity at Industrial and Ecological Chemistry Department in the frame of Faculty of Chemistry and Chemical Technology, Moldova State University Scientific activity: Metal chemistry of natural waters; involvement of metals with transition valence in self-purification processes of natural waters; methods of extracting iron and copper from aqueous solutions (including natural waters); quality assessment of surface and ground waters as supply sources for drinking water preparation. Didactic activity (Lecturer of the courses): Chemical experiment technique; Unitary operations in the chemical industry, Chemical technologies, Inorganic chemical technology, Processes and devices in chemical and environmental engineering. More than 80 scientific publications and Laureate of the Republican Prize for Youth, for the cycle of scientific works "Analysis and transformation of toxic substances in the environment and estimation of health risk" (2002).

Elena Bunduchi graduated from Moldova State University, Faculty of Chemistry, in june 1996. She holds a PhD in "Redox-Catalytic transformations of hydroquinone and glyoxalic acid in aquatic environments" in 2010 at Moldova State University. From 1997 she began her teaching activity at the Department of Industrial and Ecological Chemistry at Moldova State University, becoming associate professor in 2014. Currently she is Chief of Department of Industrial and Ecological Chemistry „Acad. Gh. Duca" at Faculty of Chemistry and Chemical Technology at Moldova State University. She's research focuses on the natural waters self-purification process assured through oxygen active forms (H2O2, OH radicals) and their role in natural waters as a living environment for hydrobiontites.

Elena Cheremnykh, Ph.D., Senior Researcher, lab. Biochemistry FSBI Scientific Center for Mental Health. The author of the method and instrument for assessing toxicity according to the reactions of the protozoa - Paramecium caudatum and Tetrahymena pyriformis. The Biolat device is used in agriculture to assess feed toxicity. Scientific interests — toxicology of compounds with antiprotozoal activity, human and animal blood complement system, automation of biological research.

Ecaterina Covaci is a doctor in chemical sciences, senior scientific researcher and Associiate Professor at the Techical University of Moldova, Department of Oenology and Chemistry. She has a scientific background in physical chemistry She is graduate from the Technical University of Moldova. Her research interests involve management and using of secondary winery wastes, wine oxidability and stabilisation depending on different factors.(ORCID: 0000-0002-8108-4810)

Victor Covaliov, Professor, Doctor in chemical sciences, coordinating scientific researcher at the Moldova State University. He had a scientific background in physical chemistry (electrochemistry) and ecological and environmental chemistry. The main field of scientific interest is the research of waste water treatment by physico-chemical and biochemical methods, chemical and photochemical transformation of water pollutants, redox processes in the aquatic environment, water quality, solid and liquid wastes management, integrated treatment technologies and reactors. He is graduate from the Moldova State University. He is author of over 350 invention patents, 12 monographs and over 500 scientific publications.(ORCID: 0000-0002-0941-3050)

Olga Covaliova has a doctor in chemical sciences, coordinating scientific researcher at the Moldova State University/ Institute of Chemistry, Chisinau, Republic of Moldova. She has a scientific background in physical chemistry (electrochemistry) and ecological and environmental chemistry. The main field of scientific interest is focused on the research of waste water treatment by physico-chemical and biochemical methods, chemical and photochemical transformation of water pollutants, redox processes in the aquatic environment, water quality, solid and liquid wastes management. She is a graduate from the State University of Moldova. She is author of over 300 scientific publications and invention patents. (ORCID: 0000-0002-0387-3195)

Igor Cretescu, f-Director of Environmental Engineering and Management Department (TUIASI) and Vice-president of the Romanian Society of Inventors, has a PhD in Chemical Engineering (2000, TUIASI, Romania) and a Postdoc in Electrochemical and Bio-Sensors (2002). He is involved in more than 60 research projects, developed in the frame of national and international research grants, and contracts with industrial companies. His scientific expertise and research directions include analysis, control and monitoring of environmental components, process engineering and diagnosis, treatment of water and gaseous fluxes, eco-technologies, innovative system design, modelling and optimization of technological processes, applied electrochemistry, etc. He holds awards in innovation, with more than 70 awarded patents (diploma and medals) at national and international Inventions Fairs, and also provides consultancy, information, knowledge and technology transfer. • Scientific portfolio: (Scopus Author ID: 6602569674; Web of Science Researcher ID: GVU-4057-2022; ORCID ID:): 220 titles (scientific articles, patents, books), including 180 papers in ISI Journals Master List (Clarivate), 25 patents, 15 books/chapters. Hirsch Index (h) = 24, based on more than 2500 citations.

Alim Ennan, professor, works as a Director Physical-Chemical Institute for Environment and Human Protection of MES Ukraine and NAS Ukraine. His research interests are physicochemical properties of aerosols, engineering of air purification from aerosols and gaseous toxicants. His current research interests are Development of sorption-filtering materials and low-temperature catalysts for neutralization of toxic gaseous compounds, means of individual respiratory protection based on them; development of theoretical bases of processes of capture and utilization of welding aerosol; development of theoretical bases of stabilization of an ecological condition and recreation of the Kuyalnitsky estuary, and also measures for protection, reproduction and rational use of its resources in the interests of Odessa and the Odessa region.

Aliona Ghendov-Mosanu is an associate professor at the TUM, Faculty of Food Technology. She holds a bachelor's degree in baking technology, confectionery and pasta, a master's degree in baking technology. The PhD was conducted in the field of Processes and Apparatus in the Food Industry in the research of the cherry fruit drying process with microwaves application. The doctoral habilitatus thesis was carried out in the field of obtaining and stabilizing dyes, antioxidants and preservatives of vegetable origin for functional foods.

Viorica Gladchi, PhD, Associate Professor, Dean of the Chemistry and Chemical Technology Faculty, Moldova State University. Research interests are focused on the studying the kinetics and mechanisms of free radical or photochemical redox catalytic processes responsible for transformation of various chemical pollutants in the model and natural aquatic systems; evaluation of surface waters self-purification capacity; applying different methods of redox state and kinetic parameters analysis, elaboration of new indicators of water quality estimation, prognosis and its ability for autopurification,as well as recommendations for their application. Holder of 10 university subjects, author of over 160 scientific works, 5 monographs, 6 methodical-didactic works for students. Participant in the realization of more than 10 international and national scientific projects.

Giuseppe Ianiri was born in Termoli, Italy in 1995. He obtained his Master degree in Food Science and Technology in 2020 from the Department of Agricultural, Environmental and Food Sciences, University of Molise, Campobasso, Italy. He is a member of the National and Molise Regional order of Food Technologists. He is currently a PhD student in Agriculture, Technologies and Biotechnologies at the University of Molise. He is working on the determination of persistent organic and inorganic pollutants in food and environmental matrices.

Tatyana Kiose has research interests which are focused on the development of supported on various carriers of metal complex catalysts for respiratory purposes. She obtained and generalized data on the influence of the phase composition, structure and morphology of fixed metal complexes on their activity in the oxidation reaction of sulfur dioxide. Metal complex catalysts based on natural and acid-modified zeolites and aluminosilicates have been studied by her. She studies the physico-chemical and structural-adsorption properties of catalyst compositions of different compositions. She developed new effective catalysts for the equipment of industrial respirators.

Marit Kolb studied food chemistry at Technische Universitaet Braunschweig in Germany and received her Ph.D. in 1993 from Technische Universitaet Braunschweig. From 1994 to 1997 she was postdoctoral fellow in the Institute of Plant Nutrition and Soil Science, Federal Agricultural Research Centre Braunschweig. Since 1998 she is senior scientist in the Institute of Environmental and Sustainable Chemistry at Technische Universitaet Braunschweig. Her research is focused on the field of analytical chemistry and ecotoxicology in the context of environmental pollution and sustainable chemistry.

Sivakumar Krishnamoorthy is a Professor in the Department of Chemistry at Sri Chandrasekharendra Saraswathi Viswa Mahavidyalaya (Deemed to be University) [SCSVMV University], Enathur, Kanchipuram-631561, Tamil Nadu, India. He has more than 22 years of experience in teaching and research. He has guided for Ph.D.s., M.Phil. Scholars and M. Sc., Students. He has published 95 research papers. His h-index is 16, and i10 index is 20. He has successfully completed research projects from various funding agencies, like DST, UGC. His current research interest involves the development of chemosensory materials for environmental and biological applications.

Ketevan Kupatadze graduated from S.S. Orbeliani State Pedagogical University, faculty of Natural Sciences, Chemistry and Biology specialization (With honors). 1997 _ graduated from the same University postgraduate studies. 1998 _ defended a Candidate (PhD) and 2003 _ Doctoral dissertations. She is the author of more than 70 published scientific works, monograph, manual as well as the author of three educational–virtual programs in Chemistry and Biochemistry. She authored over 200 scientific-popular articles in chemistry. 2017 for chemistry popularization she was awarded with the Special Medal of the Moldovan Academy of Sciences and Chemical Society. From 2017 till now every Autumn and Spring, she has active collaboration with the Science Popularization Center of Eskisehir (Turkey), where she conducts special workshops for chemistry teachers and students. She graduated from the secondary faculty Journalism specialization and published in many newspapers. 1998-2003 _ Docent (Senior teacher) of the Faculty of Natural Sciences at S.S. Orbeliani State Pedagogical University. 2003-2005 _ Professor at the same University. She was a member of the Scientific Council of the mentioned University and in 2000-2005 she served as the Head of the Department of Postgraduate Studies. At Ilia State University she works from 2006, first at the faculty of life sciences, then the faculty of Science and Art and nowadays she is Associate Professor Faculty of Natural Sciences and Engineering. She manages courses in Chemistry and Biochemistry direction (Bachelor and Master level). The main research area-chemistry, didactic of chemistry, History of chemistry. She is expert of chemistry and author in the internet journal "Maststsavlebeli" (The Teacher). She is technical editor of "Southern Brazilian Journal of Chemistry" and editorial member of "Tchê Química" and "SCIREA Journal of Chemistry". In 2021, she was named among the three best women scientists by NATO

Angela Lis has a Doctor of Philosophy in Chemical Sciences, university lecturer, scientific researcher at Moldova State University, Faculty of Chemistry and Chemical Technology, Department of Industrial and Ecological Chemistry "Acad. Gh. Duca". Specialist in the field of ecological chemistry of water and environmental protection, co-author of over 30 scientific papers. Field of scientific interests: water chemistry; buffering capacity of natural waters; surface water quality; redox-catalytic and photochemical transformations of environmental pollutants in aquatic systems; the self-purification capacity of natural waters.

Doina Lutic is an Associate Professor at the Faculty of Chemistry of "Alexandru Ioan Cuza" University of Jassy, Romania. After defending her PhD thesis at the Technical University of Jassy in 1997, approaching the synthesis and catalytic properties of aluminophosphate-based molecular sieves, she acted as a post-doctoral researcher at Laval University, Quebec (Canada), then at the University Montpellier II (France), working in the development of various types of catalytic solids for the green chemistry. Later on, she spent two years at Växjö and Linköping Universities (Sweden) working in the field of chemical and soot sensors. Her current scientific interests include the synthesis and characterization of solid micro- and mesoporous materials and their practical applications including environmental remediation and (photo)catalysis. Scientific portfolio: Web of Science Researcher ID: AAG-1922-2020, ORCID 0000-0002-3959-0891, 49 papers and 768 citations (Scopus), H index =14.

Fliur Macaev was born in 1959. He studied chemistry and biology in Bashkirian State Pedagogical Institute, gaining his PhD degree in 1990 from the Institute of Organic chemistry (Ufa) at the Ural Polytechnic Institute under the supervision of Prof. Ghenrikh Tolstikov. After postdoctoral training on synthesis of vitamin D3 analogs at the Institute of Organic Chemistry of the Polish Academy of Sciences with Prof. Jersy Wicha (1995–1996) and then on synthesis of an insect antifeedant at the Wageningen Agricultural University (the Netherlands) with Prof. Aede de Groot (1998–1999), he joined the staff of the Institute of Chemistry of Moldova Academy of Science in 1999 as the Head of the Laboratory of Organic Synthesis, where he obtained a Dr. habililtat in 2002. His research interests cover chemistry of natural compounds, stereochemistry, organic synthesis, ionic liquids chemistry, medicinal chemistry, rational drug design, computer-aided molecular design and structure-activity analysis.

Viorica Parvulescu earned the "N.Teclu" Prize of Romanian Academy on 1990 with "Contributions to theoretical base of the preparation of the bimetallic supported catalysts," cited by 2572, 221 publications (169) in journals with IF, 52 in journals without IF and proceedings) 2 books as co-author, 2 chapters in Wiley-VCH book, 1 chapter as co- author in Elsevier book, 1 chapter in IntechOpen book, 4 patents, 192 international conferences and 93 national conferences, 46 national projects manager and co-worker, 14 international bilateral projects. Hirsch index=22

Violina Popovici is lecturer within Technical University of Moldova, Faculty of Food Technology. She defended bachelor's degree in chemical technology and biotechnology, a master's degree in Restaurant Management and Catering Services. The PhD was carried out in the field of Biological and Chemical Technologies in the Food Industry, the topic of the research was about the stabilization of the vegetable oils enriched with bioactive compounds from renewable sources.

Tatyana Rakitskaya, professor, works as a Head of the Department of Inorganic Chemistry and Chemical Ecology of Odessa I.I. Mechnikov National University. Her research interests are basic research in the field of metal-complex redox catalysis and the development of theoretical foundations, methods and means of protecting humans and the environment from the effects of anthropogenic action. She researches nanochemistry, nanotechnology, coordination chemistry and metal complex catalysis of redox reactions, namely oxidation of carbon monoxide and sulfur dioxide, ozone decomposition. Metal complex catalysts based on natural and acid-modified zeolites and aluminosilicates, as well as nanocompositions based on metal oxides and the solid component of welding aerosols have been studied. She develops personal protective equipment for the respiratory system from phosphine, carbon monoxide, sulfur dioxide and ozone. Developed special courses of lectures for chemistry students (bachelors, masters, PhD-students).

Alla Truba works as an assistant professor of the Department of Inorganic Chemistry and Chemical Ecology of Odessa I.I. Mechnikov National University. Her research interests are nanochemistry, coordination chemistry and catalysis of the ozone decomposition reaction. As catalysts for low-temperature ozone decomposition, she investigated metal-complex compounds of copper(II), cobalt(II) and manganese(II), supported on natural materials (clinoptilolite, tripoli, bentonite, basalt tuff). She established the influence of the phase composition, structure and morphology of manganese and iron oxides on their activity in the reaction of low-temperature decomposition of ozone and oxidation of sulfur dioxide. She was investigating the effect of the phase solid component of the welding aerosol on the parameter of the ozone decomposition reaction. She studies the physico-chemical and structural-adsorption properties of catalyst compositions of different compositions.

Harsha Ratnaweera is a Professor in Water and Wastewater Technology at the Faculty of Sciences and Technology, Norwegian University of Life Sciences (part-time from 2001, full time from 2012). He was employed as the Director of International Projects and Innovation at the Norwegian Institute for Water Research (NIVA) 1991-2012, where he initiated and led NIVA's commercialization of research results. He has a Dr. Ing. degree in Civil Engineering (1992) from the Norwegian University of Science and Technology NTNU, and a MSc (Hons) Chemical Engineering (1987) from the National Technical University of Ukraine KPI, Ukraine. He is a Fellow of the International Water Association. Prof. Ratnaweera's research areas are on modelling and optimization of coagulation processes in water and wastewater treatment; Real-time monitoring and control of treatment processes; Holistic optimization of sewerage systems; virtual sensors and validation of measurements; membrane process; image analysis for sensor development; water resources management; and harmonisation of water-related graduate studies in Asia, Africa and Eurasia with Norway. Prof. Dr. Ratnaweera is the Norwegian representative in the Council of the European Water Association and a member of the Management Committee (MC) of the International Water Association - IWA Specialist Group "Particle Separation". He has been a member of the Norwegian National Commission for UNESCO and serves as the Chairman and a Board Member in several water and environment related organisations in Norway. He has worked as an advisor/consultant to the World Bank, UNECE, Norwegian Ministry of Foreign Affairs, NORAD and SIDA. He has worked in many countries in Asia, Africa, Eastern and Western Europe and fluent in English, Norwegian and Russian languages. Prof. Dr. Ratnaweera has initiated and leading a network of projects involving 82 universities form 52 countries (). Water Harmony () and Water and Society () are examples of projects where universities from Asia, Africa and Eurasia are involved and which are coordinated by Prof. Ratnaweera. He has also lead the organization of four international conferences in IWA Particle Separation Conference, Oslo and EWA Water Management in Cold Climates, Spitsbergenin June 2016, EWA Sustainable Urban Drainage Systems, Oslo, 2019 and2nd IWA-EWA Conference on Water Management in Cold Climates, Harbin, China in 2020., More information

Lidia Romanciuc, doctor in chemical sciences, senior scientific researcher at the Moldova State University/Institute of Chemistry, Chisinau, Republic of Moldova. She has a scientific background in physical chemistry (kinetics and catalysis) and ecological chemistry. The main field of scientific interest is the research of water self-purification mechanisms, chemical and photochemical transformation of natural water pollutants, redox processes in the aquatic environment, water quality, hydrochemistry and monitoring of natural waters. She is graduate from the Moldova State University. She passed several scientific scholarships in Europe, Japan, USA, has about 100 scientific publications and invention patents. She contributed to the organization of a series of international scientific conferences of ecological and environmental chemistry, with the participation of experienced and young scientists from the CIS, Europe, USA, etc. She contributed to the development of bilateral, regional, and international scientific cooperation relations.(ORCID: 0000-0002-4555-1194)

Iurie Scutaru graduated from the Faculty of Chemistry of the State University of Moldova, PhD thesis - at the State University "T. Shevcenco" from Kiev (Ukraine). Scientific interests are focused on the field of homogeneous kinetics and catalysis, free radicals and electron transfer, substances of natural origin with biological activity, modern instrumental methods for researching chemical composition and transformations, oenological chemistry, winemaking technologies and of the recovery of by-products.

Ludmila Semenyak was born in 1954. Professional activity is connected with the study of ecological and chemical regularity of habitation conditions of aquatic biological resources. In 1976 she graduated from the Odessa Hydrometeorological Institute with a degree in Oceanology. In 1990 she defended her dissertation for the degree of PhD, in chemistry (specialty - Environmental Protection and Rational Use of Natural Resources), in 1996 - for the degree of Doctor of Biological Sciences (specialty - Environmental Protection and Rational Use of Natural Resources). In present works at the All-Russian Research Institute of Fisheries and Oceanography (FGBNU VNIRO) as a chief researcher.

Elena Shtamm, Doctor of Chemical Sciences (1992), N.N. Semenov Federal Research Center for Chemical Physics Russian Academy of Sciences, Russian Federation. Leading Researcher. Specialist in the field of physical and ecological chemistry, environmental protection. Research interests: Environmental chemistry, problems of chemical safety of environmental objects, assessment of impact on water resources, water toxicology, environmental monitoring, industrial ecological control. Author of more than 250 scientific publications, 3 monographs, 5 patents, 41 on the problems of the quality of the natural aquatic environment and self-purification processes, 20 devoted to the issues of detoxification and disinfection of natural and waste water.

Vyacheslav Shvydkiy was born on July 11, 1958 in Dnepropetrovsk, USSR Graduated from the faculty of Geography of Moscow state University. M. V. Lomonosov 30.06.1985 Defended his thesis in physical chemistry 09.04.2008 Affiliation Position: senior researcher Research interests: ecological chemistry of water environment, water toxicology, chromatography.

Gabriela Soreanu, an Assoc. Prof. Dr. Eng, Technical University of Iasi (TUIASI), completed her PhD in Chemical Engineering at the "Gheorghe Asachi" Technical University of Iasi, Romania (2001) and performed two postdoctoral stages in the field of gas biotreatment at the Université de Sherbrooke, QC, Canada (2002-2003) and Environment Canada, Burlington, ON (2004-2007), respectively. She is the head of the BIO-SPACE LIFE laboratory (Laboratory of gas biofiltration and space technologies for life support in space) at TUIASI, director of several RDI projects awarded by the Romanian Space Agency (ROSA) and Romanian Research Authority (UEFISCDI), as well co-editor of the Elsevier book "From Biofiltration to Promising Options in Gaseous Fluxes Biotreatment: Recent Developments, New

Trends, Advances, and Opportunities" (2020). Scientific portfolio: articles (62), books/chapters (12), patents (4). Cumulated IF > 120. Hirsch (h) = 16.

Rodica Sturza, in 1989 - doctoral thesis in chemistry entitled: "Electrochemical properties of titanium-dioxymanganese anodes obtained by the electric spark method", Electrochemistry specialty, within the Institute of Chemistry and Chemical Technology of the Lithuanian Academy of Sciences (Vilnius). In 2008, habillitatus thesis entitled: "Theoretical and practical principles of food fortification with micronutrients: iodine, iron, calcium", food technology, Technical University of Moldova. Area of scientific concern: Technology of food products; food safety: reducing the risks of chemical, biochemical, microbiological food contamination; technogenic contaminants; falsification of products; non-compliant packaging materials; control of food along the food chain. She is the coordinator of the State Program "Food Security and Safety in the Period of Demographic Transition" (2018-2021), member of the Exact and Engineering Sciences section of the Academy of Sciences of Moldova (since 2018) and honorary member of the Romanian Academy of Technical Sciences (since 2019); Horizon Europe National Contact point: Food security, sustainable agriculture, marine and maritime research and the bio-economy; Biotechnology.

Natalia Sucman was born in 1983. She studied chemistry at the State University of Moldova, gaining her PhD degree in 2013 from the Institute of Chemistry (Moldova). She joined the staff of the Institute of Chemistry of Moldova Academy of Science in 2006 as a researcher. Her research interests cover the chemistry of natural compounds, stereochemistry, organic synthesis, ionic liquids chemistry, medicinal chemistry, rational drug design, computer-aided molecular design, and structure-activity analysis.

Poongodi T. has a M.Phil. degree in from SCSVMV (Deemed to be University), Kanchipuram, Tamil Nadu, India in 2018. His research interests focus on the photodegradation.

Corina Tasca is a PhD Student at the Moldova State University/ Institute of Chemistry. She is is dealing with the study of biochemical treatment of wastes and wastewaters from agroindustrial sector.

Sergey Travin Graduated Moscow Institute of Physics and Technology (MIPT), faculty of molecular and chemical physics 1976, diploma with honors. Ph.D. in Chemistry 1979 Doctor of Sciences in Economics, Professor 2001 Full member of the Russian Academy of Natural Sciences in the section "Education and sustainable development" 2002 Current position: senior scientist in N.N. Semenov Federal Research Center for Chemical Physics Russian Academy of Sciences Scope of interest: Mathematical modeling in Chemical kinetics, Chemistry and Physics, Computer science, Information systems

Ashok Vaseashta is the Chief Researcher and Chairman for the International Clean Water Institute in Virginia, USA. In addition, he serves as a Visiting Professor at Transylvania University of Brasov, Riga Technical University, Chaired Professor of Nanotechnology at the Academy of Sciences of Moldova, Academician at Euro-Mediterranean Academy of Arts and Sciences and Senior Strategic Research Advisor for several organizations. Inspired by nature and guided by societal necessities, he strives for innovations to address global challenges of the 21st century, such as environment, safety, security, sustainability and resiliency using advanced technological solution platforms. He is a scholar, dedicated futurist and visionary leader who provides strategic leadership to promote and advance research initiatives and priorities using data driven decisions. He received PhD from the Virginia Polytechnic Institute and State University, Blacksburg, VA in 1990 followed by Kobe post-doctoral fellowship. Following his PhD, he served as professor and researcher at Virginia Tech and Marshall University. He also served as the Director of Research at the Institute for Advanced Sciences Convergence and International Clean Water Institute for Norwich University Applied Research Institutes. Prior to his current position, he served as Vice Provost for Research at the Molecular Science Research Center in Orangeburg, South Carolina. He served as visiting professor at the 3 Nano-SAE Research Centre, University of Bucharest, Romania and visiting scientist at the Helen and Martin Kimmel Center of Nanoscale Science at the Weizmann Institute of Science, Israel. He served the U.S. Department of State in two rotations, as strategic S&T advisor and U.S. diplomat. His research interests span nanotechnology, environmental/ecological science, and safety and security. His research on nanotechnology has been on improving the understanding, design, and performance of nanofibers and sensors/detectors, mainly for applications such as wearable electronics, target drug delivery, detection of biomarkers and toxicity of nano and xenobiotic materials. In the security arena, he has worked on counterterrorism, countering unconventional warfare and hybrid threats, critical-Infrastructure protection, biosecurity, dual-use research concerns, and mitigating hybrid threats including fake-news. In addition, he has made numerous contributions in environmental/ecological science including directing research for International Clean Water Institute, pollution monitoring, contamination detection and remediation, and sustainability through green nanotechnology. He has authored over 270 research publications, edited/authored nine books on nanotechnology and environment and has presented many keynotes and invited lectures worldwide. He serves on the editorial board of several highly reputed international journals. He is an active member of several national and international professional organizations. He is a fellow of the American Physical Society, Institute of Nanotechnology and the New York Academy of Sciences. He has earned several other fellowships and awards for his meritorious service including 2004/2005 Distinguished Artist and Scholar award.

Caroline Vitone is a PhD student in Agricultural Technologies and Biotechnologies. Vitone has a bachelor's degree in Food Science and Technologies from University of Molise with a thesis on free-from foods and the use of nitrites and nitrates as preservatives; then obtained a master's degree from University of Molise in Food Sciences and Technologies, with a thesis on the use of grape must yeasts to produce craft beer.

Inga Zinicovscaia- head of the Sector of Neutron Activation Analysis and Applied Research, Joint Institute for Nuclear Research. Scientific interests: neutron activation analysis, air pollution, water treatment, nanotoxicology. Author of more than 150 scientific papers in Scopus and WoS journals.

Index

Q

R

S

T

U

W

www.ingramcontent.com/pod-product-compliance
Lightning Source LLC
LaVergne TN
LVHW062109300926
849161LV00010B/2408